Hinweise für die Autoren

Die Zeitschrift „Kristall und Technik" nimmt folgende Veröffentlichungen auf: Originalarbeiten, Zusammenfassungen, kurze Mitteilungen, Buchbesprechungen, bevorzugt in englischer Sprache.

Alle zum satzreifen Manuskript gehörenden Unterlagen sind beizufügen.

Das Manuskript soll in Maschinenschrift im Format A 4 (21 × 29,7 cm) auf weißem, tintenfestem Papier, einseitig, mit zweizeiligem Abstand der Zeilen und einem linken freien Rand von 3 cm geschrieben sein. Die Manuskriptblätter sollen fortlaufend ab „1" numeriert werden.

Noch nicht für den Druck überarbeitete Vortragsmanuskripte werden in der Regel nicht angenommen.

Der Kopf des Manuskriptes soll folgende Zeilen enthalten: Abgekürzte Vornamen und den Familiennamen des Verfassers, genaue Bezeichnung der Arbeitsstelle des Verfassers, Überschrift der Arbeit.

Der Arbeit ist eine Zusammenfassung in der Original- und einer Fremdsprache (deutsch, englisch, französisch oder russisch) voranzustellen.

Formeln, Zeichen, Buchstaben usw., für die keine Schreibmaschinentypen vorhanden sind, müssen deutlich lesbar mit Tinte geschrieben sein.

Nomenklatur des Chemischen Zentralblattes benutzen.

O als Null ist mit der Schreibmaschine einfach zu unterstreichen, der Großbuchstabe O nicht.

Die Absätze sind deutlich im Manuskript zu kennzeichnen (durch Einrücken von mindestrens 3 Anschlägen).

Für die Auszeichnung des Manuskriptes ist zu beachten:

Überschriften im Text und zum Beginn des Beitrages sind übersichtlich anzuordnen und nur durch eine einfache Unterstreichung kenntlich zu machen

Tabellenüberschriften sind nicht besonders kenntlich zu machen, sie werden normal gesetzt.

Auszeichnungen innerhalb des Textes werden in der Regel von der Redaktion erledigt. Es bedeuten:

hellrot ~~~~~~~~ = Kapitälchen
blau ~~~~~~~~ = kursiv

hellrot | = Kleindruck (senkrechter roter Strich am linken Manuskriptrand)

Tabellen sind gesondert als Manuskript zu schreiben und am Schluß des Beitrages einzuordnen.

Fußnoten sind auf gesondertem Blatt fortlaufend numeriert aufzuführen.

Abbildungen [„Fig." (= Figur)] sind als reproduzierbare Vorlagen dem Manuskript in gesondertem Umschlag beizulegen.

Fotos sind auf weißem Hochglanzpapier und im Mindestformat von 6 × 6 cm zu liefern. Bildausschnitte und eventuell noch vorzunehmende Beschriftungen sind auf einem auf dem Foto angebrachtem Deckblatt aus durchsichtigem Material zu vermerken. Die Größen aller Fotos und deren Verkleinerungsmaßstäbe sind nach Möglichkeit einheitlich zu halten.

Alles Nebensächliche bei Halbtonbildern bitte weglassen und von einer Vergrößerung der Vorlagen absehen.

Strichzeichnungen genügen als gut lesbare Skizzen.

Auf der Rückseite der Abbildungsvorlagen sind anzugeben: Nummer der Figur, Name des Autors bzw. der Autoren, abgekürzter Beitragstitel, gewünschte Größe der Figur im Druck in Zehntel-Angaben, z. B. 9/10, 8/10,/5/10 usw.

Bildunterschriften sind auf einem gesondertem Blatt aufzuführen.

Die Literatur soll ohne Bezifferung am Ende der Arbeit in alphabetischer Reihenfolge der Autorennamen aufgeführt werden. Im Text zitierte Autorennamen gelten als Literaturhinweis.

Die Literaturzitate sollen folgende Angaben enthalten:

bei Büchern z. B.: RAST, K.: Physikalisch-chemische Rechungen, Berlin 1958
bei Zeitschriften z. B.: KRAUSE, H. N., MÜLLER, K.: Kristall u. Technik 1, 70 (1966)

Für jede veröffentlichte Arbeit werden 50 Sonderdrucke vom Verlag kostenlos abgegeben. Sind mehrere Autoren gemeinsam an einem Beitrag beteiligt, so erhalten sie insgesamt 50 Sonderdrucke kostenlos.

Die Bestellung weiterer Sonderdrucke gegen Berechnung, jedoch nur bis zu einer Anzahl von 200 Stück, muß mit der Einsendung des Manuskriptes der Hauptschriftleitung bekanntgegeben werden.

Kristall und Techni

Crystal Research and Technolog

Akademie-Verlag · Berlin

.–3. Heft · 7. Jahrgang · 1972

TRAW 7 (1) 1–358 (1972)

1–3

Kristall und Technik

Crystal Research and Technology

Zeitschrift für experimentelle und technische Kristallographie

Kleber-Gedenkband

BAND 7 · HEFT 1–3
1972

Akademie-Verlag
Berlin

Herausgeber und verantwortlich für den Inhalt Professor Dr. HERMANN NEELS, 703 Leipzig, Scharnhorststraße 20, Fernruf: 32519, 31032 bzw. 7254 Machern, Zeititzer Weg 25, Fernruf: Brandis 452.
Verlag: Akademie-Verlag GmbH, 108 Berlin, Leipziger Straße 3 – 4,
Fernruf: 220441, Telex-Nr. 112020, Postscheckkonto: Berlin 35031.
Die Zeitschrift „Kristall und Technik" erscheint monatlich. Bestellnummer: 1082/7/1 – 3
Preis je Heft 20,– M
Gesamtherstellung: VEB Druckerei „Thomas Müntzer", 582 Bad Langensalza
Veröffentlicht unter der Lizenznummer 1526 des Presseamtes beim Vorsitzenden des Ministerrates der Deutschen Demokratischen Republik.
Printed in the German Democratic Republic

Inhalt

Buchbesprechung

Kurze Originalmitteilungen

Buchbesprechungen

Am 15. 12. 1971 wäre Professor Dr. phil. nat. habil. Will KLEBER 65 Jahre geworden. Wie bereits zu seinem sechzigsten Geburtstag war von seinen Mitarbeitern und Schülern beabsichtigt, zu diesem Anlaß ein Festkolloquium durchzuführen, auf dem eine Würdigung seiner wissenschaftlichen Verdienste erfolgen sollte und das einen großen Kreis seiner Schüler und Freunde zusammengeführt hätte. Will KLEBER, Professor mit Lehrstuhl für Kristallographie und Leiter des Bereichs Kristallographie der Sektion Physik an der Humboldt-Universität zu Berlin, verstarb am 27. 8. 1970.

Will KLEBER übernahm nach einer Tätigkeit als Hochschullehrer in Heidelberg und Bonn im Jahre 1953 die Leitung des damaligen Mineralogisch-Petrographischen Instituts und Museums der Humboldt-Universität zu Berlin, dem späteren Institut für Mineralogie und Museum und dem jetzigen Bereich Kristallographie. Es war sein großes Verdienst, hier eine Ausbildungs- und Forschungsstätte von nationaler Bedeutung und internationalem Ansehen geschaffen zu haben. Er gab mit wissenschaftlichem Weitblick und unermüdlicher Schaffenskraft wesentliche Impulse für die Entwicklung der modernen Kristallographie und dokumentierte die Bedeutung dieser Fachrichtung als interdisziplinäre Wissenschaft. So ist sein Name untrennbar mit der Entwicklung der Kristallographie in der DDR und der internationalen Anerkennung ihrer wissenschaftlichen Leistungen verbunden. Seine Verdienste als Hochschullehrer und Forscher fanden in der Auszeichnung mit dem Nationalpreis und in der Wahl als Ordentliches Mitglied der Deutschen Akademie der Wissenschaften zu Berlin und Mitglied der Deutschen Akademie der Naturforscher Leopoldina Halle ihre Würdigung.

Will KLEBER war wegen seiner hohen wissenschaftlichen Leistungen, seiner hervorragenden Darstellungsweise, seines großzügigen und verständnisvollen Verhaltens und seines edlen Charakters als Persönlichkeit von seinen Mitarbeitern, Schülern, Freunden und Kollegen im In- und Ausland geschätzt und geachtet. Sie widmen dem Gedenken an Professor Dr. phil. nat. habil. Will KLEBER die Beiträge dieser Hefte der Zeitschrift ,,Kristall und Technik", deren Mitherausgeber er war, und vertreten damit sein Anliegen um die weitere Entwicklung der Kristallographie.

Im Auftrage der Redaktion
H.-J. BAUTSCH

Im Mai 1971

Kristall und Technik	7	1–3	1972	9–16

Originalbeiträge

J. D. H. DONNAY

The Johns Hopkins University, Baltimore, Maryland, USA

and the late W. KLEBER

Sektion Physik, Bereich Kristallographie der Humboldt-Universität, Berlin, DDR

Haidingerite (I). From Crystal Morphology to Bond Assemblage

The known structural space group of haidingerite is Pcnb. The general forms, 121 and 221, indicate that the period of the bond chains along the y axis is $b/2$, half the structural period b. In planar projections: onto yz, the energetic periods are $b/2$ and $c/2$; onto zx, they are $c/2$ and $a/2$, with a primitive mesh, as compared with a centered mesh (ca).

Die bekannte Raumgruppe des Haidingerits ist Pcnb. Die allgemeinen Formen 121 und 221 zeigen, daß die Periode der Bindungsketten in Richtung der y-Achse $b/2$ ist, die Hälfte der Strukturperiode b. In der ebenen Projektion auf yz sind die energetischen Perioden $b/2$ und $c/2$, auf zx $c/2$ und $a/2$, mit einem primitiven Netz im Vergleich zum zentrierten Netz (ca).

1. Introduction

The aim of crystal-morphology studies is two-fold: (1) given complete knowledge of the crystal structure, to find an explanation for the crystal morphology; (2) given morphological observations, to infer from them useful hints as to the unknown crystal structure. Nowadays the knowledge of space group and structural cell can be taken for granted, and the second aim above can be restated as follows: knowing the structural cell and space group from x-ray-diffraction study, to find which additional halvings of net-plane spacings will account for the morphological data.

The method rests on the law of BRAVAIS (1849) and its generalizations (DONNAY and HARKER, 1937; DONNAY, DONNAY and KLEBER, 1961; KLEBER and DONNAY, 1961; DONNAY and DONNAY, 1961, 1962, 1963, 1965; DONNAY, 1962, 1966) and on the concept of periodic bond chains in crystal structures, which was introduced in crystallography by HARTMAN and PERDOK (1955) and further exploited by HARTMAN (1961). Preliminary results of the present work were presented to a meeting of the American Crystallographic Association (DONNAY and KLEBER, 1963).

The method consists of the following steps: (1) to obtain structural cell dimensions and space group by x-ray diffraction; (2) to establish the relative frequencies of occurrence of authenticated crystal forms; (3) referring the forms to the structural axial ratios, to compare, with the list of observed frequencies, the list of forms drawn up according to the frequencies predicted by the space group and to correct the discrepancies by introducing additional extinctions (besides the space-group systematic extinctions); (4) to interpret these corrections as additional halvings leading to a prediction of the cell dimensions and symmetry of the three-dimensional bond assemblage and of its planar and linear projections.

Let it be emphasized once more that crystal structure and bond assemblage are two different properties of the crystal and need not, therefore, have the same symmetry. In a structure determined by x-ray diffraction, equivalent points are chemically identical atoms or ions with the same scattering power; in the bond assemblage, which controls the morphological development, the equivalence is based on the bonding properties (for example, Na^+ and Cl^-, in NaCl, are morphologically equivalent because the bond Na^+ to Cl^- has the same strength as the next bond Cl^- to Na^+). That the structure and the morphology of a crystal species should have different lattices or different space groups should occasion no surprise.

2. Structural Cell and Space Group of Haidingerite

The formula assigned to haidingerite in the literature is $CaH(AsO_4) \cdot H_2O$. Fischer determined the following cell dimensions:

$$a = 6.946\,, \quad b = 16.14, \quad c = 7.930 \text{ Å}\,,$$

with space group $Pcmb$ and $Z = 8$. This cell was confirmed by Binas and Dornberger-Schiff, who gave the space group $Pcnb$, on which they based a partial crystal structure. Cassien, Herpin and Perminngeat have published the complete structure of haidingerite, with cell dimensions:

$$a = 6.93_{5\pm3}\,, \quad b = 16.15_{0\pm2}\,, \quad c = 7.94_{0\pm3} \text{ Å}\,;$$

they confirm the space group found by Binas and Dornberger-Schiff, to whom, however, they do not refer. The cell dimensions yield the axial ratios

$$a : b : c = 0.429 : 1 : 0.492\,.$$

3. Morphological Data of Haidingerite

The early morphologists disagreed on the choice of the parametral face (111), so that three different cells were on record in the literature before the advent of the structural cell. The new Dana (Palache, Berman and Frondel) retains Haidinger's original cell, which can be obtained from the x-ray cell by transformation 200/010/001; the cell adopted by de Schulten turns out to be the same as the structural cell (matrix 100/010/001); Goldschmidt's choice corresponds to matrix 200/010/002. In this paper the x-ray cell will be used exclusively.

The synthetic crystals described by de Schulten yield the following list of forms, arranged in observed order of decreasing frequency of occurrence:

$$b(010),\ m(120),\ a(100),\ x(121),\ t(011),\ k(101),\ y(012)\,.$$

These forms, with the exception of x and y, were reported in the original description of crystals from Joachimstal and are listed in Dana, together with four additional forms: *s*(221), *i*(201), *n*(584) and *g*(104). From these the last two are dubious and need not be considered; *g* was described as rough and slightly rounded in the original paper; the high indices of *n* are suspicious and could be a mistaken representation of (121). Form *s* has been accepted, but our conclusions would not be modified if *s* were to be rejected. Form *i*(201) is critical in that it serves to characterize the zone of the (*h*0*l*)s as a simple zone with *k* dominant. Form *i* is figured on one of the original drawings and, somehow, inspires confidence; FISCHER accepts it, and so do we.

4. Analysis of Morphological Data

The theoretical list of forms arranged according to increasing reciprocal-lattice vectors in space group *Pcnb* is as follows:

020	120	101	111	121	002	012	022	140	200	112	220	... 221 ...	402 ..
b	*m*	*k*	—	*x*	—	*y*	*t*	—	*a*	—	—	*s*	*i*

It is clear that the space-group extinctions do not account for the observed order of frequencies of occurrence of the forms. Among the first ten predicted forms, three are unknown (the 4th, 6th and 9th) and the others, except for the first two, are badly out of order. (Note ranks of *k* and *y*.) This may be taken as prima facie evidence that the symmetry (possibly even the cell) of the bond assemblage will be different from that of the crystal structure. Let us analyze, separately, the general forms, the forms of the three axial zones and the three pinacoids.

4.1 The general forms

The two general forms that have been observed, *x*(121) and *s*(221), have indices that do not obey any lattice criterion: the lattice of the bond assemblage cannot be *I*, *F*, *A*, *B* or *C*. The only criterion is "*k* even"; it indicates that the morphological lattice, which expresses the periodicity of energetically equivalent points in the bond assemblage, must have its *b* halved. We shall continue to index the forms on the structural lattice, but forms with an odd *k* index will have all their indices doubled. Example: 111 will be symbolized 222, thereby receding in the list of relative importance of the forms; *x*(121) and *s*(221), on the other hand, will remain unchanged.

4.2 The 0*kl* forms

The *c* glide plane of the space group requires the 0*kl* forms to be written with *l* even, thus demanding that *y*(012) be dominant over *t*(022), as shown in the above theoretical list. In order to account for the observed sequence, *t* dominant over *y*, we must double the indices of *y*, so that its symbol becomes *y*(024). We thereby add a new criterion, "*k* even", valid only for the bond assemblage, to the glide-plane criterion "*l* even", which holds for both the structure and the bond assemblage.

4.3 The *h*0*l* forms

The *n* glide plane requires the *h*0*l* forms to obey the criterion "$(h + l)$ even", and the projection of the structure onto the *ca* plane will have a centered net. If the bond assemblage of this projection had the same centered net as the projected structure, the resulting zone would be a double zone (in such a zone the dominant, *k*(101), would be followed by (103), which is contrary to observation. To express the dominance of *k* over *i*, it is necessary to write *k* as (202) and *i* as (402). In effect this adds a new criterion, "*h* and *l* both even", valid only for the bond assemblage, to the glide-plane criterion "$(h + l)$ even", which governs both the structure and the bond assemblage.

4.4 The *hk*0 forms

The only known *hk*0 form is *m*(120). The *b* glide plane of the structure, which demands that *k* be even, thus expresses the morphological fact. No additional criterion is needed.

4.5 The pinacoids

In accordance with the space-group criteria, the pinacoids must be symbolized *a*(200), *b*(020) and (002). The last one is unknown in haidingerite, and *b* dominates over *a*. There is no way in which we can increase the predicted importance

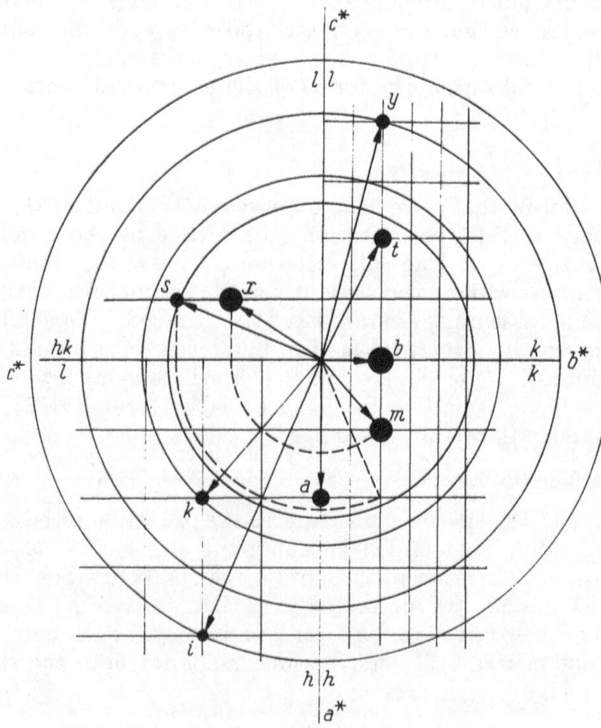

Fig. 1. Haidingerite – $CaH(AsO_4) \cdot H_2O$. – Graphical determination of the lengths of the reciprocal-lattice vectors that correspond to known crystal forms. The nets b*c* and c*a* have been brought into the plane of the net a*b* by rabattements around the axes b* and a* respectively

of *a*, since we cannot symbolize it (100). As (200) it comes after *x* and *t* in the amended list; this must be recognized as an irreducible anomaly. The absent pinacoid (002) comes ahead of *a* according to space-group. If its indices are doubled, (004) would still precede *y*(024) and *i*(402), and should be observed. Its indices cannot be tripled, to (006), because all atoms must be in general position in the haidingerite space group and this precludes any possible division of the *c* axis in six parts. The absent pinacoid will thus have to be written (008), and as such it will properly be relegated way down the list of predicted form importances.

4.6 Final theoretical list

Taking into account the additional criteria of the bond assemblage, the amended list of forms (see Fig. 1) is considerably improved. It now reads as follows:

020.	120.	121.	022.	140.	200.	220.	221 ...	202 ...	024 ...	402 .
b	*m*	*x*	*t*	—	*a*	—	*s*	*k*	*y*	*i*

The *a* pinacoid ranks 6th instead of 10th, *t* comes well ahead of *y*, *k* is behind *a* and *t*, and several unobserved forms have receded down the list, such as 111, 002, 112.

5. Inferences Concerning the Bond Assemblage

It is now possible to compare the periodicity of the bond assemblage with that of the crystal structure, in three dimensions and in projections.

We have already seen (Section 4) that the bond assemblage, in three dimensions, has cell dimensions a, $b/2$, c as indicated by the general forms: *x*(121) proves it, and *s*(221) confirms it. Another way of stating the same result is to say that $b/2$ is the repeat of a periodic bond chain (PBC of Hartman and Perdok).

The additional extinction criteria that had to be introduced properly to express the relative morphological importances of the $0kl$ and $hk0$ forms simply indicate that the *b* halving already found in the three-dimensional bond assemblage is also to be seen on the *bc* two-dimensional projection. (On the *ab* projection the space group itself halves the *b*.)

The zone of the $h0l$ forms does yield supplementary evidence: its observed simple zone tells us that, in the *ca* projection of the bond assemblage, both *c* and *a* must be halved.

Finally, as regards the one-dimensional projections on the axial directions, the information obtained from the pinacoids shows that the repeat distances will be $a/2$, $b/2$ and $c/8$. The first two are valid for both bond assemblage and structure. The last one should hold for the bond assemblage only; the projection of the structure having period $c/2$ by space-group requirement (Table 1).

Remark. — Note that, in the orthorhombic case of haidingerite, it is possible to speak of halving the axial lengths instead of the interplanar distances. In all cases where the direct-lattice vector does not coincide in direction with the reciprocal-lattice vector, doubling the indices hkl must be interpreted as requiring the halving of $d(hkl)$.

Table 1
Periodicities

	3 dimensions	2 dimensions			1 dimension		
		$b\ c$	$c\ a$	$a\ b$	a	b	c
Structure or projected structures[1])	$a\ b\ c$	$b\ \frac{c}{2}$	centered $c\ a$	$a\ \frac{b}{2}$	$\frac{a}{2}$	$\frac{b}{2}$	$\frac{c}{2}$
Bond assemblage[2])	$a\ \frac{b}{2}\ c$	$\frac{b}{2}\ \frac{c}{2}$	$\frac{c}{2}\ \frac{a}{2}$	$a\ \frac{b}{2}$	$\frac{a}{2}$	$\frac{b}{2}$	$\frac{c}{8}$

[1]) Space-group systematic extinctions only.
[2]) Space-group systematic extinctions and additional morphological extinctions.

6. Confrontation of Morphological Predictions with Structural Results

The first morphological prediction to be tested is that, in the three-dimensional bond assemblage, the period along the y axis is $b/2$, as opposed to b in the crystal structure. The structure (CASSIEN, HERPIN, PERMINGEAT) consists of two double layers of tetrahedra perpendicular to the y axis, one at $y = 0$, the other at $y = 1/2$. Within the spacing of a double layer, that is between -0.11 and $+0.11$, two calcium atoms are located, at -0.05 and $+0.05$, forming a dumbbell whose center carries a punctualized charge of $4+$; these dumbbells are placed at $x = 1/2$, $z = 0$ and $x = 0$, $z = 1/2$, between the zig-zag chains of tetrahedra that constitute the layers, at $y = 0$ and $y = 1/2$, respectively. The period of the bond chains along y is thus $b/2$; it is marked by the centers of the calcium dumbbells. Likewise, at $x = 0$, $z = 0$ and at $x = 1/2$, $z = 1/2$, bond chains along y have links that stretch from the center of an arsenic dumbell at $y = 0$ to the center of the next one at $y = 1/2$. The first prediction is fulfilled.

The second prediction to be checked is the halving of the b period in the two-dimensional projection of the bond assemblage onto plane bc. In this projection (CASSIEN et al., Fig. 3), the energetic period along the b axis is clearly apparent — it stretches from the midpoint of the projected segment Ca(8)–Ca(1) to the midpoint of Ca(3)–Ca(5) and is thus rigorously equal to $b/2$. (The next repeat, from Ca(3)–Ca(5) to Ca(8)–Ca(1), is equivalent to the first one by screw rotation around 2_1 in the x direction.) The second prediction comes true.

The third prediction is the energetic halving of both c and a in the projection of the bond assemblage onto plane ca. The ca projection of the structure shows the superposition of the four structural layers of AsO_4 tetrahedra, parallel to (010), at $y = 0 \pm 0.11$ and $y = \frac{1}{2} \pm 0.11$. The four arsenic atoms numbered 1, 3, 6 and 7 by CASSIEN et al. are projected at the corners of a rectangle about the origin; those numbered 2, 4, 5 and 8 form a similar rectangle about the center of the mesh ca. Each rectangle is the positive core of a group of 12 oxygen ions and 4 hydroxyl ions, that surround it. In the ca projection the complex anionic group at the origin, $[As_4O_{12}(OH)_4]^{8-}$, also centers the mesh. At the midpoints of the mesh edges a positive group can be recognized; it is made up of four calcium ions, at the corners of a somewhat larger rectangle, and four water molecules that are projected near the midpoints of the sides parallel with *c*; its formula can be written $[Ca_4(H_2O)_4]^{8+}$. We note that the projected structure, from the energetic point of view, looks like an atomic layer in the NaCl structure: two kinds of ions, whose charges are equal and of opposite signs, occupy the two kinds of squares on a chess-board. The ions are energetically equivalent; their punctualized charges lie at the nodes of a primitive net, of mesh $(c/2, a/2)$. The third prediction is verified. The case of barite was similar (DONNAY and DONNAY, 1961d, 1961e); haidingerite is instructive in showing how large the complex ionic groups can be.

The last check concerns the predicted period $c/8$ for the projection of the bond assemblage onto the *c* axis. If we did not know the haidingerite crystal structure, we could reason as follows. All atoms must occupy general point-positions in *Pcnb*, and the eight sites of one position will lie at heights z, $-z$, $\frac{1}{2}+z, \frac{1}{2}-z$, two on each of the levels. Consider the two cations (As and Ca) in the chemical formula, and let their *z* parameters be z' and z'', $z'' > z'$. If *c* is to be divided into eight equal parts, two necessary and sufficient conditions,

$$2z' = 1/8 , \quad z'' - z' = 1/8 ,$$

must be fulfilled. It follows that one kind of cations will be at $z' = 1/16 = 0.0625$ and the other kind, at $z'' = 3/16 = 0.1875$. Oxygen atoms can be confidently expected to be situated between the cationic planes, if bond chains are to be formed.

According to the recently published structure (CASSIEN et al.), the fractional *z* co-ordinates of As and Ca are 0.0597 and 0.1824, respectively. It follows that the fractional distances between projected As atoms is 0.1194; between projected Ca atoms, 0.1352; and from projected As to projected Ca, 0.1227. We have, of course, in one *c* distance,

$$0.1194 \times 2 + 0.1352 \times 2 + 0.1227 \times 4 = 1.0000 .$$

It may be that the projected intercationic distances, close to 0.1250 though they are, do not qualify as periods of the projected bond assemblage, since (by definition) periods must be equal. If so, we can recognize as energetic period the distance between the midpoints of projected pairs of As atoms and of Ca atoms; this distance is rigorously equal to 1/4. We would then have to index the missing pinacoid (004) after all (cp. Section 4) and consider its absence an anomaly in view of the existence of *y*(024) and *i*(402).

The difficulties we are running into in the case of the last prediction are not unexpected. This prediction was based on the absence of (001), confronted with the presence of *y* and of *i*. Rare forms are unreliable, more often than not, and it is hard to decide by which factor the Miller indices of a missing form are to be multiplied. Just how much "absent" can an absent form be? Indexed (008) the litigious pinacoid would be very unlikely ever to occur; as (004) it could presumably show up occasionally. But all we know is that it has not been reported, in a species that is rather poor in forms.

References

BINAS, H., DORNBERGER-SCHIFF, K.: Chemie der Erde **21**, 450 (1962)

BRAVAIS, A.: Compt. Rend. Acad. Sci. Paris **29**, 132–137, 143–144 (1849)

CASSIEN, M., HERPIN, P., PERMINGEAT, F.: Bull. Soc. franç. Minér. Crist. **89**, 18–22 (1966)

DONNAY, J. D. H.: Kôbutsu-gaku Zasshi **5**, 277–284 (1962a) (in Japanese)
DONNAY, J. D. H.: Seventh Int. Congress I. U. Cr., Moscow, USSR (1966) Abstracts, A 30–31
DONNAY, J. D. H., DONNAY, GABRIELLE: Compt. Rend. Acad. Sci. Paris **252**, 908–909, 1982–1983 (1961ab)
DONNAY, J. D. H., DONNAY, GABRIELLE: Amer. Cryst. Assoc. meeting, Boulder, Colorado, Program, 27–28 (abstract) (1961c)
DONNAY, J. D. H., DONNAY, GABRIELLE: Compt. Rend. Acad. Sci. Paris **253**, 291–292 (1961d)
DONNAY, J. D. H., DONNAY, GABRIELLE: Kristallografiya **6**, 840–846 (1961e) (in Russian); Crystallography **6**, 679–684 (1962) (in English)
DONNAY, J. D. H., DONNAY, GABRIELLE: ACA meeting, Villanova, Pennsylvania, Program, 57–58 (abstract) (1962a)
DONNAY, J. D. H., DONNAY, GABRIELLE: Twentieth Ann. Diffr. Conf., Pittsburgh, Pennsylvania, Program, 17 (abstract) (1962b)
DONNAY, J. D. H., DONNAY, GABRIELLE: Acta Cryst. **16**, A147 (abstract) (1963)
DONNAY, J. D. H., DONNAY, GABRIELLE: Geol. Soc. Amer., Special Paper **82**, 48 (abstract) (1965)
DONNAY, J. D. H., DONNAY, G., KLEBER, W.: N. Jahrb. Mineral., Monatsh. **1961**, 97–101 (1961)
DONNAY, J. D. H., HARKER, D.: Compt. Rend. Acad. Sci. Paris **204**, 274–276 (1937a)
DONNAY, J. D. H., HARKER, D.: Amer. Mineralogist **22**, 446–467 (1937b)
DONNAY, J. D. H., KLEBER, W.: ACA meeting, Cambridge, Massachusetts, Program, 58 (abstract) (1963)
FISCHER, E.: Chemie der Erde **20**, 162 (1960)
GOLDSCHMIDT, V.: Atlas der Krystallformen **4**, 105. Heidelberg 1918
GROTH, P.: Chemische Krystallographie **2**, 832. Leipzig 1908
HARDIE, L. A., MUNOZ, J. L., DONNAY, J. D. H., DONNAY, GABRIELLE: Amer. Mineralogist **49**, 415–420 (1964)
HARTMAN, P.: C. R. Acad. Sci. **252**, 3288 (1961)
HARTMAN, P., PERDOK, W. G.: Acta Cryst. **8**, 49–52, 521–524, 525–529 (1955)
KLEBER, W., DONNAY, J. D. H.: Z. Krist. **115**, 161–168 (1961)
PALACHE, C., BERMAN, H., FRONDEL, C.: The System of Mineralogy of James Dwight Dana and Edward Salisbury Dana, 7th ed., vol. 2, New York 1951
SCHULTEN, A. DE: Bull. Soc. franç. Minéral. **26**, 20 (1903); Z. Krist. **41**, 94 (1903)

(Received June 14, 1971)

Author's address:

Prof. J. D. H. DONNAY
The Johns Hopkins University
Baltimore
Maryland 21218
USA

| Kristall und Technik | 7 | 1–3 | 1972 | 17–20 |

J. D. H. Donnay

The Johns Hopkins University, Baltimore, Maryland, USA

and Gabrielle Donnay

McGill University, Montreal, Quebec, Canada

Haidingerite (II). From Valence Summation to Hydrogen Bonding

The haidingerite formula must be written $CaAsO_3(OH) \cdot H_2O$. The calcium co-ordination number, reported as 6, is shown to be 7 by valence summation. Chains of hydrogen-bonded arsenate groups are confirmed. The two protons of the water molecule form hydrogen bonds to neighboring chains.

Die Haidingerit-Formel muß geschrieben werden $CaAsO_3(OH) \cdot H_2O$. Die Koordinationszahl für Calcium — bisher mit 6 angegeben — beträgt, wie gezeigt wird, auf Grund der Summierung der Valenzen 7. Die Ketten der über Wasserstoffionen verbundenen Arsenatgruppen werden bestätigt. Die beiden Protonen der Wassermoleküle bilden Wasserstoffbindungen zu den benachbarten Ketten.

The known interatomic distances in the haidingerite crystal structure (Cassien et al.) enable us to confirm, by means of the valence-summation procedure (Donnay, Allmann), the presence of one hydroxyl group and one water molecule per formula unit. In addition we set out to predict the hydrogen bonding in the structure.

The tetrahedral co-ordination of arsenic is confirmed. The co-ordination of calcium, described as 6-fold by Cassien et al., is changed to 7-fold for the following reasons: the maximum Ca—O bond length, obtained from the effective ionic radii (Shannon and Prewitt), is 3.25 Å; the length of the bond to be added, Ca—O(4) = 2.692 Å, is well below this cut-off value; the mean bond length for the seven nearest neighbors, 2.466 Å, is reasonably close to the median 2.430 Å.

The valence sums of the anions (Table 1) show O(1) to be the hydroxyl group and O(5) the water molecule, in agreement with Cassien et al. The high valence sum of O(4), 2.16 v. u., must be taken as avidence of π bonding character. The low valence sums for O(2) and O(3), 1.853 and 1.637 v. u. respectively, and the high obtained for O(5), 0.307 v. u. instead of zero, imply the presence of hydrogen bonding.

Approaches (shorter than 3.10 Å) between nonbonded oxygen atoms were calculated (Table 2). It is found that O(1) and O(3) are 2.541 Å apart, a distance which corresponds to a charge transfer of 0.288 v. u. from O(1)H to O(3) (see curve adapted from Lippincott and Schroeder, in Donnay and Allmann, loc. cit.). The O(3) valence sum is thus increased to 1.93 v. u. (Table 1, last

Table 1

Valence summation for synthetic haidingerite $CaAsO_3(OH) \cdot 2\,H_2O$

Cations Anions	vii Ca^{2+} Å v. u.	iv As^{5+} Å v. u.	Σ v C(v.u.)	Anion chemistry	Corrections for H bonds
iiiO(1)		1.785 (1.043) 1.040	1.040	OH	+0.224 from O(5) −0.288 to O(3) 0.98
ivO(2)	2.350 (0.332) 0.331 2.472 (0.284) 0.283	1.691 (1.242) 1.239	1.853	O	+0.145 from O(5) 2.00
ivO(3)	2.395 (0.313) 0.312	1.651 (1.329) 1.325	1.637	O	+0.288 from O(1) 1.93
ivO(4)	2.518 (0.267) 0.266 2.430 (0.299) 0.298 2.692 (0.203) 0.202	1.621 (1.400) 1.396	2.162	O	2.16
ivO(5)	2.407 (0.308) 0.307		0.307	H_2O	−0.224 to O(1) −0.145 to O(2) −0.06
$\bar{L}$ (Å) L_{max} (Å) p	2.466 3.25 3.145	1.687 2.28 2.845			
v_i (v. u.) $\sum_A$ v (v. u.)	2/7=0.286 (2.006) 2.000	5/4=1.250 (5.014) 5.000			

The valence strength in parentheses is the one obtained directly from the equations (Donnay and Allmann). Below it is the normalized value.

column); O(1)H, the donor of this hydrogen bond, is also the acceptor of another hydrogen bond, reaching it from $O(5)H_2$, so that its valence sum becomes 0.98 v. u. A second hydrogen bond is donated by $O(5)H_2$ and accepted by O(2), whose valence sum is thereby increased to 2.00 v. u. The valence sum of the water oxygen O(5) is reduced from 0.307 to −0.06 v. u. Thus all the protons in the crystal structure are involved in hydrogen bonding.

The hydrogen bonds O(1)—H— —O(3) bind arsenate groups into chains (Fig. 1). The two hydrogen bonds from the water molecule connect arsenate groups from adjacent chains.

The chemical formula must be written $CaAsO_3(OH) \cdot H_2O$.

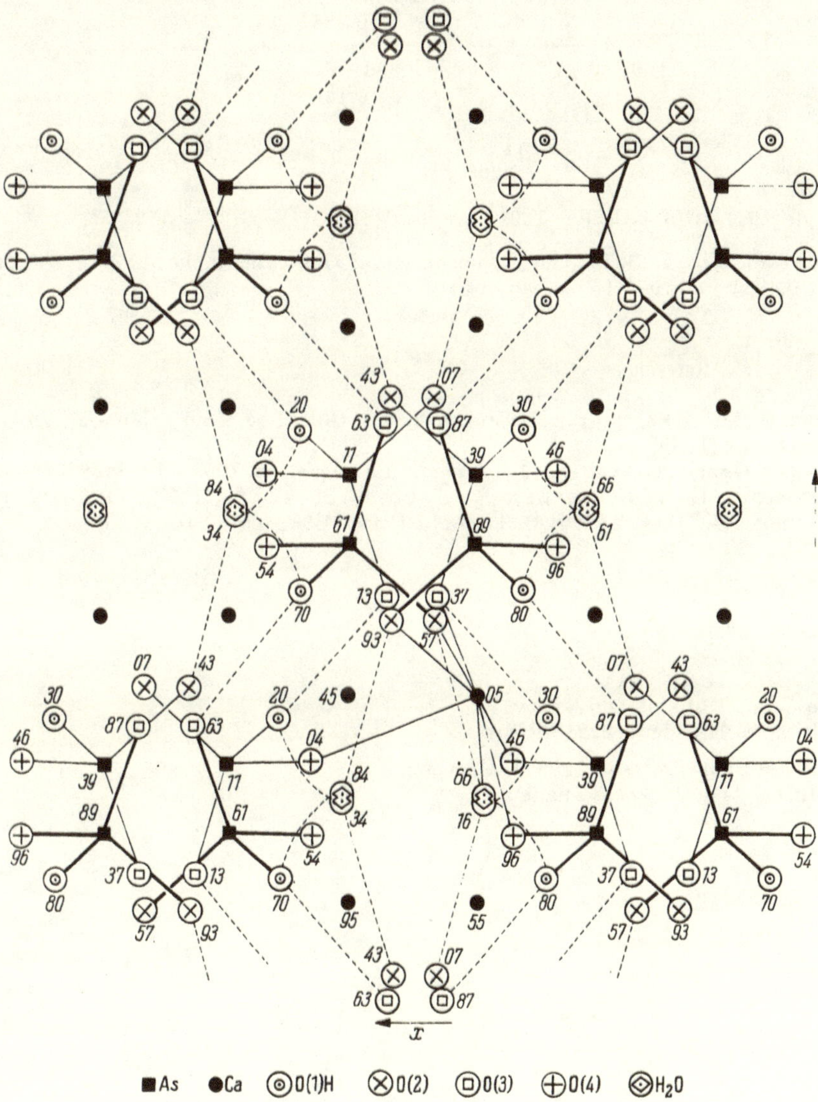

Fig. 1. Projection of haidingerite on (010). Heights in hundredths of *b*

Table 2

Approaches between nonbonded oxygen atoms in haidingerite

Donor / Acceptor	O(1)H Å v. u.	$O(5)H_2$ Å v. u.
O(1)	...	2.669 0.224
O(2)		2.856 0.145
O(3)	2.541 0.288	

Angle O(1)–O(5)–O(2) = 150°. Angle O(5)–O(1)–O(3) = 117°.

(Erratum, Fig. 1, S. 19: the y co-ordinate of the water molecule at $x \sim 0.14$ and $z \sim 0.49$ should read 16 instead of 61.)

References

CASSIEN, M., HERPIN, P., PERMINGEAT, F.: Bull. Soc. franç. Minéral. Crist. **89**, 18–22 (1966)

DONNAY, Gabrielle, ALLMANN, R.: Amer. Mineralogist 55, 1003–1015 (1970)

LIPPINCOTT, E. R., SCHROEDER, R.: J. Chem. Physics **23**, 1099–1106 (1955)

SHANNON, R. D., PREWITT, C. T.: Acta Cryst. **B25**, 925–946 (1969)

(Received June 14, 1971)

Authors' addresses:

Prof. J. D. H. DONNAY
The Johns Hopkins University
Baltimore, Maryland 21218, USA

Prof. Gabrielle DONNAY, McGill University
Montreal 110, Quebec, Canada

Kristall und Technik	7	1–3 1972	21–25

P. Hartman

Geologisch en Mineralogisch Instituut der Rijksuniversiteit, Leiden, The Netherlands

Structural Morphology of Crystals with the $K_2CuCl_4 \cdot 2\,H_2O$ Structure Type

The morphology of the $K_2CuCl_4 \cdot 2\,H_2O$ structure is deduced from the crystal structure. F forms are {011} and {110}, determined by the PBC's ⟨100⟩ and ⟨111⟩. A slice d_{110} has a peculiar structure in that it consists of an electrostatic neutral layer of K and Cl ions, and another neutral layer of Cu, Cl and H_2O. The K–Cl layer contains two PBC's while the other layer has the character of a K face.

Die Morphologie der $K_2CuCl_4 \cdot 2\,H_2O$-Struktur wird aus der Kristallstruktur abgeleitet. F-Formen sind {011} und {110}, bestimmt an den PBC-Vektoren ⟨100⟩ und ⟨111⟩. Die Ebenen d_{110} haben eine besondere Struktur, da sie aus einer elektrostatisch neutralen Schicht von K- und Cl-Ionen und einer anderen neutralen Schicht von Cu, Cl und H_2O bestehen. Die K–Cl-Schicht enthält zwei PBC-Vektoren, während die andere Schicht den Charakter einer K-Fläche hat.

1. Crystal structure

Recently the crystal structure of $K_2CuCl_4 \cdot 2\,H_2O$ was refined by means of neutron diffraction by Chidambaram and co-workers. Unit cell data are: $a = 7.477$ Å, $c = 7.935$ Å, space group $P\,4_2/mnm$, $Z = 2$. The coordinates for all atoms in the unit cell, except the hydrogens, are given in Table 1. The

Table 1

Atomic coordinates of $K_2CuCl_4 \cdot 2\,H_2O$

atom	x	y	z	atom	x	y	z
Cu(1)	0	0	0	Cl(1)	0.2161	0.2161	0
Cu(2)	1/2	1/2	1/2	Cl(2)	0.2839	0.7161	1/2
K(1)	0	1/2	1/4	Cl(3)	0.7161	0.2839	1/2
K(2)	1/2	0	3/4	Cl(4)	0.7839	0.7839	0
K(3)	1/2	0	1/4	Cl(5)	0.2262	0.2262	1/2
K(4)	0	1/2	3/4	Cl(6)	0.2738	0.7262	0
O(1)	0	0	0.2484	Cl(7)	0.7262	0.2738	0
O(2)	0	0	0.7516	Cl(8)	0.7738	0.7738	1/2
O(3)	1/2	1/2	0.7484				
O(4)	1/2	1/2	0.2516				

structure consists of copper atoms octahedrally surrounded by 4 Cl and 2 H_2O, and of potassium ions situated in the centre of a distorted cube, the vertices of which are occupied by Cl atoms.

The morphology is deduced from the structure using the PBC method (see e.g. HARTMAN and PERDOK; HARTMAN). The strong bonds in the structure are the Cu—Cl and K—Cl bonds. The bonds occurring within a unit cell are given in Table 2. As the Cu—H_2O and Cl—H_2O bonds are presumably weak they are ignored in the following considerations.

Table 2

Strong bonds in the $K_2CuCl_4 \cdot 2\,H_2O$ structure

Cu(1)	– Cl(1)	Cu(2)	– Cl(2)
	– Cl(4,1$\bar{1}$0)		– Cl(3)
	– Cl(6,0$\bar{1}$0)		– Cl(5)
	– Cl(7,$\bar{1}$00)		– Cl(8)
K(1)	– Cl(1)	K(2)	– Cl(2,0$\bar{1}$0)
	– Cl(2)		– Cl(1,001)
	– Cl(3,$\bar{1}$00)		– Cl(4,0$\bar{1}$1)
	– Cl(4,$\bar{1}$00)		– Cl(3)
	– Cl(5)		– Cl(6,0$\bar{1}$1)
	– Cl(6)		– Cl(5)
	– Cl(7,$\bar{1}$00)		– Cl(8,0$\bar{1}$0)
	– Cl(8,$\bar{1}$00)		– Cl(7,001)
K(3)	– Cl(1)	K(4)	– Cl(1,001)
	– Cl(3)		– Cl(2)
	– Cl(2,0$\bar{1}$0)		– Cl(3,$\bar{1}$00)
	– Cl(4,0$\bar{1}$0)		– Cl(4,$\bar{1}$01)
	– Cl(5)		– Cl(5)
	– Cl(7)		– Cl(6,001)
	– Cl(6,0$\bar{1}$0)		– Cl(7,$\bar{1}$01)
	– Cl(8,0$\bar{1}$0)		– Cl(8,$\bar{1}$00)

2. The [100] projection

First we consider the shortest lattice translation, ⟨100⟩. Figure 1 shows a projection along [100]. In this direction the following bond chain is found:

$$[100]:\ \mathrm{Cu(1)}-\mathrm{Cl(1)} \begin{matrix} \diagup\ \mathrm{K(3)}\ \diagdown \\ \diagdown\ \mathrm{K(2,00\bar{1})}\ \diagup \end{matrix} \mathrm{Cl(4,0\bar{1}0)}-\mathrm{Cu(1,100)}$$

This chain does not have a stoichiometric composition, so other atoms are added to it, namely $H_2O(1)$, $H_2O(2,00\bar{1})$, Cl(7) and Cl(6,0$\bar{1}$0). Together these atoms form a PBC without an electrostatic dipole moment perpendicular to its direction. An equivalent PBC occurs arround Cu(2). The PBC's are bonded to their nearest neighbours by strong bonds in the slices $d_{0\bar{1}1}$ and $d_{0\bar{1}1}$ only. Hence {011} is an F form.

In a layer d_{200} the PBC's are not bonded, so {100} is an S form in this zone. The same holds for a layer d_{002}, but then {001} is an S form in two equivalent zones [100] and [010]. This means that {001} is a K form rather than an S form. Parallel to (001) are layers of ions with alternating charges: 2 K^+ alternate with ($Cu^{++} + 4\,Cl^-$). Therefore the presence of this face may occur at high supersaturations.

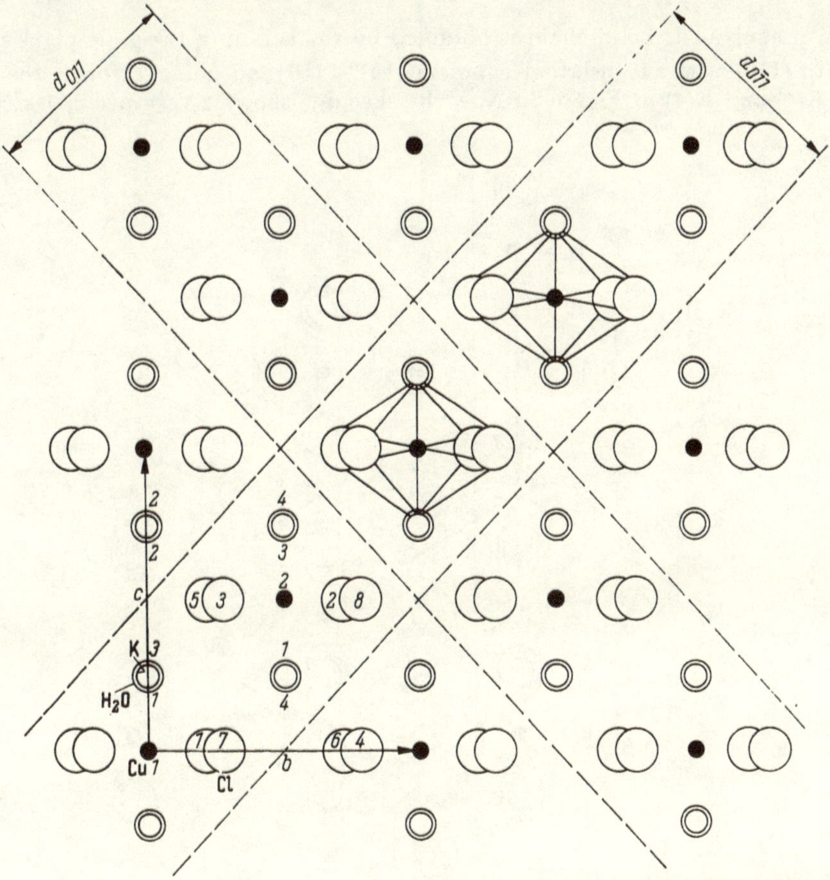

Fig. 1. Projection along [100]. Small solid circles: Cu; Open circles with increasing diameter: H_2O, K and Cl. Slices $d_{0\bar{1}1}$ and d_{011} are indicated. At the intersection of both slices a PBC [100] is shown in which Cu – Cl, K – Cl and Cu – H_2O bonds are drawn

3. The [111] projection

The F form {011} must be determined by other PBC's than ⟨100⟩ alone. A PBC [111] is easily found (Fig. 2).

$$[111]\,\alpha : \mathrm{Cu(1)-Cl(1)} \begin{matrix} \diagup \mathrm{K(1)-Cl(2)} \diagdown \\ \diagdown \mathrm{K(3)-Cl(3)} \diagup \end{matrix} \mathrm{Cu(2)} \begin{matrix} \mathrm{—K(2,010)} \diagdown \\ \mathrm{—K(4,100)} \diagup \end{matrix} \mathrm{Cl(4,001)-Cu(1,111)}.$$

By adding 4 other Cl atoms and 2 H_2O molecules the bond chain becomes stoichiometric.

These PBC's are bonded to their neighbours in the slices $d_{01\bar{1}}$ and $d_{10\bar{1}}$, which confirms their F character. In a slice d_{110} the PBC's are not bonded, so we might expect {110} to be an F form. There is, however, another possibility to construct a PBC:

$$[111]\,\beta : \mathrm{K(1)-Cl(2)-K(2,010)-Cl(7,011)-K(1,111)}$$

An equivalent bond chain is obtained by the action of the glide plane parallel to $(1\bar{1}0)$ with a translation component 1/2 [110] and going through the atoms K(1) and K(4) in Figure 2. Now the slice $d_{1\bar{1}0}$ shows a very peculiar structure.

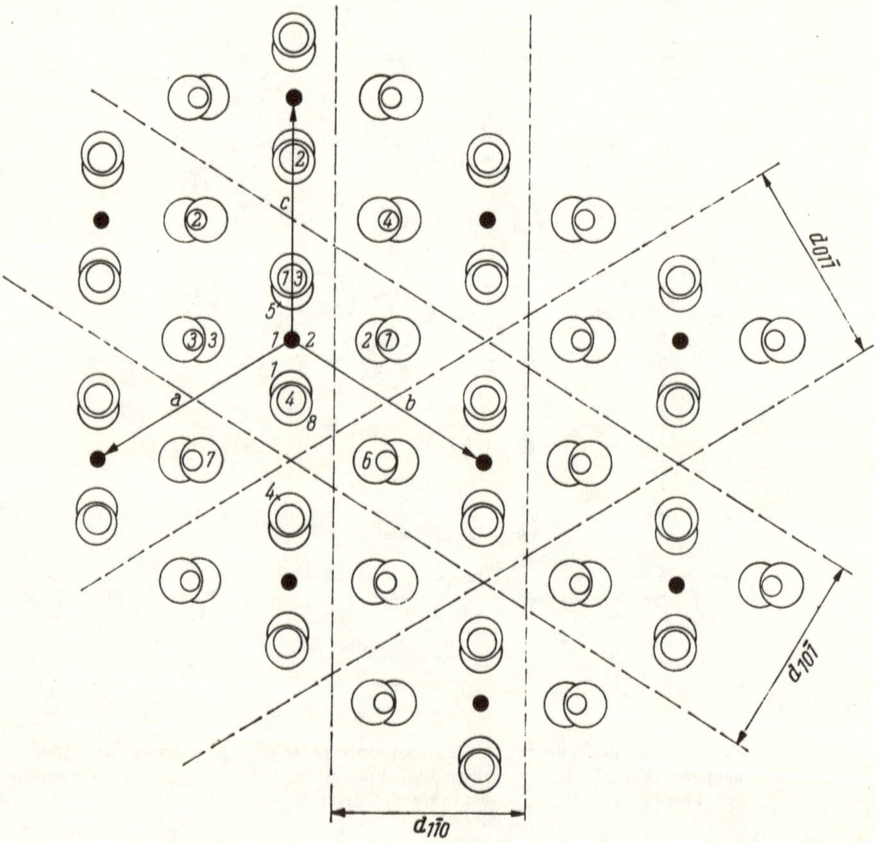

Fig. 2. Projection along [111]. Designation of atoms as in Figure 1. Slices $d_{01\bar{1}}$, $d_{10\bar{1}}$ and $d_{1\bar{1}0}$ are indicated. The PBC is found at the intersection of the first two slices

It consists of a neutral layer of coplanar atoms ($2\ Cu^{++} + 4\ Cl^- + 4\ H_2O$) and another neutral layer of almost coplanar atoms ($4\ K^+ + 4\ Cl^-$). The former layer in itself does not contain a bond chain, but the K–Cl layer does: it has the bond chains $[111]\beta$ and $[11\bar{1}]\beta$. Hence the slice $d_{1\bar{1}0}$ is non-symmetrical, but it has no electrostatic dipole moment perpendicular to itself and it contains two bond chains. These criteria suffice to classify {110} as an F form that can grow according to a layer mechanism. The surface structure is not determined as both the K–Cl surface and the $Cu–Cl_2–2\ H_2O$ surface have the same surface energy in the sense of Born. The K–Cl layer behaves as an F face, while the $Cu–Cl_2–2\ H_2O$ layer consists of isolated aggregates ($CuCl_2 \cdot 2\ H_2O$) of two types: one with short Cu–Cl bonds (2.285 Å) and the other with long bonds (2.895 Å).

This layer behaves in itself as a K face. The most stable configuration of the surface is therefore determined by kinetic factors. As the K–Cl layer, once formed, is not so easily disintegrated the surface might well be described as that of a K–Cl layer with only patches of the $Cu–Cl_2–2\,H_2O$ layer adsorbed on it.

4. Discussion

The F forms found are $\{011\}$ and $\{110\}$, determined by the PBC's $\langle 100\rangle$ and $\langle 111\rangle$. In order to account for habit changes at high supersaturation KLEBER distinguished two types of PBC's which he called "p-Vektoren" and "s-Vektoren", whether they were parallel (p) or oblique (s, schräg) with respect to a certain face. When the number of s-vectors is high, the face is likely to appear at high supersaturation as shown by KLEBER on a number of examples. In the present case the K forms with large interplanar distances are $\{111\}$ and $\{002\}$. The number of p- and s-vectors is:

$$(111) : d = 4.400\ \text{Å}\ , \quad 6\ s\text{-vectors and no } p\text{-vectors}$$

$$(002) : d = 3.967\ \text{Å}\ , \quad 4\ s\text{-vectors and } 2\ p\text{-vectors.}$$

It is therefore likely that $\{111\}$ appears at high supersaturation.

The face also has alternating positive ($2\,K^+$) and negative ($Cu^{++} + 4\,Cl^-$) layers parallel to its surface. Unfortunately facts about the influence of the supersaturation on crystals of this structure type are unknown. GROTH describes crystals of the K, Rb and NH_4 compounds. The general morphology is: $\{011\}$ large, $\{110\}$ smaller. On the NH_4 compound $\{001\}$ and $\{111\}$ have been noticed as subordinate forms.

It seems therefore that the observations are in agreement with the result of the foregoing analysis. The observed morphology is also in agreement with the law of DONNAY and HARKER, which predicts the sequence: $\{011\}$, $\{110\}$, $\{111\}$, $\{002\}$, $\{020\}$, ...

References

CHIDAMBARAM, R. et al.: Acta Cryst. **B26**, 827 (1970)
DONNAY, J. D. H., HARKER, D.: Amer. Mineralogist **22**, 446 (1937)
GROTH, P.: Chemische Krystallographie. Vol. I, 355, 356. Leipzig 1906
HARTMAN, P., PERDOK, W. G.: Acta Cryst. 8, 49, 521 (1955)
HARTMAN, P.: Z. Krist. **119**, 65 (1963)
KLEBER, W.: Z. Physik. Chem. **206**, 327 (1957)
KLEBER, W.: Z. Krist. **109**, 115 (1957)

(Received June 10, 1971)

Author's address:

Dr. P. HARTMAN
Geologisch en Mineralogisch Instituut
der Rijksuniversiteit
Garenmarkt 1B
Leiden
The Netherlands

Kristall und Technik	7	1–3	1972	27–35

I. KOSTOV

University of Sofia, Bulgaria

The Structural Patterns of Crystal Faces and Crystal Growth

Two main types of elementary slices are differentiated and the correspondence between their thickness, crystal habit variation and degree of supersaturation is indicated. A transition from spiral to accretion growth with increasing supersaturation is considered plausible and in agreement with observed habit modification.

Zwei Haupttypen von Elementarschichten sind zu unterscheiden. Die Korrespondenz zwischen ihrer Dicke, der Änderung des Kristallhabitus und der Übersättigung wird gezeigt. Ein Übergang vom Spiralwachstum zum Aggregatwachstum mit steigender Übersättigung ist sehr wahrscheinlich und in guter Übereinstimmung mit beobachteten Habitusänderungen.

Die kristallographische Orientierung einer Oberfläche fixiert auch ihre kristallchemische „Aktivität“.

Will Kleber, 1962.

Although there is a voluminous literature about the relationship between the structure and morphology of crystals, opinion on that subject still diverges, showing its very complex character. The starting point in the treatment of this relationship has always been the Law of Bravais extended by Donnay and Harker, which represents the fundamental structural-geometrical basis for further considerations. BECKENKAMP's (1929) rule for persistence of a crystal face, when coinciding with densely built up zones follows the conceptions of NIGGLI (1926a, b) and is elaborated by KLEBER (1939), and HARTMAN (1953, 1967), the latter defining the strong bonds as PBC vectors. Differentiating between *p*- and *s*-vectors KLEBER (1957a, b) underlines the significance of the number of the *p*-vectors for the stability of a crystal face, the number of *s*-vectors determining the "activity" of a face.

When the structural-geometrical aspect of that problem is considered important seem to be both the density of the lattice planes and that of the rows which are interdependent. Their dynamic behaviour during crystal growth depends largely on the prevailing conditions of crystallization, under which, to use NIGGLI's fitting terms, the phenotype develops at the expense of the genotype. When a well faceted crystal forms the significance of the structural patterns of the lattice planes for the growth of the corresponding faces is obvious. These patterns can be simple or very complicated, in the latter case greatly hindering a quantitative interpretation of the development of the crystal forms under varying conditions.

In what follows there is an attempt to generalize the structural patterns of the crystal faces of different substances and apply these generalizations qualitatively to the problem of modification of their crystal habits. The crystal habits of a substance can provide good evidence about the conditions of its crystallization, therefore they should be considered of prime importance, especially when interpretation of crystallization in nature is concerned.

1. Method for Differentiating Crystal Habits

A method by using the anisometricity of the crystal lattices is offered by the author, who presented also a rational, unified structural-genetic notation of them (Kostov 1962, 1965, 1968). The anisometricity is presented by the axial ratios c/a for the crystals with spindle symmetry, and $2\,c/(a+b)$, resp. $2b/(a+c)$ and $2\,a/(b+c)$ for the low-symmetry crystals. These ratios can be either less, approximately equal, or over one, the corresponding types being axial A, A^c, A^b or A^a, pseudoisometric (I), and planar P, P^c, P^b or P^a. The degree of axiality and planarity depends on the repetition of structural patterns therefore reductions in the values of a, b, and c parameters should be accordingly made when the structural genotype is evolved. The true isometric crystals of the I-type are conveniently denoted in concurrence with the primitive, body-centered, and face-centered lattices as I^c, I^d, I^o, corresponding to theoretical cubic, dodecahedral and octahedral habits.

The variation of the habits of the anisometric crystals (the phenotype) can be generally indicated by corresponding face or zone indices shown as subscripts. Thus for the axial A^c type the habits can be $A^c_{[uvw]}$, $A^c_{(hkl)}$, $A^c_{[uvw]\,(hkl)}$, and $A^c_{(hkl)\,[uvw]}$, i.e. elongated $[uvw]$, flattened on (hkl) or as $\{hkl\}$ habit corresponding to the crystal form with such indices, elongated $[uvw]$ and flattened on (hkl), and flattened on (hkl) and elongated $[uvw]$. Pseudoisometric such habits are denoted as A^c_i. Flattened habits specify predominant two-dimensional, elongated — one-dimensional development, hence the simplified structural-morphological notation A^c_2 and A^c_1 for the end types (Kostov, 1968).

2. Two Basic Types of Elementary Slices and Growth Behaviour

2.1 Isometric crystals

Turning to the structural patterns of the crystal faces the simple structures of NaCl and CaF_2 can be utilized in order to differentiate two basic types of elementary slices met also in complex crystal structures. One of them can be termed discrete (δ'), the other serial (δ''), the first one referred to the F-faces the second to the K-faces of Hartman (Fig. 1). Taking into consideration experimental work about the crystallization of these two substances a correlation can be made between the thickness and type of elementary slices, and supersaturation as an important factor for crystal habit modification. With increasing supersaturation the crystal habit succession for NaCl runs $\{100\}-\{110\}-\{111\}$, for $CaF_2-\{111\}-\{110\}-\{100\}$, i.e. in accordance with the thickness of the elementary slices and along the line of discrete — serial elementary slices, the thickest being of importance at lower supersaturations, those less thicker and of the serial type underlining the habits formed at higher supersaturations.

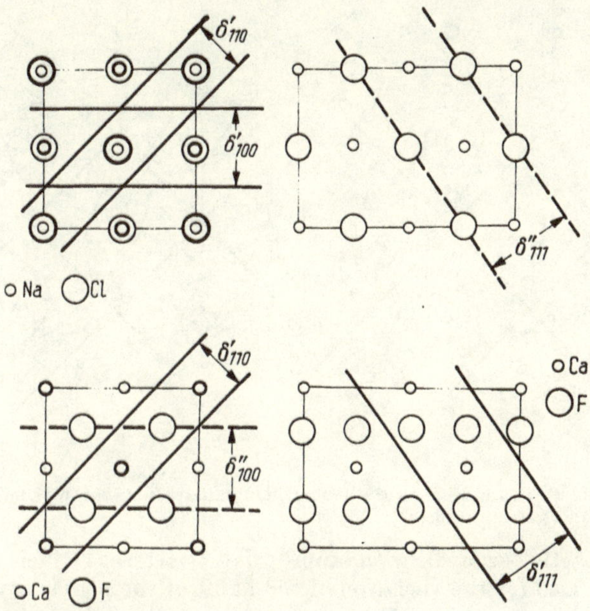

Fig. 1. Structural schemes of NaCl (above) and CaF_2 (below) projected on (001) and (110), with indicated discrete (heavy lines) and serial (dashed lines) elementary slices

2.2 Anisometric crystals

If the trend of modification of the cited isometric crystals is compared with that of the anisometric axial crystals a difference appears in that axial crystals follow an opposite tendency, i.e. higher supersaturations favour the development of crystal habits of the forms with thicker discrete elementary slices, then the habit being distinctly prismatic to acicular, in agreement with the rule of JANSEN (1933). In Figure 2 the structural schemes of stibnite Sb_2S_3 and natrolite $Na_2Al_2Si_3O_{10} \cdot 2\,H_2O$ are shown projected on (001) with indicated the thickest elementary slices, which are habit-determining at higher supersaturations. Both minerals are orthorhombic, belonging to the space groups D_{2h}^{16} – *Pbnm* resp. C_{2v}^{19} – *Fdd*2, possessing accentuated A^C axiality $(2\,c/(a + b) < 1)$ and structural patterns of the (110) faces, although rather complicated, belonging to the discrete type. Habit-determining in natrolite is the three-dimensional tetrahedral network with a pronounced chain-like concentration of the tetrahedral groups along the *c*-axis. There is a distinct pseudotetragonal screw arrangement of the tetrahedral chains, influencing the patterns of the serial elementary slices across the *c*-axis, that should be of a definite significance for the growth of the mineral along that direction. Stibnite has a perfect cleavage along (010) but the elementary slice of th's face is thinner than that of (110), which apparently accounts for its subordinate position in the stibnite crystals.

The $A_2 - A_1$ tendency in the habit modification of the axial crystals with increasing supersaturation should be explained not only in terms of short bonds but also in terms of the specific structural patterns of the crystal faces and a transition from a dislocation growth mechanism to accretion of nucleated

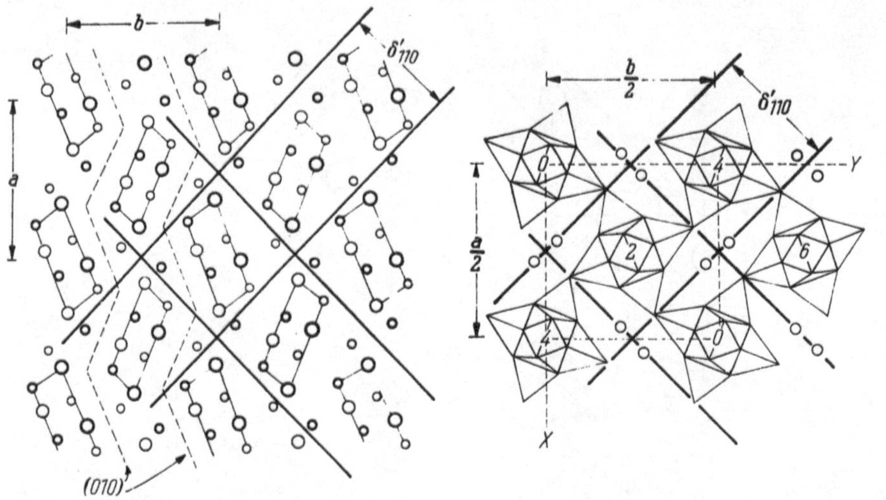

Fig. 2. Structural schemes of stibnite (left) and natrolite (right) with indicated principal discrete elementary slices

clusters (cybotactic groups) with appropriate patterns. Such explanation should be valid also for the variation of the habit of the planar crystals, which seems to follow in general lines the development of the habits of the isometric crystals, in that crystal faces with thicker discrete elementary slices persist under lower supersaturations. Thus Kern has shown that the layered structure of CdJ_2 produces thinner plates (P_2 habit type) at lower supersaturations, thicker — at higher supersaturations. Similar conclusions can be reached at from de Schulten's experiments with $Al(OH)_3$ (Groth).

3. Crystals of Minerals and their Genetic Conditions

Crystals of minerals provide additional data for the interpretation of the habit variation of the planar crystals. Molybdenite MoS_2, of the space group D_{6h}^4 — — $P6_3/mmc$, that has a typically planar c'/a ratio of 1.95, is found rather frequently in relatively high-temperature occurrences (pegmatites) as thin hexagonal plates of the $P_{(0001)}$ or P_2 type, whereas in certain low-temperature sulphide-bearing deposits it occurs as large bipyramidal crystals of the $P_{(10\bar{1}1)}$ or $P_{\hat{1}}$ type. Biotite mica from hypabyssal monzonitic intrusions rather rich in volatiles crystallizes as thin foliae of the $P^c_{(0001)}$ or P^c_2 type, whereas in volcanic rocks of the same composition appears frequently as prismatic $P^c_{[0001]}$ or P^c_1 crystals. The viscosity of the melt in the second case, owing to more rapid cooling, should be sonsidered higher than that of the former. Well-known are, on the other hand, thick tabular to prismatic books of muscovitic mica in pegmatites, obviously formed at higher temperatures and increased mobility of the components.

4. Cleavage Planes and Growth Forms

Pertinent to the low-high supersaturation habit problem is the observation of Kalb about the significance of the cleavage planes as growth forms. If {111} is the early, i.e. the low supersaturation form of CaF_2, and {100} such form for NaCl, it is logical to assume that mica, or whatever other substance with layered structure and perfect (0001) cleavage, would produce flattened crystals at lower supersaturations and prismatic at higher supersaturations. Axial crystal, however, display a reverse tendency, as shown by rutile TiO_2 with a distinct (110) cleavage, corresponding to the higher supersaturation A_1 habit type. Another striking exception from the KALB's rule is provided by the oxalic acid, which is also axial of the A^b type (a 6.12, b 3.60, c' 6.02) and crystallizes at higher supersaturations in long prismatic $A^b_{[001]}$ or A^b_1 habit. The oxalic acid has, however, (110) cleavage and corresponding faces built up of elementary slices discrete type but thinner (3.07 Å) in comparison to the elementary slices of the principal habit forms ($\delta'_{(101)} = 5.95$ Å).

5. Further Examples

Figure 3 represents schematically the structural patterns of the sole discrete and one serial elementary slices in mica and gypsum, which are typical for a relatively homogeneous (mica) and heterogeneous (gypsum) distribution of the pattern elements within the discrete elementary slices. The habit variation of mica crystals, owing to their pseudohexagonal symmetry, is along one direction mainly ($P_2^c - P_1^c$), whereas gypsum displays a variation along the b-axis (P_2^b—P_1^b) but is with a pronounced tendency towards prismatic or even acicular development along the c-axis as well ($P_{[010]}$ or P_{1c}^b type), due to columnar arrangement of the calcium and sulphate polyhedra along that direction. Gypsum crystals of the latter type are indicative of rapid crystallization and higher supersaturations, these apparently accentuating the axial motif within the structural pattern of the (010) face. Elongated [010] habits form probably also under higher supersaturations but under a different growth mechanism which might involve coalescence of cybotactic groups.

The possitibility of *conversion of a serial into a discrete multi-layered elementary slice* can be demonstrated with the structures of pyrrhotite, FeS (idealized),

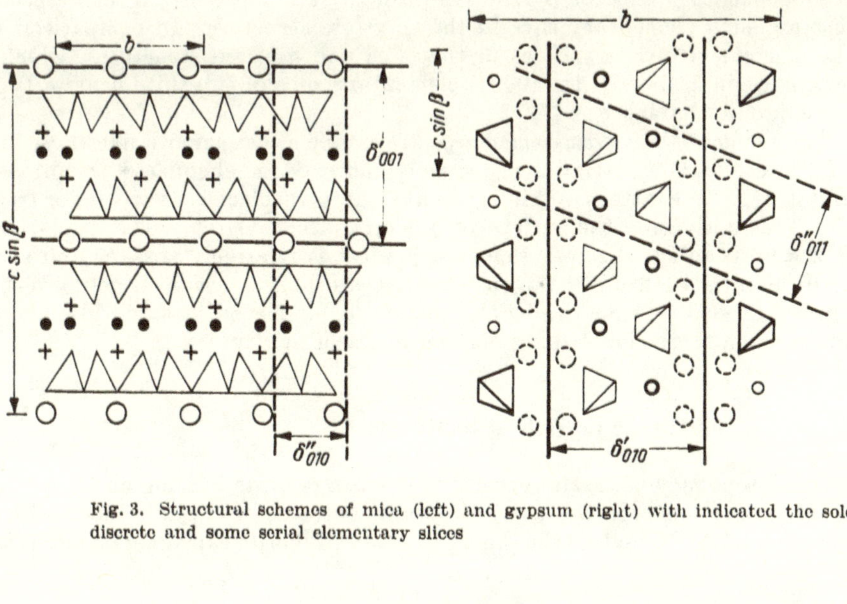

Fig. 3. Structural schemes of mica (left) and gypsum (right) with indicated the sole discrete and some serial elementary slices

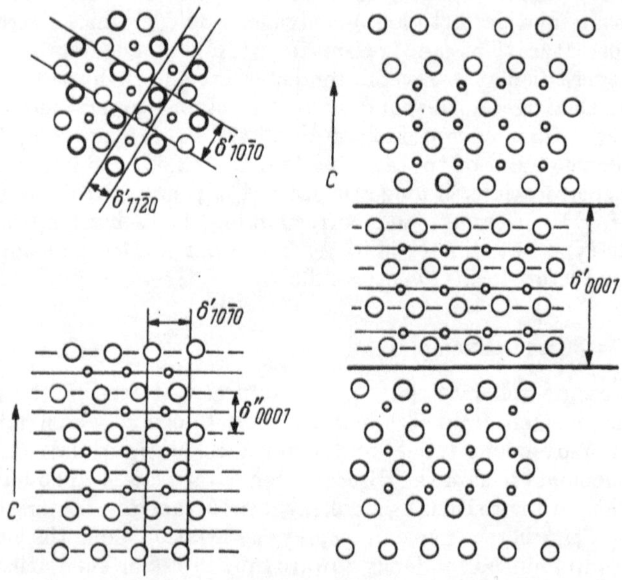

Fig. 4. Pyrrhotite (left) and smythite (right) structural schemes showing relationship between serial and discrete elementary slices

of the space group $D_{6h}^4 - P6_3/mmc$, with an axial ratio c'/a 0.83, and smythite, Fe_3S_4, of the space group $D_{3d}^5 - R\bar{3}m$, with an axial ratio c'/a 3.25 (Fig. 4). In concurrence with the pronounced layered structure of smythite are its thin platy crystals and micaceous cleavage on (0001). Pyrrhotite is of the *A*-type and occurs as columnar or thick tabular crystals devoid of cleavage, or with parting on (0001) due probably to epitaxial smythite intercalations. The serial 0001 elementary slice of pyrrhotite through stacking rearrangement becomes discrete such elementary slice in the smythite structure. In comparison with the discrete elementary slices in the mica and gypsum structures, which are heterodesmic in bonds, the (0001) elementary slice of smythite bears a typical homodesmic character.

The *surface of a discrete elementary slice* can be either smooth like those in the smythite and mica structures, or wavy, notched or chequered, as presented in Figure 5. The latter two kinds can have units displaced screw-wise or trench-like, thus activating the surface to one or another degree. The displacement of the units in the discrete elementary slices is relatively shallow and should be differentiated from analogous architecture in the serial elementary slices. These can also be smooth, deeply grooved or displaced in a rod-like manner, but in them there is a striking horizontal discontinuity.

6. Minute Crystals as Sub-Units

Another tendency of crystal growth that has a definite bearing on the problem of interdependence between the structural patterns of crystal faces and habit variation formally revives the theory of Haüy. Fluorite can again be taken as an

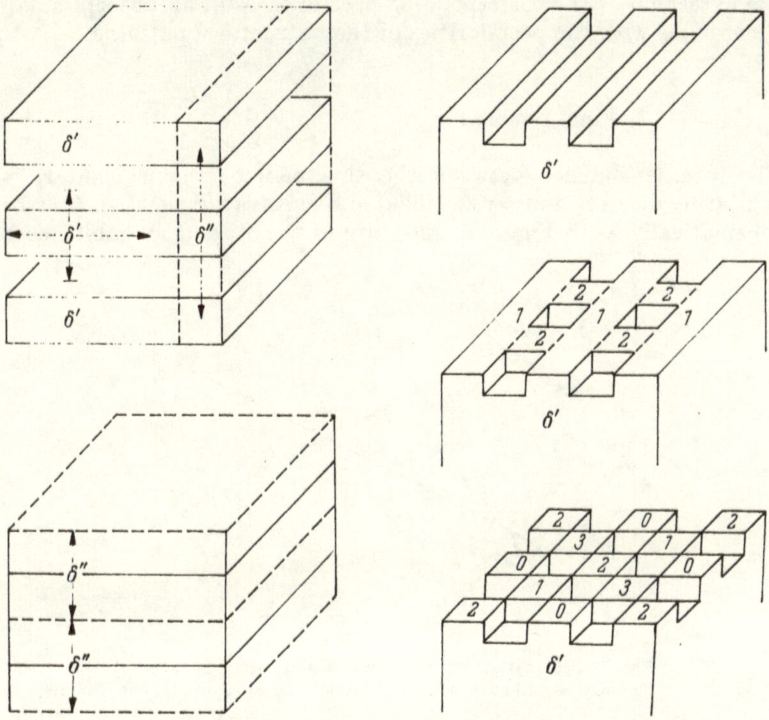

Fig. 5. Left: discrete elementary slices (above) and serial ones (below), both with smooth surfaces. Right: wavy, notched, and chequered surfaces of discrete elementary slices. Analogous can be the serial elementary slices but they bear a discontinuous horizontal arrangement as shown above left in connection with the smooth discrete slices. Numbers indicate possible displacement

example. The octahedral fluorite crystals are occasionally built up of stacked minute crystals whose habit is either octahedral, or dodecahedral, or cubic. Dodecahedral crystals of the same mineral are found formed from stacked minute dodecahedra or cubelets, and finally cubic crystals are frequently built up of aggregated small cubes. If that tendency in fluorite is coupled with the one concerning the modification of the habits under different supersaturations, it can be said that lower supersaturation crystal habits are formed by stacking of minute or submicrscopic crystals of the same or of the higher supersaturation habit types. This rule seems to hold good for the anisometric crystals as well, as exemplified by short prismatic to tabular $A_{(001)}$ or A_2 tourmaline habits composed of acicular $A_{[0001]}$ or A_1 type minute crystals.

The frequent occurrence of composite habit forms amidst crystals of various substance suggests that the *accretion* is indeed a highly probable mechanism of growth being of especial significance under the "rough" environment prevailing in nature. Such growth has been demonstrated by Shubnikov, Sheftal et al., and others. It is appropriate to recall here the thought of Frank that "growth is not really growth but a sort of stimulated re-nucleation". At higher supersaturations three-dimensional, two-dimensional, or/and one-dimensional clusters can easily form, which together with free atoms can contribute towards the extension of the corresponding crystal faces in an autoepitaxial manner in accordance with the peculiarities of their structural patterns.

7. Final remarks

The interdependence between the thickness of the elementary slice (differentiating discrete and serial kinds) and supersaturation can thus be presented schematically as in Figure 6. The site of the A_i and P_i habits corresponds to

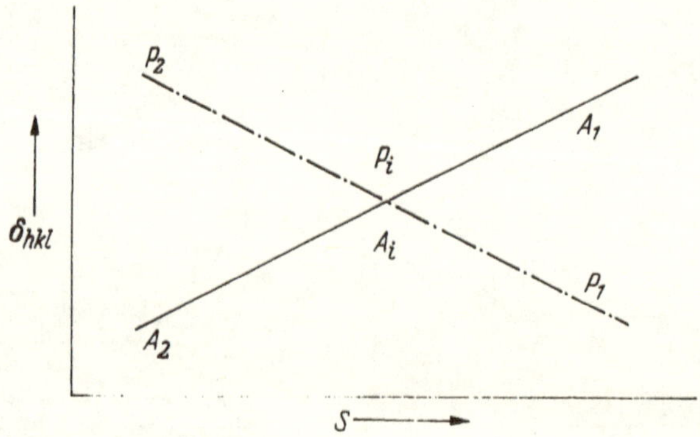

Fig. 6. Schematic representation of the interdependence between thickness of elementary slices and supersaturation for the axial and planar crystals

conditions favouring the growth of crystals richest in forms and probably of rather large size. To the right of it twin-formation appears to be stimulated. The transition between dislocation, layer, and accretion mechanisms of growth seems to follow the abscissa.

In addition to supersaturation another principal habit-modifying factor is the influence of admixtures, but as Kleber (1957) stated „führt Übersättigung zu den gleichen Effekten wie die Adsorption". Along this line can be treated for instance the change of the cubic habit of MgO into octahedral under the adsorption of OH ions. When grown in a OH-bearing medium films of brucitic $Mg(OH)_2$ composition are apparently formed on the (111) planes of periclase thus hindering their development. As known, crystals of MgO formed through sublimation are invariably of cubic habit.

References

Beckenkamp, J.: Cbl. Miner. Geol. Palaeont. Abt. A, 289 (1929)
Frank, F. C.: Proceed. Int. Conf. Cryst. Growth 1958
Groth, P.: Chemische Kristallographie **1**, 122 (1906)
Hartman, P.: Diss. Groningen 1953
Hartman, P.: In „Rost Kristallov" **7**, 8 (1967)
Jansen, W.: Z. Krist. **85**, 239 (1933)
Kalb, G.: Cbl. Miner., Abt. A, 271 (1928)
Kern, R.: Bull. Soc. franç. Minér. Crist. **78**, 461, 497 (1955)
Kleber, W.: N. Jb. Min. A Beil.-Bd., **75**, 72 (1939)
Kleber, W.: Z. physik. Chem. **206**, 327 (1957a)

KLEBER, W.: Z. Krist. **109**, 115 (1957b)
KOSTOV, I.: Mineral. sbornik, Lvov **16**, 75 (1962)
KOSTOV, I.: Bull. „Str. Dimitrov" Inst. Geol. **14**, 33 (1965)
KOSTOV, I.: IMA Papers and Proceedings, 5th General Meeting, 100 (1968)
NIGGLI, P.: Z. Krist. **63**, 49 (1926a)
NIGGLI, P.: Lehrbuch der Mineralogie II. Berlin 1926b
SHEFTAL, N. N. et al.: Izv. Akad. Akad. USSR, ser. phys. **21**, 146 (1957)
SHUBNIKOV, A. V.: Kak rastut crystali (How the crystals grow), Moscow 1935

(Received May 20, 1971)

Author's address:

Prof. Dr. I. KOSTOV
Lehrstuhl für Mineralogie und Kristallographie
Universität Sofia
Sofia/Bulgarien
Moskowska 49

Kristall und Technik | 7 | 1–3 | 1972 | 37–41

E. I. Givargizov, N. N. Sheftal'

Institute of Crystallography, Academy of Sciences USSR, Moscow

Morphology of Germanium Whiskers

The morphology of germanium whiskers grown by the VLS technique in chloride process is studies. It is shown that the whiskers can be grown in this process over a wide temperature range. The effect of surface active impurities on the morphology and the growth direction of whiskers is studied.

Es wird die Morphologie nach der VLS-Technik im Chloridprozeß gewachsener Germanium-Whiskers untersucht und gezeigt, daß die Whisker in diesem Prozeß über einen weiten Temperaturbereich wachsen können. Der Einfluß oberflächenaktiver Verunreinigungen auf die Morphologie und die Wachstumsrichtung der Whisker wird experimentell verfolgt.

1. Introduction

Against a background of numerous studies concerning whisker growth from vapor phase stand out those in which a liquid layer is situated between the vapor and the growing crystal. It is a new concept of crystal growth, so-called "vapor-liquid-solid", or VLS, growth mechanism (Wagner, Ellis). Such studies deal mainly with silicon: many papers are concerning that material (Berežkova; Wagner), while there are only a few publications on germanium whiskers grown by VLS-technique. Wagner, Doherty, Ellis and Kamadjiev, Mladjov, Velchev grew germanium whiskers by a iodide disproportionation process, and Bootsma used the pyrolysis of germane (GeH_4). It was claimed (Wagner) that the hydrogen reduction of halides, such as $GeCl_4$, could not be used for VLS-growth of germanium but it was shown later (Givargizov, Sheftal') that it is possible.

There are no detailed investigations on the morphology of micron- and submicron-size germanium whiskers while it is the micromorphology that gives information on the whisker growth mechanism.

In this paper further experiments are described performed with the aim for correlating the micromorphology with crystallization conditions for germanium whiskers.

2. Experimental Technique

Crystallization was performed in a vertical reduction cell of a deposition apparatus described elsewhere (Givargizov). Whisker growth commenced when mixtures of germanium tetrachloride and purified hydrogen at 650–900 °C were introduced into the system. In some experiments the mixtures were doped with surface-active impurities, such as arsenic.

The whiskers were grown onto the germanium (111)-orientation wafers, 0,8 mm in thickness; these wafers were mechanically polished and then etched in $HF/HNO_3/CH_3COOH$ solution for removing a surface abraded layer. A gold film 100—1000 Å in thickness was then deposited at 10^{-5} torr over the entire surface of the wafers. On heating the metallic film was alloyed with the substrate where upon the liquid alloy, under the influence of surface tension broke up into separate globules. These globules then served as the "seeds" for whisker growth.

The resulting whiskers were examined with the Reichert optical microscope and with the scanning electron microscope JSM-2 using the surface topography mode of operation.

3. Results and Discussion

Scanning electron micrographs of the as-grown germanium whiskers are shown in Figures 1 to 5.

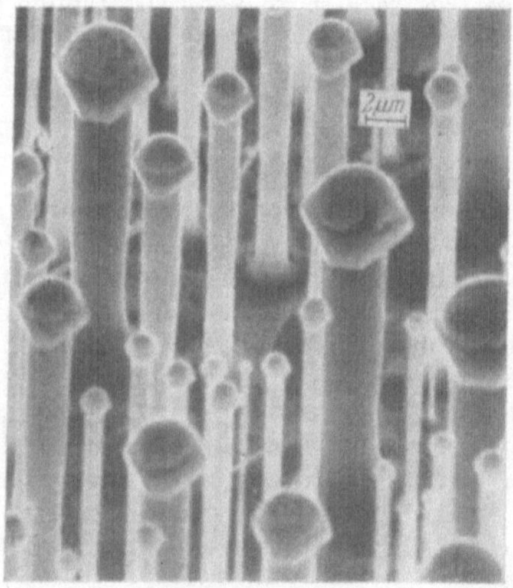

Fig. 1. Gemanium whiskers grown without deliberate addition of a surface-active impurity. Growth temperature is 900 °C. Scanning electron micrograph. Tilt angle $\alpha = 45°$

1. The whiskers prepared in optimal conditions (see Figs. 1 and 3) are normal to the (111)-surface, i.e. grow in the direction [111] to continue the corresponding direction of substrate.

2. The whiskers grown without deliberate addition of a surface-active impurity (Figs. 1 and 2) are morphologically rather different from those grown in the presence of such impurities (Figs. 3, 4 and 5). The first group of whiskers mentioned above (we shall refer to them as "type-I whiskers") has peculiar caps faced from bottom by three $\{\bar{1}11\}$-planes; the second group ("type-II") is morphologically similar to silicon whiskers (WAGNER; GIVARGIZOV).

3. Various instabilities are inherent to whiskers grown under non-optimal conditions (see Figs. 3, 4 and 5).

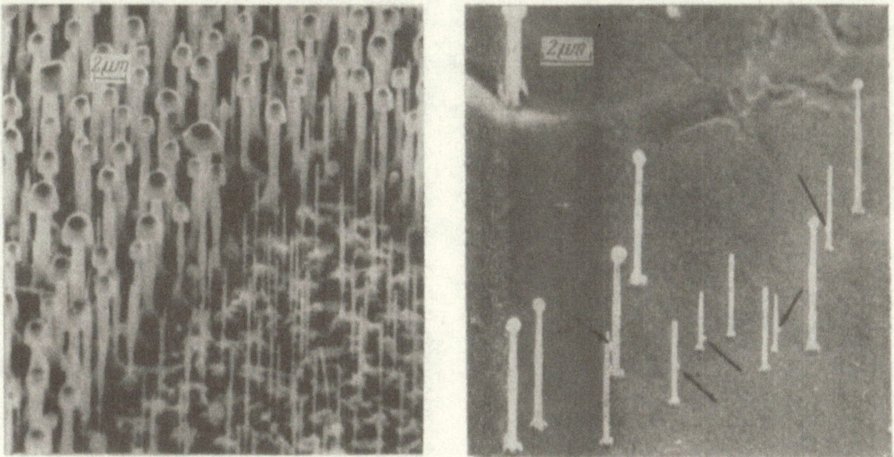

a) b)

Fig. 2a, b. Instabilities of whiskers grown without surface-active impurity. Scanning electron micrographs, $\alpha = 45°$

a) For the type-I whiskers the instability manifests itself in losing their caps, this phenomenon being mainly characteristic of the thinnest whiskers. In Figure 2a there is an area (underneath, on the right) from which the gold film previously has been partly removed. This has led to formation of relatively thin whiskers whose caps are lost. This losing is thought to have occurred during cooling, after $GeCl_4$ supply was ceased: due to the transient process and under the influence of surface tension at the liquid-solid interface the droplet-like cap slips along the side of the whisker downwards to its foot, as shown in Figure 2b. Arrows point the frozen droplets.

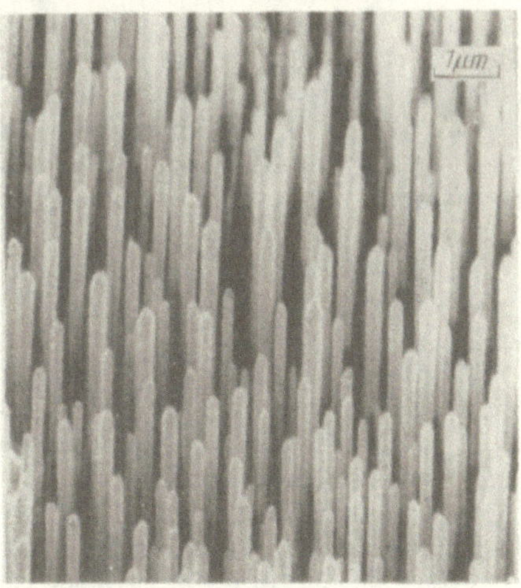

Fig. 3. Germanium whiskers grown in presence of arsenic as surface-active impurity. Growth temperature is 700 °C. Scanning electron micrograph, $\alpha = 45°$

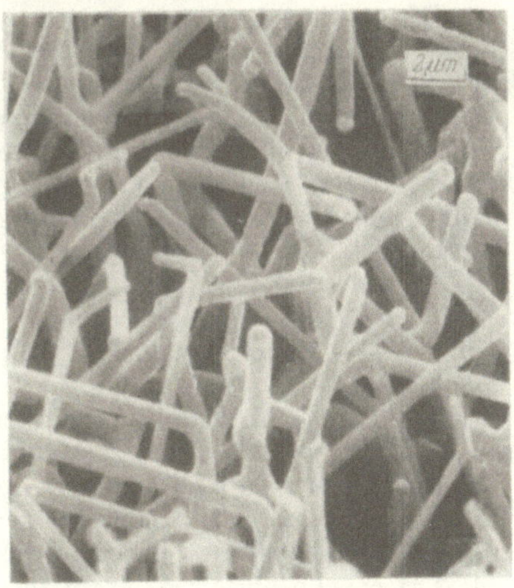

Fig. 4. Instability of whiskers grown at relatively high concentrations of surface-active impurity. Scanning electron micrograph, $\alpha = 45°$

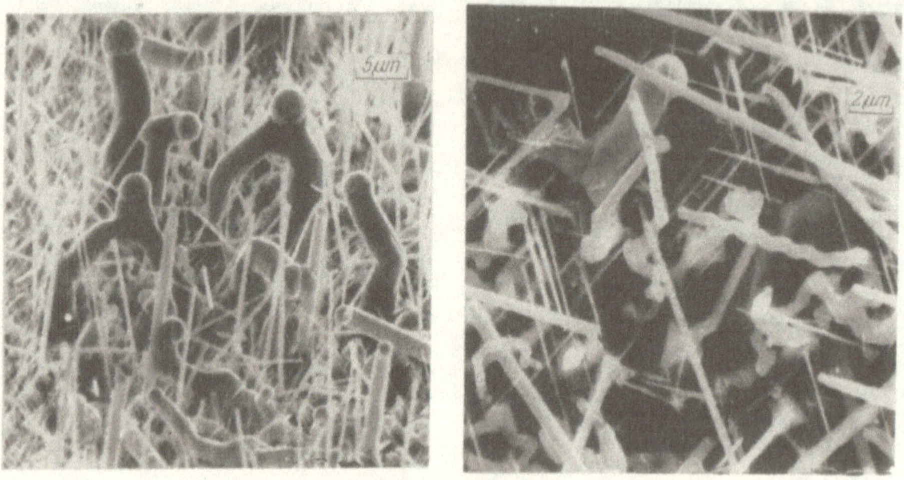

a) b)

Fig. 5a, b. Various types of instability of whiskers grown in presence of surface-active impurities. Scanning electron micrographs, $\alpha = 45°$

b) For the type-II whiskers instability manifests itself in kinking and branching (Figs. 4 and 5a), as it was observed by WAGNER, and in formation of twisted or nodule whiskers (Fig. 5b). However, in this case caps stay in their former position. As can be seen in Figure 5a, the thicker the whisker, the higher its instability. It can be associated with differences in surface tension of thin and thick whiskers. On the other hand, the whiskers, shown in Figures 3 and 4, have nearly the same diameters, but they were grown at different concentrations of surface-active impurity (the concentration was higher for the whiskers shown in Figure 4). By comparing these facts themselves and taking into account the instability of the type-I whiskers we can conclude that the surface-active impurity reduces both the specific surface energy σ_{LS} of the interface "liquid-solid" (this leads to disappearance of the instability inherent for the type-I whishers) and the energy σ_{LV} of the interface "liquid-vapor" (this leads to the appearance of the growth direction instability inherent for the type-II whiskers).

4. Conclusions

1. The micromorphology of the germanium whiskers grown by the VLS-technique in chloride process is studied. It is shown that whiskers can be grown in this process over a wide temperature range (from 650 to 900 °C).

2. The whiskers of two morphological types were observed. The differences between these types are related to the effect of surface-active impurities.

4. The conclusion is that arsenic as the surface-active impurity reduces the specific surface energy both of the "liquid-vapor" and of "liquid-solid" interfaces. The impurity affects essentially both morphology and growth direction instability of the whiskers.

We are grateful to L. N. Obolenskaya for experimental assistance.

References

BEREŽKOVA, G. V.: Nitevidnye Kristally („The Whiskers"), Moskva 1969

BOOTSMA, G. A.: Report at the Conference on Crystal Growth and Epitaxy from the Vapor Phase, Zürich, Sept. 23–26, 1970

GIVARGIZOV, E. I.: Soviet Physics – Solid State **6**, 1415 (1964)

GIVARGIZOV, E. I., SHEFTAL', N. N.: J. Crystal Growth **9**, 326 (1971)

KAMADJIEV, P. R., MLADJOV, L. K., VELCHEV, N. B.: C.r. Bulg. Acad. Sci. **19**, 779 (1966)

WAGNER, R. S.: in "Whisker Technology", Ed. A. P. Levitt, New York 1970, pp. 47–119

WAGNER, R. S., DOHERTY, C. J., ELLIS, W. C.: J. Metals **16**, 761 (1964)

(Received June 11, 1971)

Authors' address:

Dr. E. I. GIVARGIZOV, Prof. Dr. N. N. SHEFTAL'
Institute of Crystallography
Academy of Sciences USSR
Leninsky prospect 59
Moscow, B-333
USSR

| Kristall und Technik | 7 | 1–3 | 1972 | 43–73 |

S. Toschev † und I. Gutzow

Institut für physikalische Chemie der Bulgarischen Akademie der Wissenschaften, Sofia

Nichtstationäre Keimbildung: Theorie und Experiment

Vorliegende Arbeit gibt einen zusammenfassenden Bericht zur Kinetik der Keimbildung, wobei besonders der nichtstationäre Verlauf dieses Prozesses untersucht wird. Auf Grund einer eingehenden Diskussion des Keimbildungsmechanismus wird der Einfluß aktiver Unterlagen, wie auch die Art der Ausgangs- und der sich bildenden neuen Phase auf die nichtstationäre Zeitverzögerung berechnet. In diesem Zusammenhang werden auch einige Wahrscheinlichkeitsaspekte der homogenen und heterogenen Keimbildung erörtert und die Einwirkung nichtstationärer Effekte auf das Kristallwachstum und den summarischen Kristallisationsverlauf studiert. Die theoretischen Schlußfolgerungen werden mit dem Experiment verglichen, wobei hauptsächlich Nachdruck auf methodische Probleme gelegt wird. Zwei experimentelle Beispiele – die elektrolytische Keimbildung von Metallen und die angeregte Kristallisation von Gläsern werden eingehender betrachtet.

The present paper reviews some basic problems in the kinetics of nucleation giving particular attention to the non-steady state character of this process. The detailed analysis of the mechanism of nucleation makes it possible to evaluate the influence of active substrates and that of the properties of both the mother and new phases on the magnitude of the non-steady state time lag. Some probabilistic aspects of homogeneous and heterogeneous nucleation as well as the impact of non-stationary effects on crystal growth and overall crystallization rate are also studied. The theoretical predictions are compared with available experimental data giving preference to points of methodical interest. Two experimental examples – the electrolytic nucleation of metals and the catalyzed crystallization of glasses are considered in more detail.

1. Einleitung

Gewöhnlich wird die Theorie der Phasenumwandlungen in stationärer Näherung betrachtet; insbesondere gilt dies für ihre Anwendung auf verschiedene Kondensations- und Kristallisationsfälle. In den letzten Jahren erschien jedoch eine Anzahl von Arbeiten, in denen die Kinetik der Keimbildung eingehender untersucht wurde. Dabei wurde bewußt die Betonung auf den nichtstationären Verlauf dieses Prozesses gelegt. Dies ist verständlich, wenn man in Betracht zieht, daß die Bildung einer neuen Phase seiner physikalischen Natur nach ein nichtstationärer Vorgang ist, so daß sich die stationären kinetischen Kennzeichen der Keimbildung erst nach Verlauf einer gewissen Zeit einstellen. Inwiefern solche nichtstationären Effekte in einem gegebenen konkreten Falle eine maßgebende Rolle spielen, hängt naturgemäß von ihrer Größe und der benutzten Untersuchungsmethodik ab. Es sei jedoch hervorgehoben, daß beim Betrachten der nichtstationären Keimbildung keine spezielle Klasse von Phenomänen diskutiert wird, sondern ein fundamentales Merkmal jedes Phasenbildungsvorganges erster Ordnung untersucht wird. Dies wurde erstmalig in den klassisch gewordenen Arbeiten von Zeldovich und von Frenkel eindeutig festgelegt. Die logische Weiterentwicklung der Theorie verlangt somit die Betrachtung des allgemeineren, nichtstationären Keimbildungsverlaufes. Das erklärt weitgehend das Interesse, welches diesen Effekten entgegengebracht wird. Andererseits wurde aber auch die theoretische Behandlung dieser Fragen durch den Fortschritt der experimentellen Verfahren zur näheren und quantitativen Untersuchung der Anfangsstadien der Kristallisations- und Kondensationsvorgänge stark angeregt, welche gestatten, daß nicht stationäre Effekte auch in Fällen erfaßt werden konnten, wo sie im Millisekundenbereich liegen. Ein eingehendes Studium der nichtstationären Keimbildung wurde in den letzten Jahren insbesondere im Institut für physikalische Chemie der Bulgarischen Akademie der Wissenschaften durchgeführt. Es wurden dabei sowohl einige theoretische Fragen allgemeinerer Natur untersucht wie auch der Versuch gemacht, die nichtstationäre Keimbildung experimentell zu verifizieren. So gelangte man zur Klärung einer Reihe von Versuchsergebnissen auf dem Gebiet der elektrolytischen Metallabscheidung und der Kristallisation von viskosen Schmelzen. Im vorliegenden zusammenfassenden Bericht soll ein einheitlicher Überblick über die wichtigsten Ergebnisse und Schlußfolgerungen dieser Untersuchungen gegeben werden. Damit das Problem nach Möglichkeit klar dargestellt wird, sollen besonders die Grundlagen der gegenwärtigen kinetischen Keimbildungstheorie erläutert werden. Dann werden hauptsächlich methodische Fragen behandelt und im vierten Abschnitt eingehender die erhaltenen experimentellen Ergebnisse in verschiedenen Keimbildungsfällen betrachtet. Somit werden die theoretischen Ansätze soweit wie möglich mit experimentellen Ergebnissen verglichen. Es ist zu hoffen, daß die darauf folgende Diskussion die Notwendigkeit einer umfassenden und kritischen Überprüfung mancher üblich gewordener Ansichten und Vorstellungen vom Standpunkt der nichtstationären Keimbildung zeigen wird.

2. Theoretische Betrachtungen

2.1. Allgemeinere Grundlagen

In letzter Zeit wurden von Zinsmeister die Grundlagen der kinetischen Behandlung der Keimbildung eingehend zusammengefaßt. Wir möchten jedoch einige Merkmale dieses Prozesses besonders hervorheben.

Das Entstehen einer neuen Phase wird gewöhnlich als eine Kettenreaktion betrachtet, die zur allmählichen Vergrößerung von Mikroaggregaten (oder heterophase Fluktuationen nach Frenkel) führt. Wenn g die Zahl der Atome in den Aggregaten ist, gilt für die zeitliche Veränderung der Konzentration einer gegebenen Klasse von Aggregaten der allgemeine kinetische Ausdruck (s. z. B. Ree et al.; Andres, Boudart)

$$\frac{dc_g}{dt} = \sum_{m=1}^{q-1} D_{m\to g}\, c_m - \sum_{m=1}^{q} D_{g\to m}\, c_g \,, \tag{1}$$

$$g = 1, 2, \ldots, q\,, \quad D_{q\to m} = 0\,. \tag{1a}$$

Die erste Summe in der so zusammengestellten Bilanz berücksichtigt alle möglichen Reaktionen, die zur Bildung von Aggregaten der Klasse g beitragen; die zweite Summe dagegen beschreibt alle Abbaureaktionen. Mit $D_{m \to g}$, $D_{g \to m}$ sind die entsprechenden Übergangswahrscheinlichkeiten gekennzeichnet. Die Kettenreaktion wird bei einer genügend großen Atomanzahl q abgebrochen, so daß das weitere Wachstum von solchen Aggregaten hemmungslos, bis zum Erreichen makroskopischer Abmessungen verläuft. Diese Voraussetzung ist in der Bedingung (1a) enthalten. Die Keimbildungskinetik gilt als völlig beschrieben wenn die Lösung des Gleichungssystems (1), d. h., wenn die Funktionen $c_g(t)$ bekannt sind. Darin besteht das Hauptziel der Keimbildungstheorie.

Zur Vereinfachung der Aufgabe werden nur Übergänge zwischen Aggregaten betrachtet, die sich um einen einzigen Baustein unterscheiden; es wird also die eventuelle Koaleszenz der Aggregate vernachlässigt. Wenn außerdem die Ausgangskonzentration der einzelnen Atome konstant gehalten wird, gehen die Gleichungen (1) in

$$\frac{dc_g}{dt} = (D_{g-1}^+ c_{g-1} + D_{g+1}^- c_{g+1}) - (D_g^+ c_g + D_g^- c_g) \tag{2}$$

$$g = 2, 3, \ldots, q\,, \qquad D_{g \geqq q}^- = 0$$

über, wobei

$$D_g^+ = s_g z_g\,, \tag{2a}$$

$$D_g^- = s_g w_g \tag{2b}$$

die Anlagerungs- bzw. die Abbauraten eines Aggregats mit der Oberfläche s_g bezeichnen, während z_g und w_g dieselben Größen auf die Flächeneinheit bezogen bedeuten. Im Rahmen dieser Vorstellungen definiert die Bedingung

$$D_k^+ = D_k^-$$

die kritische Keimgröße g_k, welche im labilen Gleichgewicht mit der übersättigten Mutterphase steht. In allen übrigen Fällen gelten die angegebenen Ungleichheiten

$$D_g^+ \gtrless D_g^- \quad \text{bei} \quad g \gtrless g_k\,,$$

welche erkennen lassen, daß die Existenz von unterkritischen Keimen der thermodynamischen Tendenz im ganzen System widerspricht. Deswegen kann die kritische Größe nur durch örtliche Schwankungen der freien Energie erreicht und übertroffen werden.

Das Gleichungssystem (2) kann vermittels des Stromes I_g der Aggregate durch die Abmessung g ausgedrückt werden

$$I_g = D_{g-1}^+ c_{g-1} - D_g^- c_g\,, \tag{3}$$

$$\frac{dc_g}{dt} = I_g - I_{g+1}. \tag{3a}$$

Bei Abwesenheit jeglichen irreversiblen Molekültransports aus der Ausgangsphase in die neue Phase ($I = 0$), stellt sich eine Gleichgewichtsverteilung von Aggregaten ein, welche durch eine Boltzmann'sche Verteilungsfunktion wiedergegeben ist

$$\frac{n_g}{n_{g-1}} = \frac{D_{g-1}^+}{D_g^-} = \exp\left[\frac{\Delta G(g-1) - \Delta G(g)}{kT}\right]. \tag{4}$$

Hier bedeutet $\Delta G(g)$ die mit der Bildung eines g-atomigen Aggregates verbundene Änderung des thermodynamischen Potentials des ganzen Systems. Bilden sich irreversibel Keime, dann ist $I \leqq 0$ und für die Kinetik des Vorganges sind die Gleichungen (2) maßgebend. In diesem Falle hat die Gleichgewichtskonzentration n_g keinen reellen Sinn; dennoch bietet ihre Beibehaltung als Rechengröße in der Theorie gewisse Vorteile.

Der besondere Fall $I = \text{const} = I_0$ entspricht einem stationären Zustand, bei dem die Aggregatkonzentration für jede Größe zeitunabhängig erhalten bleibt ($dc_g/dt = 0$). Eine Lösung der Aufgabe unter dieser Voraussetzung ist in den dreißiger Jahren von einer Reihe von Autoren (Farkas; Kaischew, Stranski; Becker, Döring) gegeben worden, wie dies auch in der bekannten Monographie von Volmer beschrieben und von Zinsmeister sehr ausführlich behandelt worden ist. Von besonderer prinzipieller Wichtigkiet erwies sich die von Becker und Döring entwickelte rechnerische Methode, welche in darauffolgenden Arbeiten (Volmer; Kaischew 1957a) erweitert und präzisiert wurde. Das Endergebnis dieser Berechnungen lautet in unseren Bezeichnungen (siehe auch Hirth u. Pound)

$$I_0 = D_k \Gamma n_k \tag{5}$$

worin

$$D_k = s_k z , \tag{5a}$$

$$n_k = n_1 \exp\left(-\frac{\Delta G_k}{kT}\right) \tag{5b}$$

$$\Gamma = \frac{1}{g_k}\sqrt{\frac{\Delta G_k}{3\pi kT}} . \tag{5c}$$

Wie ersichtlich, ergibt sich die stationäre Keimbildungsgeschwindigkeit als das Produkt zweier thermodynamischer Faktoren — die Gleichgewichtskonzentration der kritischen Keime n_k und die als Zeldovich Faktor bezeichnete dimensionslose Größe Γ mit einem rein kinetischen Faktor — die Anlagerungsrate zum kritischen Keim D_k.

Die Formulierung der Aufgabe mittels eines Gleichungssystems, wenn auch im Prinzip einwandfrei, ist im allgemeineren Falle einer zeitabhängigen Aggregatkonzentration nicht allzu nutzbringend. Eine analytische Lösung kann gegeben werden, jedoch gelangt man zu konkreten Ergebnissen nur nach langwierigen maschinellen Auswertungen (s. z. B. Turnbull; Courtney; Zinsmeister). Neben den mathematischen und rechnerischen Schwierigkeiten erscheint dabei eine zusätzliche Komplikation rein physikalischer Natur. Es geht dabei um die Frage, wie die Abbauraten bei sehr kleinen g-Werten zahlenmäßig gewählt werden sollen. Die gewöhnlich benutzte Beziehung

$$D_g^- = \text{const}\, g^{2/3} \exp\left(\frac{a}{g^{2/3}}\right),$$

welche aus Gl. (2b) unter Zuhilfenahme der Gibbs-Thomson'schen Gleichung folgt, rührt von thermodynamischen Überlegungen her, und die Möglichkeit einer Extrapolation für sehr kleine Aggregate ist noch nicht begründet worden. Natürlich tritt diese Komplikation prinzipiell auch im stationären Fall auf; es kann aber gezeigt werden, daß sie bei nicht zu hohen Übersättigungen das Ergebnis (5) nicht beeinflußt. Wenn dagegen eine zeitabhängige Lösung gesucht wird, ist die Lage nicht so günstig.

2.2. Nichtstationäre Keimbildung

In dieser Hinsicht ist eine andere Behandlungsweise der Keimbildungskinetik fruchtbarer, die, wenn auch nicht ganz frei von den erwähnten Mängeln, doch zu bequemen, mit dem Experiment vergleichbaren Resultaten führt. Außerdem macht sie die eingehendere Betrachtung und Konkretisierung einer Reihe von zusätzlichen Problemen möglich. Es haben nämlich ZELDOVICH und auch FRENKEL gezeigt, daß solange g genügend groß ist[1]), d. h. soweit die Atomanzahl als eine kontinuierliche Größe angesehen werden darf, die Gleichungssysteme (2) und (3) durch die partiellen Differentialgleichungen

$$\frac{\partial c(g,t)}{\partial t} = \frac{\partial}{\partial g}\left\{D^+(g)\, n(g) \frac{\partial}{\partial g}\left[\frac{c(g,t)}{n(g)}\right]\right\} \tag{6}$$

bzw.

$$I(g,t) = -D^+(g)\, n(g) \frac{\partial}{\partial g}\left[\frac{c(g,t)}{n(g)}\right] \tag{6a}$$

ersetzt werden können. Hiermit fallen die Abbauraten D_g^- durch die Einführung der Gleichgewichtsverteilung $n(g)$ vollständig aus. Formell betrachtet, beschreiben obige Differentialgleichungen die eindimensionale „Diffusion" der Aggregatgröße g in einem thermodynamischen Kraftfeld. Die Aufgabe wird erst durch die Aufstellung folgender Anfangs- und Grenzbedingungen eindeutig bestimmt:

$$c(g,t) = 0\,; \quad t = 0\,, \quad g > 1 \tag{7}$$

$$\frac{c(g,t)}{n(g)} = 1\,; \quad t \geqq 0\,, \quad g = 1\,, \tag{7a}$$

$$\frac{c(g,t)}{n(g)} = 0\,; \quad t \geqq 0\,, \quad g = q\ (q > g_k). \tag{7b}$$

Der physikalische Sinn dieser Grenzbedingungen ist von LYUBOV und ROITBURD eingehend erläutert worden. Bei $I = 0$ folgt aus (6) und (6a) mit (7a) unmittelbar die Gleichgewichtsverteilung (4). Die stationäre Kinetik ergibt sich sofort bei $I = \text{const} = I_0$.

Eine exakte analytische Lösung der ZELDOVICH-FRENKEL-Gleichungen im Falle, daß die Atomanzahl g wie auch die Zeit t als unabhängige Variabeln auftreten, ist wegen der damit verbundenen mathematischen Schwierigkeiten nicht möglich. Es sind nur einige angenäherte Lösungen bekannt (REISS; WAKESHIMA; COLLINS; CHAKRAVERTY; LYUBOV, ROITBURD); zwischen diesen ist wohl Gleichung

$$I(t) = I_0\left[1 + 2\sum_{n=1}^{\infty}(-1)^n \exp\left(-n^2\,\frac{t}{\tau}\right)\right] \tag{8}$$

als die genaueste zu betrachten (KASHCHIEV 1969a).

Der Ausdruck (8), welcher die Zeitabhängigkeit der Keimbildungsgeschwindigkeit angibt, enthält außer der stationären Geschwindigkeit I_0 noch einen neuen und sehr wichtigen Parameter τ, welcher sinn- und dimensionsgemäß einer Relaxationszeit entspricht. Wir werden weiterhin τ als nichtstationären

---

[1]) Nach FRENKEL soll $g > 10$ sein.

Zeitverzug oder nichtstationäre Induktionsperiode bezeichnen. Nach Gl. (8) gilt:

$$t = 0\,, \quad I = 0\,; \quad t \to \infty\,, \quad I \to I_0\,.$$

Praktisch wird aber schon bei $t \geqq 5\,\tau$ der stationäre Zustand erreicht, wie dies die graphische Illustration in Figur 1 (Kurve 1) zeigt. Einen im Prinzip ähnlichen Verlauf zeigt die von Zeldovich selbst vorgeschlagene Näherung

$$I(t) = I_0 \exp\left(-\frac{\tau}{t}\right), \tag{8a}$$

welche mit Kurve 2 in derselben Figur wiedergegeben ist. Für τ ergibt die Theorie z. B. nach Kashchiev (1969a)

$$\tau = \frac{\varkappa\, k\, T}{D_k\, \lambda_k}\,, \tag{9}$$

wobei $\varkappa = \frac{8}{\pi^2} \cong 0{,}8$ und

$$\lambda_k = -\left(\frac{\partial^2 \Delta G}{\partial g^2}\right)_{g=g_k} > 0 \tag{10}$$

ist. Die beiden Größen im Nenner der Gl. (9) beziehen sich auf den kritischen Keim, was im Einklang mit der zur Ableitung von Gleichung (8) eingeführten Annahme steht, nämlich, daß im Gebiet des nichtstationären Zustandes die Vorgänge in der Umgebung der kritischen Größe geschwindigkeitsbestimmend sind. Die Keimbildungsgeschwindigkeit selbst wird durch den Strom der Aggregate durch die kritische Größe gegeben.

Es läßt sich leicht zeigen, daß Gl. (9) auch in der Form

$$\tau = \frac{4}{\pi^3}\,\frac{1}{\Gamma\, D_k} \tag{9a}$$

dargestellt werden kann; diese Schreibweise veranschaulicht besser die zwischen τ und I_0 existierende Beziehung.

Die Zahl der im Zeitintervall 0, t gebildeten Keime, welche eine experimentell erfaßbare Größe ist, wird definitionsgemäß durch

$$N = \int_0^t I(t)\, dt \tag{11}$$

gegeben, woraus bei $I = I_0$

$$N(t) = I_0\, t$$

folgt. Im Falle einer nichtstationären Keimbildung hat der zeitliche Verlauf von $N(t)$ einen viel komplizierteren Gang. Wird z. B. Gl. (8) benutzt, so erhält man durch Integration nach (11), wie von Kashchiev (a) gezeigt wurde, den Ausdruck

$$N(t) = I_0 \left[t - \frac{\pi^2}{6}\tau - 2\tau \sum_{n=1}^{\infty} \frac{(-1)^n}{n^2} \exp\left(-n^2 \frac{t}{\tau}\right)\right]. \tag{12}$$

Bei $t \geqq 5\,\tau$ kann der Beitrag der Summe in (12) gegen $\frac{\pi^2}{6}\,\tau$ vernachlässigt werden, so daß weiter die Keimzahl linear mit der Zeit zunimmt, d. h.

$$N(t) = I_0\,(t - b\,\tau)\,, \tag{13}$$

worin $b = \frac{\pi^2}{6}$ ist. Wie wir an anderer Stelle vermerkt haben, (GUTZOW, TOSCHEV 1970), führt die Integration von Gleichung (8a) zu dem das Exponentialintegral $\mathrm{Ei}(x)$ enthaltenden Ausdruck

$$N(t) = I_0 t \left[\exp\left(-\frac{\tau}{t}\right) + \frac{\tau}{t}\,\mathrm{Ei}\left(-\frac{\tau}{t}\right)\right], \tag{14}$$

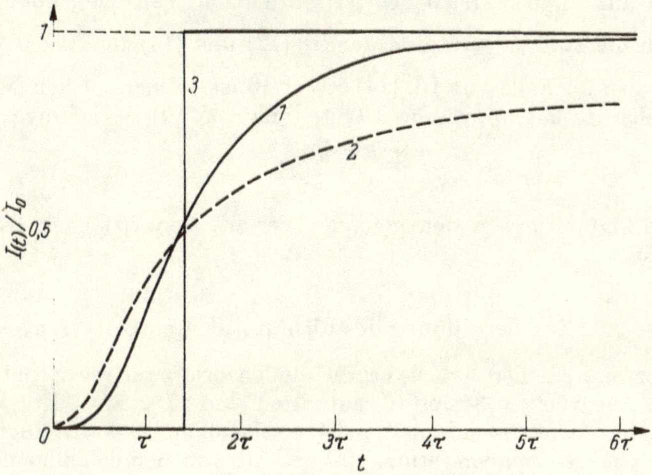

Fig. 1. Zeitlicher Verlauf der Keimbildungsgeschwindigkeit 1 – nach Gl. (8) und 2 – nach Gl. (8a). Mit 3 ist die nach Gl. (15a) für $I(t)$ folgende Approximation illustriert

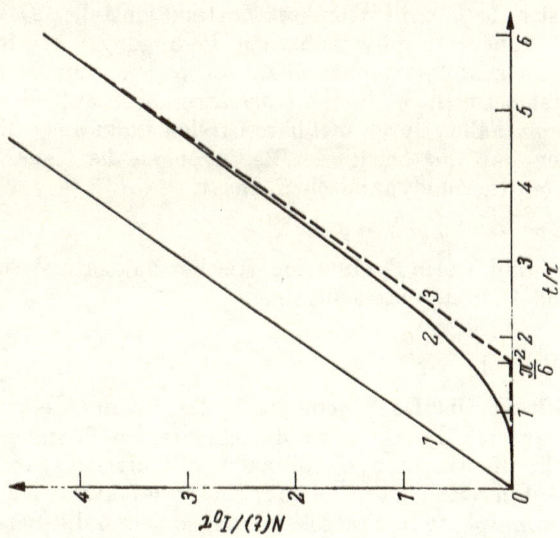

Fig. 2. Die Zahl der gebildeten Keime $N(t)$ in Abhängigkeit von der Zeit.
1 – stationäre Keimbildung
2 – nichtstationäre Keimbildung nach Gl. (12)
3 – nichtstationäre Keimbildung nach der Näherung (15) mit $b = \frac{\pi^2}{6}$

welcher erwartungsgemäß bei $t \to \infty$ und $t \to 0$ asymptotisch in (12) übergeht.

Für einige weitere Anwendungen wird es vorteilhaft sein, Gl. (12) bzw. Gl. (14) durch

$$N(t) = \begin{cases} 0 , & 0 \leqq t \leqq b\tau , \\ I_0 (t - b\tau) , & b\tau < t < \infty \end{cases} \tag{15}$$

zu approximieren, wie dies von Gutzow und Kashchiev (1970) diskutiert wurde. Wie auch aus Figur 2 ersichtlich ist, wird nach (15) der genaue Verlauf von $N(t)$ durch die zwei Asymptoten der Gl. (12) und (14) für $\frac{\tau}{t} \to 0$ bzw. $\frac{\tau}{t} \to \infty$ ersetzt, wobei im Falle von Gl. (14) $b = 1{,}46$ ist. Einer solchen Näherung entspricht folgende asymptotische Darstellung der $I(t)$-Beziehungg

$$I(t) = \begin{cases} 0 , & 0 \leqq t \leqq b\tau , \\ I_0 , & b\tau < t < \infty , \end{cases} \tag{15a}$$

welche in Figur 1 neben den genauen Verlauf von $I(t)$ nach Gl. (8) wiedergegeben ist.

2.3. Zur Berechnung des nichtstationären Zeitverzuges

Wie wir bereits gesehen haben, ergibt die Theorie zwei wesentliche Resultate, die in den Ausdrücken (8) und (9) enthalten sind. Das erste Ergebnis, welches die Keimbildungsgeschwindigkeit als eine zeitabhängige Größe angibt, ist seiner Natur nach gewissermaßen formal und deshalb von ziemlich allgemeiner Gültigkeit, insofern keine konkrete Annahmen in bezug auf die Struktur der Energieschwelle zur Bildung von kritischen Keimen vorausgesetzt werden. Die Induktionszeit τ bestimmt aber die Geschwindigkeit der inneren Umordnungsvorgänge des Systems, bis sich darin ein stationärer Zustand einstellt. Deswegen ist die Größe τ mit den konkreten physikalischen Bedingungen verbunden, unter welchen sich die Phasenbildungsphenomäne abspielen. Zur Zeit kann dieser Zusammenhang nur bei nicht zu hohen Übersättigungen aufgedeckt und somit auch τ nur in solchen Fällen durch meßbare Größen exakt ausgedrückt werden. Dies kommt daher, daß nur bei diesen Bedingungen die Keimbildungsarbeit nach dem bekannten thermodynamischen Ansatz

$$\varDelta G(g) = \beta\, v^{2/3}\, \sigma\, g^{2/3} - g\, \varDelta\mu$$

berechnet werden kann, worin β_3 ein geometrischer Faktor ist. Auf diese Weise erhält man für die Induktionszeit allgemein

$$\tau = 0{,}7\, \gamma^3 \frac{kT\, \sigma^3\, v^2}{\varDelta\mu^4\, D_k^+} \tag{16}$$

worin σ die spezifische Oberflächenenergie, v das Volumen eines Moleküls in der neuen Phase und $\varDelta\mu$ die Differenz der chemischen Potentiale der beiden Makrophasen, d. h. die thermodynamische Triebkraft des ganzen Prozesses pro Molekül, bedeuten. Hier ist $\gamma^3 = 36\,\pi$ ein Formfaktor. Wenn Tröpfchen in übersättigter Dampfphase in Frage kommen, so gelten die Beziehungen

$$\begin{aligned} \varDelta\mu &= kT \ln p/p_\infty \\ D_k^+ &= 4\,\pi\, r_k^2\, z , \\ r &= 2\,\sigma\, v/\varDelta\mu . \end{aligned}$$

Hier ist $\frac{p}{p_\infty}$ die Übersättigung und r_K der Radius des kritischen Keimes. Somit ergibt sich in diesem Falle

$$\tau = 1{,}6 \frac{\sigma}{kT\left(\ln^2 \frac{p}{p_\infty}\right) \cdot z} \tag{17}$$

Hier bedeutet z die gaskinetische Stoßzahl

$$z = \alpha\, p\, N_A\, (2\,\pi\, M\, R\, T)^{-1/2}\,,$$

wobei etwaige sterische Faktoren mit α bezeichnet sind. Eine Abschätzung der Induktionszeit nach Gl. (17) und der obigen Formel für z liefert wegen des sehr hohen Wertes der gaskinetischen Stoßzahl sehr niedrige τ-Werte, welche von der Größenordnung $10^{-6}-10^{-8}$ Sekunden sind. Daraus geht hervor, daß bei den gewöhnlich zur Untersuchung der homogenen Kondensation von Tröpfchen verwendeten Versuchsanordnungen die hier diskutierten nichtstationären Effekte praktisch keine meßbare Rolle spielen können, selbst wenn α Werte in den gewöhnlich annehmbaren Grenzen $10^{-2}-10^{-3}$ hat.

Es kann gezeigt werden, daß in kondensierten Systemen die Anlagerungsrate z pro Oberflächeneinheit durch die Viskosität η der Mutterphase nach der von der Stokes-Einsteinschen Beziehung folgenden Formel ausgedrückt werden kann

$$z = \frac{\alpha'\, kT}{\delta^5\, \eta}\,,$$

worin δ der mittlere Molekülabstand in der Schmelze ist und also eine den Moleküldimensionen nahe Größe hat. Der Koeffizient α' ist ein sterischer Faktor, welcher im Falle komplizierterer Moleküle auch Werte kleiner als 10^{-2} annehmen kann. Berücksichtigt man außerdem, daß $\Delta\mu$ von der Unterkühlung ΔT, der Schmelzwärme pro Moleküle Q_m und der Schmelztemperatur T_s nach

$$\Delta\mu = \frac{Q\, m\, \Delta T}{T_s}$$

annähernd bestimmt werden kann, so erhält man für die Induktionszeit in unterkühlten Schmelzen (sphärischer Keim angenommen)

$$\tau = 1{,}6 \frac{\sigma\, T_s^2\, \delta^5\, \eta}{Q_m^2\, \Delta T^2}\,. \tag{18}$$

Die Induktionszeit ist also bei gegebener Temperatur proportional der Viskosität. Somit sind besonders bei glasbildenden Schmelzen sehr hohe Induktionszeiten möglich, so daß τ in Stunden, sogar Tagen gemessen werden muß. In solchen Systemen wird nach Gl. (18) die Temperaturabhängigkeit der Induktionsperiode hauptsächlich vom Temperaturverlauf der Viskosität bestimmt. Der letzte läßt sich sehr oft durch die sogenannte Vogel-Fulcher-Tammann'sche Gleichung beschreiben

$$\eta = A \exp\left(\frac{B}{T - T_0}\right) \tag{19}$$

worin A, B und T_0 Konstanten sind.

Bei den meisten Metallschmelzen, Schmelzen von anorganischen Salzen, bei organischen Flüssigkeiten und Wasser ergibt die Beziehung (18) jedoch sehr geringe τ-Werte (im Bereich von $10^{-4}-10^{-6}$ Sekunden), wenn es sich um nicht zu große Unterkühlungen handelt und wenn man in Betracht zieht, daß für Metalle $\alpha' \cong 1$ sein muß. Wir werden jetzt zwei wichtige Folgerungen aus der Theorie betrachten, welche die angeführten Formeln notwendigerweise modifizieren, indem beachtenswerte und oft in der Praxis auftretende Fälle erfaßt werden können.

Die bisherigen theoretischen Überlegungen beruhen u. a. auf der von Zeldovich und Frenkel stillschweigend gemachten Annahme, daß jeder auf einem Aggregat ausgeübte Molekülstoß ein elementarer Kondensationsvorgang ist, d. h., daß das Aggregat sich damit jedesmal um ein Bildungspartikel vergrößert. Bei der Bildung von flüssigen Tröpfchen aus übersättigten Dämpfen oder bei der Entmischung von Flüssigkeiten entspricht die oben erwähnte Voraussetzung mehr oder weniger dem Tatbestand. Bei kristallinen Aggregaten ist diese Annahme nicht zutreffend, sofern die Anlagerung von Atomen auf einer Kristalloberfläche nur auf energetisch begünstigte Plätze erfolgt. Mit anderen Worten, die Anlagerungsrate D_K^+ bei der Abscheidung von geordneten Phasen muß unter gleichen anderen Bedingungen kleiner als bei der Abscheidung von isotropen Gebilden ausfallen. Wir berücksichtigen diesen Umstand, indem wir die Beziehung (5a) zusätzlich mit einem Wahrscheinlichkeitsfaktor z' multiplizieren (Gutzow, Toschev 1968)

$$D_K^+ = s_K z z' . \tag{20}$$

Die Größe z' $(0 \leqq z' \leqq 1)$ ist mit dem Mechanismus des elementaren Überganges $g_K \pm g_{K+1}$ verbunden. Hiermit ergeben sich verschiedene Möglichkeiten. Der kritische Keim kann durch den sogenannten normalen Mechanismus, oder durch die Bildung von zweidimensionalen Oberflächenkeimen weiterwachsen. Näherungsweise durchgeführte halbempirische Schätzungen zeigen, daß z' im sehr breiten Intervall von 10^{-2}—10^{-6} variieren kann. Diese Frage ist auch von Gindt und Kern behandelt worden, wobei diese Autoren zu sehr ähnlichen Schlußfolgerungen gekommen sind. Zieht man in Betracht, daß z' im Nenner des Ausdrucks für die Induktionszeit erscheint, so wird es klar, daß bei den Anfangsstadien der Kristallisation die nichtstationären Effekte viel stärker ausgeprägt sein sollten als bei der Bildung von isotropen Phasen. Die bis jetzt angeführten Formeln können somit zur Berechnung von Induktionszeiten nur dann benutzt werden, wenn mit Sicherheit $z' = 1$ gesetzt werden kann.

Wie in den oben zitierten Arbeiten ausführlicher dargelegt wurde, kann z'_∞ aus dem Verhältnis der Moleküle, die tatsächlich im Kristall eingebaut werden W/d^3, zur maximal möglichen Molekülzahl $z\,[1 - \exp(-\Delta\mu/kT)]$ ermittelt werden. Hier ist W die lineare Kristallisationsgeschwindigkeit eines unendlich großen Kristalls, welche Größe zur Berechnung von z'_∞ bei nicht zu kleinen Keimen herangezogen werden kann. Wenn man ferner in Betracht zieht, daß eigentlich direkt wachstumsfähig nur gleichgewichtsförmige Kristallkeime sind, so erweist sich, daß $z' \approx (z'_\infty)^3$ ist und somit die erwähnten sehr niedrigen Werte haben muß.

Die weitere Behandlung dieser Frage führt unmittelbar zu einer modifizierten Deutung der bekannten Ostwald'schen Stufenregel im Rahmen der nichtstationären Keimbildung. Maßgebend für die Art der sich bildenden neuen Phase ist die Keimbildungsgeschwindigkeit. Die Bildung von Keimen der metastabilen Phase bei nicht zu großen Abweichungen vom invarianten Phasengleichgewicht ist nach Stranski und Totomanow durch die niedrigere Keimbildungsarbeit der metastabilen Phase, wegen der niedrigeren σ-Werte, bedingt. Damit wird nur die stationäre Keimbildungsgeschwindigkeit berücksichtigt. Nach den hier dargelegten Betrachtungen kann man sagen, daß sich zuerst diejenige Phase abscheiden muß, welche sich durch eine kürzere Induktionszeit auszeichnet, d. h., einen höheren z'-Wert aufweist (s. hierzu GUTZOW, TOSCHEV 1968). Gewöhnlich wird dies die isotrope Phase sein (mit $z' \approx 1$), was z. B. eine allgemeinere Betrachtung der Kondensation bei der gleichzeitigen Möglichkeit zur Bildung von Tröpfchen und Kriställchen erlaubt. Somit wird die klassische Deutung der Stufenregel durch eine rein kinetische Diskussion ergänzt, wobei sich erweist, daß unter bestimmten Bedingungen die Grenzen des Existenzgebietes der metastabilen Phase erweitert werden.

2.4. Heterogene Keimbildung

Die Diskussion der Induktionszeit war bis jetzt nur auf die homogene Bildung von Keimen beschränkt, d. h. auf spontan im Volumen der Mutterphase verlaufende Keimbildung. Insofern die Rolle von Verunreinigungen, der Gefäßwände usw. nur sehr selten eliminiert werden kann, ist es wichtig, den Einfluß von fremden Unterlagen auf die nichtstationäre Keimbildungskinetik zu untersuchen.

Im stationären Fall genügt es, die Energieschwellen zur heterogenen und homogenen Keimbildung unter gleichen anderen Bedingungen zu vergleichen. VOLMER hat gezeigt, daß die Keimbildungsarbeit ΔG_K^* auf aktiver fremder Unterlage immer niedriger ausfällt als die Keimbildungsarbeit im Volumen der Mutterphase, d. h.

$$\frac{\Delta G_K^*}{\Delta G_K} = \Phi , \tag{21}$$

wobei $0 \leqq \Phi \leqq 1$ als Maß für die zwischen dem Keim und der Unterlage wirkenden Kräften eingeführt worden ist. Bei isotropen, linsenförmigen Keimen kann Φ eindeutig durch den Benetzungswinkel φ ausgedrückt werden

$$\Phi = \frac{1}{2} - \frac{3}{4} \cos \varphi + \frac{1}{4} \cos^3 \varphi , \tag{22}$$

während für Kristallkeime mit einfacher Symmetrie nach KAISCHEW (1952) Φ durch die Adhäsionsenergie β an der Kristall-Unterlage Phasengrenze ausgedrückt werden sollte

$$\Phi = 1 - \frac{\beta}{2 \sigma} \tag{23}$$

Im nichtstationären Fall muß eine ähnliche Analyse auf das Verhältnis der Induktionszeiten $\frac{\tau^*}{\tau}$ bei heterogener und homogener Keimbildung beruhen. Der allgemeine Charakter des bei der Ableitung der Beziehung (9) benutzten physikalischen Modells erlaubt direkt den Ausdruck

$$\tau^* = \frac{\varkappa k T}{\lambda_K^* D_K^*}$$

zu schreiben. Für die beiden erwähnten geometrischen Keimformen ergeben sich aus einfachen Überlegungen (TOSCHEV, GUTZOW 1967a)

$$\frac{\tau^*}{\tau} = \zeta(\varphi) = 1 - \frac{1}{2} \cos \varphi - \frac{1}{2} \cos^2 \varphi \tag{24}$$

und

$$\frac{\tau^*}{\tau} = \Theta\left(\frac{\beta}{\sigma}\right) = 3\,\frac{2 - \beta/\sigma}{5 - 2\,\beta/\sigma} \tag{25}$$

Die oben angegebenen Ausdrücke charakterisieren quantitativ die Aktivität der Unterlagen gegenüber dem Keimbildungsvorgang im Gebiet des noch nicht eingestellten stationären Zustandes. Aus diesen Betrachtungen geht hervor, daß im allgemeinen die Unterlagen die Zeit zum Erreichen der stationären Keimbildungsgeschwindigkeit verkürzen, wobei bei $\Phi = 1$ (vollständige Unbenetzbarkeit) $\tau^* \approx \tau$ ist, während bei $\Phi = 0$ (vollständige Benetzbarkeit) $\tau^* = 0$ zu erwarten ist. Wenn z. B. in ein und derselben Schmelze zwei Arten von Kondensationskeimen eingeführt sind, wobei die einen sehr aktiv verglichen mit den anderen sind, wird die Schmelze praktisch nur um die aktiveren Unterlagen auskristallisieren, ohne daß sich die inaktiven Keime am Prozeß beteiligen werden.[1]) Ein solcher Fall wurde von GUTZOW und TOSCHEV (1968) eingehend betrachtet.

3. Anwendung der nichtstationären Theorie auf einige kinetische Probleme

Die so dargelegten grundsätzlichen Beziehungen können nun auf verschiedene Fälle von Phasenumwandlungen angewendet werden. Dabei ist von besonderem Interesse zu untersuchen, inwiefern und in welcher Hinsicht die Anwesenheit von nichtstationären Effekten in der Keimbildung einen Einfluß auf die Gesamtkinetik des untersuchten Vorganges haben werden. Es sollen also einige Prozesse näher untersucht werden, welche von besonderer experimenteller oder technischer Bedeutung sind.

3.1. Kinetik der heterogenen Keimbildung

Die Kinetik des homogenen nichtstationären Keimbildungsvorganges wird durch die Gleichungen (12) bzw. (14), (15) beschrieben. Praktisch verlaufen jedoch, wie gesagt, die meisten Phasenumwandlungen heterogen, wobei von besonderem Interesse der Keimbildungsvorgang auf Kondensationszentren im Volumen der Ausgangsphase oder an aktiven Stellen der Unterlage ist. Solche Kondensationskerne oder aktiven Unterlagen mit endlicher Dimension werden in vielen Fällen bewußt, z. B. in die Schmelze eingeführt (induzierte Kristallisation). Es können dabei Bedingungen gewählt werden (geeignete Unterkühlung und genügend niedriger Φ-Wert), bei denen die Kristallisation praktisch nur auf den eingeführten Kristallisationszentren verläuft. Wenn die Abmessungen dieser Zentren nicht zu klein sind, so wird ihre Aktivität durch Gl. (22) und (23) bestimmt, wobei die Keimbildungsarbeit und die nichtstationäre Induktionsperiode durch (21) bzw. (24), (25) gegeben werden. Die von FLETCHER und von KAISCHEW und MUTAFTSCHIEW (1959) betrachteten Größeneffekte in der Keimbildung sollen hier außer Acht gelassen werden, was bei genügend großen oder sehr aktiven Zentren immer möglich ist.

---

[1]) Eine Zusammenfassung der in 3.3 und 3.4 angeführten Resultate ist von TOSCHEV 1968 gegeben worden.

Es soll N_0 die Zahl der aktiven Zentren pro Volumeneinheit der Ausgangsphase sein und die Kristallisation soll nur auf den anwesenden Kernen verlaufen. Dann wird im stationären Fall die Zahl der im Zeitintervall t, $t + dt$ gebildeten Sphärolithe definitionsgemäß durch

$$dN(t) = I_0^* [N_0 - N(t)]\, dt \tag{26}$$

gegeben, worin I_0^* die stationäre heterogene Keimbildungsgeschwindigkeit ist (auf ein Kristallisationszentrum bezogen) und $N(t)$ die Zahl der zur Zeit t an den Kristallisationszentren gebildeten Sphärolithe bedeutet. Aus der Grenzbedingung $N(t) = 0$, bei $t = 0$ folgt nach Gl. (26) direkt

$$N(t) = N_0 [1 - \exp(- I_0^* t)] \tag{27}$$

oder die momentane Keimbildungsgeschwindigkeit $I^*(t) = \frac{dN(t)}{dt}$ des ganzen Prozesses ist durch

$$I^*(t) = N_0 I_0^* \exp(- I_0^* t) \tag{28}$$

gegeben. Diese Zusammenhänge sind bereits in einer Arbeit von KAISCHEW und MUTAFTSCHIEW (1965) über die elektrolytische Keimbildung von Quecksilber auf Platin enthalten. In den anfänglichen Momenten des Prozesses ergeben Gl. (27) bzw. (28) entsprechend

$$N(t) = N_0 I_0^* t \tag{27a}$$

und

$$I^*(t) = N_0 I_0^* = \text{const}\,. \tag{28a}$$

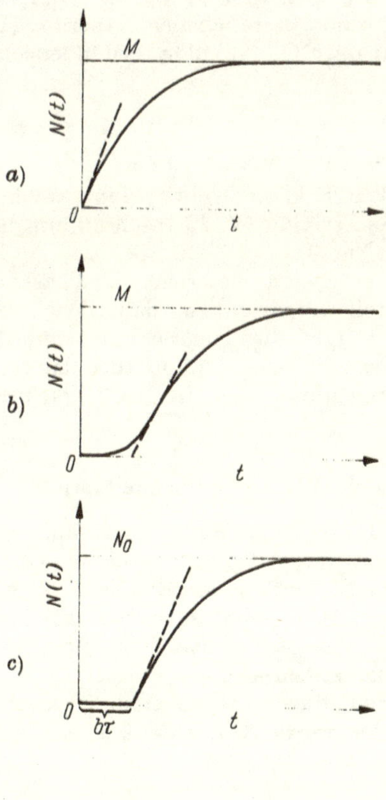

Fig. 3. Kinetik der heterogenen Keimbildung; zeitlicher Verlauf der Keimzahl-Zeit-Kurven (schematisch)
a – stationäre Keimbildung auf N_0 aktiven Stellen (s. Gl. (27)
b – nichtstationäre Keimbildung nach Gl. (29)
c – nichtstationäre Keimbildung nach der Näherung (31)

Nach Gl. (27) strebt $N(t)$ mit der Zeit exponentiell auf N_0 zu (Fig. 3a), wobei $I^*(t)$ von $N_0 I_0^*$ bis Null herabsinkt.

Der zeitliche Verlauf von $N(t)$ für den nichtstationären Prozeß ergibt sich in ähnlicher Weise (TOSCHEV, GUTZOW 1967b), wobei z. B. mit Gl. (8a) folgender Ausdruck erhalten wird

$$dN(t) = I_0^* \exp\left(-\frac{\tau^*}{t}\right)[N_0 - N(t)]\, dt\,.$$

Nach Integration unter Bezugnahme auf die gleiche Grenzbedingung ergibt sich, wenn auch Gl. (14) berücksichtigt wird, folgender Zusammenhang

$$N(t) = N_0\left[1 - \exp\left\{-I_0^* t\left[\exp\left(-\frac{\tau^*}{t}\right) + \frac{\tau^*}{t}\operatorname{Ei}\left(-\frac{\tau^*}{t}\right)\right]\right\}\right] \tag{29}$$

für den zeitlichen Verlauf der heterogenen nichtstationären Keimbildung. Hieraus folgt, daß die momentane Keimbildungsgeschwindigkeit unter diesen Bedingungen durch

$$I^*(t) = I_0^* N_0 \exp\left\{-I_0^* t\left[\exp\left(-\frac{\tau^*}{t}\right) + \frac{\tau^*}{t}\operatorname{Ei}\left(-\frac{\tau^*}{t}\right)\right] - \frac{\tau^*}{\tau}\right\} \tag{30}$$

gegeben ist. Es läßt sich zeigen, daß nach Gl. (29) $N(t)$ einen typischen S-förmigen Gang aufweisen muß, so daß bei $t \to \infty$ vom neuen $N(t) = N_0$ sein wird (Fig. 3b). Die Keimbildungsgeschwindigkeit $I^*(t)$ ist bei $t \to 0$ und $t \to \infty$ gleich Null, d. h. sie geht durch ein Maximum.

In der konkreten Handhabung sind jedoch Gl. (29) und (30) viel zu kompliziert, was wieder die Einführung der schon besprochenen Näherung (15) rechtfertigt. Mit Gl. (15) erhalten wir anstatt Gl. (29) den viel übersichtlicheren Ausdruck

$$N(t) = \begin{cases} 0\,, & 0 \leqq t \leqq b\tau \\ N_0\{1 - \exp[-I_0^*(t - b\tau)]\}\,, & b\tau < t < \infty\,. \end{cases} \tag{31}$$

Es bilden sich also bis $t = b\tau$ praktisch keine Keime; dann wächst jedoch die $N(t)$-Kurve mit einer anfänglichen Steigung $N_0 I_K^*$ rasch an, um bei $t \to \infty$ den Wert N_0 zu erreichen (s. hierzu Fig. 3c).

Keimzahl-Zeit-Kurven können bei vielen heterogenen Phasenumwandlungen genügend genau experimentell aufgestellt werden. Wie aus obigen Gleichungen und Figur 3 hervorgeht, gibt die Theorie nicht nur den prinzipiellen Verlauf der $N(t)$-Kurven, sondern sie erlaubt auch ein verhältnismäßig einfaches Kriterium zur Entscheidung über die Rolle von nichtstationären Effekten.

3.2. Die Keimbildung als ein Wahrscheinlichkeitsprozeß

Insofern die kritische Keimgröße g_K nur durch eine Reihe von zufälligen Molekülstößen erreicht und übertroffen werden kann, ist die Keimbildung ein typisches Beispiel eines stochastischen Vorganges. Größen wie die Zahl der gebildeten Keime oder die zur Bildung einer gewissen Keimzahl notwendige Zeit haben daher einen zufälligen Charakter und ihre statistische Verteilung ist den Gesetzmäßigkeiten der Wahrscheinlichkeitstheorie untergeordnet.

Da der Keimbildungsprozeß als eine Sequenz von voneinander unabhängigen zufälligen Ereignissen ist, muß insbesondere die Zahl m der in einem bestimmten

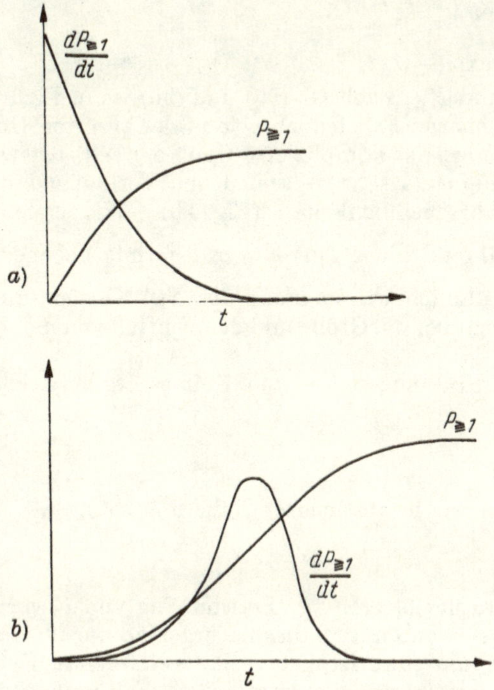

Fig. 4. Zur Wahrscheinlichkeitsanalyse der Keimbildung. Verlauf der differentiellen und integralen Wahrscheinlichkeiten zur Bildung des ersten Keimes
a – stationärer Fall
b – nichtstationärer Fall

Zeitintervall bei konstanten Bedingungen gebildeten Keime nach dem Poisson'schen Gesetz

$$P_m = \frac{N^m \exp(-N)}{m!} \tag{32}$$

um den mittleren Wert N verteilt sein. Es läßt sich zeigen (TOSCHEV, STOYANOV), daß in den Gültigkeitsgrenzen dieses Gesetzes die Wahrscheinlichkeit $dP_{\geqq 1}$ zur Bildung des ersten Keimes im Zeitintervall t, $t + dt$ durch den Ausdruck

$$dP_{\geqq 1} = \frac{dN(t)}{dt} \exp[-N(t)]\, dt \tag{33}$$

gegeben wird, worin der Mittelwert $N(t)$ durch Gl. (11) definiert wird und auch experimentell bestimmt werden kann. Das Integral

$$P_{\geqq 1} = \int_0^t dP_{\geqq 1} = 1 - \exp[-N(t)] \tag{34}$$

hingegen gibt die Wahrscheinlichkeit an, daß sich im Zeitintervall 0, t mindestens ein Keim gebildet hat. Im Falle, daß die nichtstationären Effekte vernachlässigt werden können ($I(t) = \text{const}$), ergeben die Ausdrücke (33), (34) entsprechend

$$dP_{\geqq 1} = I_0 \exp(-I_0 t)\, dt \tag{35}$$

und

$$P_{\geqq 1} = 1 - \exp(-I_0 t)\,. \tag{36}$$

Der Verlauf von $dP_{\geqq 1}$ und $P_{\geqq 1}$ nach Gl. (35) und (36) ist in Figur 4a wiedergegeben.[1]) Im nichtstationären Fall jedoch, wie in der zitierten Untersuchung gezeigt werden konnte, führt der komplizierte Gang der $N(t)$-Kurve z. B. nach Gl. (12) zu einem von Figur 4a sehr verschiedenen Verlauf der differentielle und der integralen Wahrscheinlichkeiten (Fig. 4b). Wie ersichtlich, zeigt $\frac{dP_{\geqq 1}}{dt}$ ein Maximum, während die $P_{\geqq 1}(t)$-Kurve S-förmig gebogen ist. Diese Unterschiede gestatten, ähnlich wie im Falle der $N(t)$-Kurven, auch eine experimentelle Entscheidung über die Größe und den Einfluß von nichtstationären Effekten zu machen.

Die mittlere Zeit t zur Bildung des ersten Keimes ergibt sich definitionsgemäß als

$$\bar{t} = \int_0^\infty t\, dP_{\geqq 1} \tag{37}$$

und mit Gl. (35) erhalten wir in stationärer Näherung sofort

$$\bar{t} = \frac{1}{I_0} \tag{38}$$

Diese letzte Relation wird des öfteren zur Bestimmung von I_0 verwendet. Es muß jedoch hervorgehoben werden, daß dies naturgemäß nur in solchen Fällen möglich ist, wo es sich um praktisch stationär verlaufende Keimbildungsprozesse handelt. Im allgemeineren, nichtstationären Fall muß $N(t)$ in Gl. (37) durch (12) ersetzt werden. Man gelangt jedoch mit dieser und ähnlichen komplizierten $N(t)$-Funktionen zu keinen expliziten Ausdruck für $\bar{t}$. Unter Verwendung der Näherungsformel (15) kann auch im nichtstationären Falle wenigstens die Struktur von $\bar{t}$ aufgedeckt werden. Wie wir zeigen konnten (Gutzow, Toschev 1970), erhält man mit Gl. (15) aus (37) einfach

$$\bar{t} \approx b\tau + \frac{1}{I_0} \tag{39}$$

Durch die Benutzung von Gl. (15) wird eigentlich die in Figur 5 gestrichelt gegebene $\frac{dP_{\geqq 1}}{dt}$-Funktion durch die ausgezogene Kurve ersetzt. Diese letzte ist nichts anderes als die um den Wert $b\tau$ an der Zeitachse verschobene $\frac{dP_{\geqq 1}}{dt}$

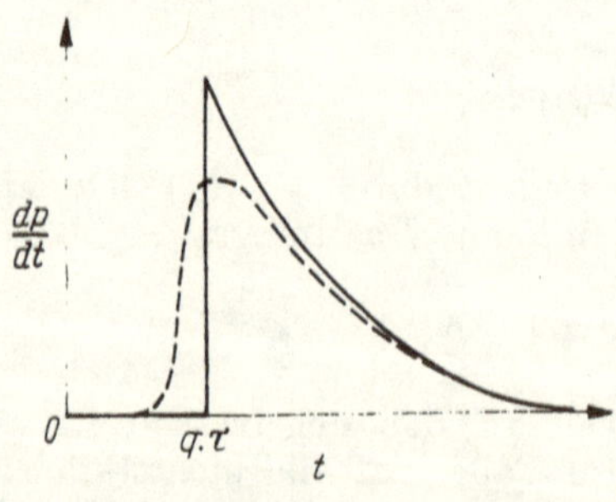

Fig. 5. Approximation des wahren Verlaufes der $\frac{dP_{\geqq 1}}{dt}$ Kurve vermittels einer der Gl. (15) analogen Näherung (schematisch) (Abszisse lies $b \cdot \tau$)

[1]) Hier ist I_0 auf das ganze zur Kristallisation verfügbare Volumen bzw. auf die ganze Oberfläche bezogen.

Funktion, welche durch Gl. (35) gegeben wird. Es ist jedenfalls aus Gl. (39) ersichtlich, daß in Fällen, wo $b\tau$ mit $\frac{1}{I_0}$ vergleichbar ist, die Bestimmung von I_0 als der reziproke Wert der zur Bildung des ersten Keimes notwendigen Zeit zu bedeutenden Fehlern führen kann. In Fällen, wo $b\tau > \frac{1}{I_0}$, was durchaus möglich ist, wird $\bar{t}$ ausschließlich durch $b\tau$ bestimmt.

3.3. Nichtstationäre Keimbildung und die Kinetik des summaren Kristallisationsverlaufes

Die Kinetik des isothermen summarischen Kristallisationsverlaufes einer Schmelze wird nach der bekannten von KOLMOGOROV und von AVRAMI gegebenen Gleichung

$$\alpha(t) = 1 - \exp(-K_n t^n) \tag{40}$$

beschrieben. Hier ist $\alpha(t)$ der bei t erreichte Kristallisationsgrad, welcher z.B. nach

$$\alpha(t) = V(t)/V_0$$

aus dem Verhältnis des zur Zeit t kristallisierten Volumens zum Ausgangsvolumen V_0 bestimmt werden kann. Der ganzzahlige Exponent n hängt von dem konkreten Charakter des Wachstumsprozesses ab (siehe z. B. AVRAMI; HOLLOMON, TURNBULL, wie auch YOUNG) und kann Werte zwischen 1 und 4 einnehmen, während in den für eine gegebene Temperatur konstanten Wert von K_n ein sterischer Faktor neben der Keimbildungsgeschwindigkeit I und der Wachstumsgeschwindigkeit W enthalten ist.

In der Ableitung wie auch in der Anwendung von Gl. (40) wurde $I = \text{const} = I_0$ angenommen, d. h., es wurde bisher nur die stationäre Näherung des summarischen Kristallisationsprozesses betrachtet. In Fällen, wo höhere τ-Werte zu erwarten sind (z. B. bei der Kristallisation hochviskoser Schmelzen) könnten aber auch bedeutende nichtstationäre Effekte in summarischen Kristallisationsverlauf auftreten. Diese Frage wurde in einer in letzter Zeit erschienen Arbeit von GUTZOW und KASHCHIEV untersucht. Es konnte dabei gezeigt werden, daß bei Annahme einer nichtstationären Keimbildungskinetik (Gl. (8) bzw. (12)) der summarische Kristallisationsverlauf in der Form

$$\alpha(t) = 1 - \exp(-K_n t^n \Psi_n) \tag{41}$$

dargestellt werden kann. Mit Ψ_n ist hier eine Korrektionsfunktion bezeichnet, welche bei $t \to 0$ den Wert $\Psi_n = 0$ annimmt, während bei $t \to \infty$ entsprechen $\Psi_n = 1$ gilt. Auf eine genaue Wiedergabe der Ψ_n-Funktion wird hier verzichtet; wir weisen nur darauf hin, daß es möglich ist, in der der Gl. (15) entsprechenden Näherung diese Funktion auch als

$$\Psi_n = \begin{cases} 0 \,, & 0 \leqq t \leqq b\tau \\ \left(1 - \frac{b\tau}{t}\right)^{n+1}, & \tau < t < \infty \end{cases} \tag{42}$$

darzustellen. In dieser Näherung führen die nichtstationären Effekte zu einer einfachen Verschiebung der $\alpha(t)$-Kurven zu höheren t-Werten (Fig. 6). Der Betrag dieser Verschiebung $b\tau$ ist nicht nur ein Maß für die Größe der nichtstationären Effekte, sondern bestimmt, wie auch in der erwähnten Publikation gezeigt wurde, die kinetische Stabilität der unterkühlten Flüssigkeit und gibt somit die Möglichkeit zur Aufstellung eines einfachen Kriteriums der Glasbildung.

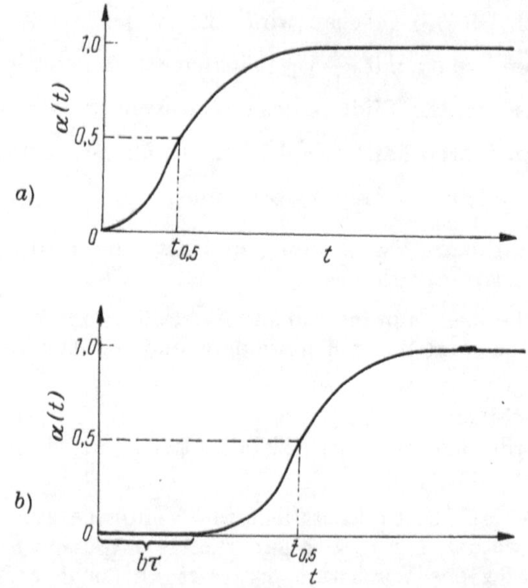

Fig. 6. Verschiebung der die summarische Kristallisation beschreibenden $\alpha(t)$-Kurven
a – stationärer Kristallisationsverlauf
b – nichtstationärer Kristallisationsverlauf nach Gl. (42)

3.4. Nichtstationäre zweidimensionale Keimbildung

In den bisher besprochenen Beispielen führte die nichtstationäre Keimbildung zu Effekten, welche eindeutig als ein Zeitverzug in der makroskopischen Kinetik des Vorganges bezeichnet werden kann, wobei sich die stationäre Kristallisationskinetik in erster Näherung erst nach Verlauf des Zeitintervalles $b\,\tau$ einstellt. Das Wachstum gleichgewichtsförmiger, ideal aufgebauter Kristallflächen muß auch über einen Keimbildungsprozeß verlaufen, welcher zur Bildung zweidimensionaler Keime führt. Auf den nichtstationären Verlauf der zweidimensionalen Keimbildung hat wohl als erster Lyubov hingewiesen. Eine eingehende Betrachtung dieses Problems wurde von Gutzow und Toschev (1970) gegeben, wobei ein von Kaischew (1957b) entworfenes Kristallisationsmodell benutzt wurde. Nach diesem Modell wird die Wachstumsgeschwindigkeit W des Kristalls durch

$$W = \frac{d}{\bar{t}_2} \tag{43}$$

gegeben, wobei $\bar{t}_2$ der Mittelwert der zur Bildung eines zweidimensionalen Keimes notwendige Zeit ist, während d der Gitterparameter des Kristalls ist. Im Prinzip gelten auch für diesen Keimbildungsfall die im vorigen Abschnitt gemachten Betrachtungen und insbesondere Gleichungen (5), (8), (9) und (12), in welche nur die dem zweidimensionalen Charakter des Prozesses entsprechenden Größen I_2, τ_2 usw. eingeführt werden müssen. Im stationären Fall erhält man auch hier

$$\bar{t}_2 = \frac{1}{I_2^0\,F} \tag{44}$$

worin F den Flächeninhalt der anwachsenden Kristallfront bedeutet, während für den nichtstationären Verlauf ähnlich wie in 3.2

$$\bar{t}_2 = b\,\tau_2 + \frac{1}{I_2^0 F} \tag{45}$$

geschrieben werden muß. Hier ist

$$I_2^0 = D_{K_2}\,\Gamma_2\,n_{K_2}\,,$$

worin die entsprechenden Größen dieselbe Bedeutung, wie in Gl. (5) haben.

Nach Gl. (43), (45) läßt sich die Wachstumsgeschwindigkeit als

$$W = I_2^0\,F\,d\left[1 + \frac{\varkappa\,b}{2\,\pi}\,F\,\frac{N_s}{\Gamma_2}\exp\left(-\frac{\Delta G_{K_2}}{kT}\right)\right] = W_0\,\xi \tag{46}$$

schreiben, worin N_s die Konzentration der auf der Kristallfläche adsorbierten Atome der Ausgangsphase ist. Wie durch den zweiten Teil dieser Gleichung angezeigt ist, unterscheidet sich der so erhaltene Ausdruck für W von der in stationärer Näherung berechneten linearen Kristallisationsgeschwindigkeit W durch den Faktor ξ. Bei kleinen Übersättigungen ist $\xi \approx 1$, während bei großen Übersättigungen

$$1 \ll \frac{\varkappa\,b}{2\,\pi}\,\frac{N_s}{\Gamma_2}\exp\left(-\frac{\Delta G_{K_2}}{k\,T}\right)$$

so daß Gl. (46)

$$W = D_{K_2}(\Gamma_2)^2\,d \tag{47}$$

ergibt, was auch als

$$W = \frac{\mathrm{const}}{2\,\pi}\,z\,\frac{\Delta\mu}{k\,T}\,d^3 \tag{48}$$

geschrieben werden kann. Gleichung (47) entspricht formell einer $W(\Delta\mu)$-Beziehung, welche für das sogenannte normale Wachstum typisch ist. Es läßt sich zeigen (GUTZOW, TOSCHEV 1970), daß im Spezialfall der Kristallisation aus der Schmelze Gleichungen (47), (48) besonders in der Nähe von $T \cong \frac{2}{3}\,T_s$ von Bedeutung sein werden – d. h. in der Nähe der Einfriertemperatur des Glases T_g.

In Gl. (46) ist die Zeit t nicht als laufende Variable enthalten; es führen also zum Unterschied von den in 3.1.–3.3. besprochenen Fällen die nichtstationären Effekte zu keiner zeitlichen Abhängigkeit des Wachstumsprozesses, sondern es ändert sich nur der Absolutwert von W und die Form der $W(\Delta\mu)$ Beziehung. Dies ist zu erwarten, wenn man in Betracht zieht, daß die – wenn auch nichtstationär verlaufende – Keimbildung letzten Endes ein sich periodisch wiederholender Prozeß ist.

Es kann außerdem gezeigt werden, daß bei nicht zu extremen Bedingungen immer

$$\tau_2 > \tau_3$$

gelten muß, wobei bei nicht zu hohen Übersättigungen $\tau_2 \cong (10^2 - 10^3)\,\tau_3$.

Das hier diskutierte Modell gilt streng genommen nur für den Fall, wo die tangentielle Wachstumsgeschwindigkeit jedes Keimes so groß ist, daß jede neue Netzebene nur von einem Keim ausgebildet wird. Die Möglichkeit, daß sich ein neuer Keim auf der wachsenden Netzebene bildet, bevor sie die ganze Kristallfläche ausgefüllt hat, wurde quantitativ von HILLIG (siehe auch NIELsen) erfaßt. Von Stoyanov wurde auch für dieses Modell der nichtstationäre Vorgang berücksichtigt, wobei den hier angeführten ähnliche Ergebnisse erhalten wurden.

3.5. Athermische Keime und nichtstationäre Keimbildung

Es wurde schon des öfteren in der Literatur der Einfluß vorgebildeter Keime auf den Gesamtverlauf von Phasenumwandlungen diskutiert. Solche Keime können sich z. B. im Volumen einer Schmelze durch vorhergehende Temperaturbehandlung unterhalb des Schmelzpunktes bilden, oder sie können das Produkt eingefrorener Fluktuationen bei Temperaturen oberhalb T_s sein (Kaischew 1937). Ähnliche Verhältnisse können auch in anderen Fällen von Phasenumwandlungen bestehen, wobei gewöhnlich der von der Kristallisation unterkühlter Schmelzen übernommene Terminus athermische Keime beibehalten wird.

Von Kashchiev (1969b) wurde eingehender der Einfluß athermischer Keime auf die nichtstationäre Keimbildungskinetik untersucht. Das Ergebnis sei hier nur graphisch illustriert (Fig. 7). Wie ersichtlich, hat $I(t)$ bei Anwesenheit

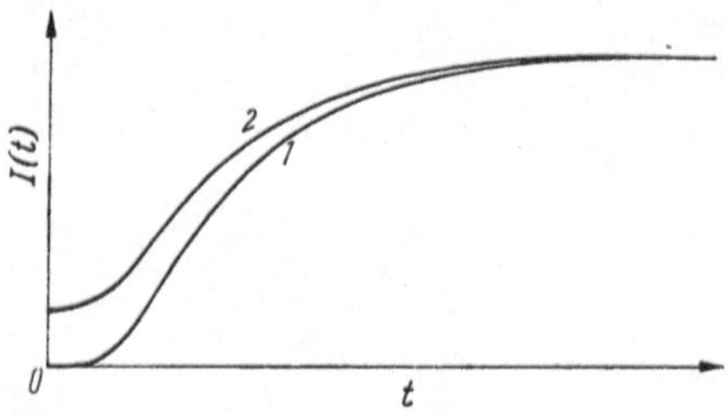

Fig. 7. Einfluß vorgebildeter Keime auf den zeitlichen Verlauf der Keimbildungsgeschwindigkeit (nach Kashchiev 1969b)
1 – $I(t)$-Kurve in Abwesenheit athermischer Keime
2 – $I(t)$-Kurve in Anwesenheit einer gewissen Anzahl athermischer Keime

athermischer Keime bei $t = 0$ einen von Null verschiedenen Wert. Diese Verschiebung der Kurve ist der Konzentration der athermischen Keime proportional (Kashchiev, Kaischew) und hängt damit von der Vorbehandlung der Ausgangsphase ab. Die stationäre Keimbildungsgeschwindigkeit bleibt dabei unverändert.

3.6. Dispersität der neugebildeten Phase und Keimbildungskinetik

Es ist zu erwarten, daß die Keimbildungskinetik auch maßgebend die Dispersität der neuen Phase beeinflussen wird, z. B. die Größenverteilung der bei einem Kristallisationsprozeß im Volumen der Schmelze gebildeten Sphärolite. Dieses Problem wurde unter Inbetrachtnahme nichtstationärer Keimbildung für den Fall untersucht (Toschev, Gutzow 1967b), daß der Prozeß im Anfangsstadium des Kristallisationsvorganges abgebrochen wird, bevor gegenseitige Berührung der Sphärolithe eingetreten ist. Somit kann noch mit genügender Genauigkeit die Abnahme des der Keimbildung zugänglichen Volumens vernachlässigt werden. Praktisch kann ein solches Modell z. B. im Anfangsstadium der induzierten Kristallisation von Glaskeramiken realisiert werden, wenn die Schmelze vor dem Erreichen eines höheren Kristallisationsgrades rasch abgeschreckt wird. In einem solchen Falle kann, wenn konstante Wachstumsgeschwindigkeit angenommen wird, die Keimbildungskinetik unmittelbar aus der Größenverteilung der Sphärolithe bestimmt werden. Dies ergibt eine zusätzliche Methode zur Erforschung nichtstationärer Effekte.

In der zitierten Untersuchung wurde das Problem insofern allgemeiner betrachtet, als auch ein Zusammenhang zwischen der wahren, im Volumen der Probe bestehenden Größenverteilung und der experimentell gewöhnlich zugänglichen Oberflächenverteilung aufgestellt wurde, welche z. B. an Bruch- oder Anschliffflächen entsteht. So ergibt sich die Möglichkeit direkte elektronenmikroskopische Messungen an halbkristallisierten Gläsern zur Bestimmung der Keimbildungskinetik und insbesondere von τ zu benutzen.

4. Vergleich mit dem Experiment

4.1. Einige methodische Probleme

Aus der vorhergehenden Darstellung ist ersichtlich, daß im Prinzip einige experimentelle Möglichkeiten bestehen, nichtstationäre Effekte in der Keimbildung messend zu verfolgen. Zwei der angedeuteten Verfahren führen direkt zum Ziel — die Aufstellung von $N(t)$-Kurven und die Analyse der Wahrscheinlichkeitsparameter der Keimbildung. Die anderen Methoden — z. B. die Untersuchung der Größenverteilung der Kristallite oder die Verfolgung des summarischen Kristallisationsverlaufes — sind im Prinzip mit bestimmten Annahmen über die Wachstumsgeschwindigkeit oder mit der Art der Kristallisation verbunden und können daher nur eine mehr oder weniger untergeordnete Rolle spielen.

Andererseits muß auch in Betracht gezogen werden, daß aus experimentellen Gründen Induktionsperioden vorgetäuscht werden können, welche physikalisch nicht als ein nichtstationärer Zeitverzug betrachtet werden dürfen. Nach Möglichkeit muß der gesuchte Wert von τ von ähnlichen Nebeneffekten differenziert werden. Solche Probleme entstehen z. B. bei der Analyse der $N(t)$-Kurven. Es ist wegen der verschwindend kleinen Dimensionen der Keime ($r_K \sim 10^{-6}$ cm) z. Z. keine direkte Keimzahlbestimmung möglich, sondern es werden nur die genügend groß ausgewachsenen Kristallisationszentren ausgezählt. Bis der Keim zu mikroskopisch sichtbaren Dimensionen auswächst, muß eine bestimmte Zeit t_w verstreichen, welche mit τ vergleichbar oder gar größer sein kann. Ähnliche Schwierigkeiten entstehen auch bei der Bestimmung der Zeit t zur Bildung des ersten Keimes. Es läßt sich jedoch zeigen, daß bei Bestimmung der integralen Wahrscheinlichkeit nach Gl. (36) die Verhältnisse günstiger liegen. Aus dieser Situation gibt es prinzipiell zwei Auswege: entweder müssen die unter bestimmten Bedingungen gebildeten Keime bei Übersättigungen „entwickelt" werden, wo die Wachstumsgeschwindigkeit genügend große Werte hat, oder es muß genau die Zeit t_w bekannt sein. Bei der ersten, noch von Tammann angegebenen Methode besteht die Gefahr, daß ein Teil der z. B. in der Schmelze schon einmal gebildeten Keime bei dem Entwickeln sich wieder auflösen würden. Wie von Toschev und Markov gezeigt wurde, ist diese Möglichkeit bei der elektrolytischen Keimbildung praktisch ausgeschlossen, wobei bei einem dem Tammannschen Verfahren nachgebildeten Prozeß alle Keime entwickelt werden können. Der zweite Weg ist nur in dem Falle realisierbar, wenn zusätzliche Messungen über die Wachstumsgeschwindigkeit bis zu den kleinsten Dimensionen vorliegen, so daß t_w möglichst genau bestimmt werden kann. Dies wurde z. B. in einer weiter unten zu besprechenden Arbeit von Gutzow, Toschev, Marinov und Popov auf Grund elektronenmikroskopischer Untersuchungen von kristallisierten Glasproben möglich.

Ein anderes Problem liegt im Absolutwert der zu bestimmenden Induktionszeiten. Es ist klar, daß nur in wenigen Fällen das experimentelle Erfassen von Zeitverzögerungen in der Keimbildung möglich sein wird, wenn sie in der Größenordnung von $10^{-5}-10^{-7}$ Sekunden liegen. Wie weiter unten gezeigt, war dies bis jetzt nur im Falle der elektrolytischen Metallabscheidung quantitativ möglich, indem eine besonders verfeinerte Untersuchungsmethode entwickeln werden mußte. Besonders muß in solchen Fällen, wo sehr kleine τ-Werte zu erwarten sind, der Umstand beachtet werden, daß auch die Zeit zur Einstellung des Sollwertes der Übersättigung mit τ vergleichbar sein kann. Andererseits sind in manchen Extremfällen, wie bei der Kristallisation von Gläsern die Induktionsperioden erwartungsgemäß so lang, daß nichtstationäre Effekte ohne besonderen experimentellen Aufwand untersucht werden können. Dabei ist der steile Temperaturverlauf der Viskosität glasbildender Schmelzen von Bedeutung, da er die Möglichkeit ergibt, durch ein rasches Abschrecken der Proben den Keimbildungsvorgang einzufrieren und dann bequem zu untersuchen.

Solche Erwägungen führen selbstverständlich zur Wahl jener Modellsysteme, in denen eine quantitative Prüfung der angeführten theoretischen Ergebnisse möglich ist. Eine Zusammenfassung der dabei erhaltenen Resultate ist weiter unten gegeben. Einige zusätzliche Beispiele, in denen nichtstationäre Keimbildung beobachtet wurde, sollen auch kurz gestreift werden.

Die experimentelle Erfassung von nichtstationären Effekten in der zweidimensionalen Keimbildung ist besonders schwierig, da der Temperaturverlauf von W bei hohen Übersättigungen mit der Theorie verglichen werden muß. Dabei ist es sehr schwer, ein eindeutiges Resultat zu erhalten, da eine Anzahl von Nebeneffekten nicht von dem gesuchten Gang zu unterscheiden sind. Eine andere und eventuell günstigere Alternative könnte wohl die statistische Verteilung der zur Bildung von Oberflächenkeimen notwendigen Zeiten ergeben, falls es möglich sein sollte, den Primärakt der zweidimensionalen Keimbildung zu registrieren. Solche Möglichkeiten bestehen beim elektrolytischen Kristallwachstum defektfreier Einkristallflächen (Kaischew, Budevski).

4.2. Nichtstationäre Keimbildung bei der elektrolytischen Abscheidung von Metallen

Das eingehende Studium der Kinetik der Keimbildung bei der elektrolytischen Abscheidung von Metallen auf artfremden Elektroden wurde durch das Benutzen einer Doppeltimpulsmethode möglich. Das Prinzip dieser Methode besteht darin, daß ⊓-förmige elektrische Impulse, gefolgt von einer Wachstumsspannung an einer elektrolytischen Zelle, angelegt werden (Kaischew, Mutaftschiew 1955; s. auch Kaischew 1967). Die Zelle besteht aus zwei in die Elektrolytlösung tauchenden Elektroden — einer Platinmikrokathode und einer Anode, welche aus dem abzuscheidenden Metall gefertigt ist. Bei geeigneter Wahl der Impulshöhe E, der Wachstumsspannung E_g und der Zeitdauer des Impulses t wird an der Kathode eine Anzahl von Keimen gebildet, welche zu sichtbarer Größe entwickelt werden. Die Zahl der während der

Zeit t gebildeten Keime kann darauffolgend mikroskopisch ausgezählt und Keimbildung und Wachstum unter definierten Bedingungen gesondert untersucht werden. Die Impulshöhe E ist ein direktes Maß der Übersättigung, bei welcher die Keime gebildet werden, da bei der elektrolytischen Metallabscheidung die Beziehung

$$\Delta\mu = \nu\, e\, E$$

gilt, worin ν die Wertigkeit der sich neutralisierenden Ionen ist und e die elementare elektrische Ladung bedeutet.

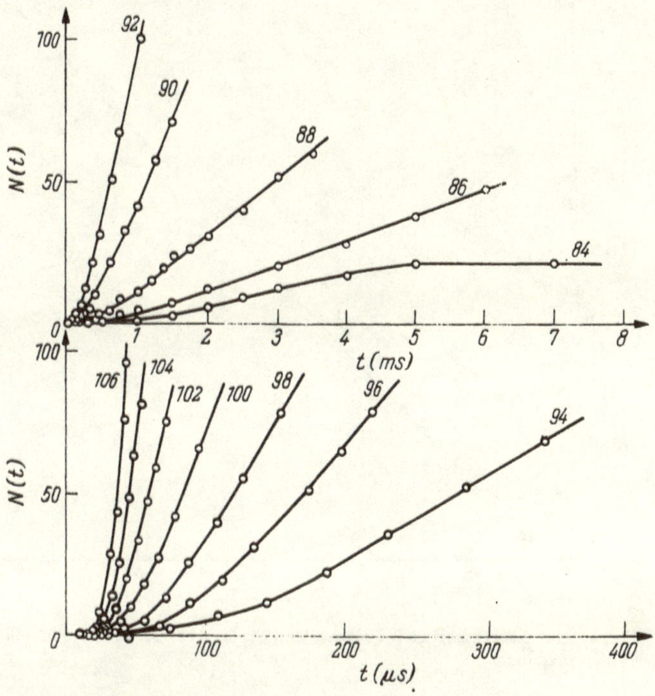

Fig. 8. Keimzahl-Zeit Kurven bei der elektrolytischen Abscheidung von Quecksilber. Die entsprechende Überspannung ist an jeder Kurve als Parameter in Millivolt gegeben

Figur 8 zeigt eine Reihe bei der elektrolytischen Abscheidung von Quecksilberkeimen nach dieser Methode bei verschiedenen Überspannungen aufgenommenen $N(t)$-Kurven. Der nichtstationäre Verlauf der Kurven ist nach dem in 3.1 Gesagten eindeutig sichtbar; da es sich dabei um einen heterogenen, auf einer Anzahl von Aktivstellen der Kathode verlaufenden Prozeß handelt (Kaischew, Mutaftschiew 1965) folgt dem linearen Teil der Kurven eine hier nur für manche Überspannungen angedeutete Sättigung (s. Gl. (29)). Die Steigung der linearen Abschnitte der $N(t)$-Kurven hängt nach Gl. (5) exponentiell von der Übersättigung $\Delta\mu$ ab, und in Koordinaten $\log I_0 \div \frac{1}{E^2}$ können ΔG_K^* und die anderen Parameter des stationären Keimbildungsprozesses bestimmt werden. Gleichung (12), welche den nichtstationären Verlauf der Keimbildung beschreibt, wurde in Koordinaten $F(x) = \frac{N(t)}{I_0^* \tau^*}$ gegen $x = \frac{t}{\tau^*}$

geprüft, wobei für jede $N(t)$-Kurve I_0^* und τ^* gesondert bestimmt wurden. Wie aus Figur 9 ersichtlich ist, werden die experimentellen $N(t)$-Kurven sehr gut von der theoretischen Beziehung (12) wiedergegeben (TOSCHEV, MARKOV 1969; TOSCHEV, MILCHEV). Die elektrolytische Keimbildung gibt auch die Möglichkeit die in 3.2 vorgenommene Analyse der Wahrscheinlichkeitsaspekte der Keimbildungserscheinungen einer Prüfung zu unterziehen. In dieser Hin-

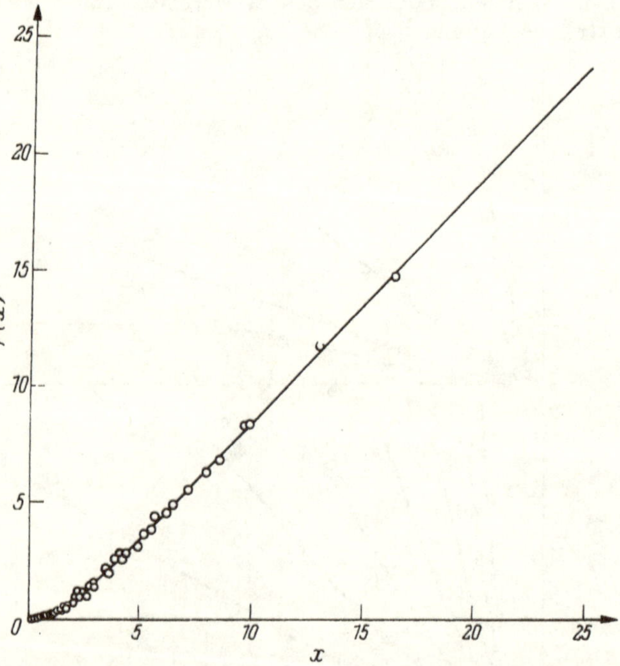

Fig. 9. Zur experimentellen Prüfung der Gl. (12) bei der elektrolytischen Keimbildung von Quecksilber

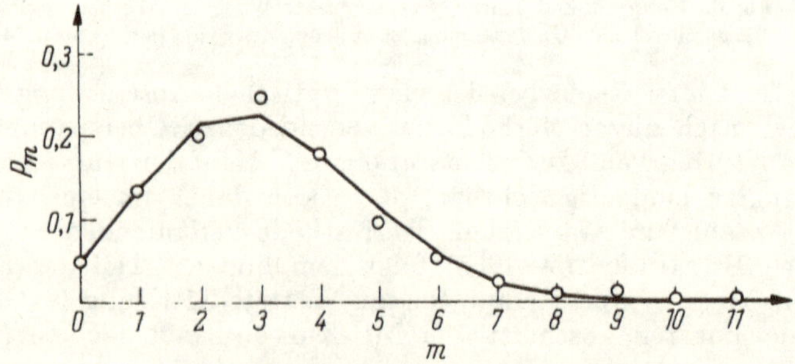

Fig. 10. Die elektrolytische Keimbildung als ein Wahrscheinlichkeits-Prozeß. Statistische Verteilung von auf einer Pt-Kathode gebildeten Quecksilbertröpfchen

Punkte – experimentelle Werte der Kembildungshäufigkeit aus 500 gleichartigen Versuchen

Polygonlinien – das Poisson'sche Gesetz mit dem experimentell gefundenen Mittelwert $\bar{m} \cong N = 3{,}14$

sicht ist der Umstand von besonderer Bedeutung, daß ein und dasselbe Experiment bei gleichen Bedingungen bequem vielfach wiederholt werden kann. So wurde gezeigt (Toschev, Milchev), daß die statistische Verteilung der während einer konstanten Zeit gebildeten Keime auch wirklich das Poisson'sche Gesetz (Gl. (32)) befolgt (Fig. 10), was die Anwendung der in 3.2 gegebenen Ansätze weitgehend rechtfertigt. Bei der elektrolytischen Keimbildung von Cadmium auf Platin konnte der nichtstationäre Charakter der Keimbildung auch auf Grund von integralen $P_{\geqq 1}(t)$-Kurven untersucht werden. Wie aus

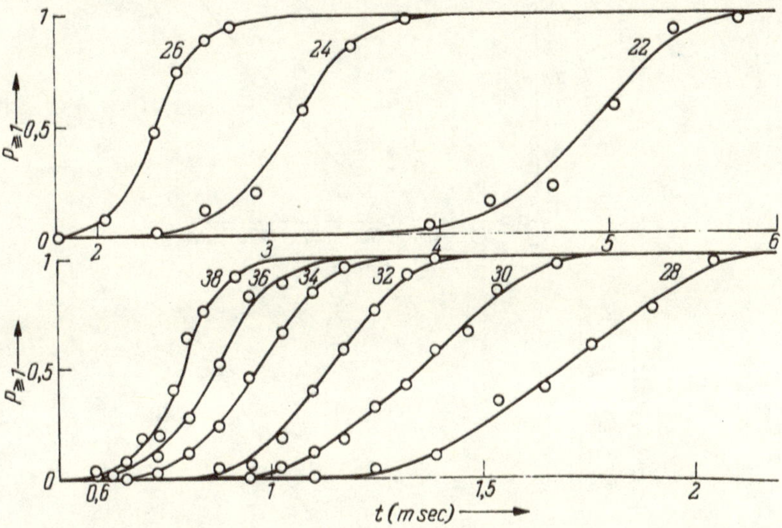

Fig. 11. Integrale Wahrscheinlichkeitskurven bei der elektrolytischen Keimbildung von Cadmium. Die Kurven geben die Wahrscheinlichkeit, daß sich in der Zeitdauer t wenigstens ein Keim gebildet hat. Zu jeder Kurve ist die entsprechende Überspannung in Millivolts angezeigt

Figur 11 hervorgeht, haben diese Kurven in Einvernehmen mit Gl. (34) und Gl. (12) (vgl. auch Figur 4b) eine typische S-förmige Gestalt (Toschev, Markov 1967). Dabei können auf Grund des von Toschev und Milchev vorgeschlagenen Auswertungsverfahren die Parameter der Keimbildung aus den Gleichungen

$$\bar{p} = 1 - \exp\left[- R(\bar{x})\right] \tag{49}$$

$$I_0^* = - \frac{\bar{\varkappa} \ln (1 - \bar{p})}{\bar{t}\, F(\bar{x})} \tag{50}$$

ermittelt werden, worin $\bar{p}$ und $\bar{t}$ die Koordinaten des Wendepunktes der $P_{\geqq 1}(t)$-Kurven sind. $R(\bar{x})$ ist eine eindeutige Funktion von $\bar{x} = \frac{\bar{t}}{\tau^*}$, von welcher I^* und τ^* bei bekannten p und t-Werten bestimmt werden können.

Somit ergab die elektrolytische Metallabscheidung ein bequemes Modell an dem sehr eingehend der nichtstationäre Charakter der Keimbildung verifiziert werden konnte. Die dabei erfaßten Induktionszeiten liegen im Bereich von 10^{-4}—10^{-1} Sekunden.

4.3. Die Kristallisation in glasbildenden Schmelzen als ein Fall nichtstationärer Keimbildung

Wie die eingehende Untersuchung der Kristallisation eines geeigneten Modellglases — des sog. Graham'schen Glases $(NaPO_3)_x$ zeigte, haben die $N(t)$-Kurven auch in diesem Falle einen typisch nichtstationären Verlauf, wovon Figur 12

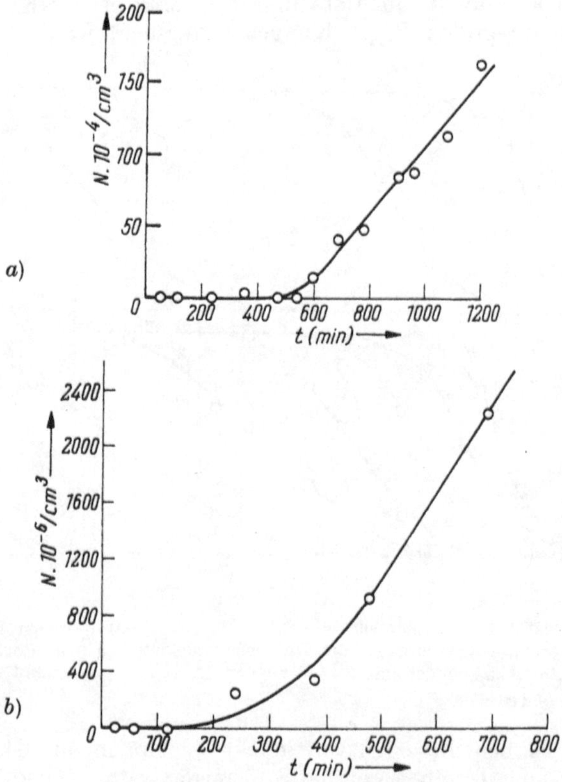

Fig. 12. Experimentelle Keimzahl-Zeit Kurven bei der initiierten Kristallisation von $Na_3P_3O_9$ in $NaPO_3$-Glas.
a — heterogene Keimbildung an Gold-Kernen bei 332 °C
b — heterogene Keimbildung an Iridium-Kernen bei 302 °C

einen Beweis ergeben (Gutzow, Toschev, Marinov, Popov). Wie gesagt, ist es kaum möglich, die Kristallisation besonders aus unterkühlten Schmelzen homogen zu führen; daher wurden in dieses Glas bewußt aktive Kristallisationskerne eingeführt. Somit wurde auch hier ein auf aktiven Unterlagen verlaufender heterogener Keimbildungsvorgang untersucht. Die $N(t)$-Kurven konnten nur in ihren Anfangsstadien verfolgt werden, da sich bei größeren Keimzahlen die um die Metallkerne gebildeten Sphärolithe gegenseitig überdecken. Somit war die Aufstellung der $N(t)$-Kurven im Sättigungsgebiet nicht möglich.

Die Unterlagen hatten eine sehr unterschiedliche Aktivität, wovon Figur 13 ein Beispiel gibt. Es wurden auf dieser Figur die an den verschiedenen Kondensationskernen (Kriställchen der Platinmetalle, wie auch Gold und Silber) bei verschiedenen Temperaturen gemessenen τ^*-Werte aufgetragen und mit den Induktionszeiten im reinen Glas verglichen. Nach Gl. (18), (19), (24) oder (25) ergeben die in Figur 13 zusammengestellten Daten in Koordinaten $\log\tau^* \div \frac{1}{T - T_0}$ eine Reihe von Geraden, welche gegenüber dem reinen Glase um den Betrag $\log\zeta$ verschoben sind. Auf diese Weise kann, wie dies in der zitierten Publikation näher gezeigt ist, die Aktivität der keimwirkenden Unterlagen quantitativ charakterisiert werden.

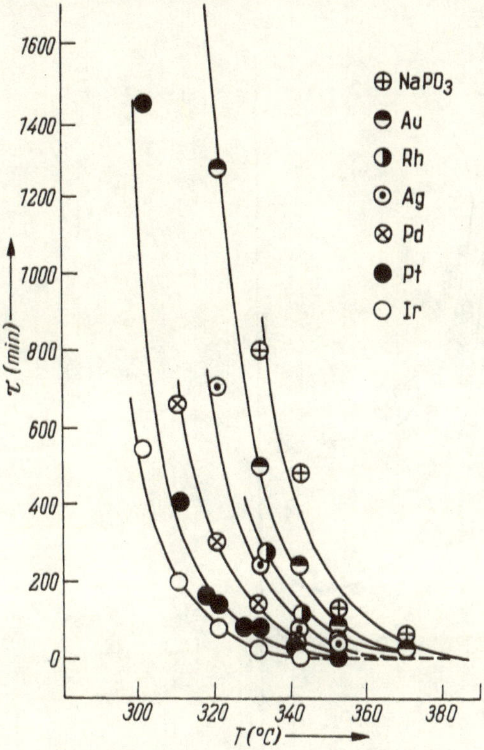

Fig. 13. Die im $NaPO_3$-Glas bei Verwendung verschiedener Initiatoren gemessenen nichtstationären Zeitverzüge τ. Die oberste Kurve gibt den Temperaturverlauf von τ im reinen $NaPO_3$-Glas wieder

Die Einführung der aktiven Unterlagen in das Modellglas, ihre Morphologie, Größenverteilung usw. sind an anderer Stelle beschrieben (Gutzow, Popov, Marinov, Toschev). Im Prinzip wurde mit dem erwähnten Glase eigentlich der Keimbildungsverlauf bei der sog. initiierten Kristallisation der Gläser modelliert, ein Prozeß, welcher in letzterer Zeit besondere technische Bedeutung erhalten hat. Eine Diskussion der initiierten Kristallisation der Gläser vom Standpunkt der nichtstationären Keimbildungskinetik findet man in einer zusammenfassenden Arbeit von Gutzow und Toschev (1971). Es sei dabei hervorgehoben, daß die experimentellen Resultate darauf hinweisen, daß die Aktivität der initiierenden Unterlagen in diesem Falle nicht als eine reine

Epitaxie im klassischen Sinne aufgefaßt werden kann, sondern auch der Einfluß der sich an der Phasengrenze Glas/Unterlage ausbildenden Spannungen eventuell von Bedeutung sein kann.

Der nichtstationäre Charakter der Keimbildung im $NaPO_3$-Glas wurde nicht nur durch die $N(t)$-Kurven demonstriert, sondern auch nach einer Reihe weiterer Messungen sichergestellt, insbesondere durch Analyse der Größenverteilung der sich gebildeten Sphärolithe (Toschev, Gutzow 1967 b) durch $\alpha(t)$-Kurven, wie auch durch eine eingehende Untersuchung des Wachstumsmechanismus der $Na_3P_3O_9$-Kristalle (Gutzow u. Konstantinov).

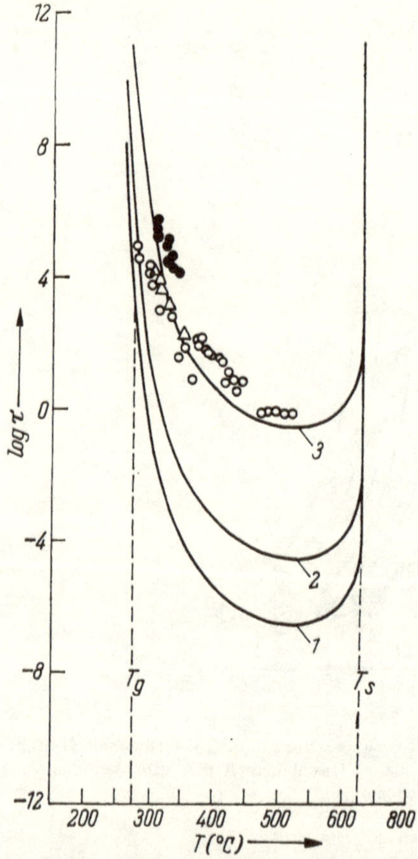

Fig. 14. Temperaturverlauf des nichtstationären Zeitverzuges bei der Kristallisation von $Na_3P_3O_9$ aus $NaPO_3$-Schmelze. Die Kurven sind nach Gl. (16), (18) und (19) berechnet, wobei 1 mit $z' = 1$, 2 mit $z' = 10^{-2}$ und 3 mit $z' = 10^{-6}$ gezogen ist.
Punkte – τ-Werte aus experimentellen $N(t)$-Kurven
Kreise – τ-Werte aus Wachstumsmessungen
Dreiecke – τ-Werte aus $\alpha(t)$ Kurven nach Daten von Westman und Krishna Murti

In Figur 14 sind die Absolutwerte der gefundenen nichtstationären Induktionszeiten mit dem theoretisch zu erwartendem Verlauf der $\tau^*(t)$-Kurve verglichen. Zur Aufstellung dieser Kurven nach Gl. (18) benutzten wir genaue

Messungen der Viskosität der $NaPO_3$-Schmelzen (GUTZOW, STRELZINA u. POPOV). Wie ersichtlich, ergibt sich eine Übereinstimmung von Theorie und Experiment erst bei $Z' = 10^{-6}$, wie auch aus groben Schätzungen unter Zuhilfenahme von $W(Z)$-Messungen für unseren Fall zu erwarten war[1]).

Im ganzen beweisen die hier nur zusammengestellten Resultate über das $NaPO_3$-Glas den nichtstationären Verlauf der Keimbildung bei der Kristallisation glasbildender Schmelzen. Daraus folgt die Notwendigkeit einer weitgehenden Überprüfung vieler in der Literatur angeführten Versuchsergebnisse und insbesondere von I_0-Bestimmungen, welche gewöhnlich einfach als $\frac{1}{t}$ ermittelt werden.

Nichtstationäre Effekte wurden neulich auch bei anderen Gläsern gefunden, so z. B. von FILIPOVICH und KALININA bei der Kristallisation von technisch sehr interessanten Li_2O-SiO_2-Zusammensetzungen. Von prinzipieller Bedeutung ist der Umstand, daß in der Kinetik der Phasentrennung oder Liquation der mehrkomponentigen Gläser auch ein nichtstationärer Verlauf festgestellt wurde. Der Beweis dafür wurde von OHLBERG und HAMMEL gegeben; dabei zeigt die Analyse der von diesen Autoren durchgeführten Messungen, daß $Z' = 1$, wie nach obigen Darlegungen (2.3) in diesem Falle zu erwarten ist. Das ist die Ursache dafür, daß meßbare τ-Werte erst bei Viskositäten von $10^{11}-10^{12}$ Poise auftraten. Ein Vergleich und eine eingehendere Zusammenstellung zwischen Liquations- und Kristallisationserscheinungen in Gläsern wurde von TASHIRO gegeben; der hier vertretene Standpunkt wurde an anderer Stelle ausführlicher auseinandergelegt (GUTZOW, TOSCHEV 1968; GUTZOW 1970).

4.4. Einige weitere Beispiele nichtstationärer Keimbildung

Es wurden von GINDT und KERN (1965) nichtstationäre Induktionszeiten auch bei der Kristallisation wäßriger Lösungen gefunden. Die dabei gemessenen τ-Werte liegen in der Größenordnung von 10^2-10^3 Sekunden und werden vorwiegend mit Wachstumshemmungen erklärt.

Eine Schätzung der bei der Tröpfchenbildung und Kristallisationserscheinungen aus der Dampfphase zu erwartenden τ-Werte ist in der schon erwähnten Arbeit von GUTZOW und TOSCHEV (1968) gegeben. Bedeutendere experimentell verifizierbare Effekte sind hier wahrscheinlich nur im Falle der direkten Kristallbildung aus der Dampfphase zu erwarten. Es sei jedoch hervorgehoben, daß nichtdestoweniger nichtstationäre Effekte bei Kondensationsprozessen von Bedeutung sein könnten, wenn man die in 2.3 erwähnte erweiterte Fassung der Ostwaldschen Stufenregel in Betracht zieht. Es ist nämlich zu erwarten, daß die Bildung von amorphen Kondensaten begünstigt ist, da bei ihnen die nichtstationären Induktionszeiten viel kleinere Werte aufweisen müssen.

Es sind hier in einer gewissen Auswahl einige grundlegende Vorstellungen und experimentelle Ergebnisse aus dem Gebiet der Keimbildungskinetik herausgegriffen worden. Wie ersichtlich, besteht schon jetzt die Möglichkeit, nichtstationäre Effekte vorauszusagen und experimentell zu erfassen. Die weitere Verfeinerung der Meßmethoden im Hinblick auf die zunehmende Tendenz, die Phasenumwandlungen in ihrer Dynamik zu verfolgen, wird u. E. die Fragen der nichtstationären Keimbildung immer aktueller machen.

---

[1]) Die τ Werte in Figur 14 sind auf homogene Keimbildung umgerechnet worden.

Literatur

Andres, R. P., Boudart, M.: J. Chem. Physics **42**, 2057 (1965)
Avrami, M.: J. Chem. Physics **7**, 1103 (1939); **8**, 212 (1940); **9**, 177 (1941)
Becker, R., Döring, W.: Ann. Physik **24**, 719 (1935)
Chakraverty, B. K.: Surface Science **4**, 205 (1966)
Collins, F. C.: Z. Elektrochem. **59**, 404 (1955)
Courtney, W. B.: J. Physic. Chem. **36**, 2009 (1962)
Farkas, L.: Z. physik. Chem. **125**, 236 (1927)
Filipovich, V. N., Kalinina, A. N.: Neorg. Materialy **6**, 351 (1970)
Fletcher, N. H.: J. Chem. Physics **29**, 572 (1958)
Frenkel, Ya. I.: Kinetic Theory of Liquids, Dover, New York 1955
Gindt, R., Kern, R.: Ber. Bunsenges. physik. Chem. **72**, 459 (1968)
Gindt, R., Kern, R.: Ber. Bunsenges. physik. Chem. **69**, 124 (1965)
Gutzow, I.: Discuss. Faraday "Vitreous State", Sept. 1970, Disc. Remarks 13A (im Druck)
Gutzow, I., Kashchiev, D.: Com. Dept. Chem. Bulg. Acad. Sci. **3** No 4, 645 (1970)
Gutzow, I., Konstantinov, I.: Comptes Rend. l'Acad. Bulg. Sci. **23**, 1107 (1970)
Gutzow, I., Strelzina, M. W., Popov, E.: Comm. Dept. Chem. Bulg. Acad. Sci. **1** No 2, 19 (1968)
Gutzow, I., Toschev, S.: Kristall u. Technik **3**, 485 (1968)
Gutzow, I., Toschev, S.: J. Cryst. Growth **7**, 215 (1970)
Gutzow, I., Toschev, S.: Proc. Symp. Nucleation and Crystallization in Glasses. Am. Ceram. Soc., Chicago April 1971 (in Druck)
Gutzow, I., Toschev, S., Marinov, M., Popov, E.: Kristall u. Technik **3**, 337 (1968)
Gutzow, I., Popov, E., Marinov, M., Toschev, S.: Kristall u. Technik **3**, 37 (1968)
Hammel, J. J.: J. Chem. Physics **46**, 2234 (1967)
Hillig, W. B.: Acta Metallurg. **14**, 1868 (1966)
Hirth, J. P., Poun, G. M.: Condensation and Evaporation, London 1963
Hollomon, J. H., Turnbull, D.: in Progress in Metal Physics vol. 4. Herausg. B. Chalmers, London 1953, S. 338
Kaischew, R.: Ann. Physik **30**, 184 (1937)
Kaischew, R.: Arbeitstagung Festkörperphysik, Dresden 1952, S. 81
Kaischew, R.: Z. Elektrochem. **61**, 35 (1957)a
Kaischew, R.: Acta Phys. Hung. **8**, 75 (1957)b
Kaischew, R.: Chem. Ing. Tech. **39**, 554 (1967)
Kaischew, R., Budevski, E.: Contemp. Physics 8, 489 (1967)
Kaischew, R., Mutaftschiew, B.: Z. physik. Chem. **204**, 334 (1955)
Kaischew, R., Mutaftschiew, B.: Bull. l'Inst. chim. Acad. Bulg. Sci. **7**, 177 (1959)
Kaischew, R., Mutaftschiew, B.: Electrochim. Acta **10**, 543 (1965)
Kaischew, R., Stranski, I. N.: Z. physik. Chem. **A170**, 295 (1934); **B26**, 317 (1934)
Kashchiev, D.: Surface Science **14**, 209 (1969)a
Kashchiev, D.: Surface Science **18**, 389 (1969)b
Kashchiev, D., Kaischew, R.: Comm. Dept. Chem. Bulg. Acad. Sci. **2**, 263 (1969)
Kolmogorov, A. N.: Bull. l'Acad. Sci. URSS (sci. math. nat.) No3, 355 (1937)
Lyubov, B. Ya.: in Rost i Nessovershenstva Metallicheskikh Kristallov, Kiev 1966, S. 5
Lyubov, B. Ya., Roitburd, A. L.: in Problemy Metallovedeniya i Fiziki Metallov, Moskva 1958, S. 91
Nielsen, A. E.: Kinetics of Precipitation, London, 1964 S. 46
Ohlberg, S. M., Hammel, J. J.: 'n Proc. 7th Intern. Congress on Glass, Brussels 28. 6.–3. 7. 1965, S. 32
Ree, F. H., Ree, T. S., Ree, T., Eyring, H.: In Advances in Chemical Physics vol. 4 Herausg. I. Prigogine, New York 1962, S. 1

Reiss, H.: J. Chem. Physics **18**, 529 (1950)
Stoyanov, S.: Comm. Dept. Chem. Bulg. Acad. Sci. **3**, 491 (1970)
Stranski, I. N., Totomanov, D.: Z. physik. Chem. **A163**, 399 (1933)
Tammann, G.: Die Aggregatzustände, 2. Aufl., Leipzig 1922
Tashiro, M.: Proc. 8th Intern. Congr. on Glass 1968, Soc. Glass Technology, Sheffield 1969, S. 113
Toschev, S.: in Kristallisation, Leipzig 1969, S. 30
Toschev, S., Gutzow, I.: Phys. Stat. Solidi **21**, 683 (1967)a
Toschev, S., Gutzow, I.: Phys. Stat. Solidi **24**, 349 (1967)b
Toschev, S., Markow, I.: Ber. Bunsenges. physik. Chem. **73**, 184 (1969)
Toschev, S., Markov, I.: Electrochim. Acta **12**, 281 (1967)
Toschev, S., Milchev, A.: Comm. Dept. Chem. Bulg. Acad. Sci. **3**, No 4, 755 (1970)
Toschev, S., Stoyanov, S.: Comm. Dept. Chem. Bulg. Acad. Sci. **3**, No 2, 295 (1970)
Turnbull, D.: Trans. AIME **175**, 774 (1948)
Volmer, M.: Kinetik der Phasenbildung, Leipzig—Dresden 1939
Wakeshima, H.: J. Chem. Physics **22**, 1614 (1954)
Westman, A. E. R,, Krishna Murti, M.: Symp. Nucleation and Crystallization in Glasses and Melts, Am. Ceramic Soc. 1962, S. 91
Zeldovich, Ya. B.: Acta Physicochim. URSS **18**, 1 (1943)
Zinsmeister, G.: Kristall u. Technik **5**, 207 (1970)

(Eingang am 7. Juni 1971)

Anschrift der Verfasser:

Dr. Svetoslav Toschev†
Dr. Ivan Gutzow
Institut für physikalische Chemie
der Bulgarischen Akademie der Wissenschaften
Sofia 13

Kristall und Technik | 7 | 1–3 | 1972 | 75–78

R. Kaischew and S. Stoyanov

Bulgarian Academy of Sciences, Institute of Physical Chemistry, Sofia

On the Distribution of Impurities during Crystal Growth

The distribution of impurities in crystals growing by spiral mechanism is obtained on the assumption that the "impure" atom mobility in the crystal is small. This distribution depends on the adsorption energy as well as on the supersaturation which is supposed to be constant along the growth front.

Найдено распределение примесей, захваченных в процессе дислокационного роста кристалла при малой подвижности примесных атомов в кристалле. Установлено каким образом это распределение зависит от адсорбционной энергии и пересыщения, которое предполагается постоянным вдоль растущей поверхности.

Let us consider a simple cubic Kossel crystal containing only one screw dislocation. At the emergence point spiral growth will take place (Fig. 1). We assume that both the supersaturation and the impurity concentration along the growth front are constant. The problem consists in finding the distribution of impurities in the crystal formed. This distribution depends on the kinetics of crystal growth if the "impure" atom mobility in the crystal is small.

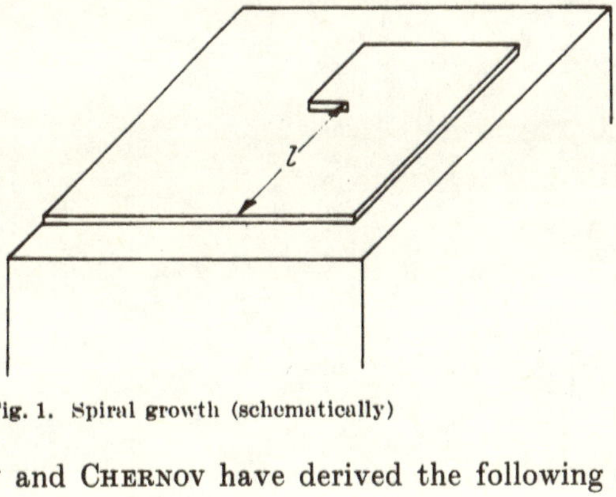

Fig. 1. Spiral growth (schematically)

Voronkov and Chernov have derived the following expressions

$$K = \frac{\mathring{K}}{1 + \frac{\eta}{S_0}} \tag{1}$$

$$K = \mathring{K} \tag{2}$$

for the distribution coefficient, considering the growth of an infinite step having orientation (01). When the adsorption energy is sufficiently high equation (1) holds good, whereas at low adsorption energy the equation (2) is ruling. Here $\mathring{K}$ is the equilibrium distribution coefficient, η and S_0 are given by

$$\eta = \frac{\omega_+}{\mathring{\omega}_+} - 1 \tag{3}$$

$$S_0 = \xi^2 \frac{u_-}{\omega_-}.$$

In the last equations ω_+ and ω_- are the frequencies of a "pure" atom addition to and detachement from the fracture respectively, while $\mathring{\omega}_+$ is the equilibrium frequency of addition and u_- is the frequency of an "impure" atom detachment. The parameter ξ depends on both the surface energy A and the work ε required to change places of an "impure" atom in the volume of the parent phase and a "pure" atom adsorbed on the crystal surface, as follows

$$\xi = \exp\left[-\frac{\varepsilon - 2A}{T}\right]$$

where the temperature T is measured in energy units.

The problem outlined above can be easily solved by using equation (1). The equilibrium frequency $\mathring{\omega}_+$ for a finite step is given by (STRANSKI, KAISCHEW)

$$\mathring{\omega}_+ = (\omega_-^{n-1}\,\omega'_-)^{1/n}, \tag{4}$$

where n is the number of atoms on the step and ω'_- is the frequency of detachment of an atom having only two nearest neighbours. It can be seen that, if n becomes infinitely large, $\mathring{\omega}_+$ tends to ω_-. Substituting (3) and (4) in (1) we obtain

$$K(n) = \frac{\mathring{K}}{1 + \dfrac{\dfrac{\omega_+}{\omega_-}\left(\dfrac{\omega_-}{\omega'_-}\right)^{1/n} - 1}{S_0}}$$

which gives for $n = n_K$

$$K(n_K) = \mathring{K}.$$

Taking into account that

$$\frac{\omega_+}{\omega_-} = 1 + S$$

$$\frac{\omega_-}{\omega'_-} = \exp\left[\frac{\psi}{T}\right]$$

we can write

$$K(n) = \frac{\mathring{K}}{1 + \dfrac{(1+S)\exp\left[\dfrac{\psi}{n\,T}\right] - 1}{S_0}} \tag{5}$$

which gives

$$K = \frac{\mathring{K}}{1 + \dfrac{S}{S_0}}$$

when $n \to \infty$. In the above expressions S represents the supersaturation and Ψ is the binding energy of the "pure" atoms in the crystal (Ψ has a negative value). Inasmuch as

$$K = \frac{c}{c_p}$$

$$\mathring{K} = \frac{\mathring{c}}{c_p}$$

from equation (5) we get

$$c(n) = \frac{\mathring{c}}{1 + \dfrac{(1+S)\exp\left[\dfrac{\psi}{n\,T}\right] - 1}{S_0}} \tag{6}$$

where $c(n)$ stands for the concentration of impurities captured by the growing step of size n, $\mathring{c}$ is the equilibrium concentration of impurities in the crystal and c_p is the impurity concentration in the parent phase.

For a complete solution of the problem it is necessary to find the dependence of n on the distance l from the step to the dislocation axis. Assuming that the velocities of all steps whose sizes are larger than n_K are equal, we obtain

$$n(l) = n_K + l \qquad 0 \leq l \leq n_K \tag{7}$$
$$n(l) = 2\,l \qquad n_K \leq l \leq \infty$$

where n_K is the linear size of the twodimensional nucleus. Finally, for the distribution of impurities whose adsorption energy is high, we find

$$c(l) = \frac{\mathring{c}}{1 + \dfrac{(1+S)\exp\left[\dfrac{\psi}{(n_K + l)\,T}\right] - 1}{S_0}} \qquad 0 \leq l \leq n_K$$

$$c(l) = \frac{\mathring{c}}{1 + \dfrac{(1+S)\exp\left[\dfrac{\psi}{2\,l\,T}\right] - 1}{S_0}} \qquad n_K \leq l \leq \infty$$

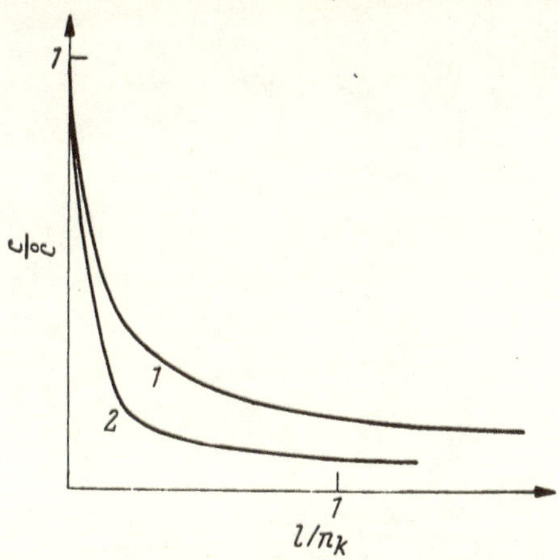

Fig. 2. The dependence of $\frac{c}{\mathring{c}}$ on l, at $\frac{S}{S_0} = 9$ (curve 1) and $\frac{S}{S_0} = 24$ (curve 2)

The dependence of $\frac{c(l)}{\mathring{c}}$ on l at $\frac{S}{S_0} = 9$ and $\frac{S}{S_0} = 24$ is shown in Figure 2 (curve 1 and curve 2 respectively).

The distribution of impurities whose adsorption energy is low is homogeneous as follows from equation (2).

References

STRANSKI, I. N., KAISCHEW, R.: Z. physik. Chem. B **26**, 100 (1934)
VORONKOV, V. V., CHERNOV, A. A.: Kristallografia **12**, 222 (1967)

(Received June 7, 1971)

Authors' address:

Prof. Dr. R. KAISCHEW
Dr. S. STOYANOV
Institut für physikalische Chemie
der Bulgarischen Akademie der Wissenschaften
Sofia — 13/Bulgarien

| Kristall und Technik | 7 | 1–3 | 1972 | 79–93 |

D. Iwanov, Chr. Nanev, R. Kaischew

Institute of Physical Chemistry
Bulgarian Academy of Sciences
Sofia

Evaporation of Zinc Single Crystals

The dislocation nature of defects leading to local evaporation of the {0001} face of zinc single crystals is proved. Two kinds of pits appear on the emergence points of dislocations, upon heating the crystals in vacuum, that is, hexagonal ones on screw dislocations and conical ones on edge dislocations (and probably on impurities). Hexagonal pits appeared at the growth points of whiskers upon evaporation. This is in excellent agreement with the assumption that whiskers grow by a screw dislocation mechanism.

Deepening and broadening of the pits were studied by interferometry. It was found that the change in profile with time can be described by the equations of Frank's kinematic theory. The same theory was used to calculate the step flux along the basal face of zinc single crystals upon evaporation at 250, 270, and 290 °C. The rate of advance of unit steps was determined from the flux and compared with the theoretically expected one. The temperature dependence of this rate made it possible to estimate the vibrational frequency of the atoms in the crystal lattice as well as the evaporation heat of zinc. The reasonable values obtained demonstrate the correctness of the interpretation of the experimental data.

Die Versetzungsnatur der Defekte, welche örtliches Abdampfen der Fläche {0001} von Zinkeinkristallen verursachen, wird gezeigt. Zwei Arten von Ätzgrübchen entstehen an den Versetzungsdurchstoßpunkten der Kristalloberfläche, wenn die Kristalle im Vakuum erhitzt werden: sechseckige Ätzgrübchen erscheinen auf den Schraubenversetzungen, wogegen kegelartige Ätzgrübchen auf den Stufenversetzungen (wahrscheinlich auch an Verunreinigungen) beobachtet werden. Wenn Kristalle mit auf der Basisfläche aufgewachsenen Whiskern abgedampft werden, entstehen sechseckige thermische Ätzgrübchen an den Whiskerstellen. Diese Tatsache stimmt gut mit der Annahme überein, daß die Whisker nach einem Schraubenversetzungsmechanismus wachsen.

Die Vertiefung und Verbreitung der kegelartigen Ätzgrübchen wird interferometrisch untersucht. Es wurde festgestellt, daß sich die Profile im Einklang mit der Frank'schen kinematischen Theorie mit der Zeit ändern. Mit Hilfe dieser Theorie wurde der Fluß der Stufen auf der Basisfläche der Zinkeinkristalle beim Abdampfen bei 250, 270 und 290 °C berechnet. Aus diesem Stufenfluß wurde die Geschwindigkeit der einzelnen Stufen bestimmt, welche mit der theoretisch erwarteten Geschwindigkeit verglichen wurde. Von der Temperaturabhängigkeit der Stufengeschwindigkeit wurde die Schwingungszahl der Atome im Kristallgitter, wie auch die Verdampfungswärme des Zinks berechnet. Die vernünftigen Werten, die dabei erhalten wurden, geben einen Hinweis für die Richtigkeit der durchgeführten theoretischen Deutung der experimentellen Ergebnisse.

1. Introduction

Theoretical calculations performed by Cabrera and Levine, Cabrera (1956) showed that the appearance of thermal etch pits during evaporation of crystals may usually be attributed to the presence of emergence points of dislocations. Chemical etching of zinc crystals (Gilman; Meleka; Rosenbaum, Saffren) with an appropriate reactant confirmed that in the case of zinc pits appear at the emergence points of dislocations. Polygonal pits were observed by Forty during evaporation of frozen zinc droplets, by Danko, Griest on cleaved zinc surfaces, and by Rais, Bromberg on zinc platelets. Ruff first carried out a comprehensive and systematic study of cleaved zinc surfaces subjected to evaporation at different undersaturations and in ultrahigh vacuum and also studied the morphology of the conical pits as a function of the undersaturation and foreign gas (oxygen) pressure. However the experimental technique used by this author does not allow continuous observation of the process.

The present paper deals with the investigation of the evaporation of basal faces of vapour-grown zinc single crystals. It is known (Nanev) that the basal faces of such crystals are very smooth. The dislocation nature of the defects leading to the appearance of pits upon evaporation was demonstrated. Furthermore it was shown that screw and edge dislocations can be differentiated. The kinetics of the evaporation process was studied.

2. Experimental

The evaporation of the basal face of zinc single crystals grown from vapours until attainment of the so-called end form (Kaischew, Nanev) was studied. We used spectrally pure zinc additionally purified by repeated sublimations under vacuum at 10^{-6} mm Hg. Zinc single crystals were obtained in an evacuated glass vessel by freezing a molten droplet so that they adhered strongly to the glass substrate. This led to the appearance of strains which generated defects (Nanev, Wilkens). Deformation was enhanced by removing the crystal from the vessel since this is done by breaking it at the location where the crystal adheres. Evaporation was performed in a Reichert Vacutherm vacuum chamber with the crystal placed in a specially devised holder. The evaporation process was followed in situ by means of a Reichert MeF-1 metallographic microscope. The use of the Nomarski interference contrast and interferometer allowed a detailed study of the morphology and the kinetics of the process. For instance interference contrast made it possible to observe pits in which the faces had a very small slope (down to 1:1000–1:10000). This is essential for observing pits on screw dislocations.

3. Determination of the nature of evaporation sources

Upon evaporation of a crystal steps are generated by its edges and corners as well as by various defects at its surface. Screw dislocations bring about non-vanishing steps on the crystal surface which are dissolved even at very small undersaturations. In this case however the pits have a very small slope so that they cannot be observed by usual techniques (Cabrera 1956). Their slope increases with undersaturation. Pits are clearly formed at undersaturations at which two-dimensional negative nuclei begin to form nearly simultaneously on screw and edge dislocations. Such nuclei are also formed on the impurities incorporated in the lattice and, at very large undersaturations on perfect regions of the crystal surface, too.

At the beginning of the evaporation process between 230 and 340 °C the basal face was covered by hexagonal flat-bottomed and conical pits. Between 230 and 280 °C the hexagonal pits predominated and were usually randomly distributed over the crystal surface (Fig. 1). Initially their edges were directed along $\langle 11\bar{2}0 \rangle$. Occasionally the pits were arranged in rows with the same direction, close to the edges of the crystal. Upon subsequent spread at a given temperature the edges of most pits changed directions and ran parallel to the direction of most rapid evaporation, i.e. $\langle 10\bar{1}0 \rangle$ (Fig. 2). This "rotation" of

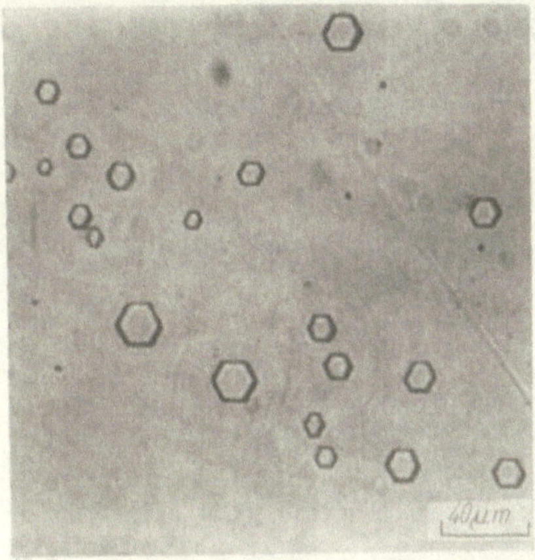

Fig. 1. Hexagonal pits on the {0001} face of a zinc single crystal obtained at 240 °C

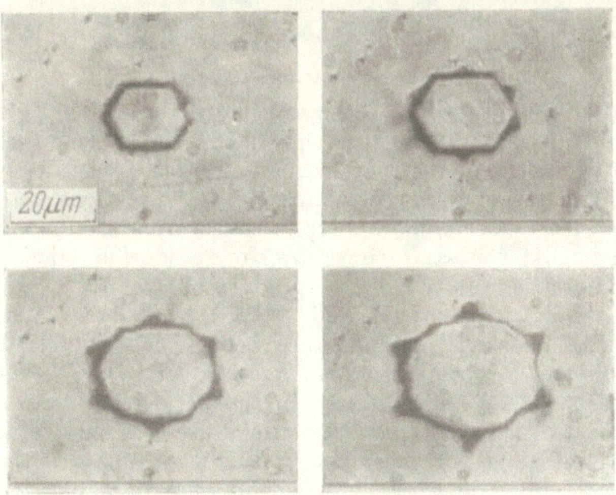

Fig. 2. "Rotation" of a hexagonal pit by 30°. Temperature 245 °C

Fig. 3. Electron micrograph of a portion of a hexagonal pit. The middle portions of the faces grow, because the initial apices are blocked

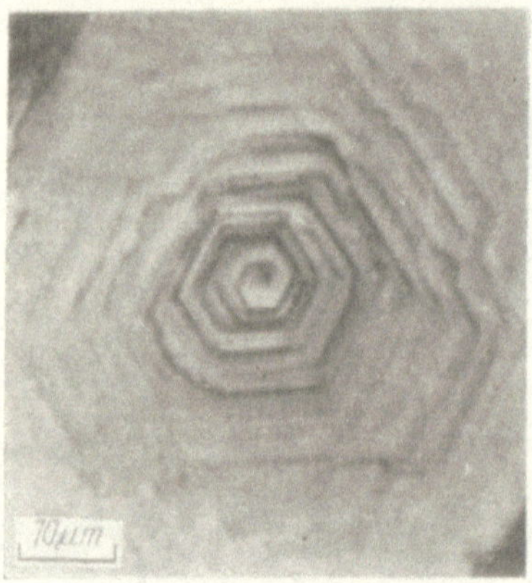

Fig. 4. Macrospiral at the bottom of a hexagonal pit. Temperature 305 °C

the pits by 30° is obviously due to the oxygen in the residual gas (dynamic vacuum of about 1.10^{-5} mm Hg). As shown by electron microscopic observations the initial apices of the pits are blocked (Fig. 3), probably by oxide which causes evaporation along new preferential directions. Actually the structure of the $\langle 11\bar{2}0 \rangle$ face is more loose so that evaporation is initially more pronounced on it. At the same time however, oxidation takes place more easily so that the initial "apices" of the pits are blocked.

If the hexagonal pits are rapidly heated to 300—340 °C by applying a thermal pulse steps form at their bottoms and they become pyramidal. Macrospirals are sometimes observed at the bottom of some pits (Fig. 4). The defects which cause the appearance of hexagonal pits on the basal face upon evaporation of the zinc single crystals deeply penetrate the bulk of the crystal and are not found at the surface only. This is shown by the deepening of the pits at temperatures above 300 °C which change to deep pyramidal pits with distinct centres. The lateral pyramidal faces of these pits have steps so that it is impossible exactly to determine their depth by interferometry. Nevertheless the latter was estimated to at least 2 to 5 times $\frac{\lambda}{2}$ (1 — 2 μm). If evaporation is arrested by cooling down to room temperature the process continues upon renewed heating without being affected by the interruption. After more prolonged interruptions (e.g. over night or by letting air in) and upon renewed heating new conical pits appeared at the bottom of most of the hexagonal pits (Fig. 5). All this indicates the presence of deep-seated defects at these points.

Conical pits appeared in the temperature range between 250 and 340 °C (Fig. 6). They are randomly distributed over the face. After more prolonged evaporation they take the shape of hexagonal pits and behave as such; the direction of preferential evaporation changes, after leaving the crystals in contact with air and upon renewed heating the conical pits are centered by new ones. The conical pits are also due to deep-seated bulk defects. The depth of the pits was interferometrically determined 3 to 10 times $\frac{\lambda}{2}$, i.e. 1,5 to 5,5 μm. Initially they have one center and their conical faces are smooth but additional centers appeared around it during subsequent evaporation.

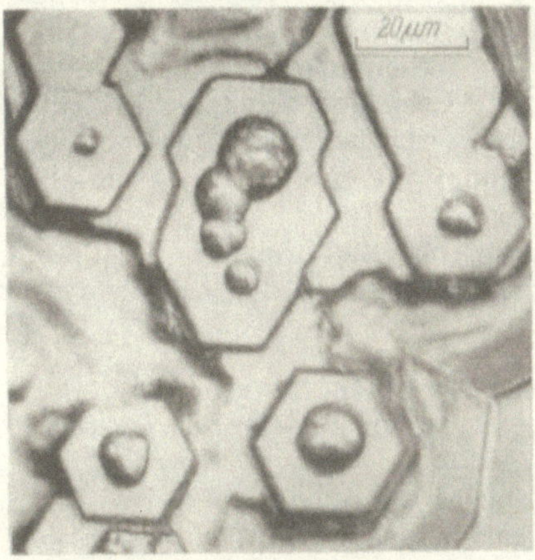

Fig. 5. Hexagonal pits obtained at 240 °C centered by conical pits (after the crystal has been cooled, left to air, evacuated and reheated at 260 °C)

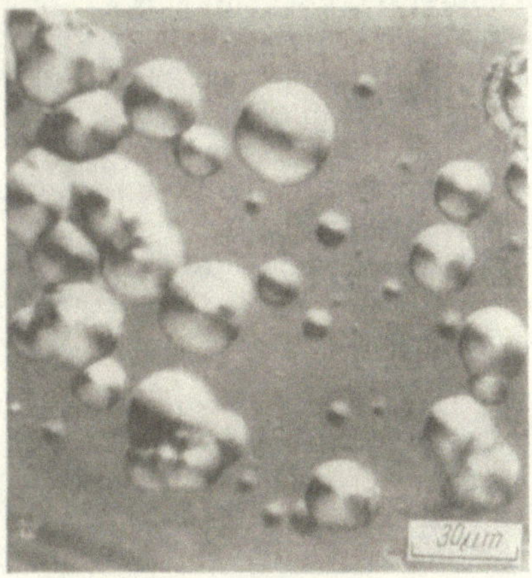

Fig. 6. Conical pits on the basal face of a zinc single crystal obtained at 300 °C

Simultaneous appearance of conical and hexagonal pits was observed between 250 and 280 °C. At a vacuum of about $1 \cdot 10^{-4}$ mm Hg hexagonal pits with a rough bottom appeared. Pits appeared also at lower temperatures in the case of more strongly deformed crystals. When macrosteps occurred on the basal faces pits appeared at higher temperature. Very strongly deformed crystals evaporated without formation of pits.

Some of the zinc single crystals were chemically etched by the method suggested by Rosenbaum and Saffren in order to confirm the dislocation nature of the defects, which brings about the appearance of pits. The crystals were etched for 30 to 60 seconds in a 1:5 mixture of 1,2 n Merck hydrochloric acid and ethanol, after which they were washed with ethanol and dried. Chemical etching

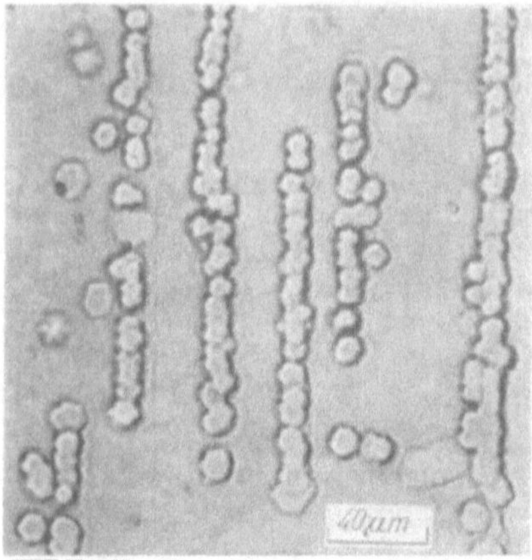

Fig. 7. Etch pits appeared in rows parallel to ⟨11$\bar{2}$0⟩

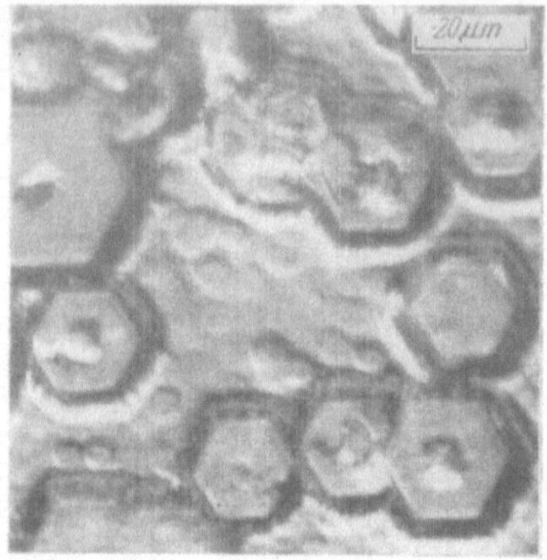

Fig. 8. Hexagonal pits (obtained from conical ones) with etch figures at the centre

of non-evaporated zinc single crystals showed once more a well-defined arrangement of some etch figures, that is, of the dislocations, in rows parallel to ⟨11$\bar{2}$0⟩, close to the edges of the basal face (Fig. 7). When evaporated zinc single crystals with pits on {0001} were additionally subjected to chemical etching chemical etch pits were formed at the centre of the hexagonal pits obtained upon evaporation (both on the initially hexagonal pits and on those which had been conical) (Fig. 8).

To summarize the above results let us point out that the pits (or at least most of them) obviously appear at the emergence points of dislocations at the crystal surface:

1. They appear on defects which deeply penetrate the bulk of the crystal. It is hardly probable that the impurities are continuously segregated with the large depth reached by the pits.

2. Upon renewed heating pits usually appear at the same sites.

3. Chemical etching showed the presence of emergence points of dislocations at points of the crystal surface where the pits appear.

4. Rows of hexagonal pits as well as chemical etch pits oriented along ⟨11$\bar{2}$0⟩ are obtained close to the edge of the basal face, where strains are largest (NANEV, WILKENS). This shows that hexagonal pits are due to dislocations.

However not all dislocations on the basal face reveal themselves upon evaporation. The hexagonal pit density on the basal face at the different temperatures varies between 4 and $10 \cdot 10^3$ cm^{-2}, while the conical pit density ranges from 2 to $10 \cdot 10^4$ cm^{-2}. Chemical etching brings about the appearance of 2 to $7 \cdot 10^5$ pits per cm^2.

According to theoretical predictions dislocations with a screw component will cause local evaporation even at low undersaturations. In our experiments the undersaturation was unknown, but it can be most generally said that an increase in evaporation temperature is equivalent to an increase in undersaturation. This makes it possible to differentiate from an energetic standpoint the defects which cause the two kinds of pits and to differentiate screw from edge dislocations. Hexagonal pits are surely due to screw dislocations since

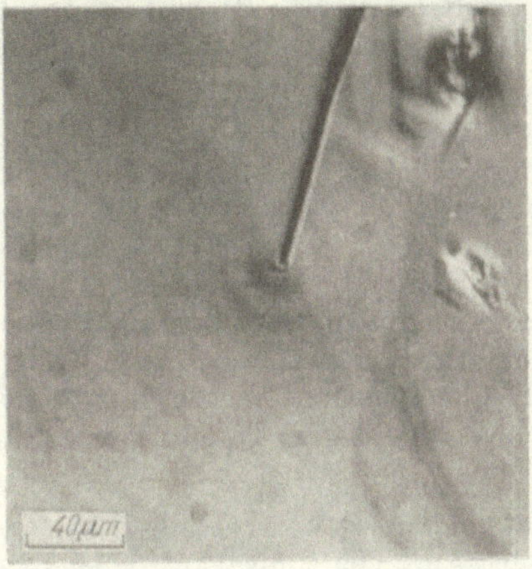

Fig. 9. Pyramidal growth at the basis of a whisker

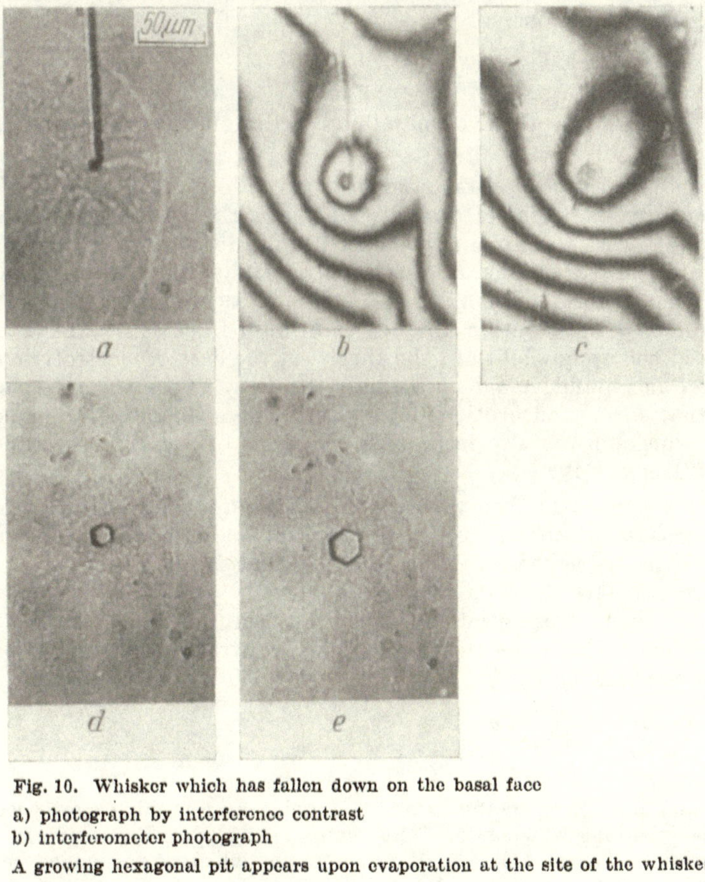

Fig. 10. Whisker which has fallen down on the basal face

a) photograph by interference contrast

b) interferometer photograph

A growing hexagonal pit appears upon evaporation at the site of the whisker (c, d, e)

1. they appear at lower temperature. The fact that they are flat-bottomed at these temperatures may be attributed to passivation of the especially active emerging point of the screw dislocation by oxygen in the residual gas.

2. Above 300 °C spirals are observed in such pits.

3. In this connection the appearance of hexagonal pits at the sites of the whiskers on the face is particularly significant. The growth of zinc single crystals under certain conditions is accompanied by growth of whiskers on their basal face in the $\langle 11\bar{2}3\rangle$ direction. (NANEV, IWANOV)[1]). The detailed investigation of the basis of the whiskers by the Nomarski interference contrast and interferometer shows that the latter have grown on a convex pyramid (Fig. 9). Heating of such crystals caused evaporation of the whiskers (which first broke down). Flat-bottomed hexagonal pits appeared at the growth points of the whiskers (Fig. 10). This is in excellent accord with the assumption that whiskers grow by a screw dislocation mechanism (NANEV, KAISCHEW).

The conical pits are obtained at higher temperatures and are probably due to edge dislocations (and to impurities). The conical pit density on the basal face is about an order of magnitude higher than that of the hexagonal pits which indicates that perhaps edge dislocations are considerably more numerous than screw dislocations.

4. Study of the evaporation kinetics of the conical pits

The evaporation kinetics was studied by an interference method. The results are interpreted in terms of FRANK's kinematic theory, which has already been successfully used to explain the macroscopic dissolution of lithium fluoride (IVES) and the experimental results obtained by BATTERMAN concerning the dissolution of germanium (FRANK, IVES). IVES and HIRTH investigated by interferometry the solution kinetics of lithium fluoride. HULETT and YOUNG confirmed the applicability of the theory in the case of electrolytic dissolution of copper and found, in agreement with the views expressed by CABRERA (1958, 1961), that surface adsorption strongly influences the process. The dissolution of iron whiskers was also interpreted in terms of the kinematic theory (SHEMENSKI, BECK, FONTANA).

FRANK's kinematic theory treats the dissolution and growth of crystals in reducing these processes only to propagation of unit steps, each of height h along a densely packed face. It is assumed that the step flux q at a given point of the crystal face depends only on the density ϱ of steps at the same point. The theory has an essential consequence, according to which the trajectories of the points with a constant step density are straight lines in the plane (x, y), whose slope is given by

$$\frac{dy}{dx} = h\left(\varrho - \frac{q}{c}\right) \tag{1}$$

[1]) A new measuring method made it possible to determine more accurately the angle between the whiskers and the vector $1/3\,\langle 11\bar{2}0\rangle$. From a large number of measurements the value 63,5 ± 5° was obtained which corresponded better to the $1/3\,\langle 11\bar{2}3\rangle$ direction (for zinc the theoretical angle is about 62°).

where

$$c = c(\varrho) = \frac{dq}{d\varrho} . \tag{2}$$

Another consequence is that the projection on the axis x of a point with a constant step density moves with a uniform velocity

$$\frac{dx}{dt} = c(\varrho) . \tag{3}$$

These theoretical relationships may be experimentally confirmed in cases similar to ours, when the evaporating surface is a curved one and when the points with constant step density describe straight-line trajectories during the successive stages of the process.

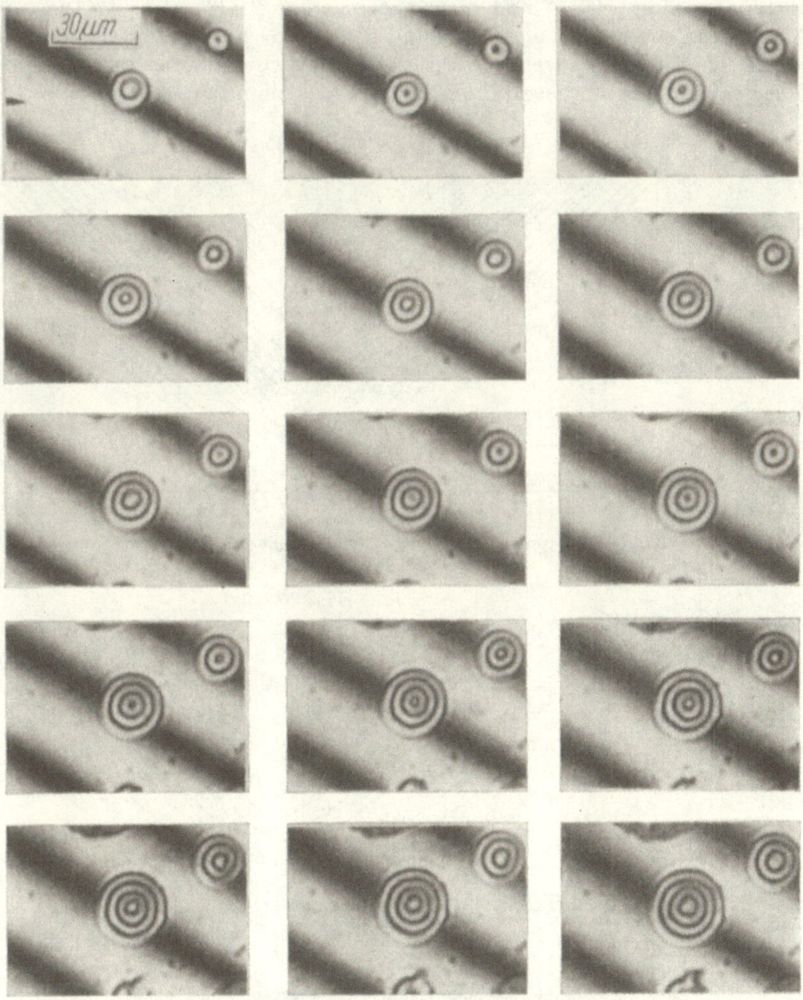

Fig. 11. Interferometric measurement of the growth and deepening of conical pits with time. Temperature 250 °C

When the zinc crystal is gradually heated the kinetics of growth and deepening of the conical pits on its basal face may be followed at a given temperature. Upon rapid heating (thermal pulse), and also above 300 °C, the evaporation proceeds so intensely that it is impossible to follow the kinetics. Between 250 and 300 °C and at a vacuum of 10^{-5} mm Hg evaporation was observed continuously by interferometry. The development of a series of pits was studied by photographs taken at definite time intervals (Fig. 11). To this effect a Nomarski interferometer and light with $\lambda = 0{,}55\ \mu m$ were used. By the method suggested by IVES and HIRTH we plotted the profiles of the pits at successive times (Fig. 12). Two kinds of profiles were observed: 1. with a nearly constant slope (e.g. the profiles obtained from the interference patterns on Figure 11) and 2. with a slope slowly increasing from the centre to the edge of the pit (e.g. Fig. 12). The slope of the first type of pits measured with respect to the basal flace, was about 4–5° and of the second one 2–3°, close to the centre, and 4–5° near the edge (i.e. an average of 3–4°). For each series of profiles the points of equal slope were determined by drawing a series of parallel tangents. The trajectories of these points in the (x, y) plane are straight lines (Fig. 13).

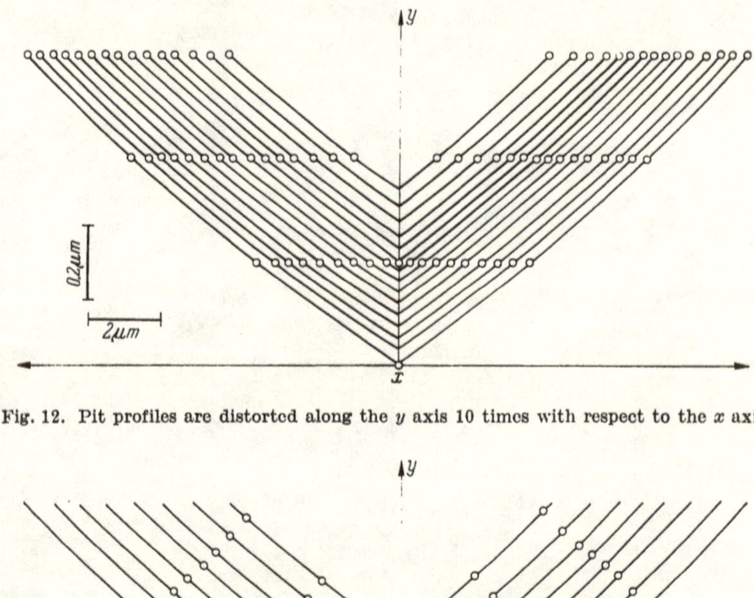

Fig. 12. Pit profiles are distorted along the y axis 10 times with respect to the x axis.

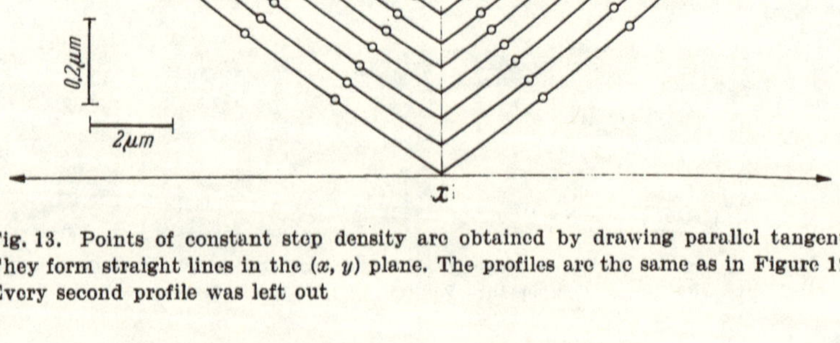

Fig. 13. Points of constant step density are obtained by drawing parallel tangents They form straight lines in the (x, y) plane. The profiles are the same as in Figure 12. Every second profile was left out

It is seen that the slope of the trajectories of the points with a constant step density $\frac{dy}{dx}$ decreases in absolute value away from the centre of the pit, in other words it decreases with increase in step density. The dependence of the distance x of the points of constant step density from the pit axis on time t is a linear one Fig. 14). The slope $\frac{dx}{dt}$ of the trajectories increases with step density.

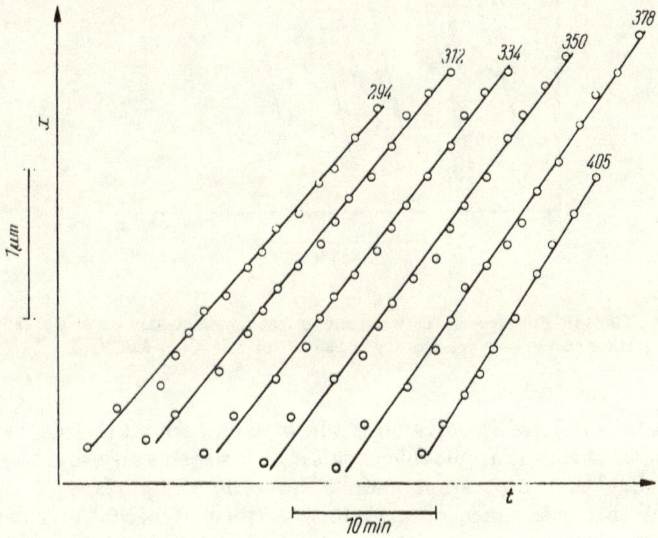

a)

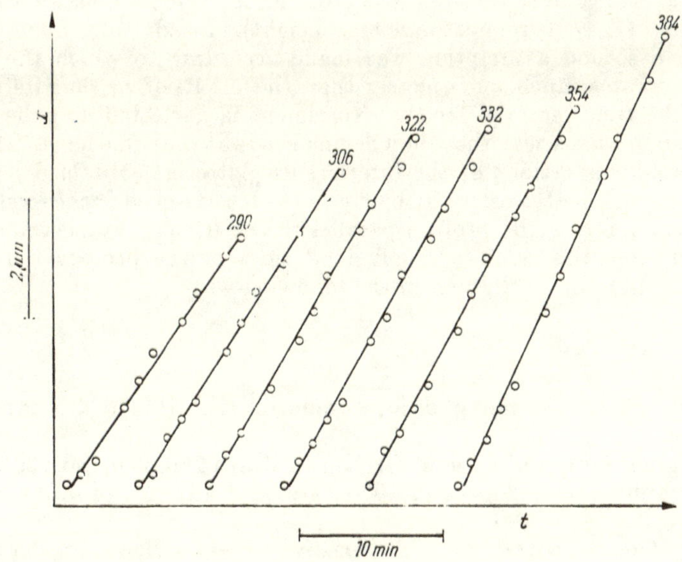

b)

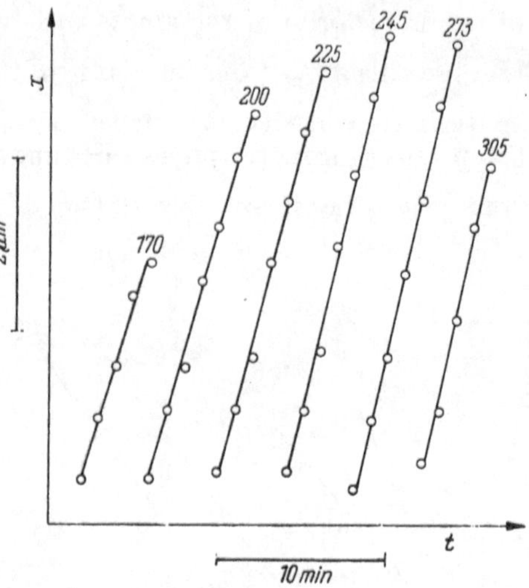

c)

Fig. 14. Relationship between x and t at constant step density. The numerals show the step number per μm at: a) 250 °C b) 270 °C c) 290 °C

The linear relationships of x on y and of x on t found in this way show that the kinematic theory is applicable in this case. It seems also that the prerequisite of the theory that unit steps should move on the surface of the crystal is fulfilled in this case since observations performed with the Nomarski interference contrast at a magnification of 400 times showed[1]) that the vicinal faces of the conical pits were smooth, at least until the appearance of new evaporation centres at the conical surface. Therefore in our calculations we assumed for h the value of 2,47 Å (monoatomic step height). In addition, upon plotting the profiles, a second assumption was made according to which the large basal face evaporates much more slowly than the pit itself so that its evaporation during the time required for the experiment is neglected and the pit profiles are drawn in such a way that they begin always at the same level. The influence of the weak distortions of the interference patterns[2]) obtained with the Nomarski interferometer perpendicularly to the interference fringes was eliminated by usually plotting the profiles parallel to the fringes. Furthermore, the data obtained from the right- and left-hand sides of the profiles were averaged.

Eq. (1) may also be written in the following way

$$q = c\,\varrho - \frac{dy}{dx}\frac{c}{h} \tag{4}$$

where all quantities required for computing the flux q are experimentally available.

The values for ϱ and q found for temperatures 250, 270 and 290 °C are illustrated graphically on Figure 15. The rate V of advance of a unit step can be

---

[1]) Interference contrast is highly sensitive toward surface irregularities.

[2]) In fact the shape of the pits is strictly conical, as observed by interference contrast.

calculated from the flux q

$$V = \frac{q}{\varrho}. \tag{5}$$

It is seen (cf. Fig. 16) that the rate of advance of the unit step strongly depends on temperature; at 290 °C it is by an order of magnitude larger than at 250 °C.

The present experiments were performed under conditions of free evaporation, i.e., the undersaturation is so large that direct evaporation from the step (without surface diffusion) may also play a noticeable role. Thus, the rate of advance of a series of parallel steps will be most generally given by the sum of two terms: the first one V_0 takes into account direct evaporation from the step, whereas V_∞ is the rate of advance of the step in presence of a surface diffusion mechanism of evaporation.

$$V = V_0 + V_\infty. \tag{6}$$

BURTON, CABRERA and FRANK give for V_∞ the expression

$$V_\infty = 2\,\lambda_S\,\sigma\,\nu\,e^{-W/kT} \cdot th\left(\frac{y_0}{2\,\lambda_S}\right) \tag{7}$$

where λ_s is the mean free path of the atoms in the adsorbed state, σ is the undersaturation, ν is the vibrational frequency of the atoms in the crystal lattice, W is the evaporation energy for one building unit, K is the Boltzmann constant, and T is the absolute temperature. The distance between the steps y_0 is equal to the reciprocal value of ϱ. For the direct evaporation rate one can also assume an exponential relationship of the form

$$V_0 = A\,e^{-W/kT} \tag{8}$$

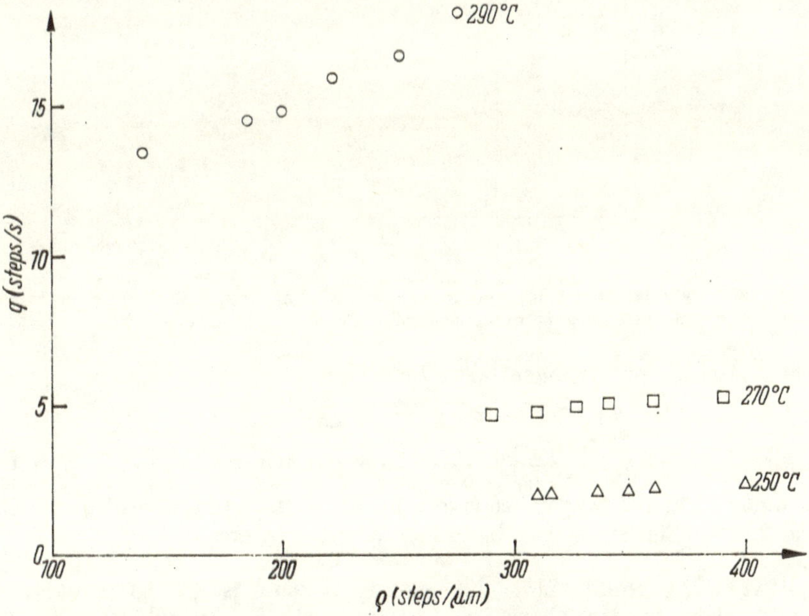

Fig. 15. Relationship between q and ϱ at: 250 °C, 270 °C and 290 °C

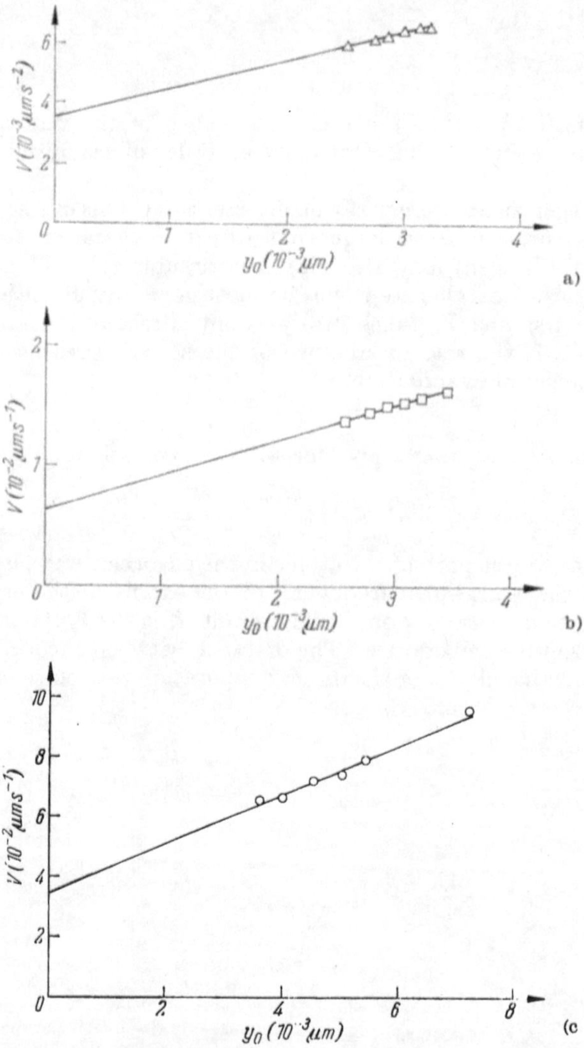

Fig. 16. Relationship between the rate V of advance of steps along the basal face and the distance y_0 between them at: a) 250 °C b) 270 °C c) 290 °C

where A is the pre-exponential factor. Thus

$$V = A\, e^{-W/kT} + \sigma\, \nu\, e^{-W/kT} \cdot y_0 \tag{9}$$

since $\lambda_S \gg y_0$[1]) i.e. $t\,h\left(\frac{y_0}{2\,\lambda_S}\right) = \frac{y_0}{2\,\lambda_S}$. In our case σ can be taken approximately as unity. Then if we represent graphically V as a function of y_0 for 250, 270 and 290 °C (Fig. 16) it can be found from the intercept V_0 and $\nu\, e^{-W/kT}$ from

[1]) At 300 °C Ruff reports for zinc $\lambda_S = 500$ Å $\gg y_0 = 25$ to 65 Å. This is confirmed by the fact that V and consequently V_∞ depend on ϱ. In turn this shows that the diffusion zones of adjacent steps overlap.

the slope. It is seen that V_0 strongly increases with temperature. On the other hand, by taking the logarithm of the slope $\nu\, e^{-W/kT}$ and plotting it as function of $\frac{1}{T}$ for the three temperatures it proved possible to estimate ν and W. The slope $\nu\, e^{-W/kT}$ used for calculating ν and W was averaged for each of the three temperatures from the data obtained on 3 or 4 pits. Thus we obtained for ν a value of about 10^{12} s^{-1} and for the evaporation heat approximately 29 kcal · $mole^{-1}$. The evaporation heat may also be calculated independently by means of the temperature dependence of V_0. A value of about 32 kcal · $mole^{-1}$ is then obtained. It is seen that reasonable values are obtained for all quantities. This demonstrates the correctness of the theoretical analysis of the experimental data.

Thanks are due to St. Stoyanov for help in discussing the manuscript.

References

BATTERMAN, B. W.: J. appl. Physics **28**, 1236 (1957)
BURTON, W. K., CABRERA, N., FRANK, F. C.: Phil. Trans. Roy. Soc. Å **243**, 299 (1951)
CABRERA, N.: J. Chem. Physics **53**, 675 (1956)
CABRERA, N.: in Growth and Perfection of Crystals, Ed. Doremus, R. H., Roberts, B. W. and Turnbull, D., New York 1958, p. 393
CABRERA, N.: in Reactivity of Solids, Ed. De Boer, J. H., Amsterdam 1961, p. 345
CABRERA, N., LEVINE, M.: Phil. Mag. **1**, 450 (1956)
DANKO, J. C., GRIEST, A. J.: J. Metals **8**, 515 (1956)
FORTY, A. J.: Phil. Mag. **43**, 949 (1952)
FRANK, F. C.: in Growth and Perfection of Crystals, Ed. Doremus, R. H., Roberts, B. W. and Turnbull, D., New York 1958, p. 411
FRANK, F. C., IVES, M. B.: J. appl. Physics **31**, 1996 (1960)
GILMAN, J. J.: J. Metals **8**, 998 (1958)
HULETT, L. D., YOUNG, F. W.: J. Phys. Chem. Solids **26**, 1287 (1965)
IVES, M. B.: J. appl. Physics **32**, 1534 (1961)
IVES, M. B., HIRTH, J. P.: J. Chem. Physics **33**, 517 (1960)
KAISCHEW, R., NANEV, CHR.: Physica Status Solidi **10**, 779 (1965)
MELEKA, A. H.: Phil. Mag. **1**, 803 (1956)
NANEV, CHR.: Physica Status Solidi **16**, 777 (1966)
NANEV, CHR., IWANOV, D.: Physica Status Solidi **23**, 663 (1967)
NANEV, CHR., KAISCHEW, R.: 3 Internationales Symposium Reinststoffe in Wissenschaft und Technik, Dresden 4–8 Mai 1970
NANEV, CHR., WILKENS, M.: Physica Status Solidi **32**, 839 (1969)
RAIS, G. B., BROMBERG, M. I.: Kristallografiya, **4**, 594 (1959)
ROSENBAUM, H. S., SAFFREN, M. M.: J. appl. Physics **32**, 1866 (1961)
RUFF, A. W.: J. Chem. Physics **41**, 1204 (1964)
SHEMENSKI, R. M., BECK, F. H., FONTANA, M. G.: J. appl. Physics **36**, 3909 (1965)

(Received June 7, 1971)

Authors' address:

D. IWANOV, CHR. NANEV
Prof. Dr. R. KAISCHEW
Institute of Physical Chemistry
Bulgarian Academy of Sciences
Sofia 13, Bulgaria

Kristall und Technik | 7 | 1–3 | 1972 | 95–106

R. Voigt, H. Neels

Karl-Marx-Universität Leipzig
Sektion Chemie, Bereich Kristallographie

Beiträge zum Ätzen von Galliumarsenid (II) Zur Kinetik der Auflösung von GaAs

Untersucht wird die Kinetik der Auflösung von {111}- und {100}-Ebenen von GaAs in NaClO- und HNO_3-Lösungen sowie im HF/CrO_3-Lösungssystem. Dazu werden Auflösungsgeschwindigkeiten bestimmt und mit Hilfe der Arrhenius-Gleichung $w = C \cdot e^{-E_A/RT}$ Aktivierungsenergien und Geschwindigkeitskonstanten angenähert ermittelt. Es ergeben sich Anhaltspunkte für den geschwindigkeitsbestimmenden Schritt (Diffusion – Konvektion, heterogene chemische Reaktion). Bei der Auflösung von GaAs können je nach Art des Mediums (Konzentrationen) Diffusion oder chemische Reaktion geschwindigkeitsbestimmend sein. In manchen Fällen läßt sich aus der Größe der Geschwindigkeitskonstanten das Erscheinen oder Nichterscheinen von Ätzgruben verstehen. Die Auflösungsgeschwindigkeit w für sich allein gestattet keine zuverlässige Aussage über die Wahrscheinlichkeit der Bildung von Ätzgruben.

The kinetics of dissolution of {111} and {100} planes of GaAs in NaClO- and HNO_3-solutions, and in the HF/CrO_3 solvent system is investigated. For that reason dissolution rates are measured, and by means of the Arrhenius equation $w = C \cdot e^{-E_A/RT}$ activation energies and rate constants approximately ascertained. Clues to the rate controlling step (diffusion – convection, heterogeneous chemical reaction) are resulting. With the dissolution of GaAs diffusion or chemical reaction can be rate controlling depending on the character of the medium (concentrations). In some cases the appearance or non-appearance of etch pits is to be understood from the value of the rate constant. The dissolution rate w for itself only permits no reliable statement on the probability of the formation of etch pits.

1. Einleitung

Die chemische Ätzung von $A^{III}B^{V}$-Verbindungen ist ein komplizierter heterogener Vorgang, der durch die gleichzeitige Oxydationsreaktion der Oberflächenatome und die Hydratation der Oxydationsprodukte mit Übergang letzterer in die Lösung beherrscht wird. Diese heterogenen Reaktionen sind kinetisch mit den Prozessen der Diffusion der Moleküle (Ionen) des Oxydationsmittels aus der Lösung zur Oberfläche der $A^{III}B^{V}$-Verbindung sowie der Diffusion der hydratisierten Oxydationsprodukte von der Kristalloberfläche in die Lösung verbunden.

Die Kenntnis der kinetischen Daten der Auflösungsprozesse ist insofern von großer Wichtigkeit, weil aus ihren Größen einmal prinzipielle Aussagen über die bestimmenden Vorgänge beim Auflösungsmechanismus möglich werden und zum anderen zwischen ihnen und den für die Ätzgrubenbildung erforderlichen Bedingungen enge Wechselbeziehungen bestehen.

Ausgangspunkt für die kinetischen Untersuchungen (s. Voigt) bildeten die experimentell als Funktion von T, Konzentration und Rührgeschwindigkeit bestimmten Auflösungsgeschwindigkeiten der einzelnen Kristallebenen.

Die Lösungsgeschwindigkeit w ist definiert als die Gewichtsabnahme in bezug auf die Einheiten der Zeit und der Oberfläche (Heimann, Willgallis):

$$w = \frac{\Delta g}{M \cdot \Delta t\, s} (\mathrm{Mol} \cdot \mathrm{cm}^{-2}\, \mathrm{s}^{-1})$$

mit M – Molekulargewicht
s – Oberfläche der Probe (cm^2),
Δg – Gewichtsverlust (g),
Δt – Zeitintervall (s).

Indem wir zunächst den Einfluß der Diffusion auf die Auflösungsgeschwindigkeit unberücksichtigt ließen, erlaubte uns die Arrhenius-Gleichung, aus den experimentell zugänglichen Größen Aktivierungsenergien und Geschwindigkeitskonstanten zu bestimmen, aus denen rückwirkend dann unter Beachtung der Korrelationen speziell zwischen errechneter Aktivierungsenergie und Rührgeschwindigkeit Rückschlüsse auf den geschwindigkeitsbestimmenden Schritt des jeweiligen Auflösungsprozesses gezogen werden konnten.

Für den Zusammenhang der Ätzgeschwindigkeit mit dem zetilichen Verlauf einer chemischen Reaktion gilt die Arrhenius-Gleichung (s. Meyer)

$$w = C \cdot e^{-E_A/RT},$$

worin C eine für jeden Verlauf charakteristische Geschwindigkeitskonstante (Moleküle $\cdot\, \mathrm{cm}^{-2}\, \mathrm{s}^{-1}$), E_A die Aktivierungsenergie (kcal $\cdot\, \mathrm{mol}^{-1}$), R die Gaskonstante (cal $\cdot$ Grad^{-1}) und T die Temperatur (°K) sind.

2. Kinetik der GaAs-Auflösung in verdünnten NaClO-Lösungen

Voraussetzung für detaillierte Studien über Baufehlernachweis, Ätzgrubenentstehung und dergleichen sind spiegelnde Oberflächen, die frei von durch mechanische Bearbeitung der Proben hervorgerufenen Störzonen sind. Erforderlich waren dafür Lösungsmittel, die die Kristallflächen relativ rasch und ohne bevorzugten Angriff entlang von Baufehlern angriffen. Verdünnte NaClO-Lösungen lieferten die besten Ergebnisse. Die Qualität der polierten Oberflächen war stark abhängig von den gewählten Versuchsbedingungen.

Eingehende Untersuchungen über den Einfluß von Konzentration, Temperatur und Rührgeschwindigkeit gestatten, folgende Aussagen zu konzipieren (s. Figur 1 und Tabelle 1): Beim Auftragen des Logarithmus der Auflösungsgeschwindigkeit gegen die reziproke Temperatur (in Kelvin-Graden) ergaben sich stets Geraden negativer Steigung, die der Gleichung

$$\lg w = -A \cdot T^{-1} + B$$

gehorchen.

Charakteristisch für das System war der Einfluß der Rührgeschwindigkeit auf die Ätzgeschwindigkeit, die bei nicht bewegten Lösungen stets 0.2–0.4 Größenordnungen niedriger lag.

Gleichzeitig ergaben unsere Rechnungen für die Aktivierungsenergie bei nicht bewegten Lösungen ebenfalls regelmäßig etwas kleinere Werte als bei Lösungen, die stark gerührt wurden.

Letzteres ist weniger als tatsächliche Veränderung der wahren Aktivierungsenergie der sich abwickelnden Reaktionen zu werten, sondern vielmehr auf eine relative Verschiebung des Verhältnisses zwischen Diffusionsgeschwindigkeit und Reaktionsgeschwindigkeit zurückzuführen.

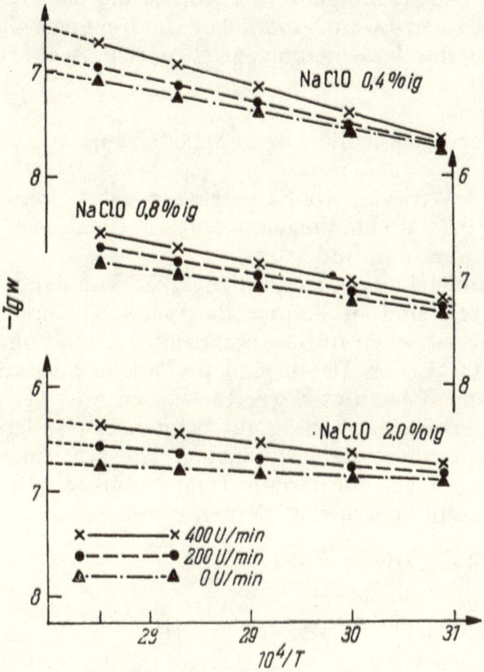

Fig. 1. Die Auflösungsgeschwindigkeit von GaAs als Funktion der Temperatur, gemessen an {100}-Flächen

Tabelle 1

Kinetische Daten der Auflösung von GaAs in verdünnten NaClO-Lösungen

Konzentration der Lösung (%)	Rührgeschwindigkeit der Lösung (U/min)	A	B	E_A (kcal · mol^{-1})	C (Molekeln · cm^{-2} · s^{-1})
0,4	0	1820	−2,06	8,32	$5{,}3 \times 10^{21}$
	200	2110	−1,14	9,64	$4{,}4 \times 10^{22}$
	400	2700	0,72	12,30	$3{,}2 \times 10^{24}$
0,8	0	1380	−3,03	6,31	$5{,}7 \times 10^{20}$
	200	1700	−1,98	7,77	$6{,}1 \times 10^{21}$
	400	1880	−1,36	8,59	$2{,}6 \times 10^{22}$
2,0	0	469	−5,46	2,14	$2{,}1 \times 10^{18}$
	200	762	−4,46	3,48	$2{,}1 \times 10^{19}$
	400	1140	−3,23	5,21	$3{,}6 \times 10^{20}$

Die beobachtete Abhängigkeit der Lösungsgeschwindigkeit von der Rührgeschwindigkeit des Ätzmittels, die Diskrepanzen in den berechneten Werten für die Aktivierungsenergie bei unterschiedlichen Umdrehungszahlen ein und derselben Lösung und die relativ kleinen Werte für die Geschwindigkeitskonstanten weisen deutlich darauf hin, daß im NaClO-GaAs-System überwiegend Diffusionsvorgänge die Geschwindigkeit der Auflösung beherrschen. Allerdings machte sich bei 0.4%iger Lösung und hoher Umdrehungszahl erwartungsgemäß bereits der Einfluß durch die heterogene chemische Reaktion bemerkbar.

3. Kinetik der GaAs-Auflösung in Salpetersäure

Wegen der oxydierenden Wirkung der Salpetersäure ist es verständlich, daß sie als Ätzmittel für $A^{III}B^{V}$-Verbindungen — oft in Gemischen mit anderen Säuren — eine breite Anwendung findet.

Die Resultate der Untersuchungen zur Abhängigkeit von der Rührgeschwindigkeit des Lösungsmittels sind in Figur 2 dargestellt, während in Figur 3 das bestehende Verhältnis zwischen Auflösungsgeschwindigkeit und Temperatur zu erkennen ist. Die kinetischen Daten sind in Tabelle 2 zusammengestellt. Die für die Figuren 2 und 3 benutzten Werte wurden an $\{111\}$-orientierten GaAs-Proben gemessen. Der Reaktionsablauf beim Lösen anders orientierter Kristallscheiben ist dem hier beschriebenen analog. Die experimentell ermittelten Abtragungsraten weisen bei konstanten Temperatur-Zeit-Parametern folgende Reihenfolge der Lösungsgeschwindigkeiten w auf:

$$w_{\{\bar{1}\bar{1}\bar{1}\}} > w_{\{100\}} > w_{\{110\}} > w_{\{111\}}$$

bzw.

$$w_{\{111\}} \approx 0.6\, w_{\{110\}} \approx 0.4\, w_{\{100\}} \approx 0.2\, w_{\{\bar{1}\bar{1}\bar{1}\}} .$$

Die für die jeweiligen Flächen berechneten Aktivierungsenergien E_A differieren um gewisse Beträge. Die Neigungen der Geraden $-\lg w = f(1/T)$ sind bei $\{\bar{1}\bar{1}\bar{1}\}$-, $\{100\}$- und $\{110\}$-Material durchweg etwas flacher als für $\{111\}$.

Für die Abscheidungsraten von As_2O_3 auf den einzelnen Ebenen ergab sich eine zu den Auflösungsgeschwindigkeiten parallele Reihenfolge.

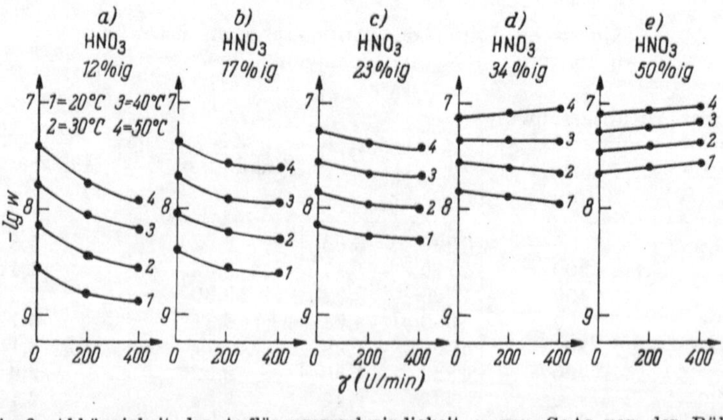

Fig. 2. Abhängigkeit der Auflösungsgeschwindigkeit w von GaAs von der Rührgeschwindigkeit γ in U/min in salpetersauren Lösungen, gemessen an $\{111\}$-Flächen

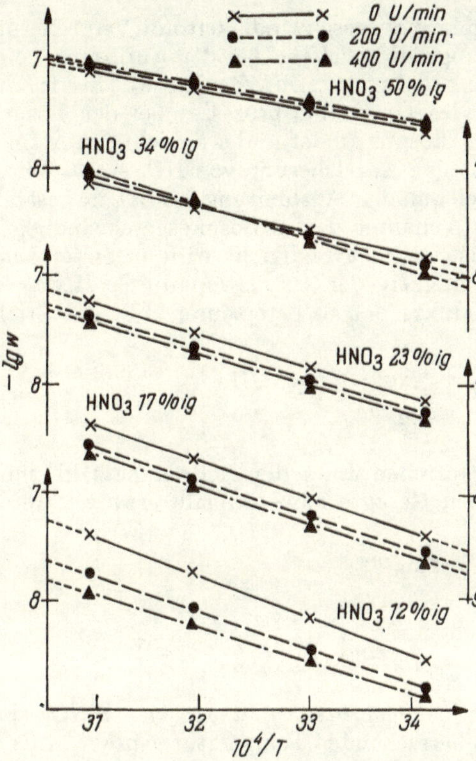

Fig. 3. Abhängigkeit der Auflösungsgeschwindigkeit w von GaAs von der Temperatur in salpetersauren Lösungen, gemessen an {111}-Flächen

Tabelle 2
Kinetische Daten der Auflösung von GaAs in HNO_3-Lösungen

Konzentration der Lösung (%)	Rührgeschwindigkeit der Lösung (U/min)	A	B	E_A (kcal · mol^{-1})	C (Molekeln · cm^{-2} s^{-1})
12	0	3630	3,83	16,6	$4{,}1 \times 10^{27}$
	200	3280	2,42	15,0	$1{,}6 \times 10^{26}$
	400	2930	1,16	13,4	$8{,}7 \times 10^{24}$
17	0	3250	2,72	14,8	$3{,}2 \times 10^{26}$
	200	3120	2,10	14,3	$7{,}5 \times 10^{25}$
	400	3120	2,06	14,3	$6{,}9 \times 10^{25}$
23	0	2870	1,63	13,1	$2{,}6 \times 10^{25}$
	200	2810	1,31	12,8	$1{,}2 \times 10^{25}$
	400	2770	1,17	12,7	$9{,}0 \times 10^{24}$
34	0	2140	−0,50	9,8	$1{,}9 \times 10^{23}$
	200	2620	1,05	12,0	$6{,}9 \times 10^{24}$
	400	2990	2,23	13,7	$1{,}0 \times 10^{26}$
50	0	1700	−1,83	7,78	$9{,}0 \times 10^{21}$
	200	1700	−1,79	7,78	$9{,}6 \times 10^{21}$
	400	1700	−1,77	7,78	$1{,}0 \times 10^{22}$

Eine Verkleinerung der Lösungsgeschwindigkeit mit Vergrößerung der Rührgeschwindigkeit der Lösung wird vielfach bei der Auflösung von Metallen in HNO_3 beobachtet (Mjuller, Orlova, Zui-Zsin'chua). Sie ist auf den autokatalytischen Charakter des Oxydationsprozesses bei der Lösung zurückzuführen. Bei nichtbewegter Lösung findet in der Schicht des Ätzmittels, die die GaAs-Oberfläche berührt, eine Anreicherung von NO_2 statt. Wird die Lösung gerührt, so hat das eine gleichmäßige Ausbreitung des NO_2 im gesamten Lösungsvolumen und damit eine Abnahme der Auflösungsgeschwindigkeit zur Folge.

Die beobachtete Auflösungsgeschwindigkeit wird in erster Linie bestimmt sein durch die Geschwindigkeit der Oxydation durch HNO_3 (w_{HNO_3}) bzw. durch die Reduktionsprodukte der Salpetersäure, z. B. des Stickstoffdioxids (w_{NO_2}):

$$w_{\mathrm{beob}} = w_{HNO_3} + w_{NO_2} = c_{HNO_3} \cdot f(\gamma) \cdot e^{-E_{HNO_3}/RT} + \frac{c_{NO_2}}{f(\gamma)} \cdot e^{-E_{NO_2}/RT}$$

worin c_{HNO_3}, c_{NO_2} Konzentrationen und γ die Umdrehungszahl $\cdot$ min^{-1} angeben. In Anlehnung an das oben Gesagte kann deshalb erwartet werden, daß bei $\gamma = 0$ U/min

$$w_{\mathrm{beob}} \approx \frac{c_{NO_2}}{f(\gamma)} \cdot e^{-E_{NO_2}/RT}$$

und für $\gamma \geq 400$ U/min

$$w_{\mathrm{beob}} \approx c_{HNO_3} \cdot f(\gamma) \cdot e^{-E_{HNO_3}/RT} .$$

Der Charakter des Auflösungsprozesses in 34%iger HNO_3 (Fig. 2d) wird sehr stark von der jeweils herrschenden Temperatur geprägt. Aus dem Verlauf der Kurven in Figur 2d kann man entnehmen, daß bei niedrigen Temperaturen (20—30 °C) die Oxydationsreaktion durch HNO_3 langsam verlaufen muß und erst bei Temperaturen >40 °C merklich zur Auflösungsgeschwindigkeit beitragen wird. Bei 40 °C ist $w_{\gamma=0} = w_{\gamma=400}$. Hier wird die beim Rühren der Lösung hervorgerufene Konzentrationsverringerung des NO_2 in der Nähe der Kristalloberfläche durch den Zufluß an HNO_3 gerade kompensiert.

Unter Berücksichtigung der erhaltenen Größen (s. auch Tabelle 2) kommt man zu dem Ergebnis, daß in Lösungen niedriger Konzentration der Lösungsprozeß im wesentlichen durch die Oxydationsreaktion der HNO_3 und des NO_2 mit GaAs beherrscht wird, während bei 50%igen Lösungen bereits Diffusionsvorgänge die bestimmende Rolle spielen. Der Übergang liegt etwa bei 34%iger HNO_3.

Die geätzten Oberflächen blieben entweder verhältnismäßig eben und spiegelnd, zeigten mehr oder weniger gut ausgebildete Ätzgruben bzw. waren unterschiedlich dicht mit As_2O_3-Abscheidungen — kristallin oder amorph — bedeckt. Die As_2O_3 -Ausscheidungen mußten zur exakten Bestimmung des Gewichtsverlustes von den Oberflächen entfernt werden. Die besten Ergebnisse liefert in dieser Hinsicht ein Gemisch aus HF und Glycerin (Mironov). Das Auftreten von Ätzgruben beschränkte sich ausschließlich auf {111}-Flächen.

Die Gründe für das Fehlen von Ätzgruben auf anderen Kristallflächen sind in der z. T. erheblich größeren Auflösungsgeschwindigkeit unter gleichzeitiger Herabsetzung der Aktivierungsenergie zu sehen. Der Einfluß der Diffusionsvorgänge auf die Geschwindigkeit der Ätzung wird hier schon bei weitaus niedrigeren Konzentrationen bemerkbar (Svešnikova et al.).

4. Zur Kinetik der GaAs-Auflösung im System HF—CrO_3

Bei unseren Untersuchungen wurde dem System CrO_3/HF besondere Aufmerksamkeit gewidmet, da hier bei der Wahl bestimmter Versuchsparameter sowohl auf {111} als auch auf $\{\bar{1}\bar{1}\bar{1}\}$, {100} und {110} Ätzgruben erhalten werden konnten. Chrom(VI)-Oxid stellt wegen der leichten Sauerstoffabgabe ein sehr kräftiges Oxydationsmittel dar und findet deshalb beim Ätzen von Substanzen mit vorwiegend kovalenter Bindung oft Anwendung.

Die Flußsäure hat einmal die Funktion, den bei der Auflösung ablaufenden Redoxvorgang in nachstehender Weise zu beeinflussen:

$$Cr_2O_7^{2-} + 6\,e + 14\,H^+ \rightleftharpoons 2\,Cr^{3+} + 7\,H_2O\,.$$

Zum anderen hängt aber von ihrer Konzentration auch die Auflösungsgeschwindigkeit der an der Oberfläche des Kristalls gebildeten Oxydationsprodukte ab.

Die kinetischen Daten der Ätzung in 8.5%igen HF-Lösungen bei verschiedenen CrO_3-Gehalten sind der Figur 4 und der Tabelle 3 zu entnehmen. Auffallend ist die mit Erhöhung der CrO_3-Konzentration relativ große Zunahme der Lösungsgeschwindigkeit, die ihrerseits zu erheblichen Unterschieden in den Geschwindigkeitskonstanten für die einzelnen Lösungen führte. Eine solche Zunahme der Auflösungsgeschwindigkeit mit Konzentrationserhöhung ist nur verständlich, solange die Diffusionsgeschwindigkeiten noch genügend groß sind.

Die Figuren 5 und 6 veranschaulichen den Einfluß von HF- und CrO_3-Konzentrationen auf die Auflösungsgeschwindigkeit. Die Kurven $-\lg w = f(\sigma_{HF})$ steigen anfangs bis zu einer vom CrO_3-Gehalt abhängigen Konzentration steil an, verflachen aber dann stark, um bei hohen HF-Konzentrationen sogar mehr oder weniger abzufallen. Die relativ abrupte Abnahme der Neigung verschiebt sich dabei mit zunehmender CrO_3-Menge in Richtung höherer Konzentrationswerte für HF. Berücksichtigt man die doppelte Funktion, die die

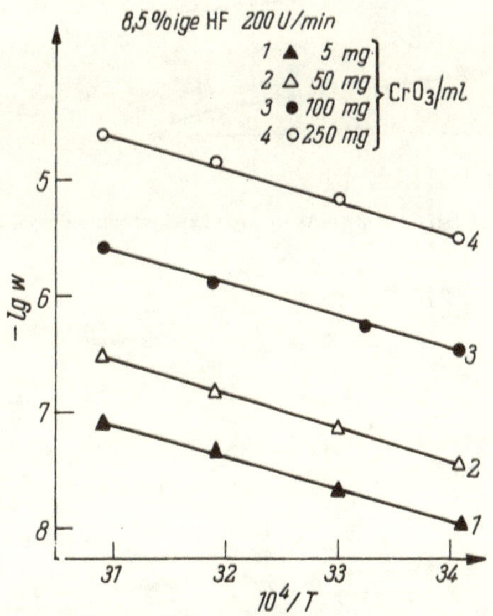

Fig. 4. Lösungsgeschwindigkeit von GaAs in CrO_3-HF-Lösungen als Funktion der Temperatur, gemessen an {100}-Flächen

Flußsäure für den Auflösungsvorgang ausübt, dazu außerdem die Löslichkeit der beim Ätzen entstehenden Abbauprodukte in der Chromsäure, so wird die Auflösungsgeschwindigkeit mit steigender HF-Konzentration so lange stark ansteigen, bis ein Konzentrationswert erreicht ist, bei dem sich Bildung und Auflösung der GaAs-Oxydationsprodukte kompensieren.

Tabelle 3

Kinetische Daten der Auflösung von GaAs in HF-CrO_3-Lösungen (Rührgeschwindigkeit der Lösung 200 U/min)

Zusammensetzung der Lösung	A	B	E_A (kcal · mol^{-1})	C (Molekeln · $cm^{-2}s^{-1}$)
8,5%ige HF + 5 mg CrO_3/ml	2840	1,74	13,0	$3,3 \times 10^{25}$
8,5%ige HF + 50 mg CrO_3/ml	2830	2,51	12,9	$1,9 \times 10^{26}$
8,5%ige HF + 100 mg CrO_3/ml	2800	3,15	12,8	$8,4 \times 10^{26}$
8,5%ige HF + 250 mg CrO_3/ml	2790	4,10	12,75	$7,8 \times 10^{27}$

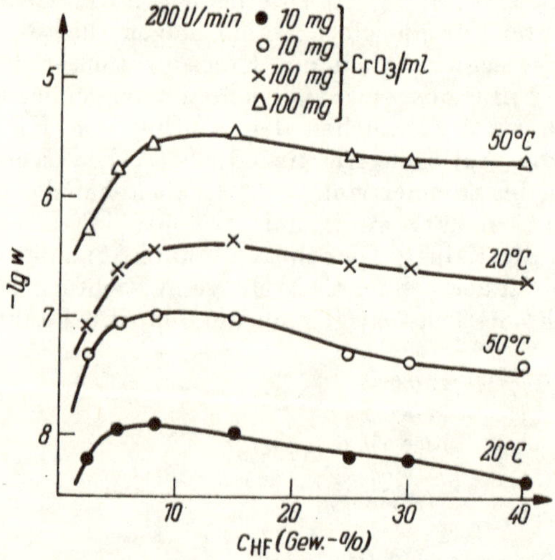

Fig. 5. Lösungsgeschwindigkeit von GaAs in Abhängigkeit von der HF-Konzentration, gemessen an {100}-Flächen

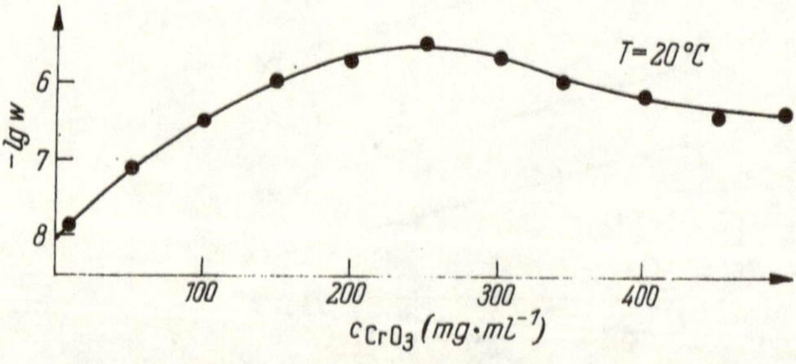

Fig. 6. Auflösungsgeschwindigkeit von GaAs in 8,5 %iger HF-Lösung bei unterschiedlichen CrO_3-Gehalten, gemessen an {100}-Flächen

Analog der Darstellung von $w = f(c_{HF})$ verlaufen die Kurven $f(c_{CrO_3})$ bei konstanter HF-Konzentration (Fig. 6). Auch hier steigt die Lösungsgeschwindigkeit mit zunehmender CrO_3-Menge bis zu einem für jede Säurekonzentration spezifischen Wert, um danach zu stagnieren bzw. allmählich abzufallen.

Die hier geäußerten Schlußfolgerungen basieren auf Untersuchungen an {100}-Flächen. Das Studium anders orientierter Kristallproben ergab im wesentlichen die gleichen Abhängigkeitsverhältnisse. Die Geschwindigkeitsraten der einzelnen Flächen standen in folgender Relation miteinander:

für die Ätzmischung

$2\ \mathrm{ml}\ H_2O_2 + 4\ \mathrm{mg}\ AgNO_3 + 1000\ \mathrm{mg}\ CrO_3 + 1\ \mathrm{ml}\ HF$

$w_{\{100\}} : w_{\{111\}} : w_{\{110\}} : w_{\{\bar{1}\bar{1}\bar{1}\}} = 1 : 0.71 : 0.69 : 0.50$

(nach ABRAHAMS, BUIOCCHI),

für die Ätzmischung

$2\ \mathrm{mg}\ H_2O_2 + 30\ \mathrm{mg}\ AgNO_3 + 150\ \mathrm{mg}\ K_3Fe(CN)_6 +$
$+ 700 - 1500\ \mathrm{mg}\ CrO_3 + 2\ \mathrm{ml}\ HF$

$w_{\{111\}} : w_{\{110\}} : w_{\{\bar{1}\bar{1}\bar{1}\}} : w_{\{100\}} = 1 : 0.98 : 0.74 : 0.71$,

für die Ätzmischung

8.5%ige HF + 10—80 mg CrO_3/ml

$w_{\{110\}} : w_{\{\bar{1}\bar{1}\bar{1}\}} : w_{\{100\}} : w_{\{111\}} : = 1 : 0.90 : 0.55 : 0.50$

bei ca. 10 mg CrO_3/ml

und 0.80 : 0.92 : 1 : 0.65

bei 80 mg CrO_3/ml .

Tabelle 4

Wirkung unterschiedlicher CrO_3/HF-Konzentrationen auf die Auflösungsstruktur von GaAs-Flächen (Orientierungen {111}, {$\bar{1}\bar{1}\bar{1}$}, {110{, {100}

Säure-konzentration (Gew.-%)	Chrom(VI)-Oxid-Gehalt ($mg \cdot ml^{-1}$)	Oberflächenbeschaffenheit[1])
<8,5	2–5	teilweise Ätzgruben, wegen äußerst geringer Vertiefung kaum zu erkennen
	5–30	Ätzgruben, sehr flach
	>30	Übergang von Kopfsteinpflasterstruktur zu reiner Polierwirkung
8,5	<5	teilweise sehr kleine Ätzgruben, teilweise grünlicher Belag
	5–100	Ätzgruben
	>100	über Kopfsteinpflaster zur Politur
16	<100	teilweise Ätzgruben, teilweise grünlicher Belag
	100–300	Ätzgruben mit teilweise unregelmäßiger Begrenzung
	>300	Politur

[1]) Die Differenz, die bei den einzelnen Flächen auftritt, ist gering und hierbei nicht berücksichtigt.

In CrO_3/HF-Lösungen geätzte Flächen hatten je nach den gewählten Bedingungen ein unterschiedliches Aussehen (Tab. 4). Die für morphologische Ätzstudien (s. Neels, Voigt) geeignetsten Ätzgruben (kontrastreich, polygonal etc.) wurden in 8.5%igen HF-Lösungen mit etwa 5–100 mg/ml Chrom(VI)-Oxid erhalten.

5. Zusammenfassung der Ergebnisse und Diskussion unter dem Aspekt spezieller Ätzuntersuchungen

Wie die vorangehenden Darlegungen gezeigt haben, kann auf Grund der gemischten Bindungsverhältnisse der Charakter der Auflösungsprozesse bei den $A^{III}B^{V}$-Verbindungen sowohl durch Diffusions-Konvektionsvorgänge als auch durch die jeweiligen heterogenen chemischen Reaktionen geprägt sein. Eine unmittelbare Abhängigkeit der sich ausbildenden Oberflächenstrukturen davon, ob nun für die Auflösung die Diffusion oder die heterogene chemische Reaktion tempobestimmend ist, war nicht zu beobachten. Die Ergebnisse weisen deutlich darauf hin, daß die experimentell erhaltene Größe der Auflösungsgeschwindigkeit w einer Kristallfläche für sich allein betrachtet keine zuverlässige Aussage, wie allgemein angenommen wird, für die Wahrscheinlichkeit, z. B. der Bildung sichtbarer Ätzgruben auf eben dieser Fläche, gestattet. Entscheidend für eine Ätzgrubenentstehung ist der Unterschied zwischen der vertikalen Auflösungsgeschwindigkeit an den Versetzungen v_v und der horizontalen Wanderungsgeschwindigkeit von Oberflächenstufen v_h bzw. der senkrechten Verschiebungsgeschwindigkeit der versetzungsfreien Oberfläche v, da erstere proportional der letzteren ist.

Die mit v bezeichnete Geschwindigkeit entspricht in erster Näherung den von uns gemessenen w-Werten.

Es kann sich dabei de facto nur um eine grobe Annäherung handeln, da eine quantitative Bestimmung der Verschiebungsgeschwindigkeit v auf Kristallflächen, auf denen im allgemeinen die spezifischen Ätzgruben entstehen, selbst unter Zugrundelegung der linearen Ausbreitungsgeschwindigkeit der Ätzgruben als Maß für das Abtragungstempo, illusorisch ist. Der Einfluß der Versetzungen auf die wirkliche Verschiebungsgeschwindigkeit der Flächen bleibt bei den uns zugänglichen w-Werten unberücksichtigt. Eventuell nähert man sich den wahren Werten mehr, wenn man $w \approx (v + v_v)\,(2)^{-1}$ schreibt. Trotz der notwendigen Einschränkung dürfen wir $w \sim v$ und damit C/C_0 proportional der Geschwindigkeitskonstanten k^* : $C/C_0 \sim k^*$ setzen. (C_0 ist die aus der Oberflächenkonzentration der Moleküle berechnete Geschwindigkeitskonstante. Für GaAs ist $C_0 \approx 8 \cdot 10^{26}$ Moleküle $\cdot$ cm^{-2} s^{-1}). Somit wird die Geschwindigkeitskonstante der Stufenwanderung weitgehend durch den Mechanismus der Auflösung bestimmt, und wir gewinnen aus dem Studium der kinetischen Daten mit C/C_0 einen für die Entstehungsprozesse der Ätzgruben äußerst wichtigen Faktor.

Hinsichtlich der erhaltenen Größe der Werte für die Aktivierungsenergie E_A könnte im ersten Moment der Eindruck erweckt werden, daß Ätzgruben bevorzugt dann entstehen, wenn der Auflösungsprozeß durch die heterogene Versetzungsreaktion beherrscht wird ($E_A \leqq 10$ kcal $\cdot$ mol^{-1}), was sich aber bei näherem Hinsehen sofort als Irrtum erweist, da, wenn auch nicht in jedem Fall, im allgemeinen mit steigender Aktivierungsenergie die C-Werte zunehmen oder mit anderen Worten, die Größe der Geschwindigkeitskonstanten. Jedoch

SCHAARWÄCHTER führte bereits an, daß bei Geschwindigkeitskonstanten $k^* = 10^2 - 10^4$ bei beliebigen Potentialdifferenzen Ätzgruben ausgeschlossen sind.

Wie man ad oculus demonstrieren kann, kommt es einzig und allein auf die Relationen der einzelnen Größen zueinander an. Nehmen wir an, die Auflösungsgeschwindigkeit an der Versetzung sei durch bevorzugte Keimbildung hervorgerufen, dann beträgt die sich daraus ergebende Geschwindigkeit v_v

$$v_v \approx a\,\nu \exp\{-(p\,\Delta G^* + \Delta H)/kT\},$$

während die Verschiebungsgeschwindigkeit der versetzungsfreien Oberfläche

$$v \approx a\,[\dot{n}\,(b^*\,k^*\,\varkappa\,\nu \exp\{-\Delta H/kT\})^2]^{1/3}$$

ist, mit

a — Stufenhöhe,
$\dot{n}$ — Keimbildungshäufigkeit/cm² s,
ν — Schwingungsfrequenz der Admolekeln in den Anlagerungsstellen pro Sekunde,
b^* — Dicke einer Atomreihe in der Oberflächenstufe,
$\varkappa$ — $= 1 - e^{-\Delta g/kT}$ = Untersättigung,
k^* — Geschwindigkeitskonstante.

Unter sehr vereinfachenden Voraussetzungen (a, $\dot{n}$, b^*, ν werden als kristallspezifische Konstanten angenommen, kT ebenfalls als Konstante) erhalten wir damit

$$v_v \sim \frac{\Delta g}{\Delta H \cdot p} \text{ und mit } w \sim v \quad w \sim v_h \sim \frac{\Delta g\,k^*}{\Delta H},$$

worin Δg die jeweilig herrschende Potentialdifferenz, ΔH die Größe der Aktivierungsenergie und p im wesentlichen den die Verminderung der Keimbildungsarbeit durch die Versetzung wiedergebenden Faktor bedeuten. Da sowohl v_v als auch v bzw. v_h proportional $\frac{\Delta g}{\Delta H}$ sind, sehen wir, welche Bedeutung für ein geeignetes Verhältnis $v_v > v$ bzw. $v_v \geqq c^* \cdot v_h$ zur Ätzgrubenbildung neben dem Faktor p der Geschwindigkeitskonstante k^* zukommt. Dabei gilt es außerdem zu berücksichtigen, daß die Potentialdifferenz Δg, die zur Bildung sichtbarer Ätzgruben erforderlich ist, eine untere Grenze $\Delta g_{\min}$ erreichen und bei $k^* > 1$ eine von k^* abhängige obere Grenze $\Delta g_{\max}$ nicht überschreiten darf. Die Lage der unteren Grenze ist im wesentlichen eine Funktion spezifischer Gitterparameter (Burgersvektor, Stufenhöhe), der Oberflächenenergie der Substanz sowie der Temperatur.

Ohne näher auf die geforderten Werte für p und Δg einzugehen, erlauben uns die gemachten Untersuchungen zur Kinetik nun eine grobe Abschätzung der für die Ausbildung der Oberflächenstrukturen in den behandelten Systemen verantwortlichen Größen sowie ein besseres Verstehen für die Verschiedenartigkeit der Ätzbilder.

Setzen wir C/C_0 in erster Näherung gleich k^*, so liegt für NaClO-Lösungen die Geschwindigkeitskonstante k^* im Bereich von etwa 10^{-8} bis 10^{-3}. Eine obere Grenze $\Delta g_{\max}$ für Ätzgruben existiert demnach hier nicht. Wenn dennoch keine beobachtet wurden, ist der Grund dafür weniger in einem zu hohen p-Wert zu suchen, da dieser als Funktion von Schubmodul, Burgersvektor und Randenergie näherungsweise konstant bleibt, sondern eher in einer zu niedrigen

Potentialdifferenz, die unter dem $\Delta g_{\min}$-Wert liegt. Wollte man mit NaClO-Lösungen Ätzgruben auf GaAs erzeugen, so wäre dies bei höheren Konzentrationen nur dann erfolgversprechend, wenn die Auflösungsgeschwindigkeit zunimmt, das Verhältnis $C/C_0 \cdot E_A$ konstant bleibt oder abnimmt. Untersuchungen dieser Art sind von uns nicht angestellt worden.

In HNO_3-Lösungen errechnet sich k^* zu $k^* \leqq 1$. Außerdem erscheinen Ätzgruben, d. h., die Mindestpotentialdifferenz $\Delta g_{\min}$ ist vorhanden, und eine maximal zulässige Potentialdifferenz folgt erst bei Konzentrationen $c_{HNO_3} \leqq 12\%$ig. Dann wird $k^* \geqq 5$, und es ist nicht weiter erstaunlich, wenn keine Ätzgruben mehr zu erkennen sind.

CrO_3/HF-Ätzungen sind trotz der hohen Lösungsgeschwindigkeiten, verglichen mit HNO_3-Lösungen, durch einen relativ großen k^*-Wert charakterisiert. Für einen CrO_3-Gehalt von 250 mg/ml in 8.5%iger HF steigt er auf fast 10, so daß Ätzgrubenbildung kaum zu erwarten ist, was auch mit angestellten Beobachtungen übereinstimmt.

Literatur

Abrahams, M. S., Buiocchi, C. J.: J. appl. Physics **36**, 2855 (1965)

Heimann, R., Willgallis, A.: N. Jb. Min., Mh. (4) 145 (1969)

Meyer, K.: Physikalisch-chemische Kristallographie, Leipzig 1968, S. 80

Neels, H., Voigt, R.: Kristall u. Technik **6**, 255 (1971)

Mironov, K. E.: Istvest. Sib. otd. Akad. Nauk SSSR, Ser. chim. nauk, Nr. 4, 169 (1967)

Mjuller, R. L., Orlova, G. M., Zui Zsin'chua: Žurn. obšč. Chim. **31**, 2461 (1961)

Schaarwächter, W.: Physica Status solidi **12**, 865 (1965)

Svešnikova, L. L., Repinskij, S. M., Orlova, G. M.: Žurn. fiz. Chim. **42**, 2815 (1968)

Voigt, R.: Dissertation Karl-Marx-Universität Leipzig 1971

(Eingegangen am 8. Juli 1971)

Anschrift der Autoren:

Prof. Dr. H. Neels
Karl-Marx-Universität
Sektion Chemie
Bereich Kristallographie
703 Leipzig, DDR
Scharnhorststr. 20

Dr. Rita Voigt
Zentralinstitut für Physikalische Chemie
der Deutschen Akademie der Wissenschaften zu Berlin
1199 Berlin-Adlershof
Rudower Chaussee 5

Ch. Ullner

Sektion Physik, Bereich Kristallographie
Humboldt-Universität zu Berlin

Clusterbildung — ein stochastischer Prozeß

Die relative Schwankung der Clusterdichte wird näherungsweise berechnet. Das verwendete Modell beschreibt einen einfachen Hochvakuum-Aufdampfvorgang und enthält lediglich die Adatomkonzentration n_1 und die Clusterdichte n_C. Im stationären Fall ist das Ergebnis Gleichung (14), in die repräsentative numerische Werte eingesetzt werden. Das numerische Ergebnis zeigt, daß in speziellen Fällen die Clusterdichteschwankungen zu berücksichtigen sind, falls Theorie und Experiment verglichen werden.

The relative fluctuations of cluster density is approximatively calculated. The used model describes a simple high vacuum deposition process and contains only adatom concentration n_1 and cluster density n_C. In the stationary case the result is equation (14) in which representative numerical quantities are substituted. Numerical result shows that in special cases the fluctuations of cluster density have to be taken into consideration, if theory and experiment are compared.

1. Die Aufgabe

Die Clusterbildung ist ein stochastischer Prozeß (Gnedenko; Ullner), d. h. die den Zustand der Clusterbildung kennzeichnenden Größen sind Zufallsvariablen. Die Form ihrer Wahrscheinlichkeitsverteilung ist grundätzlich bei genauer Kenntnis der Elementarvorgänge festgelegt, jedoch kann ihre Ermittlung enorme mathematische Schwierigkeiten enthalten.

Wird die Clusterbildung auf Oberflächen unabhängig vom örtlichen Verhalten insgesamt über eine Teilfläche betrachtet, so wird der Zustand zur Zeit t durch die Konzentration von Clustern der Größe i vollständig beschrieben. Die zufälligen Vorgänge bewirken, daß die Konzentrationen bei verschiedenen Versuchen unter genau gleichen Bedingungen zufällige Werte mit einer bestimmten Wahrscheinlichkeit annehmen können. Für einen Vergleich zwischen Theorie und Experiment ist es wichtig zu wissen, wie stark die Konzentrationen um ihren Mittelwert schwanken. Als zusätzliche Vereinfachung wird im Folgenden nur die Summe der Konzentrationen aller Cluster behandelt (Clusterdichte oder Keimdichte genannt).

Damit lautet die Aufgabenstellung, die relative Schwankung der Clusterdichte an Hand eines einfachen Modells zu berechnen.

2. Die Methode

Wird statt nach der Wahrscheinlichkeitsverteilung nur nach ihren Kenngrößen, den Momenten, gefragt, so sollte die Momente erzeugende Funktion (Bailey)

$$M(\Theta) = 1 + \sum_{r=1}^{\infty} \mu_r \frac{\Theta^r}{r!} = E[e^{\Theta X}]$$

benutzt werden. $E[e^{\Theta X}]$ bedeutet die mathematische Erwartung der aus der Zufallsvariablen X gebildeten Funktion, μ_r ist das r-te Moment zum Ursprung Null.

Die „Zufallsvariablen-Technik" (Bailey) führt auf eine partielle Differentialgleichung für die Momente erzeugende Funktion. Im zweidimensionalen Fall lautet die Differentialgleichung

$$\frac{\partial M(\Theta, \Phi, t)}{\partial t} = \sum_{j^2+k^2 \neq 0} (e^{j\Theta + k\Phi} - 1) f_{jk}\left(\frac{\partial}{\partial \Theta}, \frac{\partial}{\partial \Phi}\right) M(\Theta, \Phi, t) . \quad (1)$$

Die Lückenoperatoren $\frac{\partial}{\partial \Theta}$, $\frac{\partial}{\partial \Phi}$ gelten allein für $M(\Theta, \Phi, t)$. Die Funktion $f_{jk}(X(t), Y(t)) \, \Delta t$ gibt die Wahrscheinlichkeit dafür an, daß sich die Zufallsvariablen X und Y innerhalb des Zeitintervalls Δt um j und k ändern, falls die Zufallsvariablen zur Zeit t die Werte $X(t)$ und $Y(t)$ annehmen. Unter Verwendung der Symbolik für die bedingte Wahrscheinlichkeit [1] heißt das

$$P\{\Delta X(t) = j, \Delta Y(t) = k | X(t), Y(t)\} = f_{jk}(X, Y) \, \Delta t , \quad (2)$$

wobei j und k nicht beide zusammen Null sind.

3. Das Modell

Auf Grund rechentechnischer Schwierigkeiten muß sich das Modell auf zwei Zufallsvariable beschränken. Wir wählen die Flächenkonzentration $X(t)$ der Adatome und die Flächenkonzentration $Y(t)$ der Cluster aller Größen ($i \geqq 2$). Leider sind mit dieser unumgänglichen Festlegung auch solche kleinen Cluster in der „Keimdichte" enthalten, die im Elektronenmikroskop nicht sichtbar sind. Außerdem läßt sich nur der Fall reiner Abscheidung ($i^* < 2$) behandeln, um den Zusammenhang zwischen beiden Zufallsvariablen definieren zu können.

Das Modell berücksicht neben dem Aufdampfen und Abdampfen der Atome die Zusammenlagerungen von Einzelatomen und Clustern untereinander sowie die Anlagerung von Einzelatomen an Cluster. Die Wahrscheinlichkeit dieser Vorgänge sei proportional der entsprechenden Konzentrationen. Die Proportionalitätskonstanten sind ν (für Cluster) und λ (für Einzelatome).

Entsprechend diesem Modell lauten die $f_{jk}(X, Y)$ in Gl. (1):

a) Bei der Zusammenlagerung der Adatome unter sich entstehen aus 2 Atome ein Cluster.

$$f_{-2,1} = \lambda X^2 . \quad (3a)$$

λ ist näherungsweise gleich der vom Adatom in der Zeiteinheit überstrichenen Fläche (Zinsmeister).

b) Adatome verschwinden durch Anlagerung an Cluster oder durch Verdampfen.

$$f_{-1,0} = \lambda X Y + \mu X . \tag{3b}$$

$\tau_a = \frac{1}{\mu}$ ist die mittlere Aufenthaltsdauer eines Adatoms auf der Oberfläche.

c) Neue Adatome kommen aus dem Dampf.

$$f_{1,0} = N\downarrow . \tag{3c}$$

$N\downarrow$ ist die Aufdampfrate.

d) Um die rechnerischen Vorteile im stationären Fall (Sättigungskeimdichte existiert) ausnutzen zu können, wird die Zusammenlagerung von Clustern untereinander angenommen.

$$f_{0,-1} = \nu Y^2 . \tag{3d}$$

ν wird aus der Sättigungskeimdichte bestimmt.

e) Für alle übrigen Kombinationen von j und k gilt

$$f_{j,k} = 0 . \tag{3e}$$

Das Modell ist ähnlich dem von LEWIS und CAMBELL benutzten (LEWIS, CAMPBELL). Sie verwenden allerdings statt *d*) eine von Y abhängige Zusammenlagerungshäufigkeit λ, die für $Y = Y_s$ Null wird und so eine Sättigungskeimdichte erzeugt. Außerdem ist ihr Modell deterministisch, es existieren keine Schwankungen.

4. Der deterministische Fall

Werden die $f_{jk}(X, Y)$ nicht als Wahrscheinlichkeiten pro t aufgefaßt, sondern als mittlere Häufigkeiten des entsprechenden Elementarprozesses, so sind X und Y keine Zufallsvariablen. Wir nennen die dazugehörigen deterministischen Größen n_1 und n_C. Sie müssen nach den umgedeuteten Ausdrücken (3a) bis (3e) folgenden Gleichungen gehorchen:

$$\frac{dn_1}{dt} = -2\lambda n_1^2 - \lambda n_1 n_C - \mu n_1 + N\downarrow \tag{4a}$$

$$\frac{dn_C}{dt} = \lambda n_1^2 - \nu n_C^2 \tag{4b}$$

In Worten: Pro Zeit verschwinden Adatome durch Zusammenlagerung untereinander, durch Anlagerung an Cluster und durch Verdampfen, neue entstehen durch Aufdampfen. Die Clusterkonzentration wächst durch den Überschuß der aus Adatomen neugebildeten Cluster über die durch Zusammenlagerung untereinander verschwindenden Cluster.

Im Folgenden interessiert nur der stationäre Fall $\partial/\partial t = 0$, $n_1 = N_1$, $n_C = N_C$. Gl. (4b) ergibt wegen $\lambda, \nu, N_C > 0$

$$N_C = \sqrt{\frac{\lambda}{\nu}}\, N_1 . \tag{5}$$

Damit führt Gl. (4a) über

$$\lambda\left(2+\sqrt{\frac{\lambda}{\nu}}\right)N_1^2+\mu\,N_1-N{\downarrow}=0 \tag{6a}$$

wegen μ, $N_1 > 0$ auf

$$N_1=\frac{-\mu+\sqrt{\mu^2+4\,N{\downarrow}\,\lambda\left(2+\sqrt{\frac{\lambda}{\nu}}\right)}}{2\,\lambda\left(2+\sqrt{\frac{\lambda}{\nu}}\right)}. \tag{6b}$$

Je nachdem, ob Abdampfen oder Zusammenlagerung der Adatome vorherrschen, können zwei Näherungsen verwendet werden:

1. Fall: $4\,N{\downarrow}\lambda\,(2+\sqrt{\lambda/\nu})\ll\mu^2$ (das Abdampfen ist vorherrschend).

   Die Wurzel in Gl. (6b) wird in die Form $\sqrt{1+x^2}$ gebracht und bis zum Glied 1. Ordnung entwickelt.

$$N_1=\frac{N{\downarrow}}{\mu}. \tag{6c}$$

2. Fall: $4\,N{\downarrow}\lambda\,(2+\sqrt{\lambda/\nu})\gg\mu^2$ (das Abdampfen ist vernachlässigbar)

$$N_1=\sqrt{\frac{N{\downarrow}}{\lambda\left(2+\sqrt{\frac{\lambda}{\nu}}\right)}}. \tag{6d}$$

Die Häufigkeit ν der Clusterzusammenlagerung kann aus der Sättigungskeimdichte N_C ermittelt werden. Gl. (5) in Gl. (6d) eingesetzt ergibt näherungsweise im 2. Fall

$$\sqrt{\frac{\lambda}{\nu}}=\frac{\lambda\,N_C^2}{N{\downarrow}}. \tag{7b}$$

Dabei wurde mit $\nu\ll\lambda$ die vorausgesetzte Unterordnung der Clusterzusammenlagerung gegenüber der Adatomzusammenlagerung benutzt. Es folgt für N_1 im 2. Fall ($\lambda\,N_C\gg\mu$)

$$N_1=\frac{N{\downarrow}}{\lambda\,N_C}. \tag{6e}$$

Analog zu Gl. (7b) ergibt sich für den 1. Fall ($\lambda\,N_C\ll\mu$)

$$\sqrt{\frac{\lambda}{\nu}}=\frac{\mu\,N_C}{N{\downarrow}}. \tag{7a}$$

5. Die Normalnäherung

Werden die $f_{jk}(X, Y)$ der Gln. (3a) bis (3e) in Gl. (1) eingesetzt, so lautet die partielle Differentialgleichung

$$\frac{\partial M}{\partial t}=(e^{-2\Theta+\Phi}-1)\,\lambda\frac{\partial^2 M}{\partial\Theta^2}+(e^{-\Theta}-1)\left(\lambda\frac{\partial^2 M}{\partial\Theta\,\partial\Phi}+\mu\frac{\partial M}{\partial\Theta}\right)$$
$$+(e^{\Theta}-1)\,N{\downarrow}\,M+(e^{-\Phi}-1)\,\nu\frac{\partial^2 M}{\partial\Phi^2}.$$

Wir verwenden hier jedoch statt der Momente erzeugenden Funktion $M(\Theta, \Phi, t)$ die Häufigkeiten erzeugende Funktion $K(\Theta, \Phi, t)$, die der natürliche Logarithmus von M ist. Daher gilt:

$$M = e^K .$$

Einsetzen in die Differentialgleichung und Teilen durch $e^K = 0$ führt auf

$$\begin{aligned}\frac{\partial K}{\partial t} &= (e^{-2\Theta+\Phi} - 1)\,\lambda\left[\left(\frac{\partial K}{\partial \Theta}\right)^2 + \frac{\partial^2 K}{\partial \Theta^2}\right] \\ &+ (e^{-\Theta} - 1)\left[\lambda\left(\frac{\partial K}{\partial \Theta}\frac{\partial K}{\partial \Phi} + \frac{\partial^2 K}{\partial \Theta\,\partial \Phi}\right) + \mu\,\frac{\partial K}{\partial \Theta}\right] \\ &+ (e^{\Theta} - 1)\,N\!\downarrow + (e^{-\Theta} - 1)\,\nu\left[\left(\frac{\partial K}{\partial \Phi}\right)^2 + \frac{\partial^2 K}{\partial \Phi^2}\right].\end{aligned} \tag{8}$$

Als Lösungsansatz benutzen wir die Potenzreihenentwicklung von K:

$$\begin{aligned}K(\Theta, \Phi, t) &= K_{10}\,\Theta + K_{01}\,\Phi + \frac{1}{2}K_{20}\,\Theta^2 + \frac{1}{2}K_{02}\,\Theta^2 + K_{11}\,\Theta\,\Phi \\ &+ \frac{1}{3!}K_{30}\,\Theta^3 + \frac{1}{3!}K_{03}\,\Phi^3 + \ldots\end{aligned}$$

Hier zeigt sich der Vorteil bei Verwendung der Häufigkeiten erzeugenden Funktion, denn sie lautet für die zweidimensionale Normalverteilung einfach

$$K(\Theta, \Phi, t) = m_1\,\Theta + m_2\,\Phi + \frac{1}{2}\sigma_1^2\,\Theta^2 + \frac{1}{2}\sigma_2^2\,\Phi^2 + r\,\Theta\,\Phi . \tag{9}$$

Alle übrigen Koeffizienten sind Null. m_1 und m_2 sind die Erwartungswerte von X und Y, σ_1 und σ_2 die dazugehörenden Streuungen und r ist der Korrelationskoeffizient zwischen X und Y. Brechen wir also in unserem Lösungsansatz nach den Gliedern 2. Ordnung ab, so nehmen wir an, daß die Wahrscheinlichkeitsverteilung der Zufallsvariablen X und Y durch eine Normalverteilung hinreichend genau angenähert werden kann. Nach WITTLE ist diese Normalnäherung desto besser, je näher das Gleichgewicht erreicht ist. Allgemeine Kriterien für diese Näherung sind anscheinend noch nicht erarbeitet worden.

Die Häufigkeiten erzeugende Funktion der Normalverteilung, Gl. (9), wird in Gl. (8) eingesetzt. Die Exponentialfunktion in Gl. (8) wird ebenfalls in eine Potenzreihe entwickelt, so daß ein Vergleich der Koeffizienten vor Θ, Φ, Θ^2, Φ^2, $\Theta\,\Phi$ ein System von 5 Differentialgleichungen für die 5 Unbekannten m_1, m_2, σ_1^2, σ_2^2 und r liefert.

$$\begin{aligned}\frac{\partial m_1}{\partial t} &= \lambda\,(-\,2\,m_1^2 - 2\,\sigma_1^2 - m_1\,m_2 - r) - \mu\,m_1 + N\!\downarrow , \\ \frac{\partial m_2}{\partial t} &= \lambda\,(m_1^2 + \sigma_1^2) - \nu\,(m_2^2 + \sigma_2^2) , \\ \frac{\partial \sigma_1^2}{\partial t} &= \lambda\left(-\,4\,m_1\,\sigma_1^2 + 2\,m_1^2 + 2\,\sigma_1^2 + \frac{1}{2}\,m_1\,m_2 - m_1\,r - m_2\,\sigma_1^2 + \frac{1}{2}\,r\right) \\ &+ \mu\left(\frac{1}{2}\,m_1 - \sigma_1^2\right) + \frac{1}{2}\,N\!\downarrow , \\ \frac{\partial \sigma_2^2}{\partial t} &= \lambda\left(2\,m_{1r} + \frac{1}{2}\,m_1^2 + \frac{1}{2}\,\sigma_1^2\right) + \nu\left(\frac{1}{2}\,m_2^2 - 2\,m_2\,\sigma_2^2 + \frac{1}{2}\,\sigma_2^2\right), \\ \frac{\partial r}{\partial t} &= \lambda\,(-\,4\,m_1\,r + 2\,m_1\,\sigma_1^2 - 2\,m_1^2 - 2\,\sigma_1^2 - m_1\,\sigma_2^2 - m_2\,r) \\ &- \mu\,r - 2\,\nu\,m_2\,r .\end{aligned}$$

6. Lösung des Gleichungssystems

Wir beschränken uns auf den stationären Fall $\partial/\partial t = 0$. Die Erwartungswerte $m_1 = N_1$ und $m_2 = N_C$ liefert der deterministische Fall. Daher entsteht nach Einsetzen von Gl (5) und Sortieren nach den verbliebenen 3 Unbekannten das Gleichungssystem:

$$2\lambda\,\sigma_1^2 - \lambda\, r = -N\!\downarrow + \mu N_1 + \left(2 + \sqrt{\frac{\lambda}{\nu}}\right)\lambda N_1^2\,, \tag{10}$$

$$\lambda\,\sigma_1^2 - \nu\,\sigma_2^2 = 0\,, \tag{11}$$

$$\left.\begin{aligned}
&\left(2\lambda - \mu - 4\lambda N_1 - \lambda\sqrt{\frac{\lambda}{\nu}}\,N_1\right)\sigma_1^2 + \left(\frac{1}{2} - N_1\right)\lambda\, r = \\
&\qquad = -\frac{1}{2}N\!\downarrow - \frac{1}{2}\mu N_1 - \left(2 + \sqrt{\frac{\lambda}{\nu}}\right)\lambda N_1^2\,,\\
&\frac{1}{2}\lambda\,\sigma_1^2 + \left(\frac{1}{2} - 2\sqrt{\frac{\lambda}{\nu}}\,N_1\right)\nu\,\sigma_2^2 + 2\lambda N_1 r = \lambda N_1^2\,,\\
&2(N_1 - 1)\lambda\,\sigma_1^2 - \lambda N_1\sigma_2^2 - \\
&\qquad -\left(\mu + 4\lambda N_1 + \lambda\sqrt{\frac{\lambda}{\nu}}\,N_1 + 2\nu\sqrt{\frac{\lambda}{\nu}}\,N_1\right) r = 2\lambda N_1^2\,.
\end{aligned}\right\} \tag{12}$$

Aus Gl. (11) folgt unmittelbar:

$$\sigma_2^2 = \frac{\lambda}{\nu}\sigma_1^2\,.$$

Dann gilt für die relative Schwankung σ/m wegen Gl. (5)

$$\frac{\sigma_2}{N_C} = \frac{\sigma_1}{N_1}\,. \tag{13}$$

σ_1^2 läßt sich aus Gl. (10) und Gl. (12) ermitteln. Die übrigen Gleichungen werden nicht benötigt. Gl. (10) wird mit $\left(\frac{1}{2} - N_1\right)$ multipliziert und zu Gl. (12) addiert. Nach Umstellung ergibt sich

$$\sigma_1^2 = \frac{N_1^3\lambda\left(2 + \sqrt{\frac{\lambda}{\nu}}\right) + N_1^2(\lambda + \mu) - N\!\uparrow(N_1 - 1)}{N_1\lambda\left(2 + \sqrt{\frac{\lambda}{\nu}}\right) - \lambda + \mu}\,,$$

so daß die gesuchte relative Schwankung der Clusterdichte wegen Gl. (13) lautet:

$$\frac{\sigma_2}{N_C} = \sqrt{1 + \frac{N\!\downarrow\left(\frac{1}{N_1^2} - \frac{1}{N_1}\right) + 2\lambda}{N_1\lambda\left(2 + \sqrt{\frac{\lambda}{\nu}}\right) - \lambda + \mu}}\,. \tag{14}$$

Darin sind N_1 und $\sqrt{\lambda/\nu}$ nach Gl. (6b) und Gln. (7a, b) ebenfalls durch $N\!\downarrow$, μ, λ und N_C darzustellen.

7. Die Größenordnung der relativen Schwankung

Dazu werden representative numerische Werte von $N\downarrow$, μ, λ und N_C benötigt.

$N\downarrow = 10^{14\pm 2}$ (s · cm^{-2}) (entspricht etwa 0,02 bis 100 Å s^{-1}),

$\mu = 10^{8\pm 4}$ (s^{-1}) (entspricht einer Adsorptionsenergie von 0,1 bis 0,7 eV bei 400 bis 500 °K),

$\lambda = 10^{-3\pm 1}$ (cm^2 · s^{-1}) (entspricht einer Diffusionsenergie von $^1/_4$ bis $^1/_6$ der Adsorptionsenergie),

$N_C = 10^{12\pm 2}$ (cm^{-2}) (entspricht einer Keimbelegung von 10^{-4} bis 1).

Im Nenner von Gl. (14) wird N_1 aus Gl. (6b) eingesetzt. Außerdem ist $\lambda \ll \mu$

$$\frac{\sigma_2^2}{N_C^2} = 1 + \frac{N\downarrow\left(\frac{1}{N_1^2} - \frac{1}{N_1}\right) + 2\lambda}{-\frac{1}{2}\mu + \frac{1}{2}\sqrt{\mu^2 + 4N\downarrow\lambda\left(2 + \sqrt{\frac{\lambda}{\nu}}\right)} + \mu}.$$

Mit $\lambda\,(2 + \sqrt{\lambda/\nu})\,N_1 = -\mu + \sqrt{\mu^2 + 4N\downarrow\lambda\,(2 + \sqrt{\lambda/\nu})}$ erweitert, ergibt

$$\frac{\sigma_2}{N_C} = \sqrt{\frac{1}{N_1} + \frac{2\lambda N_1}{N\downarrow}}. \tag{15}$$

Für N_1 werden die beiden Näherungsausdrücke Gl. (6c) und Gl. (6e) eingesetzt.

1. Fall $\lambda N_C \ll \mu$ $\quad \frac{\sigma_2}{N_C} = \sqrt{\frac{\mu}{N\downarrow} + \frac{2\lambda}{\mu}} \approx 10^{-3\pm 3}$ (cm). (15a)

2. Fall $\lambda N_C \gg \mu$ $\quad \frac{\sigma_2}{N_C} = \sqrt{\frac{\lambda N_C}{N\downarrow} + \frac{2}{N_C}} \approx 10^{-2\pm 2}$ (cm). (15b)

Das numerische Ergebnis zeigt deutlich, daß die möglichen Schwankungen zu beachten sind, weil eine allgemeine Aussage über ihre grundsätzliche Vernachlässigbarkeit nicht gegeben werden kann.

8. Diskussion und Schluß

Im einfachsten Fall eines Aufdampf-Abdampf-Modells ergab der stochastische Prozeß für die Adatomkonzentration eine Poisson-Verteilung (Ullner), d. h. das Streuungsquadrat ist gleich dem Mittelwert. Die relative Schwankung ist daher reziprok zur Wurzel aus dem Mittelwert, so daß mit Erhöhung der Adatomkonzentration durch Vergrößerung der Beobachtungsfläche die relative Schwankung bis zur Vernachlässigbarkeit verringert werden kann.

Im hier benutzten, erweiterten Modell kann die relative Schwankung der Keimdichte ebenfalls durch Vergrößerung der Beobachtungsfläche beliebig verkleinert werden, denn beide Terme unter der Wurzel in Gl. (15) haben die Dimension der Fläche. Der 2. Term in den Gln. (15a, b) kann im Fall der hier benutzten üblichen Werte vernachlässigt werden. Unter Berücksichtigung der Gln. (6c, e) und (5) erkennt man, daß die relative Schwankung der Keimdichte bedeutend größer sein kann als eine der Poissonschen Verteilung entsprechenden relativen Schwankung:

$$\frac{\sigma_2}{N_C} = \frac{\sigma_1}{N_1} = \frac{1}{\sqrt{N_1}} = \left(\frac{\lambda}{\nu}\right)^{1/4} \frac{1}{\sqrt{N_C}} \gg \frac{1}{\sqrt{N_C}}, \quad \text{wegen } \lambda \gg \nu .$$

Es ist daher der Fall denkbar, trotz der Beobachtung einer sehr großen Anzahl von Keimen mit einer beträchtlichen Schwankung rechnen zu müssen, falls die mittlere Adatomkonzentration auf diesem Flächenstück nur etwa eins beträgt.

Abschließend sei nochmal darauf hingewiesen, daß eine Lösung in geschlossener Form nur für ein sehr einfaches Modell erzielt werden konnte. Die Verfeinerung des Modells, insbesondere die Erweiterung auf den nichtstationären Fall $\nu = 0$, müßte mittels Computer-Simulation erfolgen.

Literatur

Gnedenko, B. W.: Lehrbuch der Wahrscheinlichkeitsrechnung, Berlin 1968

Ullner, Ch.: Kristall u. Technik **5**, K5–K9 (1970)

Bailey, N. T. J.: The Elements of Stochastic Processes, New York 1966

Zinsmeister, G.: Beitrag zur Frenkelschen Kondensationstheorie; in Grundprobleme der Physik dünner Schichten, Göttingen 1966

Lewis, B., Campbell, D. S.: J. Vac. Sci. Techn. **4**, 209–218 (1967)

Wittle, R.: J. Roy. Statist. Soc. B **19**, 268 (1957)

(Eingegangen am 20. Mai 1971)

Anschrift des Verfassers:

Dr. Ch. Ullner
Humboldt-Universität zu Berlin
Sektion Physik, Bereich Kristallographie
104 Berlin
Invalidenstr. 43

CH. ULLNER

Sektion Physik, Bereich Kristallographie
der Humboldt-Universität, Berlin

Der Einfluß der kritischen Clustergröße auf die Clustergrößenverteilung und die Clusterdichte

Die Kondensationstheorie für den Fall $i^* = 1$ von ZINSMEISTER wird um Zerfallsprozesse kleiner Cluster erweitert. Dabei verlängert sich die stationäre Phase der Adatomkonzentration. Das von τ_a unabhängige Maximum wird größer. In der Wachstumsphase verringert sich der Abfall der Adatomkonzentrationen unabhängig von i^* (Fig. 1). Die Clusterdichte ist bei Berücksichtigung von Zerfallsprozessen kleiner und zeigt mit wachsendem i^* eine Tendenz zur Sättigung (Fig. 2). Der Maximalwert in der Clustergrößenverteilung sinkt mit wachsendem i^*, ist meist nach größeren i hin verschoben und sein Plateau ist breiter (Fig. 3).

The condensation theory for the case $i^* = 1$ from ZINSMEISTER is expanded by decomposition processes of small cluster. In this case the stationary phase of adatom concentration is growing longer. The maximum independent on τ_a is larger. In the growth phase the slope of adatom concentration is decreasing indepedently on i^* (Fig. 1). Considering decomposition processes the cluster density is smaller and tends to saturation with increasing i^* (Fig. 2). The maximum of size distribution of cluster decreases with increasing i^*, is mostly shifted to higher i and its plateau is broader (Fig. 3).

1. Definition des speziellen Kondensationsfalls und Aufstellung der Gleichungen

Wir gehen von den Bilanzgleichungen aus (siehe z. B. ZINSMEISTER 1970, Gln. (A, 39)). Die Anlagerungsraten λ_i seien unabhängig von i und proportional zur Adatomkonzentration: $\lambda_i = \omega\, n_1(t)$. Die Zerfallsraten sollen mit steigender Clustergröße sehr steil abfallen und bereits ab Cluster aus einigen Atomen vernachlässigbar gering sein. Die Zerfallsraten seien so gewählt, daß die kritische Clustergröße i^* unabhängig von der Adatomkonzentration ist, d. h. die stationäre Phase sei vorherrschend. Daher muß vorausgesetzt werden

$$\varkappa_2 > \varkappa_3 > \cdots > \varkappa_j > \varkappa_{j+1} = 0\,, \quad j < i^* < j+1\,. \tag{1}$$

Es entsteht das Gleichungssystem (2) bis (7) [siehe S. 116].
Das Gleichungssystem bis zur punktierten Linie beschreibt die zerfallsfreie Kondensation, die von ZINSMEISTER (1968, 1969) behandelt wird[1]). Den Ausdruck $\sum_{\nu=2}^{\infty} n_\nu$ in Gl. (2) erhält man durch Summation des Gleichungssystems

[1]) Die Unterschiede in unseren Gln. (2) und (3) gegenüber den entsprechenden von ZINSMEISTER (1968) um den Faktor 2 bzw. 1/2 sind im Rahmen dieser Arbeit belanglos.

$$\frac{dn_1}{dt} = N{\downarrow} - \frac{n_1}{\tau_a} - 2\,\omega\, n_1^2 - n_1\,\omega \sum_{\nu=2}^{\infty} u_\nu \quad + 2\,\varkappa_2\, u_2 + \sum_{\nu=3}^{j} \varkappa_\nu\, u_\nu\,, \quad (2)$$

$$\frac{dn_2}{dt} = \omega\, n_1^2 - u_1\,\omega\, u_2 \quad - \varkappa_2\, n_2 + \varkappa_3\, n_3\,, \quad (3)$$

$$\vdots$$

$$\frac{dn_{j-1}}{dt} = n_1\,\omega\, n_{j-2} - n_1\,\omega\, n_{j-1} \quad - \varkappa_{j-1} u_{j-1} + \varkappa_j\, n_j \quad (4)$$

$$\frac{dn_j}{dt} = n_1\,\omega\, n_{j-1} - n_1\,\omega\, n_{j-1} \quad - \varkappa_j\, n_j \quad (5)$$

$$\frac{dn_{j+1}}{dt} = n_1\,\omega\, n_j - n_1\,\omega\, n_{j+1}\,, \quad (6)$$

$$\vdots$$

$$\frac{dn_i}{dt} = n_1\,\omega\, n_{i-1} - n_1\,\omega\, n_{j+1}\,. \quad (7)$$

und anschließender Integration über t. Da unendlich große Cluster nicht in endlicher Zeit entstehen können, also $\lim_{i\to\infty} n_i = 0$, und bei $t = 0$ keine Cluster existieren, bekommen wir:

$$\sum_{\nu=2}^{\infty} u_\nu = \omega \int_0^t n_1^2\, dt \quad + \varkappa_2 \int_0^t n_2\, dt\,.$$

Damit wird aus Gl. (2)

$$\frac{dn_1}{dt} = N{\downarrow} - \frac{n_1}{\tau_a} - 2\,\omega\, n_1^2 - n_1\,\omega \int_0^t n_1^2\, dt + u_1\,\omega\,\varkappa_2 \int_0^t n_2\, dt + 2\,\varkappa_2\, u_2 + \sum_{\nu=0}^{j} \varkappa_\nu\, n_\nu\,. \quad (8)$$

2. Zeitliche Änderung der Adatomkonzentration

2.1. Die angenäherten Gleichungen

Wie die späteren Gln. (15) und (20) zeigen, hat die zeitliche Abhängigkeit der Adatomkonzentration eine zentrale Bedeutung. Sie wird aus Gl. (8) ermittelt. Diese Integrodifferentialgleichung kann für bestimmte Perioden durch einfachere Gleichungen angenähert werden, da sich die einzelnen Summanden um Zehnerpotenzen unterscheiden können.

Im Fall $j \geqq 2$ wird der Zusammenhang zwischen n_ν ($\nu \geqq 2$) und n_1 benötigt. Wir befassen uns dabei näherungsweise mit der stationären Phase von n_ν ($\nu \geqq 2$).

Aus Gl. (5) ergibt sich für $\partial/\partial t = 0$

$$n_j = \frac{n_1\,\omega\, n_{j-1}}{\varkappa_j + n_1\,\omega}\,. \quad (9)$$

Damit wird aus Gl. (4) für $\partial/\partial t = 0$

$$n_{j-1} = \frac{n_1\,\omega \cdot n_{j-2}}{\varkappa_{j-1} + \dfrac{(n_1\,\omega)^2}{\varkappa + n_1\,\omega}}\,. \quad (10)$$

Da $j < i^* < j + 1$, muß für alle n_1 die Ungleichung

$$\varkappa_j > n_1 \omega \tag{11}$$

gelten. So wird Gl. (10) näherungsweise

$$n_{j-1} = \frac{n_1 \omega}{\varkappa_{j-1}} n_{j-1} \left[1 - \frac{(n_1 \omega)^2}{\varkappa_j \varkappa_{j-1}} - \frac{(n_1 \omega)^2}{\varkappa_j^2 \varkappa_{j-1}}\right].$$

Wegen der Ungleichungen (1) und (11) können die Korrekturglieder weggelassen werden. Ebenso kann im Nenner von Gl. (9) $n_1 \omega$ vernachlässigt werden, weil uns nur die Größenordnung interessiert. Wir können also von vornherein $\varkappa_j \gg n_1 \omega$ annehmen. Eine rekursive Anwendung der angenäherten Gl. (9) von j nach 1 abwärts ergibt

$$n_j = \frac{\omega^{j-1} n_1^j}{\varkappa_2 \varkappa_3 \ldots \varkappa_j}. \tag{12}$$

Die Clusterkonzentration ist wegen der Ungleichungen (1) und (11) gegenüber derjenigen im zerfallsfreien Fall umso kleiner, je größer die kritische Clustergröße i^* ist. Dabei hat die Adatomkonzentration mit wachsendem $j < i^*$ einen immer stärker werdenden Einfluß auf die Clusterkonzentration.

Die zeitliche Abhängigkeit der Adatomkonzentration ändert sich nur quantitativ bei Einbeziehung von Zerfallsprozessen. Gl. (8) lautet mit den Clusterkonzentrationen aus Gl. (9) für $j = 2$, $n_2 \approx \frac{\omega n_1^2}{\varkappa_2}\left(1 - \frac{\omega n_1}{\varkappa_2}\right)$, und aus Gl. (12) für $2 < j < i^*$

$$\frac{dn_1}{dt} = N\!\downarrow - \frac{n_1}{\tau_a} - \frac{n_1 \omega^3}{\varkappa_2} \int n_1^3 \, dt - \\ - \frac{\omega^2 n_1^3}{\varkappa_2}\left[1 - \frac{\omega n_1}{\varkappa_3} - \frac{(\omega n_1)^2}{\varkappa_3 \varkappa_4} - \cdots - \frac{(\omega n_1)^{j-2}}{\varkappa_3 \varkappa_4 \ldots \varkappa_j}\right]. \tag{13}$$

Die Ausdrücke in der Klammer können wieder wegen (1) und (11) gegenüber 1 vernachlässigt werden. Im Vergleich zum zerfallsfreien Fall

$$\frac{dn_1}{dt} = N\!\downarrow - \frac{n_1}{\tau_a} - 2 \omega n_1^2 - n_1 \omega^2 \int_0^t n_1^2 \, dt \tag{8a}$$

lautet bei Einbeziehung von Zerfallsprozessen die angenäherte Gl. (8)

$$\frac{dn_1}{dt} = N\!\downarrow - \frac{n_1}{\tau_a} - \frac{\omega^2 n_1^3}{\varkappa_2} - \frac{n_1 \omega^3}{\varkappa_2} \int_0^t n_1^3 \, dt \,. \tag{8b}$$

Im Folgenden werden nur die Unterschiede zwischen diesen Gln. (8a) und (8b) betrachtet, d. h. auf die Änderungen gegenüber den Ergebnissen von ZINSMEISTER (1968) eingegangen.

2.2. Anfangsphase und stationäre Phase

Da sich die ersten beiden Terme der Gln. (8a), (8b) nicht unterscheiden, bleiben Anfangsphase und stationäre Phase unverändert (Fig. 1), jedoch kann τ_a größere Werte annehmen, ohne daß die Adatomkonzentration — unabhängig von τ_a — über ein Maximum wieder absinkt. Denn der 3. und 4. Term ist in Gl. (8b) gegenüber denen in Gl. (8a) um den Faktor $n_1 \omega/\varkappa_2 = 10^{-k}$ geringer, wobei $k = 1, 2, 3$ und mehr entsprechend der kritischen Clustergröße sein kann. Die Bedingung für τ_a lautet nun

$$\tau_a > 2 \cdot 10^{k/2} (2\,\omega\, N\!\downarrow)^{-1/2} .$$

Sie muß aus einem Vergleich zwischen dem 2. und dem 4. Term gewonnen werden, da der 4. Term für $t > 2\,(\omega\, N\!\downarrow)^{-1/2}$ bereits den 3. Term übertrifft. Dabei wurde stets $n_1 = N\!\downarrow t$ benutzt.

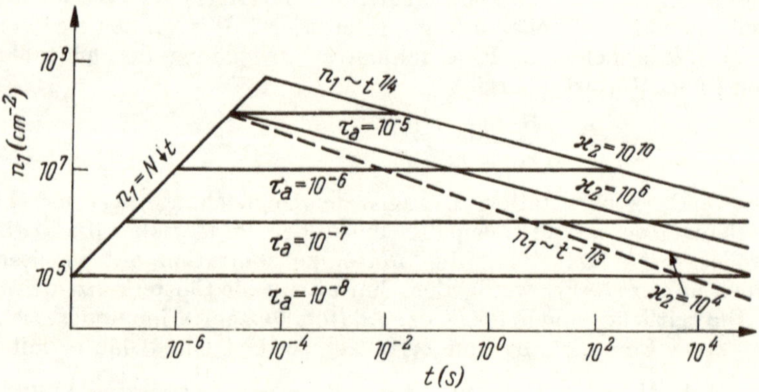

Fig. 1. Adatomkonzentration als Funktion der Zeit.
$N\!\downarrow = 10^{13}$ cm^{-2} s^{-1}
$\omega = 2 \cdot 10^{-3}$ cm^2 s^{-1}
τ_a (s)
$\varkappa_2$ (s^{-1})

2.3. Wachstumsphase

Hier tritt eine Änderung gegenüber der zerfallsfreien Kondensation ein. Es gilt

$$N\!\downarrow = \frac{n_1\,\omega^3}{\varkappa_2} \int n_1^3\, dt .$$

Durch Umstellen und Differenzieren erhält man die Differentialgleichung

$$\frac{dn_1}{dt} = -\frac{\omega^3}{N\!\downarrow\,\varkappa_2} n_1^5 ,$$

deren Lösung

$$n_1 = \left[\frac{4\,\omega^3}{N\!\downarrow\,\varkappa_2} (t - t_2) + n_1(t_2)^{-4}\right]^{-1/4} \tag{14}$$

beträgt, wobei $t_2 \approx 10^k\,\tau_a^{-2}\,(\omega\, N\!\downarrow)^{-2}$ die ungefähre Grenze zur stationären Periode angibt. Die zeitliche Abnahme der Adatomkonzentration ist hier mit $n_1 \sim t^{-1/4}$ geringer als im zerfallsfreien Fall, wo $n_1 \sim t^{-1/3}$ ist (Fig. 1).

3. Zeitliche Änderung der Clusterdichte

Die Dichte aller überkritischen Cluster (stabile Clusterdichte genannt) beträgt

$$G_j(t) = \sum_{\nu=j+1}^{\infty} n_\nu = \omega \int_0^t n_1\, n_j\, dt . \tag{15}$$

Die Anlaufphase bleibt unberücksichtigt.

Für die stationäre Periode gilt mit $n_1 = N\!\downarrow \tau_a$ und Gl. (12)

$$G_j(t) = \frac{\omega^j (N\!\downarrow \tau_a)^{j+1}}{\varkappa_2\, \varkappa_3 \ldots \varkappa_j}\, t\,. \tag{16}$$

Die stabile Clusterdichte ist um den Faktor $\frac{\omega\, n_1}{\varkappa_2} \frac{\omega\, n_1}{\varkappa_3}, \ldots, \frac{\omega\, n_1}{\varkappa_j}$ geringer als die gesamte Clusterdichte.

Zur Berechnung der stabilen Clusterdichte in der Wachstumsphase benutzen wir für n_j ebenfalls Gl. (12). Dadurch wird $G_j(t)$ in Wirklichkeit größer sein, weil der Abfall der Clusterkonzentrationen gegenüber dem der Adatomkonzentration verzögert verlaufen wird.

$$G_j(t) = \frac{\omega^j}{\varkappa_2 \ldots \varkappa_j} \int \left[\frac{4\, \omega^3}{N\!\downarrow \varkappa_2} (t - t_2) + n_1(t_2)^{-4}\right]^{-\frac{j+1}{4}} dt\,.$$

Die Auswertung für $j = 3$ ergibt:

$$G_j(t) = \frac{\omega^{j-3}\, N\!\downarrow}{\varkappa_3 \ldots \varkappa_j (3 - j)} \left[\frac{4\, \omega^3}{N\!\downarrow \varkappa_2} (t - t_2) + n_1(t_2)^{-4}\right]^{\frac{3-j}{4}} + \\ + G_j(t_2) - \frac{\omega^{j-3}\, N\!\downarrow n_1(t_2)^{j-3}}{\varkappa_3 \ldots \varkappa_j (3 - j)} \tag{17}$$

Insbesondere ergibt sich für $i = 3$

$$G_j(t) = \frac{N\!\downarrow}{4\, \varkappa_3} \ln \left[\frac{4\, \omega^3}{N\!\downarrow \varkappa_2} (t - t_2) + n_1(t_2)^{-4}\right] + G_j(t_2) + \frac{N\!\downarrow}{\varkappa_3} \ln n_1(t_2)\,. \tag{18}$$

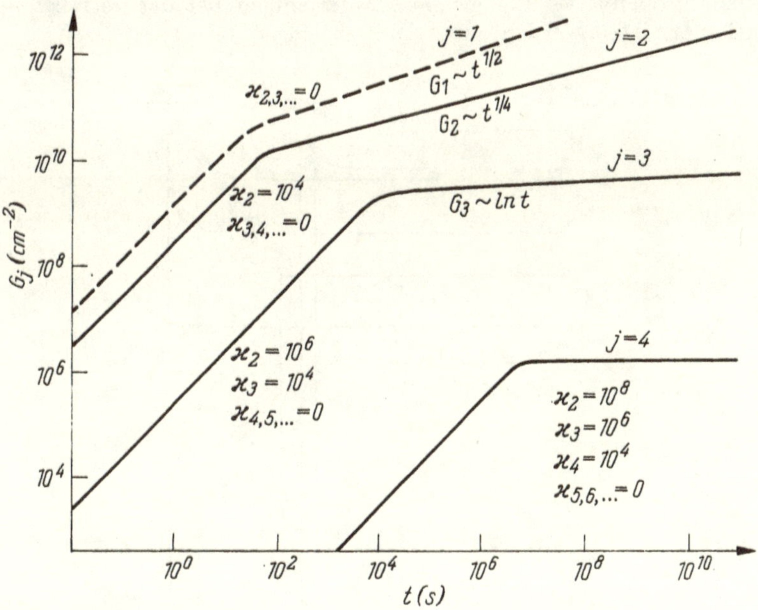

Fig. 2. Clusterdichte als Funktion der Zeit.
$N\!\downarrow = 10^{13}\ \mathrm{cm^{-2}\,s^{-1}}$
$\omega = 2 \cdot 10^{-3}\ \mathrm{cm^{1}\,s^{-1}}$
$\tau_a = 10^{-7}\ \mathrm{s}$
$\varkappa_j\ (\mathrm{s^{-1}})$

$G_j(t)$ ist in Figur 2 dargestellt. Man sieht, daß der zeitliche Verlauf von $G_2(t)$ etwa proportional der vierten Wurzel aus der Zeit ist, d. h. bei Berücksichtigung von Paarzerfall wächst die stabile Clusterdichte weniger als im zerfallsfreien Fall. Wird zusätzlich der Zerfall von Dreierclusten berücksichtigt, wächst die stabile Clusterdichte nur noch logarithmisch mit der Zeit. Noch höhere Zerfallprozesse führen zur Existenz einer Sättigungsdichte. Sie liegt nur unwesentlich über der stabilen Clusterdichte bei t_2, da der letzte Term in Gl. (17) klein gegenüber $G_j(t_2)$ ist. Da wir für n_j den stationären Ausdruck benutzt haben, ist die Existenz einer Sättigungsdichte nicht sicher, doch dürfte eine Tendenz zu einer Sättigung mit wachsendem i^* vorhanden sein. Das stimmt auch mit dem Ergebnis von LOGAN überein.

4. Clustergrößenverteilung

Die Clustergrößenverteilung muß analog zu ZINSMEISTER (1969) aus dem Gleichungssystem (6) . . . (7) ermittelt werden. Mit der Zeittransformation

$$x = \omega \int_0^t n_1(t)\, dt \tag{19}$$

entsteht als Näherungslösung

$$\begin{aligned} n_{i+1} &= n_j\,[x - (i - j)] \quad &\text{für } (i - j) < x\,, \\ n_{i+1} &= 0 &\text{für } (i - j) \geqq x\,. \end{aligned} \tag{20}$$

Wird für n_j der stationäre Wert nach Gl. (12) benutzt, kann das Verhalten der Clustergrößenverteilung gegenüber derjenigen bei der zerfallsfreien Kondensation ermittelt werden.

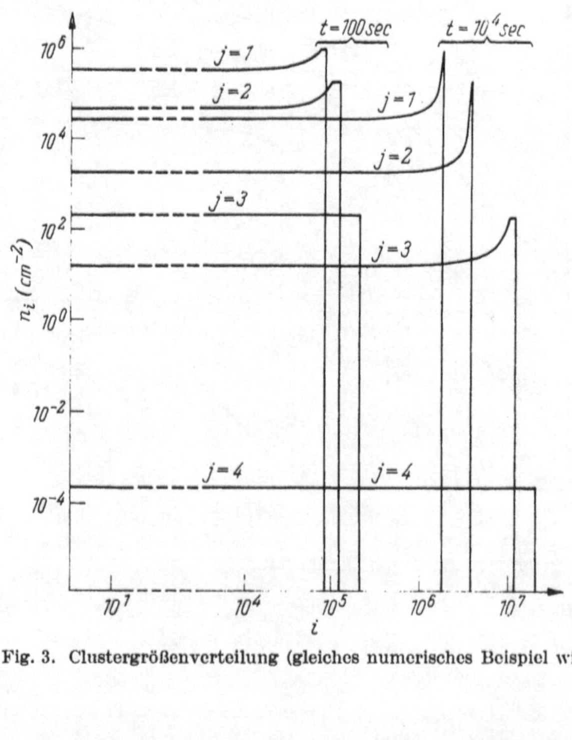

Fig. 3. Clustergrößenverteilung (gleiches numerisches Beispiel wie in Figur 2)

Das Maximum der Clustergrößenverteilung ist um den Faktor $\frac{\omega n_1}{\varkappa_2} \frac{\omega n_2}{\varkappa_3} \cdots \frac{\omega n_1}{\varkappa_j}$ geringer als bei der zerfallsfreien Kondensation. Die Höhe des Maximums ändert sich wegen der j-ten Potenz sehr empfindlich mit der Adatomkonzentration, insbesondere mit der Aufdampfrate $N\downarrow$. Die Breite der Größenverteilung müßte wegen der verlängerten stationären Phase von $n_1(t)$ für größere i^* wachsen. Allerdings bleibt die Form des Anstieges unbekannt, da wir uns auf die stationären Werte beschränkt haben. In Figur 3 ist die Gl. (12) formal benutzt worden. Der Anstieg der Clustergrößenverteilung wird in Wirklichkeit nicht so steil verlaufen, während der Abfall unverändert steil bleibt. Wegen der unterschiedlichen zeitlichen Abhängigkeit der $n_1(t)$ bei verschiedenen j (siehe Figur 1) verschieden sich die Maxima infolge der Zeittransformation, Gl. (19).

5. Schlußbemerkung

Über die Brauchbarkeit eines Modells entscheiden stets die experimentellen Ergebnisse. Ein Vergleich ist zur Zeit kaum möglich, da die vielen möglichen Störungen beim Aufdampfvorgang mit ihren enormen Auswirkungen noch nicht ausreichend unter Kontrolle gebracht werden konnten. Es erscheint uns daher auf dieser Entwicklungsstufe wenig sinnvoll, gedachte Modelle vollständig durchzurechnen und mit den bisherigen experimentellen Ergebnissen zwecks Bestätigung zu vergleichen, denn das experimentelle Material bezieht sich jeweils nur auf einen sehr kleinen Bereich innerhalb der großen Zahl von Variationsmöglichkeiten der äußeren Bedingungen.

Deshalb haben wir mit einfachsten Abschätzungen versucht, das charakteristische Verhalten der Clustergrößenverteilung und der Clusterdichte bei Einbeziehung von Zerfallsprozessen kleiner Cluster zu analysieren. Das könnte Teil eines Weges sein, um solche Bereiche zu finden, innerhalb derer ein Vergleich von Theorie und Experiment sinnvoll wird. Bei Anwendung rechentechnischer Methoden zur detaillierten Behandlung sind dann solche Abschätzungen sowieso wünschenswert.

Literatur

Logan, R. M.: Thin Solid Films **3**, 59 (1969)
Zinsmeister, G.: Kristall und Technik **2**, 207 (1970)
Zinsmeister, G.: Thin Solid Films **2**, 497 (1968)
Zinsmeister, G.: Thin Solid Films **4**, 363 (1969)

(Eingegangen am 25. Juni 1971)

Anschrift des Autors:

Dr. Ch. Ullner
Sektion Physik, Bereich Kristallographie
der Humboldt-Universität
104 Berlin
Invalidenstr. 43

Kristall und Technik	7	1	1972	123–131

I. Mietz und H. Sprenger

Humboldt-Universität, Berlin
Sektion Physik, Bereich Kristallographie

Ein Beitrag zur Erzeugung dünner orientierter Antimonfilme auf Einkristallsubstraten durch Hochvakuumkondensation

In Abhängigkeit von der Substrattemperatur, der aufgedampften Schichtdicke und der Aufdampfrate wurde die Kondensation von Antimon auf NaBr-Epitaxieschichten und auf Muskovit bei 10^{-5} Torr untersucht. Auch bei Substrattemperaturen oberhalb 1/3 T_s (T_s Schmelzpunkt) erfolgt bis zu einer kritischen Substrattemperatur (für Muskovit $>$150 °C, für NaBr-Epitaxieschicht $<$150 °C) die Kondensation über die amorphe Phase. Oberhalb dieser Substrattemperatur bildet sich die kristalline Phase direkt aus der Dampfphase.

Bei einer bestimmten Schichtdicke kristallisiert die amorphe Phase. Die dabei auftretende Orientierung ist schichtdickenabhängig.

The condensation of antimony on NaBr-epitaxy layer and mica was revealed in high vacuum in dependence on substrate temperature (20–250 °C), average thickness of the film (200–600 Å) and evaporation rate (10–140 Å/s). Also for substrate temperatures above 1/3 T_m (melting point) up to a critical substrate temperature the condensation of antimony includes the transition amorphous/crystalline (for mica $>$150 °C, for NaBr-epitaxy layer $<$150 °C). Above the critical substrate temperature antimony is deposited directly in the crystalline phase. Above a defined film thickness the amorphous phase crystallized. The orientation is influenced by the film thickness.

1. Einleitung

Kondensationsmechanismus, Kristallisationsvermögen und die damit verbundenen Textur- und Orientierungsphänomene in dünnen Schichten sind in entscheidendem Maße von der kristallchemischen Struktur und dem konvalenten Bindungsanteil des Dünnschichtmaterials abhängig. Im Falle metallischer Bindung und hoher Packungsdichte des Deposits erfolgt die Keimbildung nach dem Frenkel-Mechanismus (gradual nucleation), d. h. die aus dem Dampfstrahl auf das Substrat auftreffenden Atome bilden unter bestimmten Bedingungen Keime, die im weiteren Verlauf des Aufdampfprozesses zu Kristalliten auswachsen. Bei den Elementen mit kovalentem Bindungsanteil und entsprechend geringerer Packungsdichte und Koordinationszahl gegenüber den echten Metallen erfolgt die Kristallisation über ein intermediäres Stadium, das der flüssige oder amorphe Zustand sein kann. Bei einer kritischen „Schichtdicke" des intermediären Stadiums erfolgt spontane Keimbildung (spontaneous nucleation, Zwischenphasenmechanismus).

Ausgehend von den Arbeiten von SEMENOW führten PALATNIK und Mitarbeiter zahlreiche Experimente zur Klärung des Kondensationsmechanismus von Metallen durch. Am Beispiel der Wismutkondensation konnten PALATNIK und KOMNIK die Existenz zweier kritischer Kondensationstemperaturen nach weisen, bei denen beide Mechanismen möglich sind. Oberhalb 2/3 T_S (T_S Schmelzpunkt) erfolgt die Metalldampfkondensation bis zur kritischen Kondensationstemperatur über die flüssige Phase. Unterhalb dieser Temperatur bis zur Temperatur 1/3 T_S erfolgt die Kristallkeimbildung auf dem Substrat direkt in den kristallinen Zustand. Unterhalb dieser Temperatur findet die Kristallisation wiederum über eine amorphe Zwischenphase statt. Das Auftreten der metastabilen Zwischenphasen (amorph oder flüssig) kann mit der Ostwald'schen Stufenregel gedeutet werden (KLEBER), für ihre Gültigkeit sind die Differenzen in den Dichten der kristallinen, amorphen und flüssigen Phase wesentlich. Dieser Zwischenphasenmechanismus war in den letzten Jahren Gegenstand zahlreicher theoretischer und experimenteller Untersuchungen (BERNDT; PALATNIK, GLADKICH).

Die über die Schichtbildung von Antimon erschienenen Arbeiten beziehen sich größtenteils auf die Existenzbedingungen für die amorphe Phase und den Übergang in den kristallinen Zustand.

2. Ältere Arbeiten über die Bildungsbedingungen von dünnen Antimonfilmen

Die Ausbildung amorpher dünner Antimonfilme auf Substraten ist von den Aufdampfparametern wie Substrattemperatur, Aufdampfrate und Restgasdruck sowie in entscheidendem Maße von der Art des Substrats abhängig. Autoren älterer Arbeiten benutzten vorwiegend amorphe Substrate (RUEDL; CREMER, RUEDL; CONDAS; CONDAS, WOOTEN; GÖTZBERGER; SAGANUMA, TANAKA; KOIKE, KUROKOWA; SHEFTAL; PALATNIK). PALATNIK und KOSSEWITSCH zeigten sehr instruktiv, daß die Struktur der abgeschiedenen Antimonfilme eine Funktion des Bindungstyps, der im Substrat vorliegt, darstellt. Im Falle metallischer Bindung des Substrats kristallisiert das Antimon ohne Zwischenschaltung der amorphen Phase direkt aus der Dampfphase und unabhängig von dem vorliegenden Misfit. Bei Benutzung von Ionenkristallen, Oxidfilmen und Glimmer als Substrat tritt bis zu einer kritischen Schichtdicke die amorphe Phase auf. Oberhalb dieser Schichtdicke kristallisiert das Antimon über die Ausbildung von Sphärolithen. SUGANUMA und TANAKA konnten ebenfalls zeigen, daß im Falle von Collodium- oder Kohlenstoffilmen als Substrat und einer Abscheidung des Antimon bei Raumtemperatur oberhalb einer kritischen Dicke (180 Å) die Kristallisation radial von Zentren ausgehend erfolgt. Infolge der Stressbildung beim Kristallisationsprozeß, die auf die Volumenkontraktion bei der Phasenumwandlung zurückzuführen ist, treten sogenannte „faltige" Strukturen des Antimonfilms auf. Die Autoren berichten ferner, daß die Größe der kritischen Dicke für den Phasenübergang vom Filmgebiet abhängig ist. Bei Untersuchungen zur Kristallisation amorpher Antimonfilme unter dem Einfluß des Elektronenstrahles stellten BAHADUR und CHAUDHARY fest, daß die Orientierung des kristallisierten Antimonfilmes empfindlich von der Schichtdicke des ursprünglich amorphen Filmes abhängt.

RUEDL konnte zeigen, daß die kritische Dicke für den Übergang amorph/kristallin bei Benutzung hoher Aufdampfraten (>100 Å/s) erniedrigt werden kann.

Damit in Übereinstimmung ist das Ergebnis von CONDAS und WOOTEN, daß bei einem Restgasdruck von 10^{-8} Torr die Wahrscheinlichkeit für die Bil dung heterogener Antimonfilme mit geringer werdender Aufdampfrate fällt. Der Einfluß der Substrattemperatur auf die Struktur der gebildeten Antimonfilme wurde nur in wenigen Arbeiten (z. B. PALATNIK, KOSSEWITSCH, MOSKALEW) untersucht, die meisten Untersuchungen wurden bei Raumtemperatur des Substrats durchgeführt. PALATNIK, KOSSEWITSCH, ANTONOWA konnten die theoretische Voraussage, daß für Substrattemperaturen oberhalb $1/3\ T_S$ die Kondensation aus der Dampfphase direkt in die kristalline Phase erfolgt, bestätigen.

Die Aussage von SOMMER, daß sich Antimonfilme in ihren elektrischen Eigenschaften unterscheiden, die von elementarem Antimon und von einer Antimonlegierung als Quellenmaterial hergestellt wurden, konnte von ROBBINS und THOMAS nicht völlig bestätigt werden.

3. Versuchsdurchführung

Die Untersuchungen wurden bei einem Restgasdruck von einigen 10^{-5} Torr durchgeführt. Als Quelle diente ein Quarztiegel. Die Aufdampfratenbestimmung erfolgte durch Schichtdickenmessungen nach dem Vielstrahlinterferenzverfahren von TOLANSKY. Die Aufdampfrate betrug 10 Å/s. Für Aufdampfraten von 140 Å/s wurde ein Molybdänschiffchen als Verdampfungsquelle benutzt. Die Reinheit des verwendeten Antimon betrug 99,9999%. Die Substrattemperatur wurde zwischen Raumtemperatur und 250 °C variiert, die Schichtdicke zwischen 200 und 600 Å. Als Substrate wurden Muskovit und NaBr-Epitaxieschichten auf Muskovit benutzt. Die Untersuchung der Schichten erfolgte im Transmissionselektronenmikroskop SEM 3 des Werks für Fernsehelektronik Berlin in Abbildung und Beugung.

4. Ergebnisse

4.1. Substrat: NaBr-Epitaxieschicht auf Muskovit

Anfängliche Untersuchungen zeigten, daß auf erhitzten (100)-Spaltflächen von Alkalihalogeniden — entsprechend der vierzähligen Achse dieser Substratoberfläche — die (001)-Auflage von Antimon in symmetrieäquivalenten Stellungen aufwächst (Fig. 1).

Zur Vermeidung dieser symmetriebedingten Stellungen wurde die (111)-Fläche von NaBr bevorzugt, die aus Symmetriegründen als Substrat für die hexagonale (001)-Auflage des Antimons geeigneter ist.

Auf diese NaBr-Schicht wurde bei einer Substrattemperatur von 50 °C und mit konstanter Aufdampfrate von 10 Å/s Antimon aufgedampft. Mit zunehmender Schichtdicke konnten die Stadien amorph, polykristallin ohne

Fig. 1. Symmetrieäquivalente Stellungen von Antimon auf NaCl-Spaltflächen. Substrattemperatur: 240 °C, Aufdampfrate: 110 Å/s
Schichtdicke: 300 Å

Vorzugsorientierung, polykristallin mit steigendem Texturanteil bis zur azimutalen Orientierung hoher Güte mit Hilfe der Elektronenbeugung identifiziert werden (Fig. 2a—c).

Die abgeschiedenen Antimonpartikel zeigen gerundeten Habitus. Der Übergang von der amorphen zur kristallinen Phase erfolgt bei einer Schichtdicke zwischen 200 und 250 Å und ist schlecht reproduzierbar. Bei dieser Schichtdicke besitzt der größte Teil der Partikel keinen Kontakt zu Nachbarpartikeln, so daß die Umwandlung in die kristalline Phase innerhalb eines Partikels bei einer kritischen Radiengröße erfolgt. Die kristallinen Partikel sind zunächst regellos angeordnet (Fig. 2a). Mit zunehmendem gegenseitigem Kontakt setzt sich die (001)-Orientierung des Antimon durch, die bei Schichtzusammenschluß (ca. 600 Å) maximale Güte besitzt (Fig. 2c).

Bei einer Substrattemperatur von 100 °C liegen in qualitativer Hinsicht analoge Verhältnisse vor. Wird die Substrattemperatur auf 150 °C erhöht, so zeigt das Antimon bei der geringsten untersuchten Schichtdicke (200 ± 50 Å) bereits kristalline Struktur. Bei dieser Dicke befindet sich die Antimonschicht noch im Inselstadium, die Kristallite besitzen zunächst keine kristallographischen Formen (Fig. 3a) und sind regellos angeordnet. Mit Erhöhung der Schichtdicke tritt eine [00.1]-Textur auf, die von einem ausgeprägten hexa-

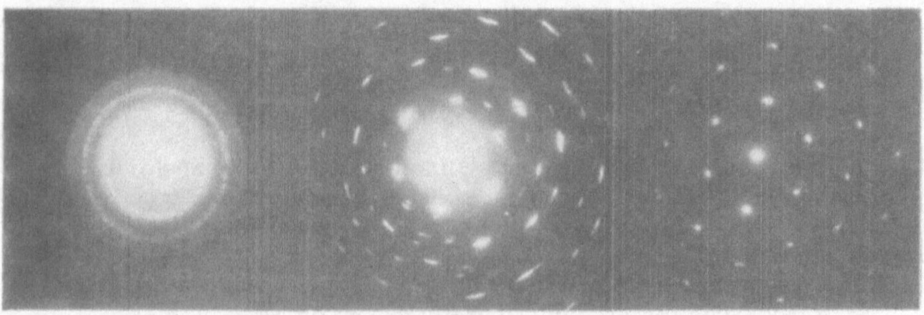

a) b) c)

Fig. 2. a—c. Abhängigkeit der Orientierung von der Schichtdicke für Antimonabscheidungen auf einer NaBr-Epitaxieschicht.
Substrattemperatur: 50 °C, Aufdampfrate: 10 Å/s,
Schichtdicke: a) 250 Å, b) 420 Å, c) 800 Å

a)

b)

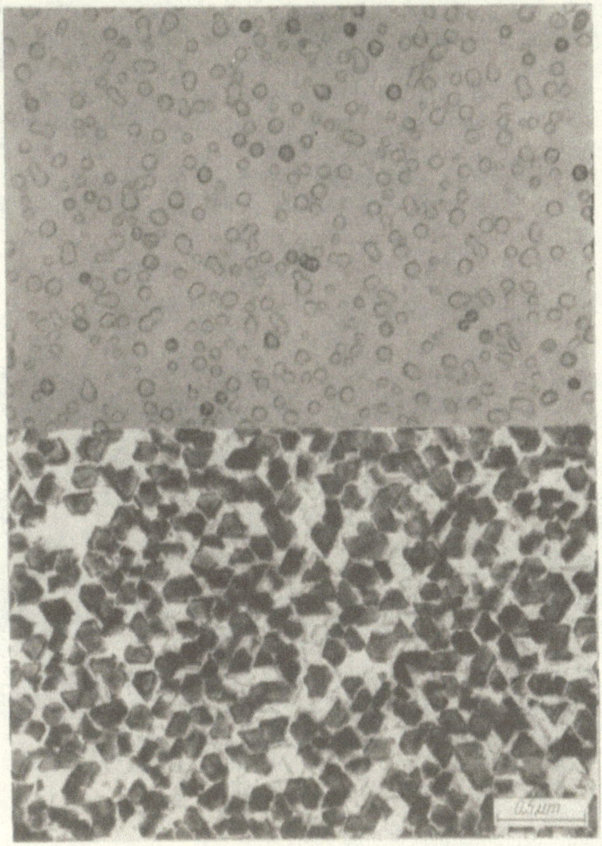

Fig. 3.a – b. Wachstumsstadien von Antimonschichten auf einer NaBr-Epitaxieschicht. Substrattemperatur: 150 °C, Aufdampfrate: 10 Å/s
Schichtdicke: a) 250 Å, b) 400 Å

gonalen Habitus der Kristallite begleitet ist (Fig. 3b). Der prozentuale Anteil der Kristallite in azimutaler Orientierung ist proportional zur Schichtdicke und ist maximal bei 500 Å. Eine weitere Erhöhung der Substrattemperatur bringt keine Verbesserung der Orientierungsgüte bei vergleichbaren Schichtdicken mit sich. Der Kondensationskoeffizient von Antimon beginnt oberhalb von 150 °C erheblich von 1 abzuweichen.

4.2. Substrat: Muskovit

Im wesentlichen ergeben sich bei Verwendung von Muskovit als Substrat für die Antimonkondensation unter sonst gleichen Bedingungen analoge Verhältnisse. Lediglich die amorphe Phase ist bei einer Schichtdicke von 200 ± 50 Å (z. T. auch bei 300 Å) noch bei einer Substrattemperatur von 150 °C stabil. Stark nach [00.1] texturierte bis azimutal orientierte Antimonfilme bilden sich bei 20 °C oberhalb 300 Å (Fig. 4a und b), bei einer Substrattemperatur von 200 °C bereits bei 200 Å aus.

Der Einfluß der Aufdampfrate auf die Schichtbildung wurde unter Verwendung einer Substrattemperatur von 20 °C untersucht. Die Schichten, die mit Aufdampfraten von 10 bzw. 140 Å/s erzeugt wurden, zeigen bei Verwendung vergleichbarer Schichtdicken charakteristische Unterschiede. Die Steigerung der Aufdampfrate ist mit einem Anwachsen des durchschnittlichen Kristallitdurchmessers von einigen hundert Å bis zu 5 µm für die größte Ausdehnung der nunmehr leistenförmigen Kristallite verbunden, die sich bei Verwendung einer Aufdampfrate von 140 Å/s ausbilden (Fig. 5). Neben der (001)-Orientierung, die bei 10 Å/s ausschließlich und mit hoher Güte auftritt, werden bei 140 Å/s andere Auflageflächen ((101) und (120)) bevorzugt.

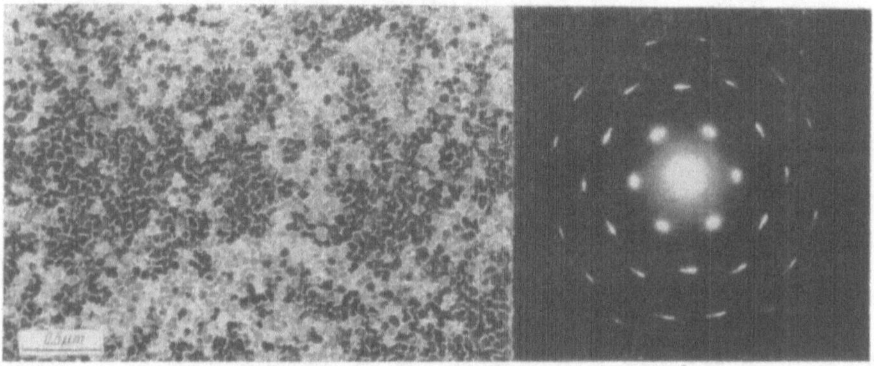

a) b)

Fig. 4. a–b. Wachstumsstadien einer Antimonschicht auf Muskovit.
Substrattemperatur: 20 °C, Aufdampfrate: 10 Å/s
Schichtdicke: 500 Å

a) Elektronenmikroskopische Abbildung
b) Elektronenbeugung

Fig. 5. Vergrößerung der durchschnittlichen Kristallitgröße in Antimonfilmen auf Muskovit als Folge einer Erhöhung der Aufdampfrate auf 140 Å/s
Substrattemperatur: 20 °C (vgl. Fig. 4a)

5. Diskussion

Die schlechte Reproduzierbarkeit der Angaben über die Größe der kritischen Dicke für den Übergang amorph/kristallin ist als Folge der Stabilisierung der amorphen Phase durch den druckabhängigen Einbau von Restgasatomen in den Antimonfilm anzusehen. Die gleiche Erklärung wird für die Tatsache gegeben, daß bei einem Restgasdruck von 10^{-5} Torr auch im Bereich der Substrattemperatur von $1/3\ T_S$ bis $2/3\ T_S$ die Existenz der amorphen Phase nachgewiesen werden konnte. Für das Auftreten der amorphen Phase in diesem Temperaturbereich dürfte unter den genannten Bedingungen die Bildung einer Oxidphase des Antimon von Einfluß sein. Auf ihre Entstehung führt RUEDL die Ausbildung heterogener Antimonfilme bei geringen Aufdampfraten und hohen Restgasdrucken zurück. Als Beweis für die Richtigkeit dieser Annahme ist der experimentelle Befund zu werten, daß das Wachstum heterogener Antimonfilme bei 10^{-5} Torr durch Aufdampfraten von >100 Å/s vermieden werden konnte und daß bei 10^{-9} Torr auch bei geringen Aufdampfraten keine heterogenen Filme beobachtet wurden (CREMER und RUEDL).

Der beobachtete Unterschied in den Grenztemperaturen für den Phasenübergang amorph/kristallin auf dem NaBr-Epitaxiesubstrat und auf Muskowit dürfte weniger durch Differenzen in der Gitterenergie als durch die unterschiedliche Oberflächenstruktur der Substrate gedeutet werden. Die (111)-orientierte NaBr-Schicht weist zahlreiche Stufen, Ecken, Korngrenzen u. dgl. auf, die die Kristallisation des amorphen Films beeinflussen dürften.

Bei der Diskussion des Phasenübergangs amorph/kristallin ist zwischen der Kristallisation einer amorphen geschlossenen Schicht und der Umwandlung eines isolierten amorphen Teilchens bestimmter Größe zu unterscheiden. Der erstere Fall kann z. B. bei Raumtemperatur vorliegen und ist dickenabhängig. Die Existenz des letzteren Falles ist durch die vorliegenden Untersuchungen wahrscheinlich gemacht (SPRENGER). Der kritische Radius für den Phasenübergang fällt mit steigender Substrattemperatur bis schließlich die Kondensation direkt in die kristalline Phase erfolgt. Zur Ermittlung der kritischen Parameter für den Phasenübergang und ihrer Abhängigkeit von den experimentellen Faktoren sind weitere Untersuchungen unter UHV-Bedingungen

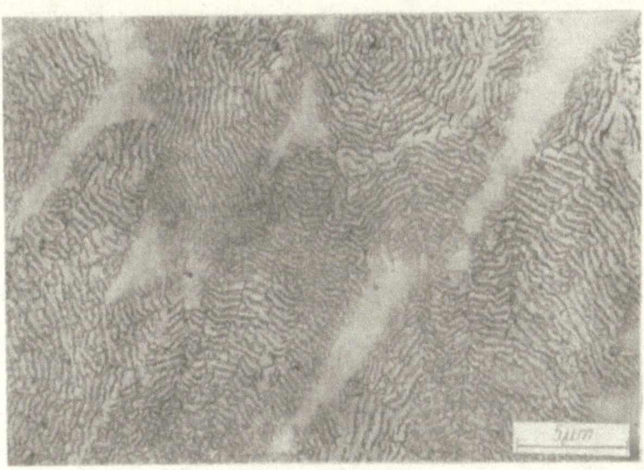

Fig. 6. Kristallisation einer geschlossenen amorphen Antimonschicht

notwendig. Interessant wäre hier eine elektronenmikroskopische in situ Untersuchung des Wachstumsmechanismus unter gleichzeitiger Registrierung der elektrischen Leitfähigkeit. Wie die Untersuchungen zeigten, ist die Intensität der (001)-Orientierung, die sich bei Substrattemperaturen von ca. 20 °C ausbildet, eine Funktion der Filmdicke. Aus den elektronenmikroskopischen Abbildungen geht hervor, daß die steigende Intensität der (001)-Orientierung an den stärkeren gegenseitigen Kontakt der Kristallite gebunden ist. Die Orientierung ist also eindeutig wachstumsbedingt. Es ist vorstellbar, daß sich beim gegenseitigen Kontakt eines (größeren) bereits kristallisierten Partikels mit einem (kleineren) noch amorphen Partikel die Kristallisation des letzteren unter Ausbildung der gleichen azimutalen Orientierung vollzieht.

Ist die amorphe Phase noch stabil, wenn eine geschlossene Schicht vorliegt, dies kann bei Raumtemperatur aber auch bei erhöhter Temperatur der Fall sein, so geht analog den Untersuchungen von Palatnik, Kossewitsch, Antonowa sowie Suganuma, Tanka die Kristallisation von Zentren radialstrahlig unter Ausbildung sogenannter „faltiger" Oberflächenstrukturen vor sich (Fig. 6).

Die Substrattemperatur ist nur von geringem Einfluß auf die Orientierungsgüte und diese muß stets in ihrer Schichtdickenabhängigkeit diskutiert werden.

Die Versuche mit Aufdampfraten von 140 Å/s brachten zwar eine erhebliche Vergrößerung der Kristallite mit sich. Es werden jedoch auch neben der (001)-Orientierung andere Auflageflächen des Antimon gleich wahrscheinlich. Nach neuesten Untersuchungen (Mietz, Thieme) kann die Vergrößerung des Kristallitdurchmessers durch einen neuartigen Mechanismus erklärt werden. Zur Herstellung einkristallin orientierter Filme mit geringer Zahl der Korngrenzen pro Flächeneinheit erscheint es notwendig, die Filmbildung unter UHV-Bedingungen bei Verwendung geringerer Aufdampfraten durchzuführen.

Literatur

Bahadur, K., Chaudhary, K. L.: Appl. Phys. Lett. **15**, 9, 277 (1969)

Berndt, K.: J. appl. Physics **37**, 3841 (1966)

Condas, G. A.: Rev. Scient. Instr. **33**, 9, 987 (1962)

Condas, G. A., Wooten, F. O.: J. appl. Physics **32**, 323 (1961)

Cremer, E., Ruedl, E.: in Auwärtner, M. Ergebnisse der Hochvakuumtechnik und der Physik dünner Schichten, Stuttgart 1957, S. 49

Götzberger, A.: Z. Physik **142**, 182 (1955)

Kleber, W.: Forschen u. Wirken, Festschrift zur 150-Jahrfeier der Humboldt-Universität, Bd. II, 271, Berlin 1960

Koike, R., Kurokowa, H.: Jap. J. appl. Physics **6**, 546 (1967)

Mietz, I., Thieme, H.: unveröffentlicht

Palatnik, L. S., Komnik, J. F.: DAN **4**, 124, 808 (1959)

Palatnik, L. S., Kossewitsch, W. M.: Kristallografija **4**, 42 (1959)

Palatnik, L. S., Gladkich, N. T.: DAN **3**, 140, 567 (1961)

Palatnik, L. S., Kossewitsch, W. M., Moskalew, W. M.: Fis. Metall. **16**, 723 (1963)

Palatnik, L. S.: in Grundprobleme der Physik dünner Schichten, Göttingen 1966, S. 92

Palatnik, L. S., Kossewitsch, W. M., Antonowa, W. A.: in Grundprobleme der Physik dünner Schichten, Göttingen 1966, S. 104

Robbins, K. G., Thomas, J. M.: NATURE **217**, 1251 (1968)

Ruedl, E.: VACUUM VII u. VIII, 56 (1957/58)

Semenow, N. N.: Zh. Obšč. Fiz. Chim. **62**, 33 (1930)

SHEFTAN, R. N.: Kristall u. Technik **3**, (4) K 65 (1968)
SPRENGER, H.: Diplomarbeit, Humboldt-Universität Berlin 1970
SUGANUMA, R. u. TANAKA, Y.: Jap. J. appl. Phys. **4**, 232 (1965)

(Eingegangen am 20. Mai 1971)

Anschrift der Verfasser:

Dr. I. MIETZ
Dipl.-Krist. H. SPRENGER
Humboldt-Universität zu Berlin
Sektion Physik, Bereich Kristallographie
104 Berlin
Invalidenstr. 43

Kristall und Technik	7	1–3	1972	133–140

A. N. Stepanova, N. N. Sheftal'

Institute of Crystallography, Academy of Sciences, Moscow, USSR

The Influence of $AsCl_3$ Impurity on Growth Mechanism and Structure Perfection of Autoepitaxial Germanium Films

The changes of film morphology and microtwin concentration in these films caused by $AsCl_3$ impurity have been studied. The correlation of growth mechanism and structure perfection of the films has been revealed.

Es werden die durch Verunreinigungen von $AsCl_3$ verursachten Veränderungen der Filmmorphologie und der Häufigkeit der Mikrozwillinge in diesen Filmen untersucht und die Beziehungen zwischen Wachstumsmechanismus und Strukturperfektion der Filme aufgezeigt.

1. Introduction

The problem of the effect of impurities on growth mechanism of crystals is extremely complex and poorly understood. In spite of some important contributions the problem is far from being solved. The object of the present research is to trace the gradual change of growth mechanism of germanium films due to impurity. Structure perfection (respecting microtwins) of the films was also examined. It is certainly possible to avoid appearance of defects in films by careful treatment of the substrate surface before film growing. However, the observation of growth figures associated with defects in our mind can favour the understanding of growth mechanism. For that reason the substrates were in some experiments deliberately oxidized to obtain a certain quantity of defects in the films.

2. Experimental technique

The germanium films were grown by means of hydrogen flow reduction of $GeCl_4$ in a vertical reactor (Theuerer). $AsCl_3$ as an impurity was introduced into the gas phase from a separate saturator. The mole fraction of the impurity in relation to $GeCl_4$ was changed from 0,02 to 0,2. All experiments were carried out under the following conditions: the mole fraction of $GeCl_4$ in hydrogen was $2{,}6 \cdot 10^{-3}$, the substrate temperature was 850° or 880 °C, the gas flow rate in the cold part of the tube was $1{,}4\ \mathrm{cm \cdot s^{-1}}$. Morphology and structure defects were examined by interferential contrast light microscopy and by electron microscopy.

3. Experimental Results

3.1. Structure Defects

The influence of the $AsCl_3$ impurity on microtwin concentration has been studied (for the growth temperature 880 °C).

(111)-films.

Two types of microtwins are observed in these films: microtwin lamellae (with inclined {111} twin boundaries (Booker) and second order twins (the first twinning occurs on the substrate surface and the second twinning occurs on inclined {111} planes). The latter results in tripyramids on the film surface (Booker). The lamellae concentration is usually far less than that of the tripyramids. Therefore the total microtwin concentration is essentially the tripyramid one. Under the influence of $AsCl_3$ the tripyramids disappear. This phenomenon is more pronounced in the case of deliberately oxidized substrates, as seen in Figure 1, where the dependence of tripyramid concentration on $AsCl_3$ mole fraction in the gas phase is shown.

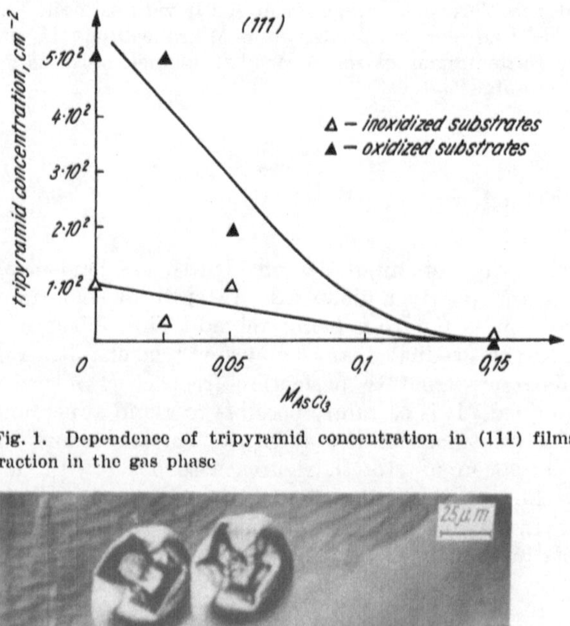

Fig. 1. Dependence of tripyramid concentration in (111) films on the $AsCl_3$ mole fraction in the gas phase

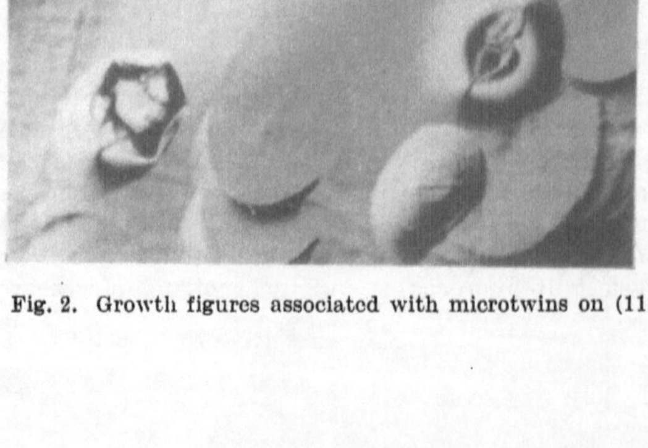

Fig. 2. Growth figures associated with microtwins on (110) undoped films

(110)-films.

The specific growth figures resulting from some defects are observed on the surface of (110) films (Fig. 2). Investigation by transmission electron diffraction reveals that these figures result from first order twins with inclined {111} twin boundaries. The dependence of microtwin concentration on the mole fraction of $AsCl_3$ in the gas phase is shown in Figure 3 for the cases of oxidized and inoxidized substrates. The higher $AsCl_3$ concentration, the higher the microtwin concentration, this increase being especially high in the case of inoxidized substrates. The lamellae concentration is far less, therefore the total concentration in Figure 2 is essentially the concentration of the above mentioned twins.

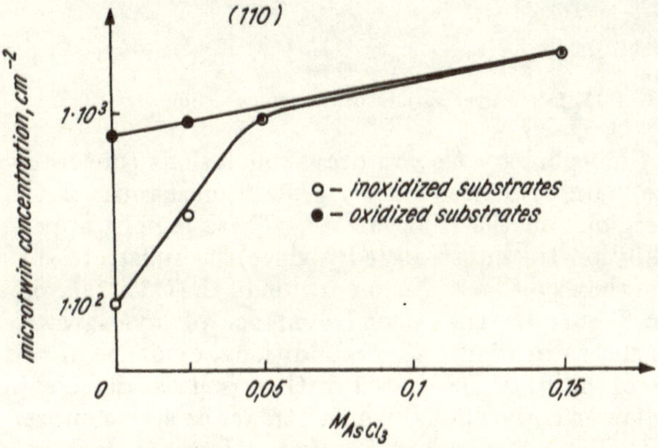

Fig. 3. Dependence of microtwin concentration in (110) films on the $AsCl_3$ mole fraction in the gas phase. (The points at 0.05 and 0.15 M_{AsCl_3} are superposed, i.e., valid for oxidized and inoxidized substrates)

(100)-films.

In these films microtwin lamellae with {111} twin boundaries were observed (Kostjuk). Structural etching shows that these lamellae have different widths. One can assume that the thinnest lines correspond to stacking faults. However, we failed confidently to distinguish them. Therefore one can only claim about (100) films that the total concentration of microtwins and stacking faults decreases with increase of impurity concentration; however, microtwins are the main part of this sum.

3.2. Growth mechanism

In the theory of crystal growth two major mechanisms are distinguished: layer-growth (or lateral) mechanism when the growth proceeds by lateral motion of the layers (steps), and the normal-growth mechanism when the surface advances normal to itself without needing steps. We also believe, according to Cahn that intermediate cases are possible, wherein there are contributions of both of these mechanisms to the whole growth mechanism. Therefore we can regard the changes of each mechanism contribution to the total growth mechanism when affected by changes of growth conditions, in particular, by impurities.

Fig. 4. (111) face on the surface of the undoped film

On the base of morphology we can draw conclusions concerning the growth mechanism. The main evidence of layer-growth mechanism is the appearance of smooth face regions on the film surface. These regions appear at the pole of the face, usually on the substrate edge since the substrate surface deviates within 1—2° from the exact face. Such a region of the (111) face on the undoped film is shown in Figure 4. The second evidence of layer-growth mechanism are growth steps shown in Figure 4. An additional evidence of this mechanism is the existence of hillocks associated with structure defects (for example, microtwins lamellae or tripyramids) which can act as step sources. Hillocks on the undoped (111) film are shown in Figure 5. However, lack of hillocks does not yet mean lack of layer-growth mechanism. With a certain relation of nucleation frequency to step spreading rate the layer arising at the defect reaches the substrate edges before arising of a second layer, and as a sequence the hillock does not form.

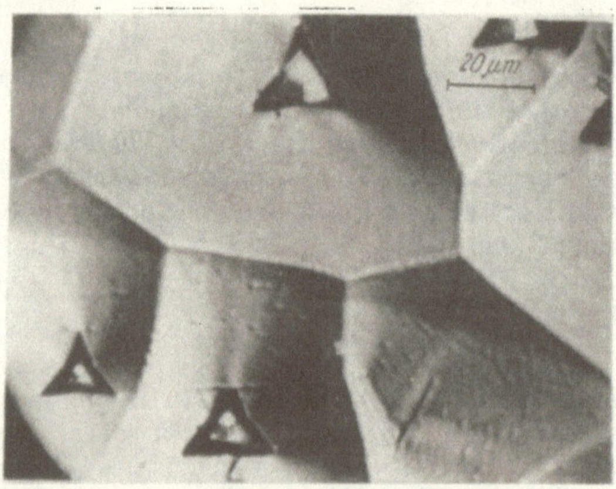

Fig. 5. Hillocks associated with tripyramids and microtwin lamellae on undoped (111) film

Consider the influence of $AsCl_3$ impurity on growth mechanism of (111), (110) and (100) germanium films:

(111)-films.

The morphology indicates that in absence of impurities the (111) face grows by layer-growth mechanism (Fig. 4 and 5). The mechanism is essentially the same in the presence of $AsCl_3$; the predominance of the lateral over normal growth seems to become even stronger. The latter can be demonstrated as follows. As mentioned above tripyramids do not form in the presence of $AsCl_3$. We performed a two-stage process. At first we grew a film in absence of $AsCl_3$ and, in this case, the film contained tripyramids. Then the film was overgrown in presence of $AsCl_3$ and in this case tripyramids extended into the doped films. But the hillocks associated with tripyramids become lower and have a plateau. It means, in our mind, that lateral growth becomes stronger on this face.

$AsCl_3$ also causes anisotropy of spreading of growth steps over this face. The smooth face region has a rounded shape on the undoped film (Fig. 4) and a triangular one on the doped film (Fig. 6).

Fig. 6. (111) face on the surface of doped film. The mole fraction of $AsCl_3$ (M_{AsCl_3}) in the gas phase is $4 \cdot 10^{-2}$ (in relation to $GeCl_4$)

(110)-films.

On the surface of the undoped (110) films all features of layergrowth mechanism are seen. There are regions of smooth face (110), growth steps

Fig. 7. (110) face on the surface of the slightly doped film. $M_{AsCl_3} = 4 \cdot 10^{-4}$

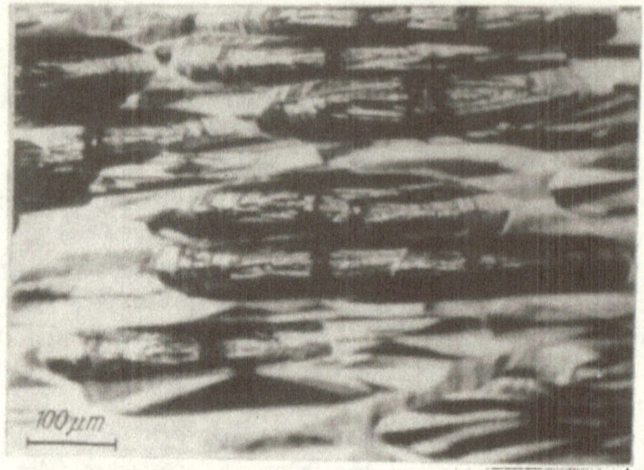

Fig. 8. Growth figures caused by microtwins on the (110) doped film. $M_{AsCl_3} = 0{,}15$

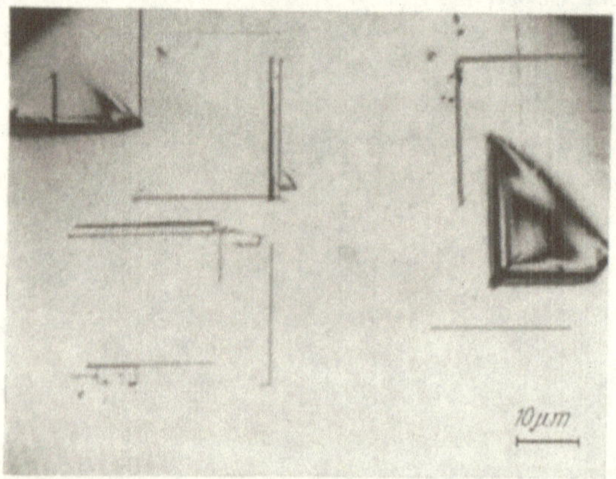

Fig. 9. Microtwin lamellae on the (100) undoped film

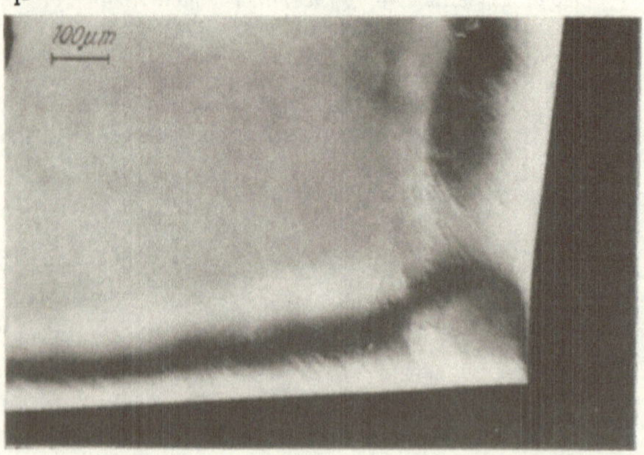

Fig. 10. (100) face on the surface of the doped film. $M_{AsCl_3} = 4 \cdot 10^{-4}$

and hillocks associated with microtwins (Fig. 7[1]) and 2). $AsCl_3$ causes gradual transition from layer-growth to chain-growth mechanism. Morpho-

[1]) In Figure 7 the smooth (110) face on the surface of the slightly doped film is shown. The faces have essentially a similar shape on the surface of undoped films.

logically this is confirmed by increasingly elongation of the hillocks associated with microtwins (Fig. 8) and by the lack of regions of the smooth face (110) on the films grown at high concentrations of $AsCl_3$.

(100)-films.

On (100) films, grown without impurity all features of layergrowth mechanism are absent: 1) the smooth face does not appear even on thick films (up to about 200 μm whereas the usual thickness of the films was about 20 μm); 2) growth steps do not reveal even if using an electron microscope decoration technique; 3) the hillocks around microtwin lamellae are absent (Fig. 9). However, all these features appear if $AsCl_3$ was introduced into the gas phase. If the $AsCl_3$ concentration is small the (100) smooth face has a rounded shape (Fig. 10). At large $AsCl_3$ concentrations anisotropy for the steps spreading

Fig. 11. (100) face on the surface of the doped film. $M_{AsCl_3} = 4 \cdot 10^{-2}$

Fig. 12. Growth figures caused by microtwin lamellae on the (100) doped film. $M_{AsCl_3} = 2{,}5 \cdot 10^{-2}$

appears as in the case of (111) and (110) faces. Consequently the face takes a square shape (Fig. 11). Hillocks associated with microtwin lamellae on the films, grown in the presence of $AsCl_3$, are shown in Figure 12. So, when the growing Ge film is affected by $AsCl_3$ it follows:

1) Layer-growth mechanism of the (111) face remains valid and seems even to enhance.

2) Layer-growth mechanism of the (110) face changes into chain-growth mechanism.

3) Layer-growth mechanism of the (100) face enhances so that all its features appear in the presence of $AsCl_3$, whereas they are absent on the films grown without impurity.

4) Anisotropy for the step spreading on these faces appears. The ansisotropy doesn't impede the layer growth of the (111) and (100) faces, only substituting rounded steps by straight ones in two or three directions for the (100) or (111) respectively. However, it is the anisotropy that causes the change of layer growth to chain growth of the (110) face according to the unique pronounced direction of the periodic bond chains (HARTMAN, PERDOK) on this face along ⟨110⟩ (taking into account the bonds between neighbours of the first and second order).

4. Discussion

Comparison of the influence of $AsCl_3$ impurity on growth mechanism of the faces and on their structure perfection reveals the following. If the layer growth enhances by the impurity, the microtwin concentration decreases (as for the (111) and (100) faces). If, however, the layer growth weakens, the microtwin concentration increases (as for the (110) face). This conclusion is confirmed by experiments with PCl_3 impurity. For example, PCl_3 does not change morphology of the (110) film and, hence, does not change its growth mechanism (at least up to mole fraction $PCl_3 \simeq 0{,}25$). As a result the microtwin concentration in these films also does not change.

5. Conclusion

It is found that there exists a correlation between the change, due to impurity ($AsCl_3$), of growth mechanism and structure perfection (respecting microtwins) for (111), (110) and (100) germanium films grown in chloride process.

The authors express thanks to N. D. ZAKHAROV and V. I. MURATOVA for electronmicroscopic investigations and to L. N. OBOLENSKAJA for experimental assistance.

References

BOOKER, G. R.: Disc. Farad. Soc. N 38, 298 (1964)

BOOKER, G. R.: Phil. Mag. 11, 1007 (1915)

CAHN, J. W.: Acta Metallurgica 8, 554 (1960)

HARTMAN, P., PERDOK, W. G.: Acta Cryst. 8, 49 (1955)

KOSTJUK, JU. G.: The All-Union Symposium on the processes of growth and synthesis of semiconductor compound crystals and films (Report abstracts), Novosibirsk 1969, p. 136

THEUERER, H.: J. Electrochem. Soc. 108, 649 (1961)

(Received June 11, 1971)

Authors' address:

Prof. N. N. SHEFTAL', A. N. STEPANOVA
Institute of Crystallography
Academy of Sciences
Leninsky prospect 59
Moscow, B-333
USSR

| *Kristall und Technik* | 7 | 1–3 | 1972 | 141–158 |

G. AMTHAUER, M. GEBHARDT, A. NEUHAUS

Mineralogisch-Petrologisches Institut der Universität Bonn

Phasen und Epitaxien in gasnitrierten polykristallinen Fe-Blechen und einkristallinen Fe-Folien[1])

Es wurden die Phasen- und Epitaxiebeziehungen nitrierter Eisenbleche und -Folien mittels Röntgen- und Elektronenbeugungsmethoden untersucht. Dabei wurden die folgenden, in zeitlicher Bildungsfolge aufgeführten Nitrierphasen gefunden: α Fe(N) (als feste Lösung) – Fe_4N (γ'-Phase) – ε-Phasen (mit ε-Fe_3N und ε-Fe_2N) – ζ-Fe_2N.

Alle genannten Nitrierphasen gehen, laufende N-Zufuhr vorausgesetzt, durch epitaxisch initiierte Gitterumbaureaktionen auseinander hervor (= Topotaxie).

Es konnten epitaxische Verwachsungen aller in der obigen Bildungsfolge benachbarten Fe-Nitride miteinander und über Epitaxiekette auch mit αFe nachgewiesen werden und zwar: αFe mit $Fe_4N(\gamma')$; $Fe_4N(\gamma')$ mit ε-Fe_3N; ε-Fe_3N mit ε-Fe_2N und ε-Fe_2N mit ζ-Fe_2N. Alle zugehörigen Verwachsungsgesetze konnten festgelegt werden. Von besonderer Aussagekraft erwiesen sich hierbei nitrierte Fe-Einkristallfolien von $\approx$600 Å Dicke, die durch Vakuumaufdampfung auf (001)-NaCl bei 475 °C gezüchtet worden waren.

The phase relations and the rules for epitaxial intergrowth of nitrogenized Fe sheets and thin Fe films have been studied by x-ray and electron diffraction. In the course of this study the following sequence of nitrogenized phases has been found: αFe(N) (as a solid solution) – Fe_4N (γ'-phase) – ε-phases (with ε-Fe_3N and ε-Fe_2N) – ζ-Fe_2N.

If there is sufficient nitrogen available, all nitrogenized phases listed above are formed from one another by epitactically initiated lattice-reconstructions (= topotaxy).

We found epitaxial intergrowth of all Fe-nitrides of the above series of phases with respect to the next member, and by epitaxychain with αFe, namely: αFe with $Fe_4N(\gamma')$; $Fe_4N(\gamma')$ with ε-Fe_3N; ε-Fe_3N with ε-Fe_2N and ε-Fe_2N with ζ-Fe_2N. The rules governing all these epitaxies are given. Especially valuable informations could be gained from thin nitrogenized Fe-films of $\approx$600 Å thickness, which were grown by vacuum evaporation at 475 °C onto (001) faces of NaCl.

1. Einführung

Für viele technische Verwendungsmöglichkeiten der Metalle und ihrer Legierungen sind die Eigenschaften ihrer Oberflächenzone von weit größerer Wichtigkeit als ihre Massiveigenschaften. Für solche Verwendungszwecke wurden chemisch bzw. mechanisch wenig widerstandsfähige, billige Konstruktionsmetalle daher voll brauchbar werden, wenn es gelänge, ihre Oberflächenzone geeignet zu veredeln, sie zu „vergüten". Zu diesen Vergütungsverfahren gehört u. a. das hier interessierende „Nitrieren". Man versteht darunter nach DIN 17014 die Behandlung des Werkstoffs in Stickstoff abgebenden Reagentien

[1]) Nach Experimenten von G. AMTHAUER, Diplomarbeit, Bonn 1970.

zum Erzielen einer mit Stickstoff angereicherten Oberflächenschicht (Koch). Obwohl das Nitrieren von Eisen und Eisenlegierungen dem seit langem geübten „Carburieren" sehr verwandt ist, erlangte es als Vergütungsmethode für die Eisen- und Stahl-Metallurgie erst in neuester Zeit Bedeutung (Paranjpe, Cohen, Bever, Floe). Zugleich mit der schnell wachsenden technischen Bedeutung des „Nitrierens" wurde auch das Zweistoffsystem Fe—N Gegenstand vieler Untersuchungen.

2. Bisheriges Schrifttum zum Nitrieren von Eisen und zum Zweistoffsystem Fe—N

2.1 Nitrieren und Nitrierungsschichten des Eisens

Wie schon Hägg und Lehrer fanden, reagiert α-Fe nur träge mit molekularem, aber spontan mit atomarem Stickstoff. Der benötigte atomare Stickstoff wird dabei durch Zerlegung von N-haltigen Verbindungen, insbesondere NH_3 und N_2 gewonnen. Die unter katalytischer Mitwirkung der Fe-Oberfläche ablaufende Reaktion des Fe mit atomarem N läßt sich durch folgenden Ablauf beschreiben:

$$Fe + NH_3 \rightarrow Fe_xN + H_2 .$$

Die Reaktion verläuft bei höherer Temperatur (z. B. 500 °C) lebhaft nach rechts unter Eindiffundierung des Hauptteils des freiwerdenden atomaren N in die Fe-Oberfläche und Bildung von Nitridphasen und festen Fe(N)-Lösungen.

Zur Gewinnung der gewünschten Nitrierschichten werden in der heutigen Technik vor allem vier Verfahren benutzt, das Gasnitrieren (im NH_3- bzw. N_2—H_2-Gasgemisch bei 490 °C bis 520 °C), das Badnitrieren (in CN-haltigen Salzschmelzen bei z. B. 570 °C), das Glimmnitrieren (Erzeugung von atomarem N durch Glimmentladung z. B. in einem N_2—H_2-Gasgemisch von einigen Torr Gasdruck bei 500 bis 600 °C) und ganz neuerdings das Plasma-Nitrieren (durch Zerlegung von NH_3 bzw. eines N_2—H_2-Gasgemisches unter Plasmabedingungen).

Bei allen vier Verfahren dringt der aktive Stickstoff wechselnd tief in die Oberflächenzone des Metalls ein und verändert sie erheblich, wie schon ein Anschliffbild zeigt. Es lassen sich i. allg. zwei sehr unterschiedliche Nitrierzonen erkennen, eine obere, helle, stark gehärtete und gegenüber dem Ausgangsmetall strukturell weitgehend veränderte, aus Nitridphasen bestehende Zone („white layer" des angelsächsischen Schrifttums) und darunter eine zweite Zone, die sich nach Schliffbild und Mikrohärte nur mäßig vom unveränderten Ausgangsmetall unterscheidet. Wir wollen die erstere Zone, gemäß Schrifttum, als Nitrid-Zone, die letztere als Diffusionszone bezeichnen.

Beide Zonen sind röntgenographisch mehrfach auf ihre Phasenzusammensetzung untersucht worden (Finnern; Bierwirth; Noren, Kindborn; Meyer, Hohbrock). Hiernach besteht die obere Nitrierzone, die Nitrid-Zone (white layer) für reine Eisenpräparate, etwa nach dem Gas- oder Glimm-Verfahren erzeugt, hauptsächlich aus ε-Nitriden; nur untergeordnet tritt auch Fe_4N (γ'-Phase) auf. Anders bei der Nitrierung mittels des Badverfahrens. Bei ihm bilden sich, wegen des hohen CN-Anteils der üblichen Bäder, neben Nitriden stets auch Carbide. In der vorliegenden Arbeit interessieren nur reine Nitrierungen.

Die zweite Nitrierzone, die Diffusionszone besteht im wesentlichen aus Stickstoff-Ferrit; untergeordnet kann auch in dieser Zone Fe_4N (γ'-Phase) auftreten. Das gilt nach FINNERN besonders für den Kontaktbereich mit der Nitrid-Zone.

2.2. Kristallphasen des reinen Zweistoffsystems Fe—N

Die älteren Arbeiten zum Phasendiagramm Fe—N lieferten zwar recht unterschiedliche Ergebnisse (HÄGG; FRY; SAWYER; EPSTEIN, CROSS; LEHRER; EISENHUT, KAUPP), dennoch führten sie bereits zu einem recht detaillierten und wohl begründeten Phasendiagramm (LEHRER). LEHRERS Phasendiagramm wurde zwar durch spätere Arbeiten vor allem von JACK (1950, 1951) und PARANJPE et al. über Teilgebiete des Zweistoffsystems Fe—N noch mehrfach überprüft und ergänzt, blieb in seinen Grundzügen aber erhalten. Im folgenden wird das Phasendiagramm LEHRER—JACK in der Fassung von JACK (1952) zugrunde gelegt (vgl. Abb. 1). Es befindet sich in gutem Einklang mit den Ergebnissen dieser Untersuchung.

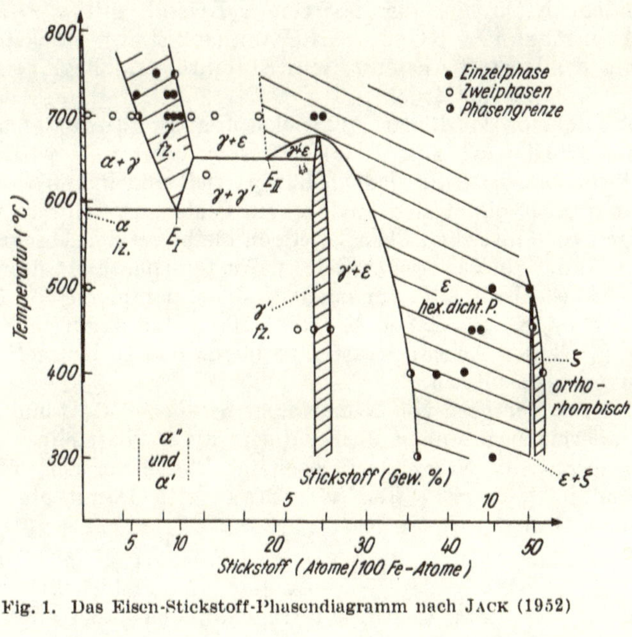

Fig. 1. Das Eisen-Stickstoff-Phasendiagramm nach JACK (1952)

Das Phasendiagramm Figur 1 zeigt zwei eutektische Punkte, E_I und E_{II} (von den Autoren dieser Arbeit hinzugefügt). E_I liegt bei 591° $\pm$ 4 °C und 2,35 $\pm$ 0,1 Gew.-% Stickstoff (LEHRER), E_{II} bei 650° $\pm$ 4 °C und 4,55 $\pm$ 0,2 Gew.-% Stickstoff (LEHRER; PARANJPE et al.). Die erstere eutektische Reaktion entspricht der Umwandlung $\gamma \rightleftharpoons \alpha + \gamma'$ bei 591 °C mit γ = Stickstoff-Austenit, α = Stickstoff-Ferrit und $\gamma' = Fe_4N$. Die zweite eutektoide Reaktion entspricht nach LEHRER und PARANJPE der Umwandlung $\varepsilon \rightleftharpoons \gamma + \gamma'$ (650 °C). Hierbei bezeichnet ε ein Eisennitrid von hexagonaler Struktur; γ bzw. γ' sind wiederum Stickstoff-Austenit bzw. Fe_4N. Der Diagrammpunkt der direkten, d. h. ohne Phasenzerfall erfolgenden Umwandlung der γ'-Phase in die ε-Phase liegt nach LEHRER (gemäß Figur 1) bei $T = 672° \pm 8$ °C und $5{,}7 \pm 0{,}2$ Gew.-% Stickstoff.

Im folgenden sollen anhand des Phasendiagramms (Fig. 1) die vorstehend aufgeführten und einige weitere bekannt gewordene Fe—N-Phasen kurz beschrieben werden. Namen und Bezeichnungen dieser Fe—N-Phasen sind denen des Zweistoffsystems Fe—C weitgehend analog, ihre Strukturen mit ihnen z. T. auch isotyp.

α-Fe(N) = Stickstoff-Ferrit: Bei dieser Phase handelt es sich um eine feste Lösung von atomarem N im iz-kubischen α-Fe. N besetzt dabei oktaedrische Gitterlücken des Fe-Gitters. Die Gitterkonstante dieses interstitiellen Mischkristalls ändert sich nur wenig mit steigendem N-Gehalt und zwar steigt a_0 des reinen α-Fe mit steigendem N-Gehalt (PARANJPE et al.; DIJKSTRA). Die Löslichkeit des α-Fe für N nimmt mit steigender Temperatur zu und erreicht beim eutektischen Punkt E_I mit $\sim$0,1 Gew.-% N ihr Maximum.

Auch γ-Fe(N) = Stickstoff-Austenit ist eine feste Lösung von atomarem Stickstoff in fz-kubischem γ-Fe mit N statistisch verteilt in oktaedrischen Lücken des Fe-Gitters. Hierbei steigt die Gitterkonstante des reinen γ-Fe mit wachsendem N-Einbau vom N-freien γ-Kristall mit $a_0 = 3{,}564$ Å auf $a_x = 3{,}635$ Å für maximal 2,5 Gew.-% N (PARANJPE et al.). Stickstoff-Austenit ist isotyp mit Kohlenstoff-Austenit, sein Stabilitätsfeld liegt gemäß Figur 1 bei $T \geqq 591$ °C (für Normaldruck).

α'-Phase = Stickstoff-Martensit: An Stickstoff stark übersättigtes, tetragonal deformiertes α-Fe-Gitter (JACK 1951). Hierbei besetzt N wiederum oktaedrische Gitterlücken des α-Fe, jedoch nur partiell und in tetragonaler Gruppierung. N-Martensit bildet sich analog dem C-Martensit durch Abschrecken der zugehörigen austenitischen Phase. Jedoch bleibt bei der Abschreckung von N-Austeniten i. allg. ein gewisser Anteil an Austenitphase erhalten. Die Umwandlung $\gamma \rightarrow \alpha'$ ist also i. allg. nur partiell. Zellenparameter von N-Martensit für $x = 3{,}5$ Gew.-% N: $a_x = 2{,}850$ Å, $c_x = 3{,}106$ Å und damit $c/a = 1{,}09$. Für $x \leqq 2{,}4$ Gew.-% N des Ausgangsaustenits bleibt das Gesamtgefüge des Abschreckproduktes austenitisch.

α''-Phase: Durch Tempern von N-Martensit bei $T < 250$ °C bildet sich nach JACK (1951) abermals eine neue Phase, die er als α'' bezeichnet. Sie erwies sich als metastabile, mit N-Martensit isocheme Phase, die sich erst in sekundärer Reaktion in N-Ferrit + $Fe_4N(\gamma')$ umwandelt. Damit ergibt sich folgende Gesamtreaktion:

$$\alpha' \xrightarrow{\text{isocheme U.}} \alpha'' \xrightarrow{\text{Zerlegungs-U.}} \alpha + \gamma' \quad \text{(für } T < 250\text{ °C und Normaldruck).}$$

Die α''-Phase ist, wie die α'-Phase, iz-tetragonal, weist aber eine Elementarzelle mit 8 leicht verzerrten iz. Struktureinheiten des Ausgangsmartensits auf. α'' entsteht nach DIJKSTRA auch beim Anlassen abgeschreckter übersättigter N-Ferrite.

γ'-Phase = Fe_4N: Sie besitzt nach JACK (1951) eine kubische Struktur gemäß Figur 2 mit Fe als fz-kub. Teilgitter und atomarem N in den zentralen Oktaederlücken der Fe-Zelle. Dieser Strukturvorschlag wurde von PARANJPE et al.; DVORIANKINA et al. und NAGAKURA bestätigt. Die γ'-Phase besitzt ein schmales Homogenitätsgebiet zwischen 5,7 und 6,1 Gew.-% N (Fig. 1).

ε-Phasen: Strukturen und Homogenitätsgebiete dieser Phasen überdecken gemäß Figur 1 ein breites, stark temperaturabhängiges Zustandsfeld zwischen $\approx$4 und $\approx$11 Gew.-% N. All diese ε-Phasen weisen ein hexagonal-dicht ge-

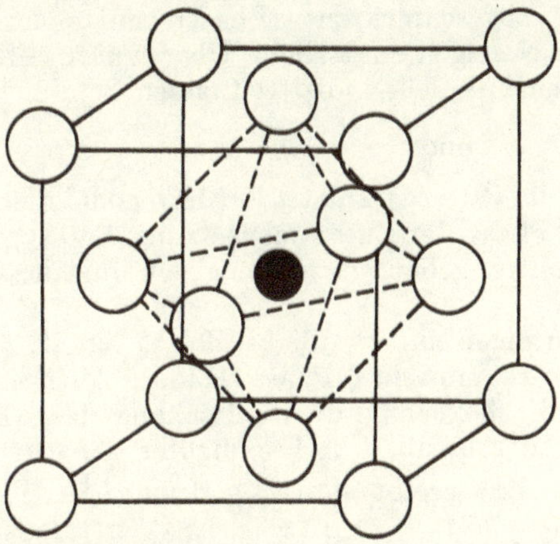

Fig. 2. Struktur von Fe_4N (γ'-Phase)

packtes Fe-Teilgitter mit atomarem N in Oktaederlücken auf (Jack 1952; Hendricks et al.). Der maximale N-Gehalt des ε-Feldes beträgt $\approx$11 Gew.-% = 33,1 Mol-%, sein minimaler N-Gehalt hängt, wie Figur 1 zeigt, stark von der Temperatur ab. Er beträgt für $T \approx 300$ bis 400 °C $\approx$ 8 Gew.-% N, oberhalb $\approx$650 °C hingegen nur $\approx$4 Gew.-% N. Innerhalb des Zustandsgebietes der ε-Phasen liegt bei 7,72 Gew.-% N = 25,0 Atom-% N die Phase ε-Fe_3N sowie bei 11,1 Gew.-% N = 33,33 Atom-% N die Phase ε-Fe_2N.

Einen Strukturausschnitt letzterer 2 Phasen in Form einer Projektion zweier einander in z-Richtung folgender N-Ebenen auf (0001) mit A-Schicht in Höhe $z = 1/2$ und B-Schicht in Höhe $z = 0$ zeigt Figur 3. Alle 3 N-Positionen dieser Abbildung entsprechen Oktaederlücken des hexagonal-dicht gepackten Fe-Gitters. Von ihnen sind bei ε-Fe_2N alle 3 N-Positionen besetzt, bei ε-Fe_3N nur 2 (ausgefüllte + schraffierte Kreise). Das ergibt für die N-Schichten von ε-Fe_3N eine derart geordnete Drittelbesetzung aller vorhandenen Oktaederlücken, daß die 6 unmittelbaren Nachbarpositionen eines N innerhalb der Schicht, sowie die Positionen unmittelbar über und unter dem Bezugs-N leer bleiben.

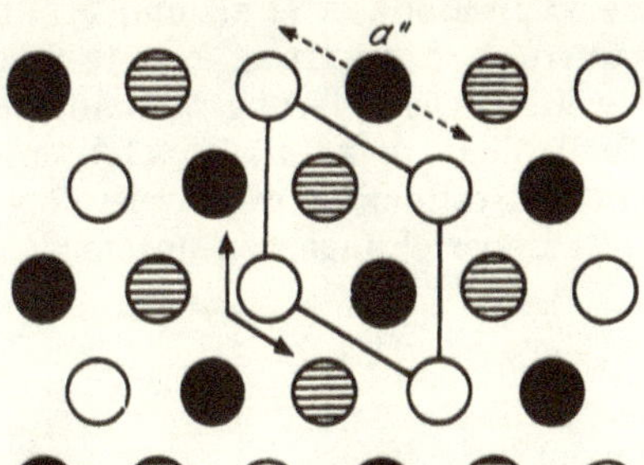

Fig. 3. Lagen der Stickstoffatome in ε-Fe_3N und ε-Fe_2N nach Jack;

ausgefüllte Kreise in A-Ebenen:	besetzt für Fe_2N und Fe_3N
schraffierte Kreise in B-Ebenen:	besetzt für Fe_2N und Fe_3N
offene Kreise in B-Ebenen:	besetzt für Fe_2N aber unbesetzt für Fe_3N

Röntgendiagramme von abgeschreckten ε-Phasen mit einer Zusammensetzung nahe Fe_4N und Fe_2N zeigen zusätzliche Überstrukturreflexe, die sich mit Hilfe der folgenden Elementarzelle indizieren lassen:

$$a_0'' = 2\sqrt{3} \cdot a_0 \qquad \text{und} \qquad c_0'' = c_0 .$$

Hierbei bedeuten a_0 und c_0 die Gitterparameter der hexagonal dichten Packung der Fe-Atome. Für eine ε-Phase der Zusammensetzung $Fe_{24}N_{10}$ konnte JACK (1952) einen Strukturvorschlag geben, nicht hingegen für das ε-Nitrid der Zusammensetzung $\approx Fe_{24}N_6$.

ζ-Fe_2N: Bei Fe—N-Legierungen mit $x > 33{,}24$ Mol-% N tritt abermals eine neue Struktur auf, die ortho-rhombische ζ-Phase (HÄGG). Auch in ihr behalten die Fe-Atome angenähert die hexagonal-dichte Packung der ε-Phasen, doch sind die N-Atome entlang der o-rhb. b-Achse dichter gepackt als sich für $a\sqrt{3}$ hex. errechnen würde. Das ergibt a(o-rhb.) gleich a(hex.) + 0,1% aber b(o-rhb.) deutlich größer als $a\sqrt{3}$ (hex.) (+ 1%). Diese Ergebnisse von JACK sind zwar mehrfach bestätigt worden (PARANJPE et al.; BRIDELLE; BURDESE; NAGAKURA et al.), doch gibt es auch abweichende Deutungen (PINSKER).

2.3. Epitaxiebeziehungen

Schon FRY beobachtete, daß die Fe_4N-Nitridnadeln sich häufig nach den kristallographischen Richtungen der ferritischen Grundmasse ausscheiden. Er konnte aber noch keinen genauen Orientierungszusammenhang angeben. Das gelang erst auf Grund besserer Kenntnis der Strukturen der Kristallphasen (PITSCH; BOOKER et al.; MEHL et al.; SCHNEIDER).

Nach Untersuchungen von PITSCH wird durch Anlassen von abgeschreckten übersättigten Ferriten unterhalb von 250 °C α''-$Fe_{16}N_2$, oberhalb dieser Temperatur γ'-Fe_4N ausgeschieden. Er findet dabei folgende Orientierungen:

1) Für α''/α (001)-$\alpha''//(001)$-α mit $[100]$-$\alpha''//[100]$-α

2) Für γ'/α a) $(\bar{1}1\bar{2})$-$\gamma'//(102)$-α mit $[1\bar{1}0]$-$\gamma'//[010]$-α

b) $(\bar{1}11)$-$\gamma'//(201)$-α mit $[1\bar{1}0]$-$\gamma'//[010]$-α

Die Orientierung 1 stimmt mit den von BOOKER et al. und MEHL et al. angegebenen Verwachsungsbefunden überein. SCHNEIDER konnte auch die Orientierung 2b bestätigen. Da über weitere Orientierungszusammenhänge in Nitrierungsschichten in der Literatur[1]) nichts bekannt ist, stellten sich die Autoren dieser Arbeit die Aufgabe, die Kristallphasen des Zweistoffsystems Fe—N auf mögliche und tatsächliche Epitaxiebeziehungen zu untersuchen und deren Orientierungsgesetze festzulegen.

3. Eigene Untersuchungen

3.1. Probenherstellung

Für die Probenherstellung sind naturgemäß zwei Arbeitsgänge erforderlich, die Präparation der zu nitrierenden Fe-Bleche bzw. Fe-Folien und das Nitrieren.

[1]) Vgl. jedoch NEUHAUS und GEBHARDT, 1968.

Präparation der Fe-Bleche und Fe-Folien

Während für die Röntgenuntersuchungen handelsübliche Eisenbleche benutzt werden konnten, mußten für die Versuche mit transmittierender Elektronenbeugung durchstrahlbare Eisenfolien besonders hergestellt werden. Das geschah durch Aufdampfen von metallischem Fe auf geeignete Trägeroberflächen in einer Hochvakuum-Aufdampfanlage Ba 360 der Firma Balzers bei 10^{-5} Torr. Das verwendete Fe besaß einen Reinheitsgrad von 99,99%. Für alle Durchstrahlungsaufnahmen erwies sich eine Foliendicke von ≈600 Å als gut geeignet.

Für die Präparation der **polykristallinen** Fe-Folien wurde Fe im obigen Hochvakuum teils auf normal temperierte Kollodiumfolien, teils auf (001)-NaCl bei tiefen Temperaturen aufgedampft. Um **einkristalline** Fe-Folien züchten zu können, wurde die Hochvakuum-Aufdampfung von Fe nach dem Prinzip der epitaxischen Impfung durchgeführt, d. h. durch Hochvakuum-Aufdampfung von Fe auf einen strukturell günstigen und geeignet aktivierten Trägerkristall (NEUHAUS 1950). Als geeigneter Träger bietet sich, gemäß Literatur (BRÜCK; NEUHAUS 1950 u. 1952) (001)-NaCl an, als günstige Aktivierungstemperatur eine Temperatur zwischen ≈400 °C und ≈500 °C. Um letztere Experimentalforderung erfüllen zu können, mußte zunächst ein besonderer, definiert heizbarer Präparathalter gebaut werden (vgl. Figur 4). Er besteht im wesentlichen aus einer definiert aufheizbaren Kupferplatte, die durch vier Stahlblechhal-

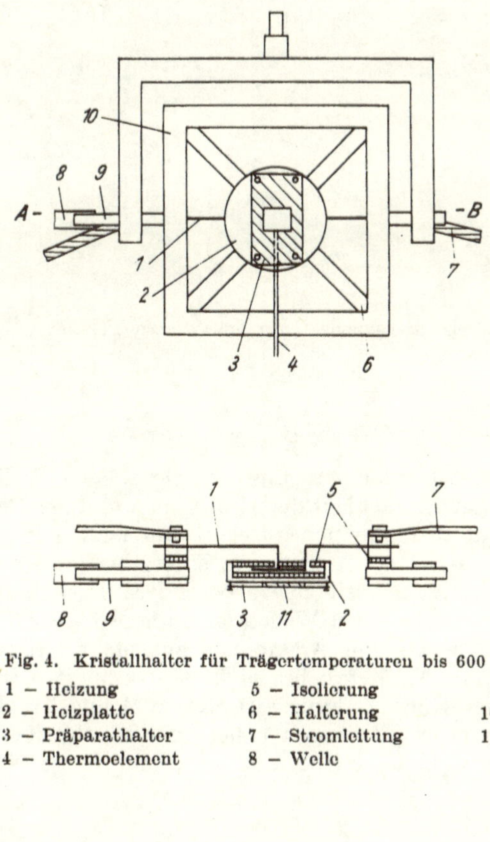

a) Schnitt senkrecht zur Aufdampfrichtung

b) Querschnitt A – B

Fig. 4. Kristallhalter für Trägertemperaturen bis 600 °C

1 – Heizung
2 – Heizplatte
3 – Präparathalter
4 – Thermoelement
5 – Isolierung
6 – Halterung
7 – Stromleitung
8 – Welle
9 – Achse
10 – Rahmen
11 – Kristall

terungen im Zentrum eines nach zwei Richtungen drehbaren Stahlrahmens fixiert ist. Die Heizung erfolgt über eine Wolframspirale mit einer Maximalbelastung von 40 Watt. Die optimale Aktivierungstemperatur für die Erzeugung einkristalliner Fe-Aufdampfschichten auf (001)-NaCl wurde mittels einer besonderen Temperatur-Versuchsreihe festgelegt. Hiernach wurden die besten Einkristallorientierungen bei Trägertemperaturen zwischen ≈ 425 °C und ≈ 450 °C erzielt. Ein Beispiel hierzu gibt das Transmissions-Elektronenbeugungsdiagramm Figur 5, das bei einer Aktivierungstemperatur des Trägers von 450 °C erhalten wurde. Wie das Diagramm zeigt, sind die Reflexpunkte sehr scharf und ohne Andeutungen von Texturverschmierungen mit (001)-Fe parallel zur Trägeroberfläche. Die epitaxisch gewachsene Fe-Folie ist also gut einkristallin. Diese Beobachtungen stimmen mit denen von Brück überein, doch ist die von letzterem Autor angegebene Orientierungstemperatur von 500 °C für heutige Vakua, wegen der gesteigerten Abdampfung von NaCl, zu hoch.

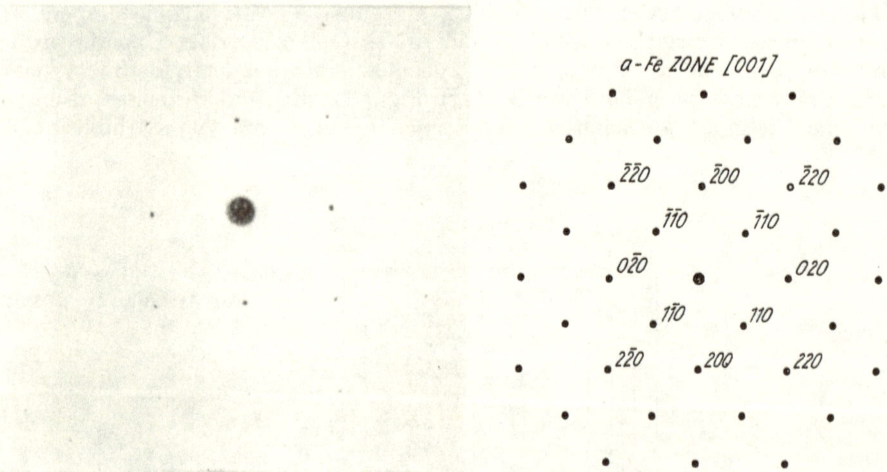

a) b)

Fig. 5. Elektronenbeugungsdiagramm einer epitaxischen Fe-Schicht mit nur einer Orientierung
a) Aufnahme b) Indizierung

Herstellung der Nitrierungsschichten

Die Nitrierung der im vorstehenden Abschnitt beschriebenen Fe-Bleche und Fe-Folien erfolgte, wegen der Einfachheit der Reaktionen und der Definiertheit der Reaktionsbedingungen, ausschließlich durch Gasnitrieren mit Ammoniak. Will man hierbei reine Fe—N-Phasen erzeugen, so ergibt sich als erste Aufgabe die Darstellung von extrem reinem, insbesondere H_2O- und O-freiem Ammoniak. Das gelingt auf folgendem Wege (dazu Figur 6, siehe S. 149).

Man leitet eine begrenzte Portion Ammoniak aus einer handelsüblichen NH_3-Druckflasche in ein mit Na-Metall beschicktes, evakuiertes Druckgefäß, das sich seinerseits in einer Kühlmischung aus Methanol und Trockeneis von $T = -68$ °C befindet. Hierbei bildet sich nebeneinander flüssiges NH_3, das infolge Lösens von Na eine blaue Farbe annimmt, und festes NaOH (evtl.

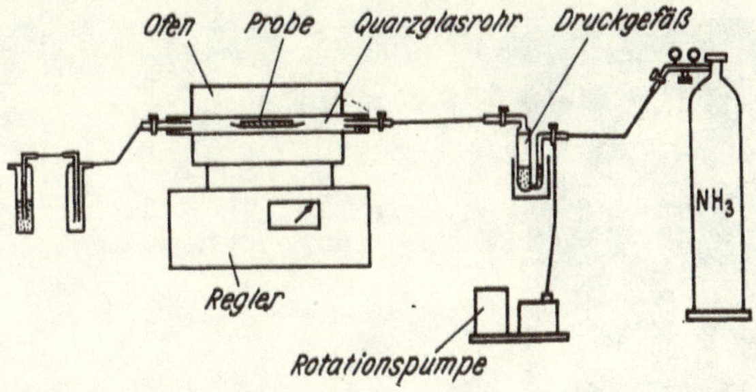

Fig. 6. Nitrierapparatur

auch Na_2O). Entfernt man nunmehr die Kältemischung, so dampft reinstes, insbesondere H_2O-freies NH_3-Gas ab, das gemäß Figur 6 in das Nitrierrohr mit der zu nitrierenden Probe geleitet wird.

Die Nitrierung der Fe-Bleche erfolgte hauptsächlich bei den technisch gebräuchlichen Temperaturen 470 °C bzw. 520 °C bzw. 570 °C, die Nitrierung der Fe-Folien bei Temperaturen von 200 °C bis 450 °C. Die Aufheizgeschwindigkeit betrug 15 °C · min^{-1}. Die jeweilige Nitriertemperatur (Temperatur im Nitrierrohr) wurde mit einem Pt/PtRh-Thermoelement gemessen und mittels einer Regelanlage auf ± 5 °C konstant gehalten. Die Nitrierzeiten betrugen zwischen 30 und 180 min für Fe-Bleche und zwischen einigen Sekunden und wenigen Minuten für Fe-Folien. Die letzteren sehr kurzen Reaktionszeiten verstehen sich aus den sehr kurzen Diffusionswegen in den Fe-Folien ($\sim$300 Å).

Die Nitrierung der auf NaCl abgeschiedenen poly- bzw. einkristallinen Fe-Folien erfolgte zunächst an der freien Folie, also abgelöst vom Trägerkristall. Elektronenbeugungsdiagramme an solchen nitrierten Folien wiesen jedoch stets mehr oder weniger deutliche Homogenitätsstörungen auf. Das gilt in besonderem Ausmaß für die Einkristall-Folien. Wesentlich bessere Ergebnisse lieferten Präparate, die **auf dem Träger** nitriert und erst dann von ihm abgelöst worden waren. Bei den folgenden Untersuchungen sind daher i. allg. Objekte benutzt worden, die nach dem letzteren Präparationsverfahren hergestellt worden sind.

3.2. Beugungsuntersuchungen an Kristallphasen Fe–N

3.2.1. Bestimmung der Kristallphasen

α) Nitridbildung in ≈600 Å dicken Fe-Folien

Die Untersuchungen an diesen sehr dünnen Folien mußten grundsätzlich mittels Elektronenbeugung erfolgen. Sie ergaben bei Nitriertemperaturen von 200° bis 300 °C und sehr kurzen Nitrierzeiten (einige Sekunden) Fe_4N (γ'-Phase) neben α-Fe. In längerzeitig (einige Minuten) nitrierten Proben ist die N-Aufnahme bereits so beträchtlich, daß sich schon N-reiche Phasen bilden, insbesondere ε-Fe_3N (vgl. Figur 7 und Tabelle 1).

Bei Nitrierungstemperaturen $\gtrsim 300° - 450$ °C bilden sich bei Nitrierung im NH_3-Strom schon nach wenigen Sekunden ε-Nitride neben Fe_4N (γ'). In den meisten Fällen tritt daneben auch bereits ζ-Fe_2N auf. Mit zunehmender Nitrierungsdauer dominiert ζ-Fe_2N mehr und mehr, während ε-Fe_3N und Fe_4N an Menge rasch abnehmen.

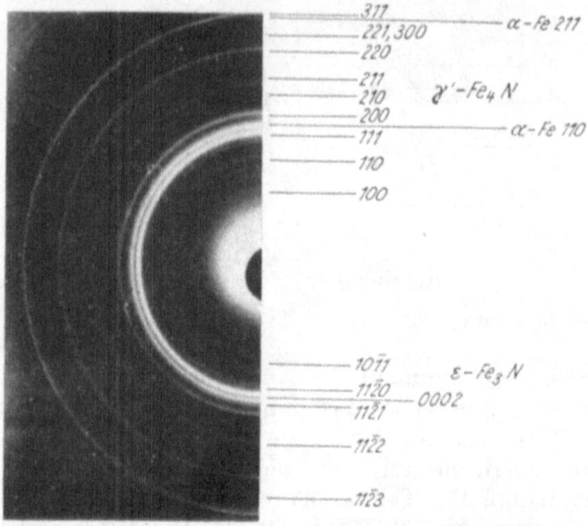

Fig. 7. Elektronenbeugungsaufnahme einer Nitrierungsschicht bestehend aus F_4Ne (γ'-Phase) und ε-Fe_3N neben α-Fe

Tabelle 1

Auswertung der Elektronenbeugungsaufnahme einer 40 s bei 250 °C nitrierten polykristallinen Fe-Folie, dazu Figur 7

Aufnahme			$Fe_4N(\gamma')$ [1]		ε-Fe_3N		α-Fe[1]	
r (mm)	*I* (vis)	*d* (Å)	*d* (Å)	*hkl*	*d* (Å)	*hkil*	*d* (Å)	*hkl*
5,14	s	3,79	3,79	100	—	—	—	—
6,49	s	3,00	—	—	2,991	10$\bar{1}$1	—	—
7,2[illegible]	s	2,68	2,684	110	—	—	—	—
8,25	s	2,36	—	—	2,358	11$\bar{2}$0	—	—
8,89	ssst	2,19	2,191	111	2,197	0002	—	—
9,38	sst	2,08	—	—	2,077	11$\bar{2}$1		
9,58	s	2,03	—	—	—	—	2,026	180
10,25	st	1,90	1,897	200	—	—	—	—
11,51	ss	1,69	1,697	210	—	—	—	—
12,12	m	1,61	—	—	1,607	11$\bar{2}$2	—	—
12,56	sss	1,55	1,549	211	—	—	—	—
13,63	sss	1,43	—	—	—	—	1,433	200
14,53	st	1,34	1,342	220	—	—	—	—
15,65	m	1,25	1,265	221	1,244	11$\bar{2}$3	—	—
				200				
16,73	ss	1,17	—	—	—	—	1,170	211
17,50	st	1,14	1,144	311	—	—	—	—
17,78	s	1,09	1,095	222	—	—	—	—

ssst = sehr sehr stark, sst = sehr stark, st = stark, m = mittel, sss = sehr sehr schwach, ss = sehr schwach, s = schwach.

[1]) aus ASTM-Karte.

Bei Nitriertemperaturen von ≈ 450 °C bildete sich bei ausreichender Nitrierzeit schließlich nur noch das N-reichste der bisher bekannten Fe-Nitride, das ζ-Fe_2N. Doch ist die Unterscheidung von ε- und ζ-Nitrid nicht leicht, da sie nur mittels der schwachen, nur vom N-Teilgitter herrührenden Reflexe getroffen werden kann, z. B. durch (0001) von ε-Fe_2N und (101) oder (111) von ζ-Fe_2N. Ein Auswertungsbeispiel zeigen Tabelle 2 und Figur 8.

Tabelle 2

Auswertung einer Elektronenbeugungsaufnahme einer 120 s bei 400 °C nitrierten Fe-Folie (dazu Figur 8)

Aufnahme			ζ-Fe_2N		ε-Fe_2N	
r (mm)	*I* (vis)	*d* (Å)	*d* (Å)	*hkl*	*d* (Å)	*hkil*
4,43	s	4,41	—	—	4,420	0001
5,68	s	3,44	3,45	101	—	—
6,47	st	3,02	—	—	3,024	$10\bar{1}1$
6,96	s	2,80	2,804	111	—	—
8,18	st	2,39	2,404	020, 210	2,394	$11\bar{2}0$
8,87	st	2,20	2,207	002	2,210	0002
9,30	ssst	2,10	2,110	021, 211	2,105	$11\bar{2}1$
10,46	s	1,87	—	—	1,877	$20\bar{2}1$
12,08	sst	1,62	1,626	022, 212	1,624	$11\bar{2}2$
13,28	s	1,47	—	—	1,477	$21\bar{3}1$
					1,473	0003
14,18	st	1,38	1,385	321, 230	1,388	$10\bar{1}3$
				400		$30\bar{3}0$
14,87	ss	1,31	—	—	1,319	$30\bar{3}1$
15,62	st	1,25	1,255	023, 213	1,255	$11\bar{1}3$
16,37	s	1,19	1,203	040, 420	1,200	$20\bar{2}3$
					1,197	$22\bar{4}0$
16,72	st	1,17	1,1668	232, 402	1,172	$30\bar{3}2$
				331, 041		
				421		
16,96	st	1,15	—	—	1,155	$22\bar{4}1$
17,75	s	1,10	1,104	004	1,113	$31\bar{4}1$
					1,105	0004
18,30	sss	1,07	—	—	1,073	$21\bar{3}3$
18,59	ss	1,05	1,055	042, 422	1,052	$22\bar{4}2$
19,53	s	1,00	1,003	024,	1,009	$40\bar{4}1$
				214	1,008	$30\bar{3}3$
					1,003	$11\bar{2}4$
21,12	m	0,92	0,9299	333, 043,	0,930	$32\bar{5}1$
				151, 423	0,929	$22\bar{4}3$

Die Identifizierung der ζ-Phase bietet auch sonst besondere Schwierigkeiten, da sie sich schon während des Abkühlens unter N-Abgabe partiell in ein N-ärmeres Fe-Nitrid umwandelt. Es ist daher sehr schwer reine ζ-Präparate herzustellen.

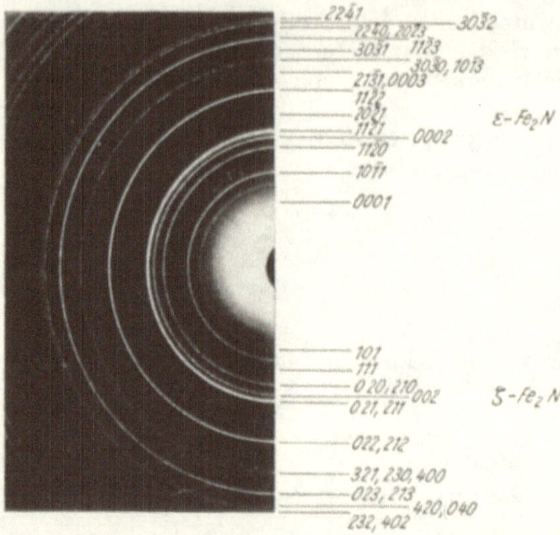

Fig. 8. Elektronenbeugungsaufnahme einer Nitrierungsschicht bestehend aus ε-Fe$_2$N und ζ-Fe$_2$N

Tabelle 3

Auswertung einer Elektronenbeugungsaufnahme einer 180 s bei 475 °C nitrierten polykristallinen Fe-Folie

Aufnahme			ζ-Fe$_2$N		Fe$_4$N (γ')[1]		ε-Fe$_3$N-Fe$_2$N	
r (mm)	*I* (vis)	*d* (Å)	*d* (Å)	*hkl*	*d* (Å)	*hkl*	*d* (Å)	*hkil*
4,43	s	4,40	—	—	—	—	4,409	0001
5,18	s	3,77	—	—	3,79	100	—	—
5,66	s	3,45	3,45	101	—	—	—	—
6,46	st	3,02	—	—	—	—	3,001	20$\bar{2}$1
6,94	s	2,81	2,804	111	—	—	—	—
7,76	m	2,52	—	—	—	—	—	—
8,13	st	2,40	2,404	020, 210	—	—	2,380	22$\bar{4}$0
8,41	sss	2,32	—	—	—	—	2,332	30$\bar{3}$1
8,82	st	2,21	2,207	002	2,19	111	2,204	0002
9,26	ssst	2,11	2,110	021, 211	—	—	2,094	22$\bar{4}$1
10,37	ss	1,88	—	—	1,89	200	1,867	40$\bar{4}$1
12,01	sst	1,62	1,626	022, 212	—	—	1,617	22$\bar{4}$2
13,23	ss	1,47	—	—	—	—	1,470	42$\bar{6}$1
							1,469	0003
15,58	st	1,25	1,255	023, 123	—	—	1,250	22$\bar{4}$3
16,29	s	1,20	—	—	—	—	1,200	40$\bar{4}$3
							1,190	4480
16,63	st	1,17	1,166	232, 402	—	—	1,166	60$\bar{6}$2
				331, 041				
				421				
16,92	st	1,15	—	—	—	—	1,151	4481

[1]) Aus ASTM-Kartei.

Bei der thermischen Zersetzung der N-reichsten Nitride bei ~500 °C wurde die Bildung von Fe_4N (γ') neben einem weiteren Nitrid beobachtet, das einen Reflex $d = 2.32$ Å aufweist. Nach den röntgenographischen Untersuchungen dieses 2.32 Å-Nitrids durch JACK (1952) (vgl. Tabelle 3) handelt es sich bei diesem um eine ε-Phase mit einer Zusammensetzung zwischen Fe_3N und Fe_2N (ε-Fe_3N-Fe_2N). Tab. 3 zeigt neben der letzteren Phase auch die 2. Zerfallskomponente Fe_4N (γ') sowie die Ausgangsphase ζ-Fe_2N.

β) Nitridbildung in Fe-Blechen

Da die Diffusionsgeschwindigkeit von Stickstoff in Eisen bei geeigneten Temperaturen sehr groß ist, die verfügbaren Diffusionswege in den dünnen Fe-Folien aber sehr klein sind, ließ sich die zeitliche und räumliche Bildungsfolge der verschiedenen Nitridphasen durch die Untersuchung der Folien allein nicht ausreichend sichern. Es wurden daher auch Eisenbleche der üblichen Dicke nitriert, wobei, in Anlehnung an die industrielle Praxis, als Nitriertemperaturen 470 °C, 520 °C und 570 °C gewählt wurden. Alle nitrierten Proben wurden im Ammoniakstrom von der jeweiligen Versuchstemperatur auf Zimmertemperatur abgekühlt, da sich die gebildeten Nitride sonst weitgehend in Fe_4N (γ') umwandelten. Die Untersuchungen erfolgten zweckmäßig mittels Röntgenbeugung. Die an diesen Fe-Blechen nachgewiesenen Kristallphasen sind in Tabelle 4 als Zeit/Temperatur-Tafel zusammengefaßt worden. Hierbei wurde zwischen den verschiedenen ε-Phasen nicht unterschieden, da das Verhältnis der Streuintensitäten von Fe und N für Röntgenstrahlen, im Gegensatz zu Elektronenstrahlen, zu ungünstig ist (s. Fig. 9a und 9b).

Tabelle 4

Phasenanalyse nitierter Fe-Bleche als Zeit-Temperatur-Tafel dargestellt (Nach Nitrierung im NH_3-Strom abgekühlt)

Temp. (°C)	Zeit (Min.)			
	30	60	120	180
470	α	$\alpha + \gamma'$	$\alpha + \gamma' + \varepsilon$	$\varepsilon + \zeta$
520	$\alpha + \gamma' + \varepsilon$	$\varepsilon + \zeta$	$\varepsilon + \zeta$	–
570	$\varepsilon + \zeta$	$\varepsilon + \zeta$	$\varepsilon + \zeta$	–

Das Auftreten der ζ-Phase bei allen Temperaturen der Tabelle 4 legt nach dem Phasendiagramm Figur 1 die Vermutung nahe, daß sich diese Phase erst bei der Abkühlung gebildet hat. Es wurden daher weitere Untersuchungen durchgeführt, bei denen die nitrierten Bleche sogleich nach Versuchsende in Wasser abgeschreckt worden waren. Hierbei ließ sich eine teilweise Oxidation der Proben nicht vermeiden (Bildung von Fe_3O_4 und α-Fe_2O_3). Die Phasenanalyse der nitrierten Bleche führte zu den Ergebnissen der Tabelle 5, wobei die Oxide für die Betrachtungen zur Bildung der Fe–N-Phasen unberücksichtigt bleiben.

Das Fehlen des ζ-Fe_2N in Tabelle 5 bei allen Versuchstemperaturen zeigt, daß die obere Grenze des Stabilitätsfeldes von ζ-Fe_2N in Diagramm 1 zu hoch angesetzt wurde.

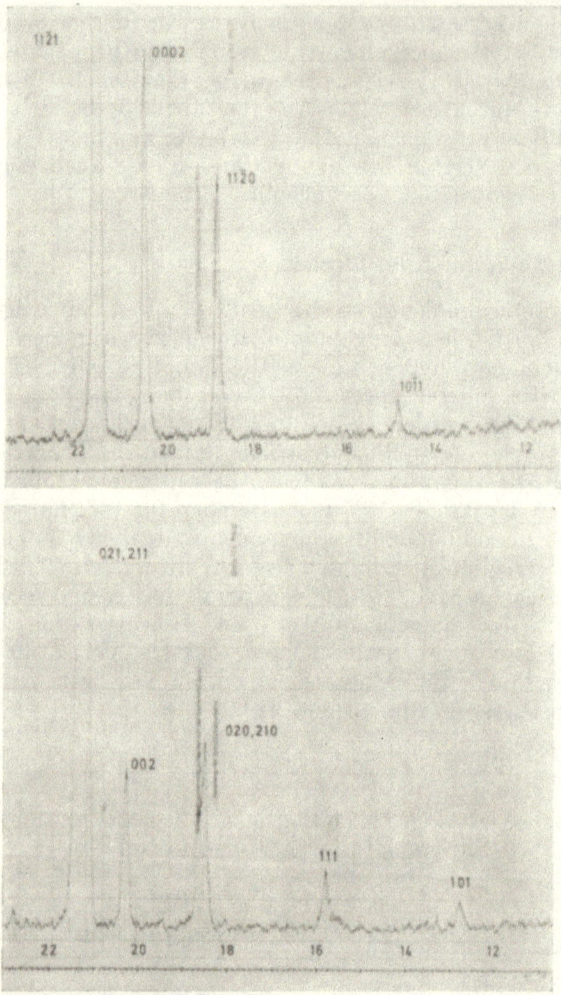

a)

b)

Fig. 9. Röntgendiffraktometerdiagramm der ε- und ζ-Phase (Cu-K$_\alpha$; 40 kV, 20 mA; 100 Imp. s^{-1}, 1/4 °Θ/min)
a) ε-Phase b) ζ-Phase

Tabelle 5

Phasenanalyse nitrierter Fe-Bleche als Zeit-Temperatur-Tafel dargestellt

(Nach Nitrierung in kaltem Wasser abgeschreckt)

Temp. (°C)	Zeit (min)		
	30	60	180
470	$\alpha + \gamma'$	$\alpha + \gamma'$	ε
520	$\alpha + \gamma'$	ε	ε
570	ε	ε	ε

3.2.2. Orientierungsbeziehungen (Epitaxien, Topotaxien) im System Fe—N

Bei den engen partiell-isomorphen Gitteranalogien (NEUHAUS 1950) der Fe-Teilgitter aller gefundenen Nitride waren von vornherein vielfache Orientierungsbeziehungen (Epitaxien) dieser Kristallphasen untereinander und mit dem Trägergitter α-Fe zu erwarten. Für die Suche nach solchen Epitaxien eigneten sich vorzüglich die weiter oben beschriebenen nitrierten Fe-Einkristallfolien, da auf vielen ihrer Diagramme Reflexe zweier einkristalliner Phasen auftreten.

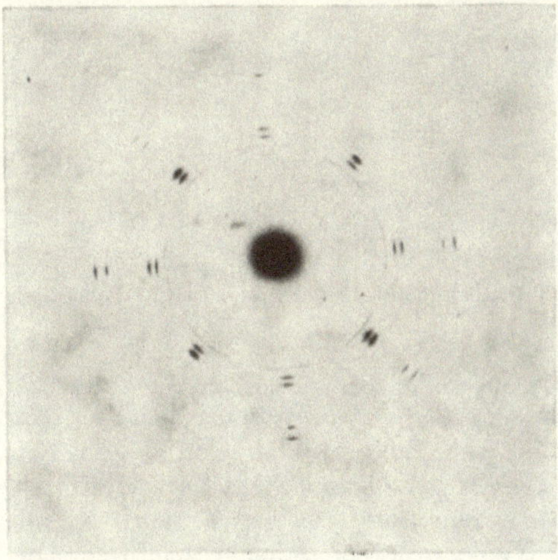

Fig. 10. Elektronenbeugungsaufnahme einer epitaktischen Verwachsung aus α-Fe und Fe_4N (γ'-Phase)

Beispiele:

α) Figur 10 zeigt das leicht texturierte Beugungsdiagramm einer Nitrierungsschicht aus Fe_4N (γ') und α-Fe. Die Auswertung dieses Diagramms ergibt zwei unterschiedlich stark ausgeprägte, um 45° gegeneinander verdrehte Orientierungen von α-Fe als Wirtgitter sowie zu jedem dieser zwei Fe-Gitter ein gleich orientiertes γ'-Gitter. Das ergibt folgendes Orientierungsgesetz:

$$(001)\text{-Fe}//(001)\text{-Fe}_4\text{N}\,(\gamma') \qquad \text{und} \qquad [110]\text{-}\alpha\text{-Fe}//[100]\text{-Fe}_4\text{N}\,(\gamma')\,.$$

Der „misfit“ zwischen a_0-Fe_4N (γ') $= 3{,}79$ Å und [110]-αFe $= 4{,}05$ Å beträgt $-6{,}4\%$ bezogen auf α-Fe.

β) Elektronenbeugungsaufnahmen von Nitrierschichten, die Fe_4N (γ') und ε-Fe_3N enthielten, zeigen zunächst große Ähnlichkeit mit Faserdiagrammen, mit einer Vielzahl an Orientierungen (Fig. 11). Fe_4N (γ') besitzt in dieser Aufnahme zwei gegeneinander um 45° verdrehte Stellungen mit (001) parallel zur Folienoberfläche. Wird nunmehr die sehr naheliegende Annahme gemacht, daß (111)-Fe_4N (γ') und (0001)-Fe_3N (ε) epitaxisch miteinander verwachsen sind, so läßt sich Figur 11 leicht interpretieren. Da in diesem Falle zwei γ'-Kristallit-Stellungen vorliegen, so ergeben sich für die Verwachsung von γ' mit ε aus Symmetriegründen $2 \times 4 = 8$ unterschiedlich orientierte (111)-Ebenen mit folgendem Orientierungsgesetz:

$$(111)\text{-Fe}_4\text{N}\,(\gamma')//(0001)\text{-Fe}_3\text{N}\,(\varepsilon) \quad \text{mit} \quad [1\bar{1}0]\text{-Fe}_4\text{N}\,(\gamma')//[21.0]\text{Fe}_3\text{N}\,(\varepsilon).$$

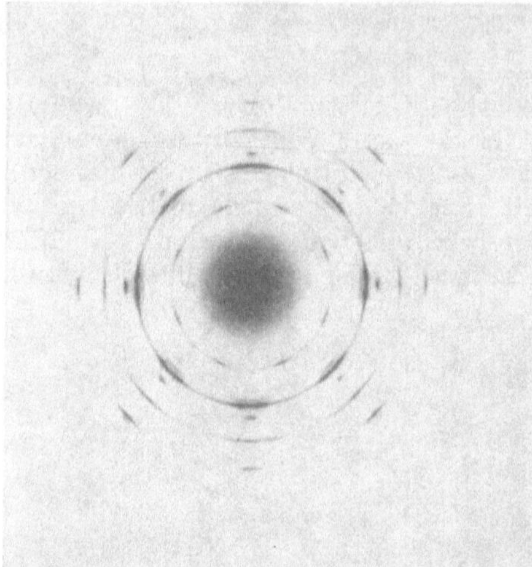

Fig. 11. Elektronenbeugungsaufnahme einer orientierten Folie bestehend aus Fe_4N (γ'-Phase) und ε-Fe_3N

Der „misfit" beträgt in diesem Falle nur $\approx 1{,}4\%$. Die zunächst verwirrende Vielzahl an Orientierungen wird also gut verständlich.

γ) Es wurden schließlich noch Beugungsdiagramme von Nitrierschichten erhalten, die ε-Fe_2N und ζ-Fe_2N in epitaxischem Verwachsungsverband ausweisen. Sie ähneln sehr stark denen, die aus den beiden ε-Nitriden Fe_3N und Fe_2N bestehen. Auf ihre Wiedergabe sei daher verzichtet. Die rhombische Symmetrie des ζ-Fe_2N sollte die äquivalenten Verwachsungsmöglichkeiten gegenüber der hexagonalen Symmetrie verdreifachen. Doch ist die Abweichung von der hexagonalen Symmetrie so gering, daß sie im Elektronenbeugungsdiagramm nicht nachgewiesen werden konnte. Damit ergibt sich die folgende Orientierung:

$$(0001)\text{-}\varepsilon//(001)\text{-}\zeta \quad \text{und wahrscheinlich} \quad [21.0]\text{-}\varepsilon//[100]\text{-}\zeta.$$

Der „misfit" beträgt $\approx 0\%$.

4. Zeitliche Bildungsabfolge, räumliche Anordnung und Bildungsmechanismus der Nitridphasen in Nitrierungsschichten des Eisens

Entsprechend der Zunahme an freier Enthalpie der Nitrierreaktionen am Fe (metallisch) und dank dem kleinen Wirkungsbereichs-Radius des N dringt atomarer Stickstoff leicht und schnell in das αFe-Gitter ein und bildet dort eine feste Lösung αFe(N) mit N in [6]-koordinierten Lückenpositionen des αFe-Gitters. Da das αFe-Gitter jedoch nur 0,1 Gew.-% N in fester Lösung aufnimmt, wandelt sich diese erste Nitrierphase bei weiterer N-Zufuhr, durch epitaxisch initiierten und gelenkten, also topotaxischen Gitterumbau (Neuhaus, Gebhardt 1968) sogleich in die N-reichere Nitridphase Fe_4N (γ') um. Diese Umwandlung beginnt an der Phasengrenze Gas/Festoberfläche und schreitet, wie vorher schon die Umwandlung $\alpha Fe \rightarrow \alpha Fe(N)$, unter Ausbildung einer geschlossenen Nitrid-Deckschicht nach der Tiefe zu fort.

Da die Diffusion von atomarem N in der Nitrid-Deckschicht wesentlich gehemmter ist als im αFe-Gitter bzw. im αFe(N), so bewirkt die Herausbildung dieser Deckschicht und ihre Verdickung nach der Tiefe zu eine fortschreitende, starke Verschlechterung des N-Transports in die unter der Nitridzone liegende unveränderte αFe-Zone. Der Reaktionsmechanismus der Nitrierungsreaktion ändert sich also mit fortschreitender Nitrierung von einer wesentlich reaktionsbedingten, in der Primärschicht, zu einer wesentlich diffusionsbedingten Reaktion nach Ausbildung einer Nitridzone. Das ergibt einen Stickstoff-Stau in der Nitridschicht und abnehmende N-Zufuhr zu der unter der Nitridschicht folgenden Zone des primären α-Fe (Diffusionszone).

Weiter anhaltende N-Zufuhr bewirkt nunmehr eine Umbildung des primären Fe_4N (γ') in N-reichere Nitride und zwar zunächst in die ε-Phase und schließlich in ζ-Fe_2N, dem N-reichsten der bisher in Nitrierungsschichten gefundenen Nitride. Auch diese Umwandlungsreaktionen sind sämtlich epitaxisch initiierte und gelenkte also topotaxische Gitterumbau-Reaktionen (Neuhaus, Gebhardt 1968).

Aus dieser **zeitlichen** Bildungsfolge der Nitridphasen läßt sich ihr **räumlicher** Aufbau ableiten, wenn man einbezieht, daß 1) alle Nitride der obigen Bildungsreihe einander topotaxisch verdrängen und 2) daß die N-Zufuhr mit steigender Nitriertiefe, infolge diffusiver Selbsthemmung, steigend abnimmt.

Aus Punkt 1) folgt unmittelbar eine räumliche Anordnung der stark ineinander geschoben zu denkenden Einzelnitride gemäß Figur 12 und damit die Herausbildung einer besonderen, hoch vergüteten „Deckschicht" neben der darunter folgenden, nur schwach nitrierten Zone des α-Fe(N).

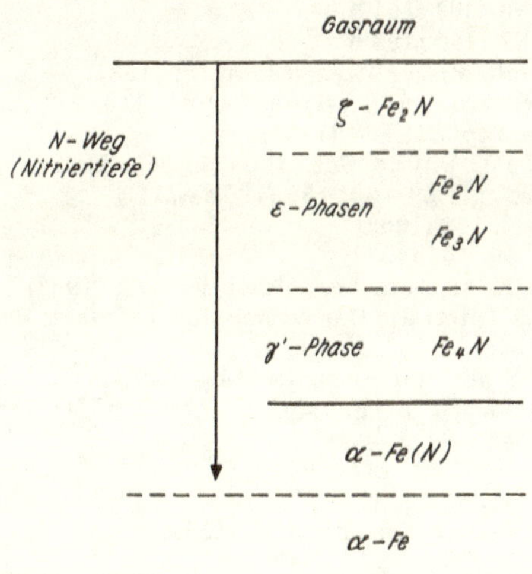

Fig. 12. Endanordnung der Nitridphasen in einem nitrierten Eisenblech

Bei der gegenüber αFe sehr geringen Durchlässigkeit der „Nitridschicht" für atomares N bedeutet steigende Dicke der Nitridzone entsprechend Punkt 2) starke Selbsthemmung der Zufuhr von atomarem Stickstoff zu den unter der „Nitridzone" liegenden Bereichen des αFe, so daß sie nur noch zur Bildung einer schmalen Zone von α-Fe(N) mit maximal 0,1 Gew.-% N ausreicht.

Damit dürfte die Koordinierung der zeitlichen Bildungsfolge der Nitrierphasen in Nitrierungsschichten und ihre räumliche Endanordnung in den fertigen Nitrierschichten im wesentlichen verständlich werden.

Literatur

BIERWIRTH, G.: Siemens-Zeitschrift **5**, 365 (1968)
BOOKER. G. R., NORBURG, J., SUTTON, A. L.: J. Iron Steel Inst. **187**, 205 (1957)
BRIDELLE, R.: Ann. Chim. **10**, 824 (1955)
BRÜCK, L.: Ann. d. Physik **3**, 233 (1936)
BURDESE, A.: Metallurg. Ital. **49**, 195 (1957)
DIJKSTRA, L. J.: Trans. AIME **185**, 252 (1949)
DVORIANKINA, G. G., PINSKER, Z. G.: Sov. Physics **3**, 439 (1958)
DVORIANKINA, G. G., PINSKER, Z. G.: Sov. Physics **5**, 236 (1960)
EISENHUT, O., KAUPP, E.: Z. Elektrochem. **37**, 392 (1930)
EPSTEIN, S., GROSS, H. C.: Bur. of Stand. Inl. of Res. **3**, 1005 (1929)
EPSTEIN, S., GROSS, H. C., GROESBECK, E. C., WYMORE, J. J.: Bur. of Stand. Inl. of Res. **3**, 1005 (1929)
FINNERN, B.: Bad- und Gasnitrieren, München 1965
FRY, A.: Stahl und Eisen **43**, 1271 (1923)
GEBHARDT, M.: Protection against Corrosion by Metal Finishing, Zürich 1967, S. 323
HÄGG, G.: Z. Physik Chem. **8**, 355 (1930)
JACK, K. H.: Proc. Roy. Soc. A **195**, 439 (1948)
JACK, K. H.: Proc. Roy. Soc. A **208**, 200 (1951)
JACK, K. H.: Proc. Roy. Soc. A **208**, 216 (1951)
JACK, K. H.: Acta Cryst. **5**, 404 (1952)
KOCH, M.: Werkstatt und Betrieb **95**, 594 (1962)
LEHRER, E.: Z. Elektrochem. **36**, 383 (1930)
LEHRER, E.: Z. Elektrochem. **37**, 460 (1930)
MEYER, O., HOHBROCK, R.: Arch. Eisenhüttenwes. **5**, 251 (1931)
MEHL, R. F., BARRETT, C. S., SUTTON, A. L.: Trans. AIME **113**, 211 (1934)
NAGAKURA, S.: J. Physic. Soc. Jap. **25**, 488 (1968)
NAGAKURA, S., TANEHASHI, K.: J. Physic. Soc. Jap. **25**, 840 (1968)
NEUHAUS, A.: Fortschr. Mineralogie **29–30**, 136 (1950–51)
NEUHAUS, A.: Angew. Chemie **64**, 158 (1952)
NEUHAUS, A.: Z. Elektrochem. **56**, 543 (1952)
NEUHAUS, A., GEBHARDT, M.: Nova Acta Leopoldina **24**, 208 (1968)
NEUHAUS, A., GEBHARDT, M.: Interface Conversion for Polymer Coatings New York 1968, S. 91
NOREN, T. M., KINDBORN, L.: Stahl und Eisen **78**, 1881 (1958)
PARANJPE, V. G., COHEN, M., BEVER, M. B., FLOE, C. F.: Trans. AIME **188**, 261 (1950)
PINSKER,
PITSCH, W.: Arch. Eisenhüttenwes. **32**, 497 (1961)
SAWYER, C. B.: Trans. AIME **69**, 798 (1923)
SCHNEIDER, H. G.: Freiberger Forschungsheft B **71** (1964)

(Eingegangen am 14. 6. 71)

Anschrift der Verfasser:

Dipl. Min. G. AMTHAUER
Dr. M. GEBHARDT
Prof. Dr. A. NEUHAUS
Mineralogisch-Petrologisches Institut der Universität Bonn
53 Bonn
Poppelsdorfer Schloß

O. W. Flörke

Institut für Mineralogie, Ruhr-Universität Bochum

Transport and Deposition of SiO_2 with H_2O under Supercritical Conditions

Transport and deposition of silica in different forms: quartz, chalcedony, cristobalite, tridymite and hyalite were investigated between 400–800 °C and 200 to 2000 bar. The conditons of formation of the forms are discussed. The crystallization of quartz whiskers is reported and ideas of the formation of hyalite are given. Conclusions are drawn to natural occurences.

Transport und Abscheidung von SiO_2 in seinen verschiedenen Formen: Quarz, Chalzedon, Cristobalit, Tridymit und Hyalit wurden zwischen 400–800 °C bei 200–2000 bar untersucht. Es werden die Bildungsbedingungen für die verschiedenen Formen diskutiert. Dabei wird auch über die Bildung von Quarz-Whiskern berichtet, sowie eine Vorstellung über die Entstehung und die Struktur von Hyalit entwickelt. Beziehungen zu entsprechenden SiO_2-Vorkommen in der Natur werden hergestellt.

1. Introduction

Vapour deposited silica occurs in nature as well as under laboratory or technical conditions. Many volcanic rocks bear enormous masses of SiO_2 which obviously have been transported through the gas phase. With H_2O the rate of volatilisation of SiO_2 increases strongly with increasing pressure and temperature (for references see: Gillingham; Heitmann; Anderson and Burnham; Kirsch). Under these conditions the occurrence of the different silica-polymorphs sometimes is very puzzling with respect to the phase diagram. Quartz crystals can be associated with crystals of syngenetic tridymite or cristobalite. Exampels of non-equilibrium paragenetic growth are not rare in nature. An example for vapour transported silica is the paragenesis of tridymite and quartz in the trachyte of Perlenhardt-Hill, Siebengebirge, Germany (BRD). Though chemical transport in many cases serves for quick adaption of equilibrium (Schäfer; Nitsche) this is not the case in any circumstance and for any kind of chemical bond. In pores of basalts from the Blau Kuppe near Eschwege, Germany (BRD) vapour deposited silica crystals were observed by Ramdohr which showed the habit of Tridymite but the properties of cristobalite. According to the phase diagram a paramorphic transformation of tridymite into cristobalite far below the equilibrium temperature and within the stability field of tridymite had to be assumed. Paramorphs of quartz after tridymite or cristobalite are known from other localities (references see: Flörke), and they are easy to understand. But the basalt of Blaue Kuppe at the time of the deposition of these crystals and afterwards cannot have had much higher temperatures than about 950 °C (H. U. Schmincke, personal communication). From our knowledge of silica transformations a paramorphic transformation of tridymite into cristobalite, even under non-equilibrium conditions has to be excluded. At the section „Tridymite" a simple solution of this problem shall be given. Other topics of interest for our investigations were the conditions of formation of chalcedony, lussatite and hyalite. — To avoid confusions with these terms, they shall be defined as follows:

Chalcedony: microcrystalline quartz, commonly with fibrous texture; average H_2O-content 1—3%

Lussatite: microcrystalline, very strongly disordered cristobalite-tridymite, commonly with fibrous texture; average water content 4—10%; the x-ray powder pattern is markedly different from strongly disordered water-free cristobalite-tridymite

Hyalite: amorphous silica; average water-content 3—6%; its structure formes a continuous random network similar to silica glass; like silica glass hyalite doesn't show x-ray low angle scattering[1])

Opal: amorphous silica; average water-content 4—10%; discontinuous structure, giving rise to low angle x-ray scattering[1]).

According to the work of many scientists (references see Anderson, Burnham) the solubility of SiO_2 in water vapour increases with temperature and pressure. The nature of the gaseous species which serves as volatile compound C in the transport reaction:

$$A, s + B, g = C, g; \qquad -\Delta H; \qquad T_2 \to T_1; \qquad T_1 < T_2$$

s = solid, g = gaseous, ΔH = heat of reaction; T = Temperature) is not well known. From thermodynamic considerations exists a certain probability for $Si(OH)_4$ as getting the most important species with increasing pressure (Brady). On the other hand Wendlandt and Glemser concluded that this is valid only at low densities.

2. Experimental

Based in experiences gained with growth of single crystals of quartz, closed bombs with Bridgman-seal, machined from an austenitic cobalt-base superalloy, were used. Gold capsules (12 cm length, 10 mm diameter) were loaded with nutrient and water and sealed by welding. The capsules were placed in the bombs (Fig. 1/A), and water was added for counterbalancing the pressure in the capsule. Denting the capsules a bit before welding helped in balancing inner and outer pressure. The bombs were hanging vertically in furnaces with variable temperature gradients. The cap had the lowest temperature. Nevertheless the only practicable lubricant for cap threads was pyrophyllite with vaseline. In some runs the capsules were subdivided into several chambers by pinching (Fig. 1/B). About 100 runs were carried out at 400—1000 °C and 200—2000 bars. The durations were 10—30 days.

[1]) Low angle diffraction experiments by courtesy of Dr. A. A. W. A. van Eulen, Vlaardingen, Netherlands.

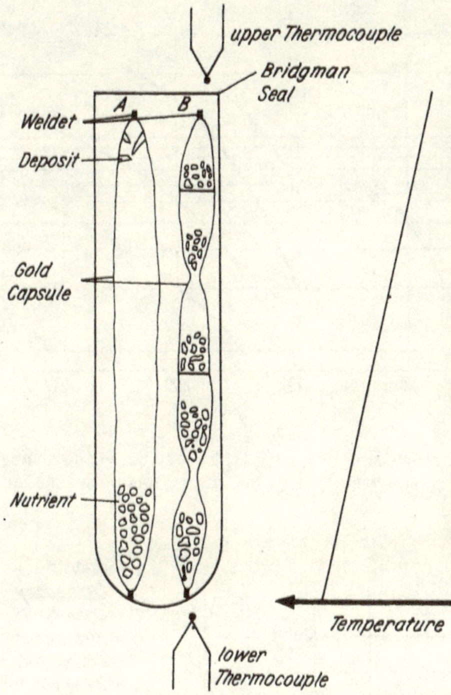

Fig. 1. Scheme of bombs used for vapour transport

3. Results

Depending on the condition of the runs, quartz, cristobalite, tridymite, disordered structural mixtures of cristobalite-tridymite and hyalite deposited. — Lussatite under supercritical conditions never was obtained. In another series of experiments it was prepared at very low pressures and temperatures (100—200 °C, 50 bar), sometimes associated with SiO_2-X and crystallized silicic acids. These results shall be published in another paper.

A draft of transport rates at 2000 bar and different temperatures is given in Figure 2. It shows the increase of rate with temperature and temperature-gradient, i.e. of supersaturation. The rate of transport decreases rapidly with decreasing pressure. For comparison, with $\Delta T = 200$ °C at $T_1 = 700$ °C and $p = 200$ bar the transport rate was approximately 5 mg/day. Figure 3 shows the different forms of deposited silica depending on transport rate, deposition temperature and temperature gradient.

Quartz was deposited in three different forms. At sufficient high temperatures and pressures it was obtained in well developed crystals, showing an acute rhombohedral habit, which gives the crystals the shape of barley corns. This habit as a non-equilibrium form develops when quartz is growing at high temperatures and high transport rates (Figs. 4 and 8). Choosing the right balance in p, T, ΔT and with seeds properly cut, the supersaturation at T_1 will be high enough for quick single crystal growth and low enough to prevent

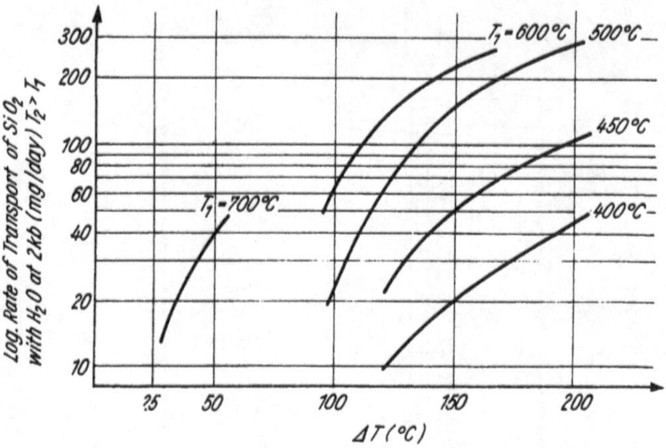

Fig. 2. Log. rate of transport of SiO_2 with H_2O at 2 kbar versus temperature gradient ΔT. Nutrient: rock crystal from swiss alps, grain fraction 5 – 0,5 mm

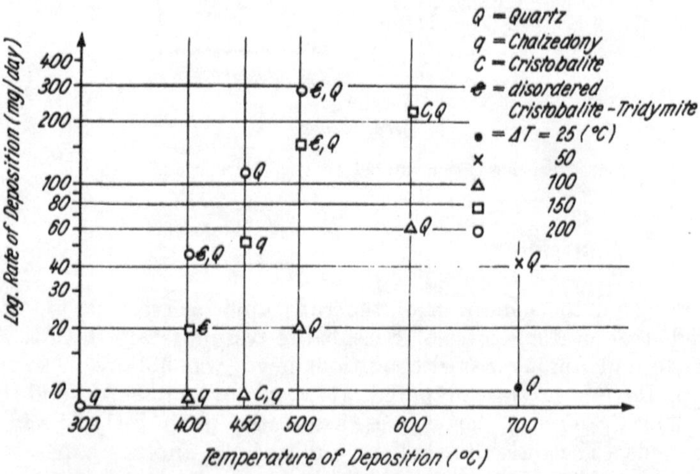

Fig. 3. Forms of silica deposited at 2 kbar, depending on rate of deposition, deposition temperature (temperature of the substrate) and temperature-gradient between nutrient (T_2) and deposit (T_1); ($T_2 > T_1$)

wild growth. On z-cut seeds we were able to grow quartz crystals of several grams weight under 2 kbars with $T_2 = 600$, $T_1 = 500$ °C within 20 days (growth rate 160 mg/day).

At low temperatures with high pressures and high transport rates quartz deposited in the chalcedony variety (Figs. 3 and 5). With decreasing deposition temperature the crystallite size of chalcedony decreases from about 5000 Å at $p = 2$ kbar, $T_1 = 500$ °C, $\Delta T = 150$ °C to about 700 Å at se same pressure, $T_1 = 300$ °C (in this case deposition from the liquid!) and $\Delta T = 200$ °C (Fig. 6).

At low pressures (200–500 bars) and temperatures between 700 and 900 °C frequently the formation of quartz-whiskers was observed (Fig. 7). They can

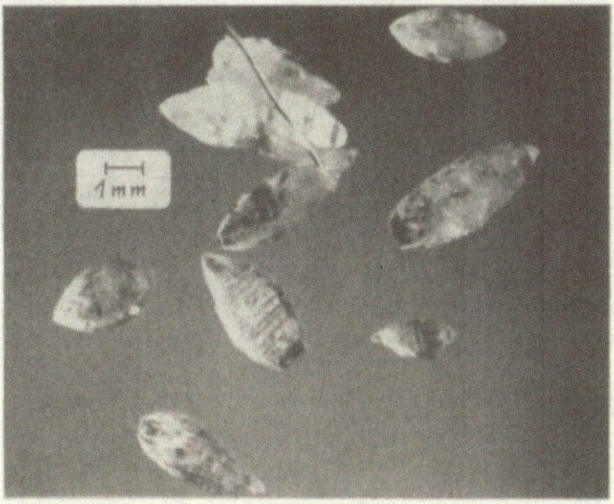

Fig. 4. Quartz crystals showing acute rhombohedral habit; $T_2 = 750$ °C, $T_1 = 550$ °C, $p = 2$ kbar

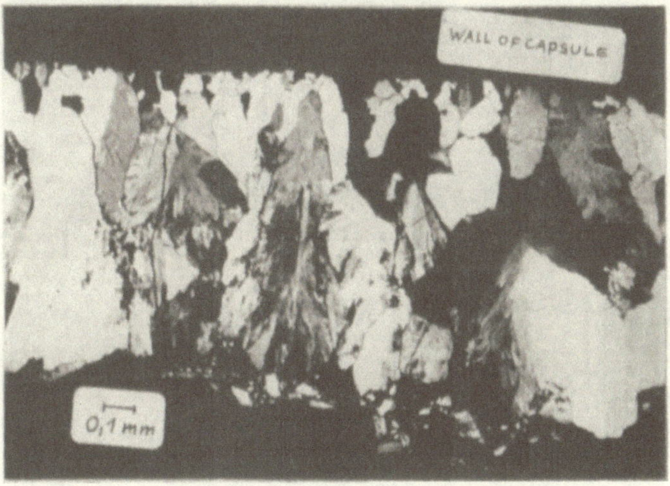

Fig. 5. Thin section (nicols crossed) of quartz and chalcedony deposit; $T_2 = 700$ °C, $T_1 = 500$ °C, $p = 2$ kbar. Initially at the wall of capsule coarse quartz with quartzite-like texture deposited. Thereafter deposition of chalcedony-like texture predominates

be sharply bent without breaking, thus showing a remarkable strength (Blaschke, Flörke). Hitherto the occurrence of quartz whiskers in nature is not yet reported.

Cristobalite with well ordered structure deposited in submicroscopical crystallites with fibrous texture at temperatures above 500 °C, high transport rates and at high pressure (2 kbar). Below 500 °C at high transport rates it showed the tendency of stacking disorder. At low rates of transport, however, it developed a well ordered structure even at low temperatures (Figs. 3 and 8).

At lower pressures (200—500 bar) and temperatures between 700—900 °C single crystals of varying degree of disorder deposited.

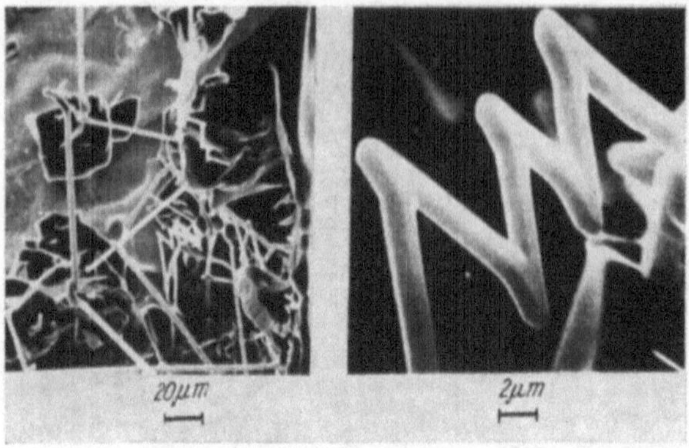

Fig. 6. Quartz-whiskers grown at 750 °C and 300 bar; Scanning electronmicroscope

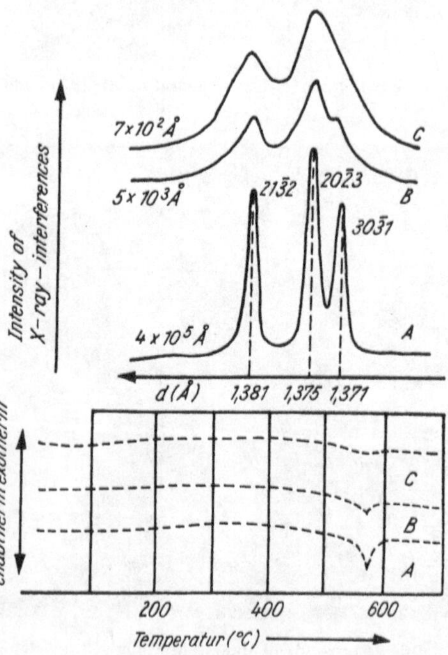

Fig. 7. d = 1,38 – 1,37-section of the X-ray powder patterns (photometer traces) and DTA-curves of
A: quartz deposited at 500 °C
B: chalcedony deposited at 500 °C
C: chalcedony deposited at 300 °C

Tridymite, like cristobalite crystals and quartz whiskers a typical product of gas transport at low densities, never was obtained at high pressures. At low pressures (200 – 500 bars) and temperatures of about 600 – 900 °C it was the dominating deposit. It crystallized in well shaped single crystals and twin groups. — In this paper only preliminary results on the statistic of structure variation with conditions of deposition can be given. In the average the degree of stacking disorder increases with decreasing deposition temperature.

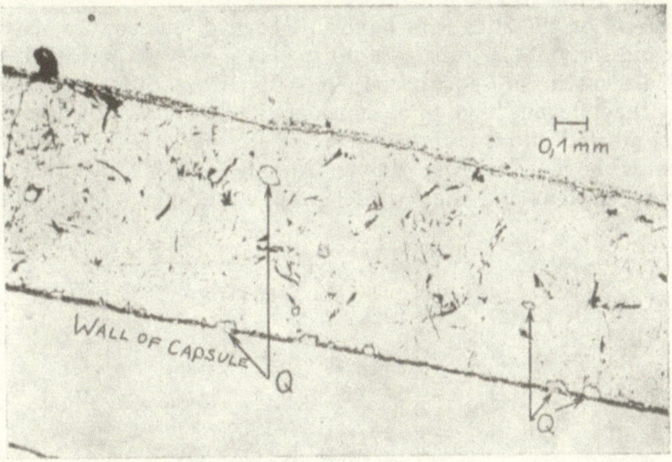

Fig. 8. Thin section of deposit of disordered cristobalite with sporadic quartz crystals of barley corn habit: $T_2 = 650\ °C$, $T_1 = 500\ °C$, $p = 2$ kbar

Assumed a well ordered cristobalite crystal, representing the cubic packing sequence "c", the probability of finding the sequence is $P(c) = 1$. The probability of errors in the c-sequence according to the hexagonal "h"-sequence of tridymite shall be called ε. Then the probability of c-sequences in any given crystal equals $1 - \varepsilon$. With $\varepsilon < 0{,}5$ the cristobalite structure dominates the packing sequence, whereas with $\varepsilon > 0{,}5$ the tridymite structure prevails.

Due to growth inhibition effects (cristobalite octahedra are closed forms with flat faces of low growth rate (Flörke 1966)) cristobalite crystals with $\varepsilon \neq 0$ show the tendency of crystallization in hexagonal platelets instead of octahedra, i.e. with the habit of tridymite but with the properties of cristobalite. With increasing ε the properties show a continuous variation towards tridymite. So what would be adapted at the first glance as "typical" tridymite habit cannot serve as a criterion of discrimination between cristobalite and tridymite. This explaines the paradoxon from Blaue Kuppe mentioned above. The crystals are not paramorphs of cristobalite after tridymite but have been deposited as cristobalite with an appreciable ε of about 0,2 to 0,3, thus showing the habit of tridymite, but the properties are more on the side of cristobalite.

Tridymite crystals grown by vapour deposition do not show any remarkable tendency of paramorphic transformation into quartz. They can be used as seed crystals in another run and they continue growing well below 870 °C.

Hyalite is considered to be a variety of opal. Its deposition from the vapour phase as well as its structure, however, distinguishes it markedly from opal. Opal and lussatite separate out from liquids and show discontinuous structure (Jones and Segnit). In our experiments hyalite formed at the end of the runs during cooling of the bombs in air. At this state the supersaturation increases strongly. Crystallization of the separated SiO_2 cannot take place if the temperature of the substrate is low enough, but the mobility of the Si-O-bond with the aid of water vapour is still high enough to form a continous random network structure (Fig. 9).

In conclusion it may be said that the formation of the different silica forms under water vapour conditions depends on the temperature of deposition as well as on the rate of transport and the degree of supersaturation. Possible matrix effects of the substrate have not been discussed. In nature the temperature and kind of the substrate must play a very important role. This role we shall discuss in subsequent papers. In nature different transport gases, especially the halogens have to be considered. This shall be the topic of another study just started in our laboratory.

The valuable help of Miss Helene Novara and Mr. Adolf Roubal in the X-ray investigations and the hydrothermal experiments is gratefully acknowledged.

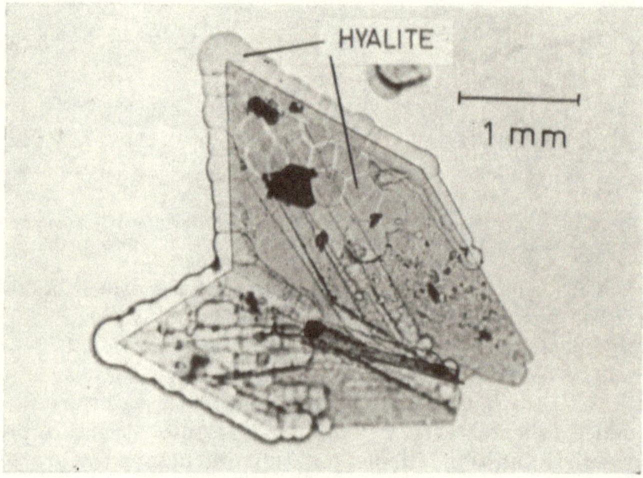

Fig. 9. Twin crystal of tridymite, grown at 750 °C with seam and cover of hyalite, deposited during cooling of the bomb

References

Anderson, G. M., Burnham, C. W.: Amer. J. Sci. **263**, 494 (1965)
Blaschke, R., Flörke, O. W.: Beiträge zur Elektronenmikroskopischen Direktabbildung von Oberflächen **4/2**, 489 (1971)
Brady, E. L.: J. Physic. Chem. **57**, 706 (1953)
Flörke, O. W.: Z. Krist. **112**, 126 (1959)
Flörke, O. W.: Kristall u. Technik **1**, 405 (1966)
Gillingham, G. E.: Econ. Geol. **43**, 241 (1948)
Heitmann, H. G.: Diss. TH Karlsruhe 1963
Jones, J. B., Segnit, E. R.: J. Geol. Soc. Australia **18**, 57 (1971)
Kirsch, H.: N. Jb. Mineral. Abh. **106**, 287 (1967)
Nitsche, R.: J. Cryst. Growth **9**, 238 (1971)
Ramdohr, P.: Jb. Preuß. Geol. Landesanst. **11**, Tl. 1, H. 2, 184 (1919)
Schäfer, H.: Chemische Transportreaktionen, Weinheim 1962
Wendlandt, H. G., Glemser, O.: Naturwissenschaften **50**, 352 (1963)

(Received May 20, 1971)

Authors' address:

Prof. Dr. O. W. Flörke
Institut für Mineralogie, Ruhr-Universität
46 Bochum

L. Däweritz

Humboldt-Universität zu Berlin
Sektion Physik, Bereich Kristallographie

Wahrscheinlichkeit der Zwillingsbildung in dünnen Schichten von Halbleitersubstanzen mit Diamant- und Zinkblendestruktur

Zur Abschätzung der Wahrscheinlichkeit der Zwillingsbildung in Epitaxieschichten halbleitender Materialien mit Diamant- und Zinkblendestruktur werden die Differenzen zwischen den Keimbildungsarbeiten von Keimen in Normalorientierung und Keimen in Zwillingsorientierung berechnet. Im Falle gleicher Versuchsbedingungen (Substrattemperatur, Übersättigung) erhöht sich die Wahrscheinlichkeit der Zwillingsbildung in der Reihenfolge SiC – Si – Ge – ZnS – ZnSe, CdS – CdSe – ZnTe – CdTe. Die Ergebnisse werden im Zusammenhang mit Experimenten diskutiert.

In order to estimate the probability of twinning in epitaxial films of semiconducting materials having diamond and zincblende structure the differences between the nucleation energy of nuclei in normal orientation and the nucleation energy of nuclei in twin orientation are calculated. In the case of equal experimental parameters (substrate temperature, supersaturation), the probability of twinning increases in the order of SiC – Si – Ge – ZnS – ZnSe, CdS – CdSe – ZnTe and CdTe. The results are discussed in connection with experimental work.

1. Einleitung

Im Hinblick auf ihre Anwendung in der Elektronik und Mikroelektronik kommt den Elementhalbleitern Silizium und Germanium sowie den Verbindungshalbleitern – insbesondere den zur Gruppe der $A^{III}B^{V}$- bzw. $A^{II}B^{VI}$-Verbindungen gehörenden – große Bedeutung zu. Besondere Aufmerksamkeit wird hierbei Untersuchungen zur Herstellung von Epitaxieschichten möglichst hoher Perfektion gewidmet.

Bei der Wahl geeigneter Substrat-Deposit-Kombinationen und entsprechender Versuchsbedingungen ist es möglich, durch Hochvakuumaufdampfung sehr gut orientierte Schichten zu erhalten. Problematisch ist jedoch die Vermeidung von Strukturdefekten infolge Zwillingsbildung, der abwechselnden Aufeinanderfolge von Bereichen mit kubischer und hexagonaler Stapelfolge sowie des Auftretens von Stapelfehlern. Sehr deutlich zeigt sich das beispielsweise bei Untersuchungen zur CdSe-Epitaxie (Kleber, Mietz, Neumann; Yasuda; Däweritz 1971b).

Die $A^{III}B^{V}$-Verbindungen kristallisieren mit Ausnahme der Nitride in der kubischen Zinkblendestruktur. Jedoch ist in dünnen Schichten in untergeordnetem Maße auch das Auftreten hexagonaler Stapelfolgen möglich. Das wurde beispielsweise für mit Hilfe der Flash-Technik hergestellte GaAs-Schichten

(FARUKHI, CHARLSON) und durch HCl-Transport erhaltene InP-Schichten (HOLLOWAY) nachgewiesen. Bei den $A^{II}B^{VI}$-Verbindungen ist das Auftreten hexagonaler Stapelfolgen bzw. von Schichten rein hexagonaler Struktur wesentlich häufiger.

Da die physikalischen Eigenschaften in einem engen Zusammenhang mit der Realstruktur der Schichten stehen, sind Aussagen über die Wahrscheinlichkeit des Auftretens der oben genannten Strukturdefekte von großem praktischen Interesse. In einer früheren Arbeit (DÄWERITZ 1971a) wurde bereits die relative Stabilität der Zinkblende- und Wurtzitstruktur von $A^{II}B^{VI}$-Verbindungen untersucht. In der vorliegenden Arbeit soll die Wahrscheinlichkeit des Auftretens der Zwillingsbildung in Epitaxieschichten halbleitender Substanzen mit Diamant- und Zinkblendestruktur, also ein weiterer wesentlicher Strukturdefekt, betrachtet werden.

2. Die Theorie von BOROVINSKY und VORONZOVA zur Wahrscheinlichkeit der Zwillingsbildung

BOROVINSKY und VORONZOVA berechneten die Ausmaße und die Bildungsarbeit zweidimensionaler einschichtiger Keime in Normal- und Zwillingsorientierung eines Kristalls mit Diamantstruktur auf den {111}-Flächen der arteigenen Unterlage. Dabei wird von der Abhängigkeit der Bindungsenergie von der Umgebung der gebildeten Aggregate ausgegangen, wobei nur die Wechselwirkung nächster Nachbarn berücksichtigt wird. Die Energie einer Bindung von Bausteinen, die nur abgesättigte Valenzen besitzen, sei Φ. Für die Bindungsenergie von Bausteinen mit einer oder mehreren freien Valenzen wird Φ_1 angesetzt. Zwischen Bausteinen mit einer freien Valenz und Bausteinen mit mehreren freien Valenzen wird nicht unterschieden, da der Anteil von Bausteinen, die mehr als eine freie Valenz aufweisen, sehr klein ist. Das gilt zumindest ab einer bestimmten Keimgröße. Zunächst sollen die Verhältnisse für die Stapelfolge ... $A\alpha$ $B\beta$ $C\gamma$ $A\alpha$ $B\beta$ $C\gamma$ ... betrachtet werden. In Figur 1 ist eine formal mögliche Aufeinanderfolge der Atomanlagerung dargestellt. Die Unterlagenebene sei A mit Atomen auf den Positionen 1, 2, 4, 5, 7, 9, ... Bei der Anlagerung des Atoms 1 der Ebene α wird eine Bindung zum darunterliegenden Oberflächenatom ausgebildet. Die Bindungsenergie beträgt Φ_1, da

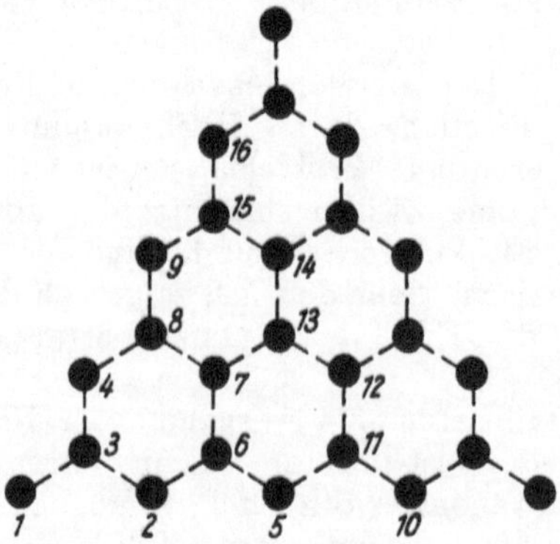

Fig. 1. Formale Anlagerungsschritte bei zweidimensionaler Keimbildung auf einer (111)-Fläche der arteigenen Unterlage mit Diamantstruktur

die drei übrigen Valenzen nicht abgesättigt sind. Die Valenzen des Unterlagenatoms hingegen sind jetzt alle abgesättigt; deshalb ändert sich die Energie von drei Bindungen von Φ_1 auf Φ. Für die freiwerdende Energie ergibt sich insgesamt $\Phi_1 + 3(\Phi - \Phi_1)$. Das gleiche gilt für die Anlagerung eines Atoms in der Position 2. Wird ein Atom auf der Position 3 in der Ebene B angelagert, so wird je eine Bindung Φ_1 zu den Atomen 1 und 2 der Ebene α ausgebildet. Bei Einbau des Atoms 4 in der Ebene α schließlich werden eine Bindung Φ_1 zu Atom 3 der Ebene B und eine Bindung Φ_1 zu dem auf Position 4 befindlichen Oberflächenatom ausgebildet.

Da bei der Anlagerung eines Teils der Atome nur die Absättigung einer Valenz erfolgt, ist das Keimwachstum mit der Überwindung von Energiebarrieren verbunden. Die Wahrscheinlichkeit des Wachstums eines Keims wird die Wahrscheinlichkeit seines Zerfalls erst nach Erreichen einer kritischen Größe überschreiten.

Eine systematische Betrachtung der Anlagerungsschritte führt zu den in Tabelle 1 angegebenen Ergebnissen.

Tabelle 1

Freiwerdende Energie bei zweidimensionaler Keimbildung auf einer (111)-Fläche der arteigenen Unterlage mit Diamantstruktur

Anlagerungsschritt	freiwerdende Energie
1	$\Phi_1 + 3(\Phi - \Phi_1)$
2	$\Phi_1 + 3(\Phi - \Phi_1)$
3	$2\Phi_1$
4	$2\Phi_1 + 3(\Phi - \Phi_1)$
5	$\Phi_1 + 3(\Phi - \Phi_1)$
6	$2\Phi_1$
7	$2\Phi_1 + 3(\Phi - \Phi_1)$
8	$2\Phi_1$
9	$2\Phi_1 + 3(\Phi - \Phi_1)$
10	$\Phi_1 + 3(\Phi - \Phi_1)$
11	$2\Phi_1$
12	$2\Phi_1 + 3(\Phi - \Phi_1)$
13	$2\Phi_1 + 1(\Phi - \Phi_1)$
14	$2\Phi_1 + 3(\Phi - \Phi_1)$
15	$2\Phi_1$
16	$2\Phi_1 + 3(\Phi - \Phi_1)$

Für einen dreieckigen Keim mit n Atomen in der Randreihe ergibt sich die Bindungsenergie W_0 durch Summation über die Anlagerungsschritte zu

$$W_0 = (2n^2 - n + 3)\,\Phi - 3\,\Phi_1\,. \tag{1}$$

Die Bildungsarbeit für einen zweidimensionalen Keim A_2 beträgt

$$A_2 = -\mu n^2 - W_0\,, \tag{2a}$$

$$A_2 = -(\mu + 2\,\Phi)\,n^2 + \Phi\,n + 3\,(\Phi_1 - \Phi)\,, \tag{2b}$$

wobei μ das chemische Potential der Gasphase bedeutet.

Unter Gebrauch der modifizierten Gibbs-Thomson'schen Gleichung

$$2\,\Phi + \mu = kT \ln p/p_0 \tag{3}$$

ergibt sich

$$A_2 = -\,n^2\,kT \ln p/p_0 + \Phi\,n + 3\,(\Phi_1 - \Phi)\,, \tag{4}$$

wobei p den tatsächlichen Dampfdruck und p_0 den Gleichgewichtsdampfdruck über der Unterlage bedeuten.

Die Keimbildungsarbeit als Funktion von n weist für ein bestimmtes n ein Maximum auf. Durch dieses Maximum wird der kritische Keim n^* bestimmt.

$$n^* = \frac{\Phi}{2\,kT \ln p/p_0}\,. \tag{5}$$

Folglich beträgt die Bildungsarbeit des kritischen Keims

$$A_2^* = \frac{\Phi^2}{4\,kT \ln p/p_0} + 3\,(\Phi_1 - \Phi)\,. \tag{6}$$

Es ist möglich, daß beispielsweise nach der Doppelschicht $C\gamma$ nicht eine Schicht $A\alpha$, sondern eine Schicht $B\beta$ aufwächst, also Zwillingsbildung auftritt. Eine Abschätzung der Energie der Ebene der Zwillingsbildung ist möglich, wenn man davon ausgeht, daß die Änderung der Anordnung der Atome vergleichbar ist mit der Änderung der gegenseitigen Anordnung der Atome bei innerer Drehung in den Molekülen gesättigter Kohlenwasserstoffe und ihrer Analogen von Silizium und Germanium. Man kann annehmen, daß die Energie der Ebene der Zwillingsbildung im Diamant — bezogen auf eine Bindung, die diese Ebene schneidet — annähernd von der gleichen Größe sein wird wie die Potentialschwelle für die innere Drehung um die zentrale Bindung im Hexamethyläthan-Molekül $(CH_3)_3C{-}C(CH_3)_3$. Die Keimbildungsarbeit für die Zwillingsorientierung ist gegenüber der für die Normalorientierung um den Betrag $\gamma/2\,(n^2 + n)\,\Phi$ größer. Hierbei bedeutet γ die in Einheiten der Bindungsenergie Φ ausgedrückte Grenzenergie der Ebene der Zwillingsbildung in bezug auf eine Bindung. γ ist folglich dimensionslos. $1/2\,(n^2 + n)$ ist die Anzahl der Bindungen zur Unterlage für einen dreieckigen Keim mit n Atomen in der Randreihe.

Die Werte für γ können aufgrund theoretischer Betrachtungen und aus experimentellen Werten für die innere Drehung an den oben genannten organischen Verbindungen abgeschätzt werden. Für den Diamant beträgt $\gamma \approx 0{,}07$. Für Silizium und Germanium, für die nur ungenügende bzw. keine Angaben über die Potentialschwelle der inneren Drehung in den entsprechenden organischen Verbindungen vorliegen, kann wegen der im Vergleich zum Diamant größeren Bindungslänge zwischen den Atomen ein Wert von 0,02 bis 0,03 angenommen werden.

Für die Bildungsarbeit eines Keims in Zwillingsorientierung $A_{2\,\mathrm{Zw}}$ ergibt sich

$$A_{2\,\mathrm{Zw}} = n^2\,(\gamma/2\,\Phi - k\,T \ln p/p_0) + n\,(\gamma/2\,\Phi + \Phi) + 3\,(\Phi_1 - \Phi)\,. \tag{7}$$

Notwendige Bedingung für die Bildung zweidimensionaler einschichtiger Keime in Zwillingsorientierung ist die Existenz eines Maximums in Beziehung (7). Das ist erfüllt für

$$kT \ln p/p_0 > \frac{\gamma\,\Phi}{2} \tag{8}$$

bzw.

$$\ln p/p_0 > \frac{\gamma\, \Phi}{2\, k\, T} \tag{9a}$$

bzw.

$$T > \frac{\gamma\, \Phi}{2\, k \ln p/p_0}. \tag{9b}$$

Für den kritischen Keim in Zwillingsorientierung n^*_{Zw} ergibt sich

$$n^*_{Zw} = \frac{\gamma/2\, \Phi + \Phi}{2\, k\, T \ln p/p_0 - \gamma\, \Phi}. \tag{10}$$

Die Bildungsarbeit des kritischen Keims in Zwillingsorientierung A^*_{2Zw} beträgt

$$A^*_{2Zw} = \frac{(\gamma/2\, \Phi + \Phi)^2}{4\, k\, T \ln p/p_0 - 2\, \gamma\, \Phi} + 3\, (\Phi_1 - \Phi)\,. \tag{11}$$

Für die Beurteilung der Wahrscheinlichkeit des Auftretens von Keimen in Zwillingsorientierung ist die Differenz der Bildungsarbeiten von Keimen in Zwillingsorientierung und solchen in Normalorientierung

$$\Delta A^*_2 = \frac{(\gamma/2\, \Phi + \Phi)^2}{4\, kT \ln p/p_0 - 2\, \gamma\, \Phi} - \frac{\Phi^2}{4\, kT \ln p/p_0} \tag{12}$$

von Interesse.

3. Berechnungen zur Wahrscheinlichkeit der Zwillingsbildung

Die Betrachtungsweise von Borovinsky und Voronzova soll auf Verbindungen vom Typ AB ausgedehnt werden. Es wird angenommen, daß die Oberflächenebene der Unterlage aus B-Atomen besteht.[1]) Die Energie einer Bindung sei sowohl für A-Atome als auch für B-Atome, die nur abgesättigte Valenzen besitzen, $\bar{\Phi}$. Bie einer exakten Betrachtung müßten Werte angenommen werden, die für die eine Atomart $> \bar{\Phi}$ und für die andere Atomart $< \bar{\Phi}$ sind. Durch die Summation über die Anlagerungsschritte ist jedoch der resultierende Fehler für hinreichend große Keime zu vernachlässigen. Die Bindungsenergie von A-Atomen mit freien Valenzen sei Φ'_1 und die Bindungsenergie von B-Atomen mit freien Valenzen Φ'_2. Somit erhält man für die Differenz der Bildungsarbeiten von kritischen Keimen in Zwillings- und in Normalorientierung

$$\Delta A'^*_2 = \frac{[\gamma/2\, \bar{\Phi} + \bar{\Phi} + 5/2\, (\Phi'_2 - \Phi'_1)]^2}{4\, k\, T \ln p/p_0 - 2\, \gamma\, \Phi - 2\, (\Phi'_2 - \Phi'_1)} - \frac{[\bar{\Phi} + 5/2\, (\Phi'_2 - \Phi'_1)]^2}{4\, k\, T \ln p/p_0 - 2 (\Phi'_2 - \Phi'_1)}\,. \tag{13}$$

Eine numerische Auswertung nach dieser Beziehung kann jedoch nicht vorgenommen werden, da die Größen Φ'_1 und Φ'_2 unbekannt sind. Es sollten jedoch abschätzende Betrachtungen möglich sein, wenn man gewissermaßen über die Bindungsanteile der A- und B-Atome mittelt und die Beziehungen (1) bis (12) verwendet.

Die auf ein Mol bezogene Bindungsenergie kann nach der Beziehung

$$\Phi = \frac{L_{Sb} + 1/2\, D_{B_2}}{4} \tag{14}$$

aus der Sublimationswärme L_{Sb} der Verbindung und der Dissoziationswärme D_{B_2}

[1]) Wird die Oberflächenebene der Unterlage aus A-Atomen gebildet, so müssen im folgenden nur Φ'_1 und Φ'_2 vertauscht werden.

der B_2-Moleküle ermittelt werden. Diese Beziehung ergibt sich aus der von Höschl und Koňák angegebenen Formel zur Abschätzung der für die Entfernung eines A- bzw. B-Atoms aus der Kristalloberfläche erforderlichen Energie. Dabei werden im Mittel zwei Bindungen getrennt. Auf eine Bindung entfällt folglich die halbe Energie.

Die Sublimations- und Dissoziationswärmen sind in den Tabellen 2 und 3 zusammengestellt. Unter Verwendung dieser Werte ergeben sich nach Beziehung (14) die in Tabelle 4 angeführten Bindungsenergien.

Tabelle 2

Sublimationswärmen[1]) L_{Sb} von $A^{II}B^{VI}$-Verbindungen

Verbindung	L_{Sb} (kcal/Mol)	Literatur
ZnS	69,0	Kubaschewski, Evans
ZnSe	65,0	Korneyeva et al. 1960 b
ZnTe	48,7	Korneyeva et al. 1960 a
CdS	51,0	Kubaschewski, Evans
CdSe	50,1	Korneyeva et al. 1960 b
CdTe	43,5	Korneyeva et al. 1960 b

[1]) Die angeführten Sublimationswärmen beziehen sich auf unterschiedliche Temperaturen. Jedoch sind die Angaben für abschätzende Betrachtungen ausreichend, da die Temperaturabhängigkeit nicht groß sein sollte.

Tabelle 3

Dissoziationsenergien D_{B_2} von B_2-Molekülen

Molekül	D_{B_2} (kcal/Mol)	Literatur
S_2	102,0	Colin et al. 1960; 1964; Marsden; Brewer
Se_2	75,4	Colin, Goldfinger
Te_2	52,2	Porter; Colin

Tabelle 4

Bindungsenergien Φ von $A^{VI}B^{II}$-Verbindungen in (eV)

	S	Se	Te
Zn	1,3	1,1	0,81
Cd	1,1	0,95	0,75

Bei der Wahl des Wertes für die Grenzenergie der Ebene der Zwillingsbildung γ kann man von der Tatsache ausgehen, daß γ in einem engen Zusammenhang mit der Bindungsenergie Φ steht. Und zwar wird γ mit steigendem Φ ebenfalls zunehmen. Nimmt man nach Pauling (1948, 1960) für die Bindungsenergie im Diamant den Wert 58,6 kcal/Mol, für Silicium 42,2 kcal/Mol, für Germanium 37,6 kcal/Mol und nach Borovinsky und Voronzova für die entsprechenden γ-Werte 0,07, 0,03 und 0,02 an, so läßt sich der Zusammenhang zwischen γ und Φ in (eV) gut durch die Beziehung

$$\gamma = a\,\Phi^b \tag{15}$$

mit $a = 0{,}00584$ und $b = 2{,}654$ beschreiben. Diese Funktion ist in Figur 2 graphisch dargestellt. Unter Verwendung der in Tabelle 4 angegebenen Φ-

Werte ergeben sich für die $A^{II}B^{IV}$-Verbindungen die in Tabelle 5 zusammengestellten γ-Werte. Durch die Multiplikation dieser Werte mit der Bindungsenergie Φ und der Anzahl der Bindungen pro Flächeneinheit gelangt man zu den in Tabelle 6 angeführten Stapelfehlerenergien.

Tabelle 5

Grenzenergien der Ebene der Zwillingsbildung γ für $A^{II}B^{IV}$-Verbindungen

	S	Se	Te
Zn	0,012	0,0075	0,0033
Cd	0,0075	0,0051	0,0027

Tabelle 6

Stapelfehlerenergien für A^{II}-B^{VI}-Verbindungen in (erg· cm^{-2})

	S	Se	Te
Zn	20	10	3
Cd	9	5	2

Figur 3 zeigt die nach Beziehung (9b) ermittelte kritische Temperatur der Zwillingsbildung als Funktion des Übersättigungsverhältnisses für ZnS, CdSe, CdTe und Silizium. Gewöhnlich wird bei der Hochvakuumaufdampfung mit Übersättigungsverhältnissen $\gg 10$ gearbeitet, und die Substrattemperatur liegt in den meisten Fällen über 200 °C. Daraus folgt, daß die Zwillingsbildung

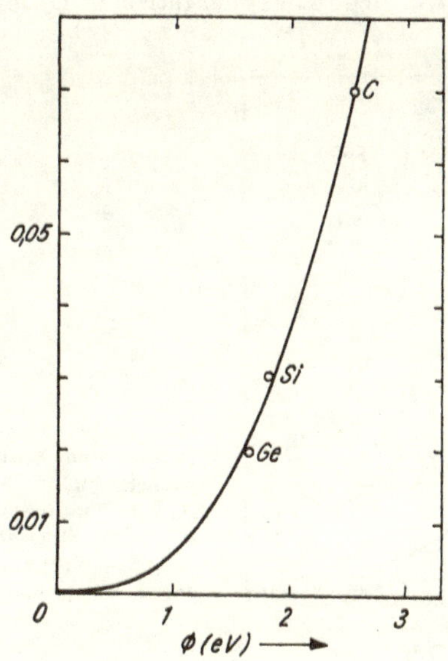

Fig. 2. Grenzenergie der Ebene der Zwillingsbildung γ als Funktion der Bindungsenergie Φ

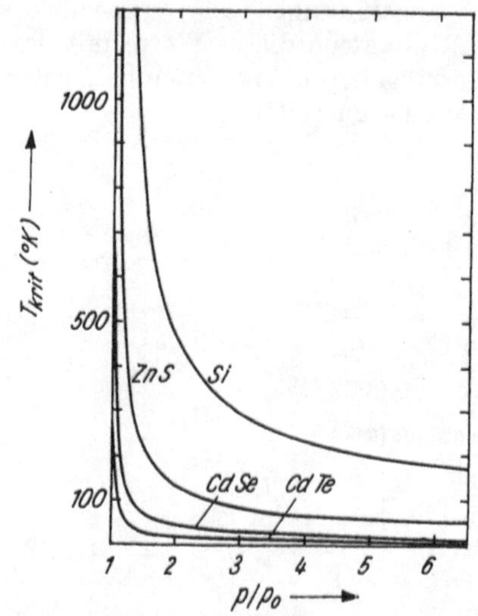

Fig. 3. Kritische Temperatur $T_{krit.}$ der Zwillingsbildung als Funktion des Übersättigungsverhältnisses p/p_0 für ZnS, CdSe, CdTe und Si

prinzipiell für alle $A^{II}B^{VI}$-Verbindungen und auch die Elementhalbleiter Silizium und Germanium zu erwarten ist.

In Figur 4 ist die Differenz der Keimbildungsarbeiten für die Zwillings- und Normalorientierung für die $A^{II}B^{VI}$-Verbindungen, Silizium, Germanium und Siliziumcarbid bei einem Übersättigungsverhältnis von 10^3 in Abhängigkeit von der Temperatur dargestellt. Die Wahrscheinlichkeit der Zwillingsbildung erhöht sich in der Reihenfolge SiC – Si – Ge – ZnS – ZnSe, CdS – CdSe – ZnTe – CdTe und ist bei den $A^{II}B^{VI}$-Verbindungen um ein Vielfaches wahrscheinlicher als bei den Elementhalbleitern. Bei den üblichen Substrattemperaturen und einem Überstätigungsverhältnis von 10^3 liegt die energetische Differenz zwischen

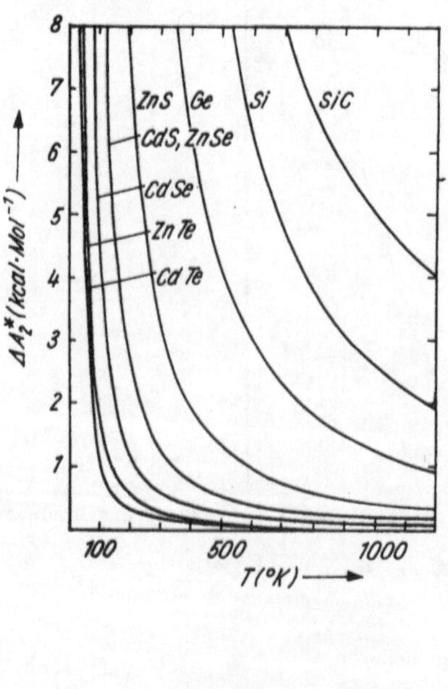

Fig. 4. Differenz zwischen den Bildungsarbeiten kritischer Keime in Zwillings- und in Normalorientierung ΔA_2^* als Funktion der Temperatur T bei einem Übersättigungsverhältnis von $p/p_0 = 10^3$ für $^{II}AB^{VI}$-Verbindungen, Si, Ge und SiC

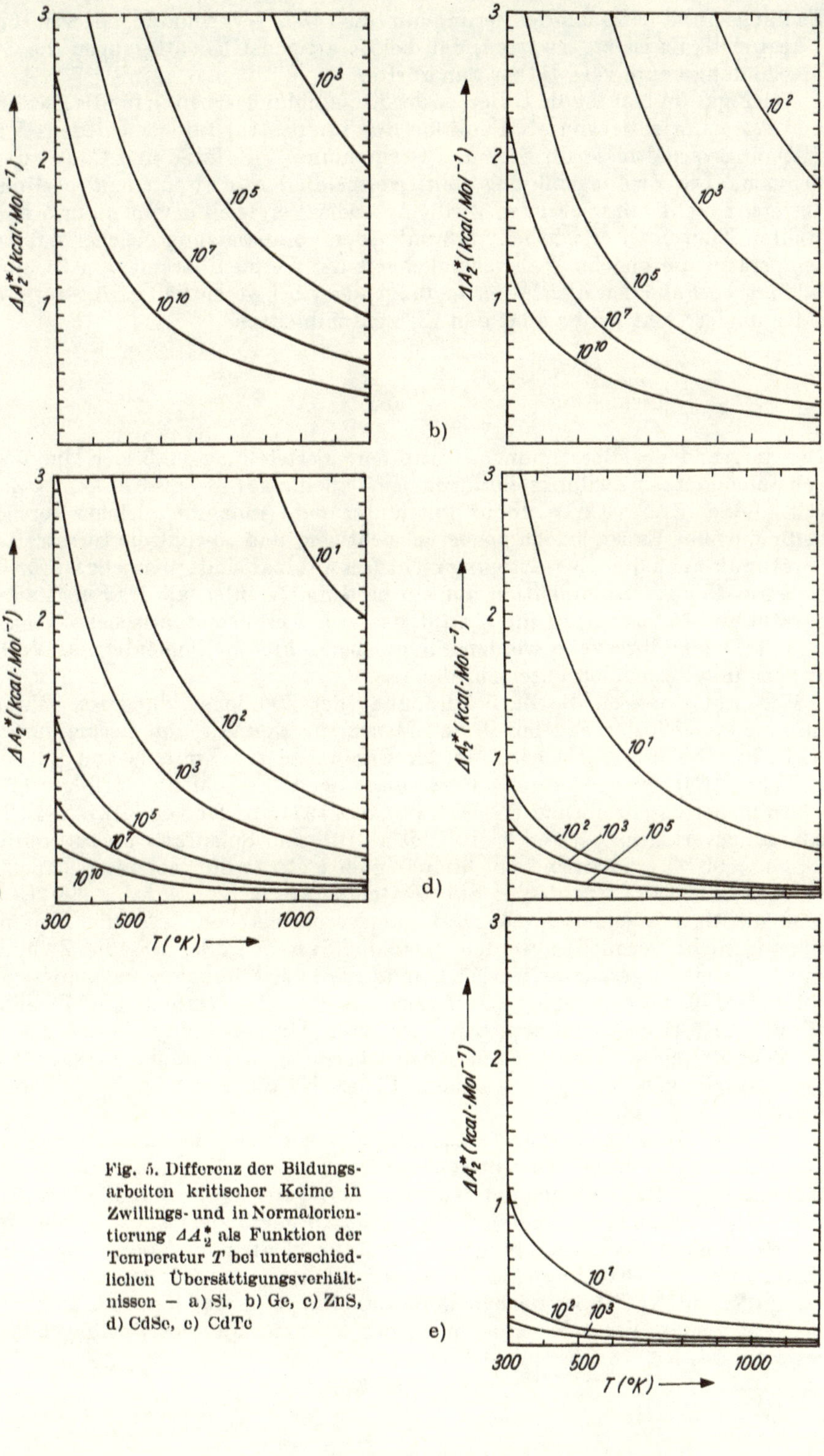

Fig. 5. Differenz der Bildungsarbeiten kritischer Keime in Zwillings- und in Normalorientierung ΔA_2^* als Funktion der Temperatur T bei unterschiedlichen Übersättigungsverhältnissen – a) Si, b) Ge, c) ZnS, d) CdSe, e) CdTe

Zwillings- und Normalorientierung für die $A^{II}B^{VI}$-Verbindungen weit unter 1 kcal/Mol. Es ist zu erwarten, daß bei derartigen Übersättigungen die Zwillingsbildung kaum vermeidbar sein dürfte.

Aus Figur 5a bis e ist die Differenz der Keimbildungsarbeiten für die Zwillings- und Normalorientierung als Funktion der Temperatur für verschiedene Übersättigungsverhältnisse für Silizium, Germanium, ZnS, CdSe und CdTe zu entnehmen. Die Zwillingsbildung sollte vermeidbar sein, wenn eine bestimmte Differenz nicht unterschritten wird. Aus dem Vergleich der in Figur 5 dargestellten Kurven ist ersichtlich, daß unter der Voraussetzung gleicher Substrattemperatur die gleiche Wahrscheinlichkeit der Vermeidbarkeit von Zwillingsbildung im Falle der $A^{II}B^{VI}$-Verbindungen erst bei wesentlich geringerer Übersättigung erreicht wird als bei den Elementhalbleitern.

4. Diskussion

Die aufgrund der Berechnungen getroffene Feststellung, daß sich die Wahrscheinlichkeit der Zwillingsbildung in der Reihenfolge SiC–Si–Ge–ZnS–ZnSe, CdS–CdSe–ZnTe–CdTe erhöht, gilt unter der Annahme gleicher Versuchsbedingungen. Es ist jedoch zu berücksichtigen, daß sowohl die Substrattemperatur als auch die Übersättigung nicht frei wählbar sind; vielmehr ist für jede Substrat-Deposit-Kombination nur ein bestimmtes Intervall der Substrattemperatur und Aufdampfrate interessant, da nur in diesem epitaktisches Wachstum erfolgt. Diese Intervalle werden für die betrachteten Elemente und Verbindungen im allgemeinen unterschiedlich sein.

Wie problematisch die Zurückdrängung der Zwillingsbildung ist, geht aus einem Vergleich der Ergebnisse von Mietz zur Epitaxie von Germanium auf NaF- und NaCl-Spaltflächen und der Ergebnisse zur Epitaxie von CdSe auf CaF_2-Spaltflächen hervor. Die Herstellung der CdSe-Schichten erfolgte in der schon früher beschriebenen Weise (Däweritz 1971b). Die von Mietz bei Übersättigungsverhältnissen von $5 \cdot 10^{13}$ bis $4 \cdot 10^{21}$ und Substrattemperaturen von 280 bis 500 °C erhaltenen Schichten zeigten stets Zwillingsbildung. Auch im System $CdSe/CaF_2$ konnte bei Substrattemperaturen von 425 bis 600 °C und dem niedrigsten erzielten Übersättigungsverhältnis von ≈ 40 die Zwillingsbildung nicht vermieden werden. Daß im System $CdSe/CaF_2$ die Zwillingsbildung nicht ausgeschaltet werden konnte, wird verständlich, wenn man berücksichtigt, daß die energetische Differenz zwischen der Normal- und Zwillingsorientierung trotz der wesentlich geringeren Übersättigung in der gleichen Größenordnung liegt wie bei den hohen Übersättigungen in den Versuchen zur Ge-Epitaxie von Mietz. In beiden Fällen ist die energetische Differenz $<0,3$ kcal/Mol (vgl. Fig. 5).

Bei einer Diskussion von Versuchsergebnissen nach den oben dargelegten Vorstellungen ist zu berücksichtigen, daß die Zwillingsbildung als mit der zweidimensionalen Keimbildung auf der arteigenen Unterlage verbunden betrachtet wird. Die Ergebnisse können jedoch wesentlich modifiziert werden, wenn man die Möglichkeit der Ausbildung von Strukturdefekten im Koaleszenzstadium (Matthews) in Betracht zieht. Die Untersuchungen von Yasuda zur Epitaxie von CdSe auf NaCl-Spaltflächen weisen allerdings darauf hin, daß zumindest bei bestimmten Versuchsbedingungen der Mechanismus der Zwillingsbildung während des Keimwachstums dominiert.

Auch bei Vernachlässigung der in jedem Versuchssystem gegebenen Besonderheiten sollten die dargelegten Betrachtungen zur Wahrscheinlichkeit der Zwillingsbildung abschätzende Aussagen erlauben.

Literatur

Borovinsky, L. A., Voronzova, R. P.: Kristallografiya **12**, 575 (1967)
Brewer, L.: J. Chem. Physics **31**, 1143 (1959)
Colin, R.: Ind. Chim. Belg. **26**, 51 (1961)
Colin, R., Goldfinger, P.: Proceedings of the Int. Symp. on Condensation and Evaporation, Dayton, Ohio 1962 (Herausgeber E. Rutner, P. Goldfinger und J. P. Hirth), S. 165
Colin, R., Jeunehomme, M., Goldfinger, P.: Nature **187**, 408 (1960)
Colin, R., Jeunehomme, M., Goldfinger, P.: Trans. Faraday Soc. **60**, 306 (1964)
Däweritz, L.: Kristall und Technik **6**, 101 (1971a)
Däweritz, L.: Kristall und Technik **6**, 281 (1971b)
Farukhi, M. R., Charlson, E. J.: J. appl. Physics **40**, 5361 (1969)
Holloway, H.: J. Phys. Chem. Solids **29**, 1977 (1968)
Höschl, P., Koňák, C.: phys. stat. sol. **9**, 167 (1965)
Kleber, W., Mietz, I., Neumann, W.: Kristall und Technik **4**, 345 (1969)
Korneyeva, I. V., Belyayev, A. V., Novoselova, A. V.: Ž. Neorg. Chim. **5**, 3 (1960a)
Korneyeva, I. V., Sokolov, V. V., Novoselova, A. V.: Ž. Neorg. Chim. **5**, 241 (1960b)
Kubaschewski, O., Evans, E. L.: Metallurgische Thermochemie, Berlin 1959
Marsden, D. G.: J. Chem. Physics **31**, 1144 (1959)
Matthews, J. W.: Phil. Mag. **4**, 1017 (1959)
Mietz, I.: Dissertation Humboldt-Universität 1967
Pauling, L.: The Nature of the Chemical Bond, Ithaca 1948, S. 53
Pauling, L.: The Nature of the Chemical Bond, Ithaca 1960, S. 85
Porter, R. F.: J. Chem. Physics **34**, 583 (1961)
Yasuda, Y.: Japan. J. appl. Physics **7**, 1171 (1968)

(Eingegangen am 20. Mai 1971)

Anschrift des Verfassers:

Dr. Lutz Däweritz
Humboldt-Universität zu Berlin
Sektion Physik, Bereich Kristallographie
104 Berlin
Invalidenstr. 43

Kristall und Technik	7	1–3	1972	179–182

J. Bohm, S. Stähr und K.-Th. Wilke

Deutsche Akademie der Wissenschaften zu Berlin,
Zentralinstitut für Optik und Spektroskopie

Epitaxie von Platin auf Strontiumbariumniobat

Bei der Züchtung von Einkristallen aus Strontiumbariumniobat (SBN) aus Platintiegeln findet eine Epitaxie von Platinpartikeln auf den Prismenflächen des SBN statt gemäß $\{111\}_{Pt}//\{hk0\}_{SBN}$ und $\langle 110\rangle_{Pt}//\langle 001\rangle_{SBN}$.

Crystals of $Sr_{1-x}Ba_xNb_2O_6$ (SBN), pulled from platinum crucibles, are covered with epitaxial grown platinum particles showing the orientation $\{111\}_{Pt}//\{hk0\}_{SBN}$ and $\langle 110\rangle_{Pt}//\langle 001\rangle_{SBN}$.

Zur Züchtung von Einkristallen hochschmelzender oxidischer Verbindungen aus ihren Schmelzen oder auch aus Schmelzlösungen werden meist Tiegel sowie andere Bauteile im Züchtungsraum aus Platin oder auch aus anderen Platinmetallen und entsprechenden Legierungen verwendet (vgl. z. B. Cockayne). Unter den dabei gegebenen physikalisch-chemischen Bedingungen finden gewisse Wechselwirkungen mit den betreffenden Substanzen statt, als deren Ergebnis bereits von Shappirio et al., Cockayne et al. sowie Wallace et al. Einlagerungen von Edelmetallpartikeln in die Kristalle festgestellt wurden.

Bei der Züchtung von Einkristallen aus Strontiumbariumniobat ($Sr_{1-x}Ba_xNb_2O_6$; SBN) nach der Czochralski-Methode (Ballman und Brown) scheiden sich bei Verwendung von Platin-Tiegeln unter normaler Atmosphäre auf der Oberfläche des wachsenden Kristalls zahlreiche kleine Platinpartikel ab. Als Transportmechanismus kommt die Bildung und Zersetzung von flüchtigem Platin(IV)-oxid gemäß $Pt + O_2 \leftrightarrow PtO_2$ infrage (Schäfer).

In der gewählten Anordnung werden die SBN-Kristalle unter Verwendung eines Keimkristalls in c-Richtung aus der Schmelze (Schmelzpunkt ca. 1480 °C)

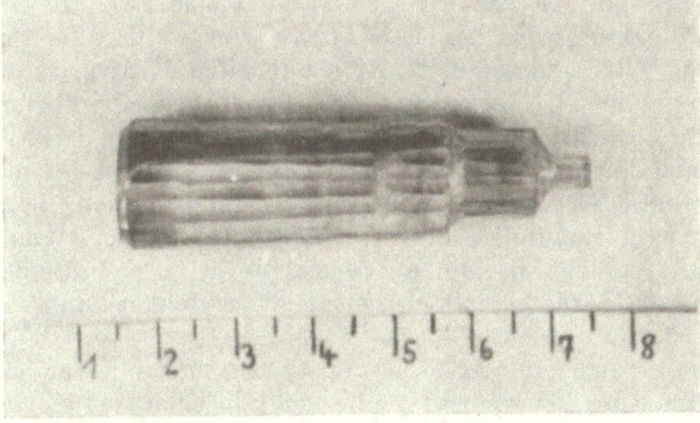

Fig. 1. SBN-Kristall in c-Richtung mit Prismenflächen

gezogen. Es ist eine Eigentümlichkeit des SBN, daß die Kristalle dabei mit glatten Mantelflächen wachsen, die in scharfen Kanten zusammenstoßen (Fig. 1). Es handelt sich um Prismenflächen parallel *c*, die zu den Formen {100}, {110}, {210} und {310} gehören, so daß die Kristalle von insgesamt 24 Prismenflächen umgrenzt werden. Das nach der Züchtung auf den Kristallen befindliche Platin läßt zwei Fraktionen unterscheiden. Ein Teil läßt sich leicht und vollständig – z. B. durch Reiben mit einem Tuch – von den Kristallen entfernen; diese Teilchen bedecken alle Flächen gleichmäßig, sind nicht orientiert und sind offenbar auf Gebiete niedergeschlagen worden, die bereits stärker abgekühlt waren. Der andere Teil der Platinpartikel läßt sich mechanisch nicht und auch mit Königswasser nur unvollständig entfernen; dieser Teil ist offenbar auf heiße Gebiete nahe der Schmelzoberfläche abgeschieden worden, die Partikel sind deutlich orientiert (Fig. 2) und besetzen die Flächen der verschiedenen Formen in ver-

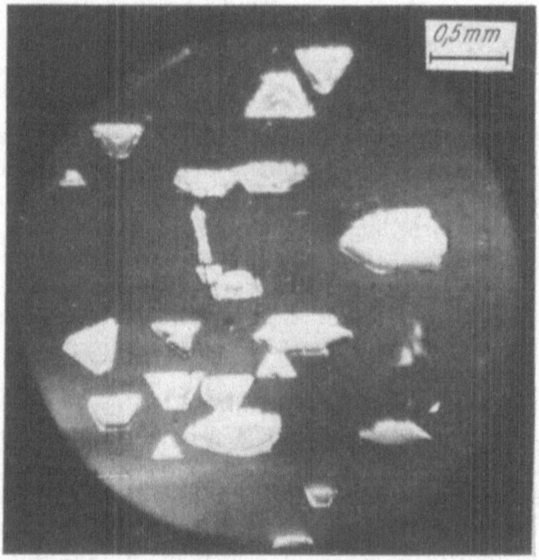

Fig. 2. Orientiert abgeschiedene Platinpartikel auf einer (110)-Fläche von SBN

schieden starkem Maße, was sowohl die Anzahl als auch die Größe der abgeschiedenen Platinpartikel betrifft. Es handelt sich also um eine Epitaxie von Pt auf SBN. Dabei wird die (111)-Fläche vom Pt parallel zur jeweiligen (*hk*0)-Fläche vom SBN und die [110]-Richtung vom Platin parallel zur [001]- bzw. *c*-Richtung vom SBN orientiert. Bei Kristallen mit einem Kationenverhältnis Sr:Ba = 0,75:0,25 ist die Abscheidung auf dem {110}-Prisma am stärksten; danach folgen das {210}- und dann das {100}-Prisma; auf dem {310}-Prisma schließlich sind nur noch eine geringe Menge sehr kleiner Partikel abgeschieden. Diese Rangfolge ändert sich mit einem steigenden Ba-Gehalt der Mischkristalle, so daß bei einem Verhältnis Sr:Ba = 0,5:0,5 das {210}-Prisma deutlich am stärksten mit Pt belegt ist; es folgen dann {110}, {100} und {310} (Fig. 3).

Bei der Epitaxie von Pt auf SBN handelt es sich im Sinne der Klassifikation von Neuhaus um eine Verwachsung eines metallischen mit einem ionogenen Gitter. Zum Vergleich der Gitterparameter sei auf die Gitterkonstante von Pt

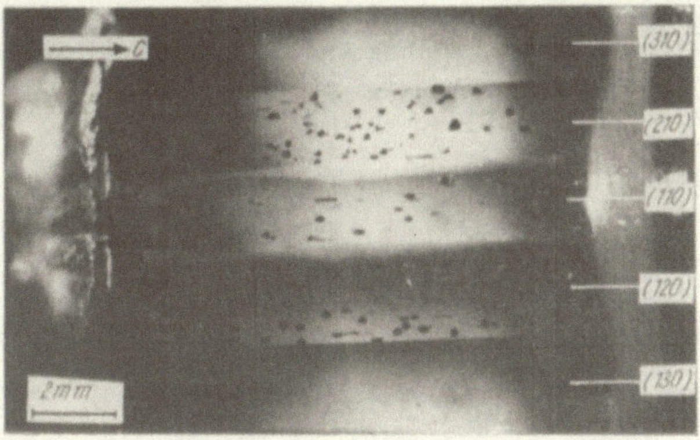

Fig. 3. Epitaxie von Platin auf unterschiedlichen Prismenflächen von SBN bei $x = 0,5$

(Raumgruppe Fm3m) bei 20 °C, $a_0 = 3,92$ Å, und auf die Gitterkonstanten von SBN (Raumgruppe P4bm) bei 25 °C für eine Zusammensetzung mit $x = 0,25$, $a_0 = 12,43$ Å und $c_0 = 3,94$ Å (Jamieson et al.), zurückgegriffen. Da die Epitaxie bei höheren Temperaturen stattfindet, bei denen weder die Gitterkonstanten noch die Raumgruppe von SBN genau bekannt sind, ist beim Vergleich der Gitterparameter mit einem gewissen Fehler zu rechnen. Entsprechend der gefundenen Orientierung ergibt sich eine Korrespondenz von 5 Elementarmaschenlängen des SBN in c-Richtung — 19,7 Å — mit 3,5 Elementarmaschenlängen des Pt in Richtung [110] — 19,4 Å. Die lineare Differenz würde hierbei nur 1,5% entsprechen. Zu diskutieren wäre evtl. auch eine Korrespondenz von 4 Elementarmaschenlängen des SBN in c-Richtung — 15,8 Å — mit 3 Elementarmaschenlängen des Pt in Richtung [110] — 16,6 Å. Im letzten Fall entspräche die lineare Differenz einem Wert von 5%. Diese Korrespondenzen parallel zur c-Richtung des SBN gelten für alle Formen von Prismenflächen gleichermaßen. Senkrecht zur c-Richtung von SBN lassen sich folgende Korrespondenzen angeben) die nun für die einzelnen Formen spezifisch sind:

Für {110} korrespondiert eine Elementarmaschenlänge des SBN $\perp c$ — 17,6 Å — mit 3,5 Elementarmaschenlänge des Pt in Richtung [$2\bar{1}\bar{1}$] — 16,8 Å — mit einer linearen Differenz entsprechend 4,5%;

für {210} korrespondiert eine halbe Elementarmaschenlänge des SBN $\perp c$ — 13,9 Å — mit 3 Elementarmaschenlängen des Pt in Richtung [$2\bar{1}\bar{1}$] — 14,4 Å — mit einer linearen Differenz entsprechend 3,6%;

für {100} korrespondiert eine Elementarmaschenlänge des SBN $\perp c$ — 12,4 Å $= a_0$ — mit 2,5 Elementarmaschenlängen des Pt in Richtung [$2\bar{1}\bar{1}$] — 12,0 Å; lineare Differenz entsprechend 3,2%;

auch für {310} läßt sich eine gute Korrespondenz einer Elementarmaschenlänge des SBN $\perp c$ — 39,3 Å — mit 8 Elementarmaschenlängen des Pt in Richtung [$2\bar{1}\bar{1}$] — 38,4 Å — mit einer linearen Differenz entsprechend 2,3% finden.

Der Habitus der aufgewachsenen Platinpartikel ist meist sechseckig und stellt einen (111)-Schritt durch einen Oktaeder dar.

Auffällig ist die Betonung der c-Richtung des SBN durch den Habitus der Pt-Partikel (vgl. Fig. 3). Nach Kleber und Ickert ist die Usrache solcher Erscheinungen in der Realstruktur der Oberfläche zu suchen. Durch die Züchtung werden nun aber hinsichtlich der Realstruktur durch Rillen- und

Streifenbildung infolge der Kristallrotation gerade die Richtungen senkrecht c_{SBN} besonders betont. Es ist noch ungeklärt, aus welchem Grund die Epitaxie des Pt hingegen parallel c_{SBN} ausgeprägt ist.

Literatur

Ballman, A. A., Brown, H.: J. Crystal Growth **1**, 311 (1967)
Cockayne, B.: 'Platinum Metals Rev. **12**, 16 (1969)
Cockayne, B., Cheswas, M., Gasson, D. B.: J. Mat. Sci. **2**, 7 (1967)
Jamieson, P. B., Abrahams, S. C., Bernstein, J. L.: J. Chem. Physics **48**, 5048 (1968)
Kleber, W., Ickert, L.: Z. physik. Chem. **224**, 364 (1963)
Neuhaus, A.: Fortschr. Mineralogie **29/30**, 136 (1950/51)
Schäfer, H.: Chemische Transportreaktionen (S. 39, 40, 74), Weinheim/Bergstr. 1962
Shappirio, J. R., AuCoin, T. R., Gualties, J. G.: J. appl. Physics **34**, 3643 (1963)
Wallace, C. A., White, E. A. D.: Proc. Internat. Conf. on Crystal Growth, Boston 1966, S. 431

(Eingegangen am 20. Mai 1971)

Anschrift der Verfasser:

Dr. habil. J. Bohm
Dipl.-Min. S. Stähr
Dr. habil. K.-Th. Wilke
Deutsche Akademie der Wissenschaften zu Berlin
Zentralinstitut für Optik und Spektroskopie
1199 Berlin-Adlershof
Rudower Chaussee 6

Kristall und Technik	7	1–3	1972	183–186

F. Bernhard, U. Müller-Jahreis

Sektion Physik der Humboldt-Universität zu Belrin
Bereich Atomstoßprozesse der Festkörperphysik

Die Kombination der Schwingquarzmethode mit Energieverlustmessungen leichter Ionen zum Nachweis von Adsorptionsschichten an extrem dünnen freitragenden Folien

Die Meßgenauigkeit der Schichtdicke dünner Aufdampfschichten mit Hilfe der Schwingquarzmethode wurde in den letzten Jahren soweit erhöht, daß dieses Verfahren heute zu den genauesten Dickenmeßmethoden gezählt werden kann (Kato). Da jedoch auch gegenwärtig noch unterschiedliche Ansichten über die Zuverlässigkeit der Schwingquarzmethode bestehen, ist jeder Vergleich mit einer anderen, vergleichbar genauen Methode nützlich.

Eine Möglichkeit des unmittelbaren Vergleichs ergab sich bei uns in Zusammenhang mit der Messung von Energieverlusten von Lithiumionen im Energiebereich von 10 bis 100 keV in freitragenden amorphen und polykristallinen Folien verschiedener Elemente im Dickenbereich von 100 bis 600 Å (Bernhard, Müller-Jahreis, Rockstroh, Schwabe), (Apel, Müller-Jahreis, Rockstroh, Schwabe), was Atom-Flächendichten von etwa 6 bis 50×10^{16} Atome je cm² bzw. Massenbelegungen von 3 bis 50 µg je cm² entspricht.

Zwischen der Massenbelegung $\varrho = M\,N\,\Delta R$ (M — Atommasse, N — Atomzahldichte, ΔR — Schichtdicke) und dem Energieverlust ΔE der Ionen in der Folie besteht die Proportionalität

$$\Delta E = S \cdot \varrho / M = S\,N\,\Delta R\,. \tag{1}$$

Der für jede Ionen-Target-Kombination spezifische Bremsquerschnitt S ist von der Energie der Ionen abhängig, nicht dagegen von den Größen N und ΔR bzw. ϱ.

Bei der Schwingquarzmethode wird die Proportionalität der Frequenzänderung eines bedampften Schwingquarzes zur Massenbelegung ϱ ausgenutzt:

$$\Delta f = -\,C\,\varrho = -\,C\,M\,N\,\Delta R \tag{2}$$

(C enthält nur die Daten des Schwingquarzes) (Sauerbrey). Damit ist ein unmittelbarer Vergleich beider Meßmethoden möglich. Für die Messungen standen uns eine Apparatur zur Bestimmung von Energieverlusten mit einer Genauigkeit $<1\%$ und eine Schwingquarz-Meßapparatur mit einer Genauigkeit von 1,5% zur Verfügung (Wronski). Mit beiden Meßapparaturen konnten relative Schichtdickenänderungen von einer monoatomaren Schicht sicher nachgewiesen werden.

Bei den Messungen wurden die Folien im Hochvakuum (Restgasdruck $<10^{-5}$ Torr) sowohl auf mit NaCl beschichtete Glasträger als auch auf den Schwingquarz aufgedampft. Danach erfolgte die Messung der Massenbelegung mittels der Schwingquarzmethode. Anschließend wurden die Folien im Wasserbad abgelöst und in der Energieverlust-Meßapparatur (Vakuum ca. 10^{-6} Torr) ausgemessen.

Bezeichnen wir den gemessenen Energieverlust der Ionen mit ΔE^*, so erhalten wir mit dem durch die Schwingquarzmethode ermittelten Wert $N\,\Delta R$ die Größe

$$S^* = \Delta E^*/N\,\Delta R\,, \tag{3}$$

die in Figur 1 für Aluminium- und Kupferfolien dargestellt ist. Da die Größe S^* eine deutliche Abhängigkeit von $N\,\Delta R$ zeigt, kann sie nicht mit dem Bremsquerschnitt identifiziert werden. Der gemessene Kurvenverlauf läßt sich jedoch

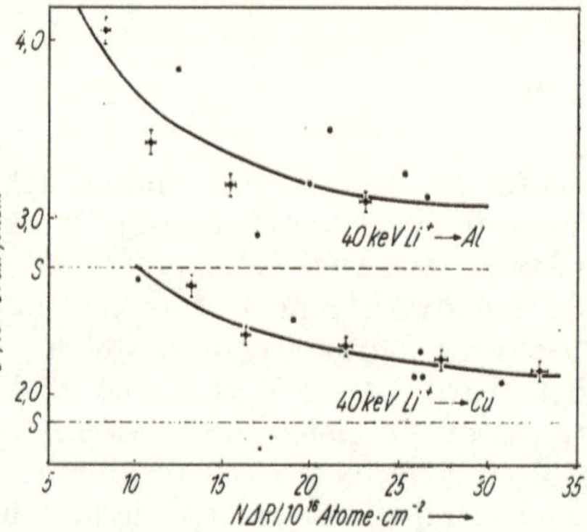

Fig. 1. Bremsquerschnittmessungen an Aluminium- und Kupferfolien verschiedener Dicke. ——— S*, – – – S

verstehen, wenn man sich den gemessenen Energieverlust ΔE^* der Ionen aus einem Energieverlust ΔE bedingt durch das Folienmaterial und einen durch Adsorptionsschichten bedingten zusätzlichen, näherungsweise konstanten Energieverlust $\widetilde{\Delta E}$ zusammengesetzt denkt:

$$\Delta E^* = \Delta E + \widetilde{\Delta E}\,. \tag{4}$$

Damit wird

$$S^* = (\Delta E + \widetilde{\Delta E})/N\,\Delta R = S + \widetilde{\Delta E}/N\,\Delta R\,. \tag{5}$$

Das heißt, zum Bremsquerschnitt S, der von $N\,\Delta\,R$ unabhängig ist, wird ein durch Adsorptionsschichten bedingter Term addiert, der bei dünnen Schichten (d. h. kleinen $N\,\Delta R$) zunehmenden Einfluß gewinnt. Physikalisch würde das eine durch die Folienpräparation bedingte Adsorptionsschicht konstanter Dicke bedeuten, die sich bei dünnen Folien prozentual stark bemerkbar macht und bei sehr dicken Folien vernachlässigbar wird. Aus dem Kurvenverlauf in Figur 1 läßt sich durch eine Ausgleichsrechnung ein mittleres $\widetilde{\Delta E}$ ermitteln. Die um diesen Wert korrigierten Kurven, die den Bremsquerschnitten von Aluminium und Kupfer entsprechen, sind in Abb. 1 gestrichelt gezeichnet worden. Aus der starken Streuung der Meßwerte ist jedoch zu schließen, daß die Adsorptionsschichten auf den einzelnen Folien durchaus unterschiedliche Stärke haben.

Da die Annahme nahe lag, daß die Adsorptionsschichten neben dem jeweiligen Oxid vorwiegend aus Wasser bestehen, wurden Messungen bei unterschiedlichen Folientemperaturen ausgeführt. Figur 2 zeigt die an einer Cu- und einer

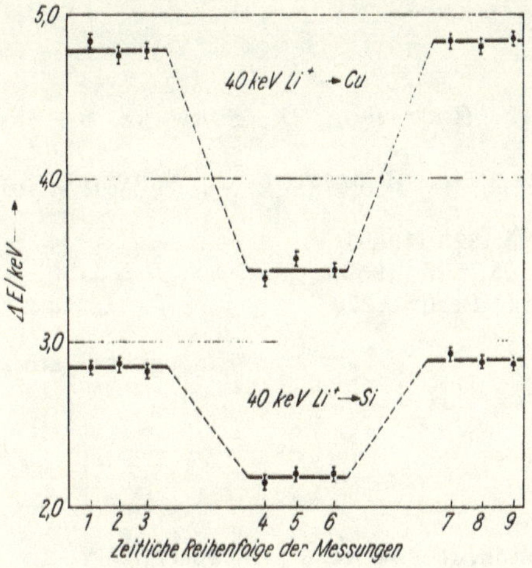

Fig. 2. Energieverlustmessungen an einer Kupfer- und einer Siliziumfolie bei verschiedenen Temperaturen. Messungen 1 – 3 und 7 – 9 bei 200 °C, Messungen 4 – 6 bei 500 °C

Si-Folie vorgenommenen Energieverlustmessungen. Die Temperatur der Folie betrug während der Messungen 1 bis 3 und 7 bis 9 etwa 200 °C und bei den Messungen 4 bis 6 500 °C. Zwischen der 6. und 7. Messung wurden die Folien in eine mit Wasserdampf gesättigte Atmosphäre gebracht und anschließend wieder im Hochvakuum ausgemessen. Durch die Aufheizung der Folien auf 500° werden bekanntlich Wasserschichten im Hochvakuum desorbiert. Dies führte zu einer erheblichen Verringerung des mittleren Energieverlustes. Die Erhöhung des Energieverlustes auf den Ausgangswert bei den Messungen 7 bis 9 ist unserer Meinung nach ein Beweis dafür, daß der überwiegende Teil der Adsorbatschicht aus Wasser besteht. Ein vorübergehendes Belüften der Meßapparatur nach der Desorption der Wasserschicht ergab ebenfalls einen Anstieg des Energieverlustes, wobei jedoch der Ausgangswert nicht erreicht wurde.

Schätzt man den Bremsquerschnitt der Adsorbatschicht zu 2×10^{-14} eVcm²/Atom ab, so ergeben sich mittlere Adsorbatbelegungen von etwa 5×10^{16} Moleküle/cm² für Cu und $4{,}5 \times 10^{16}$ Moleküle/cm² für Al, was einer Wasserschicht normaler Dichte von etwa 60 Å auf jeder Seite der Folie entsprechen würde (Fehler $\pm$ 50%). Das Unerwartete an den Meßergebnissen ist die erhebliche Stärke der Adsorbatschicht. Wir glauben nicht, daß alles Wasser an den Oberflächen angelagert ist, sondern vermuten, daß infolge der stark fehlgeordneten Struktur der Aufdampfschicht das Wasser zu einem Teil in oberflächennahen Bereichen an inneren Oberflächen adsorbiert wird.

Die Messungen zeigen, daß man bei der Beurteilung der Zuverlässigkeit der Schwingquarzmethode mögliche Veränderungen an der Schichtoberfläche ausschließen bzw. berücksichtigen muß. Wir halten eine zuverlässige Messung einzelner Atomlagen mit Hilfe der Schwingquarzmethode für gegeben und glauben, daß manche Zweifel an der Zuverlässigkeit dieser Methode auf nichterkannte Adsorptionsschichten zurückzuführen sind.

Die Autoren danken den Herren Th. Wronski und P. Apel für ihre Hilfe bei der Ausführung der Messungen.

Literatur

Apel, P., Müller-Jahreis, U., Rockstroh, G., Schwabe, S.: phys. stat. sol. (a) **3**, K 173 (1970)
Bernhard, F., Müller-Jahreis, U., Rockstroh, G., Schwabe, S.: phys. stat. sol. **35**, 285 (1969)
Kato, S.: Nucl. Instr. Meth. **75**, 293 (1969)
Sauerbrey, G.: Z. f. Physik **155**, 206 (1959)
Wronski, T.: Diplomarbeit, HU Berlin 1970

(Eingegangen am 20. Mai 1971)

Anschrift der Verfasser:

Prof. Dr. F. Bernhard
Dr. U. Müller-Jahreis
Humboldt-Universität zu Berlin
Sektion Physik, Bereich Atomstoßprozesse der Festkörperphysik
104 Berlin
Invalidenstr. 42

Kristall und Technik	7	1–3	1972	187–195

E. Wäsch

Humboldt-Universität zu Berlin

Einkristallzüchtung von Titandioxid durch chemische Transportreaktion

Es wird ein Überblick über wichtige physikalische Eigenschaften von Titandioxid gegeben und die Abhängigkeit dieser Eigenschaften von der Stöchiometrie betrachtet. Mit Hilfe der chemischen Transportreaktion $TiO_2 + 4\,HCl = TiCl_4 + 2\,H_2O$ wurden in geschlossenen Quarzglasampullen zwei TiO_2-Modifikationen synthetisiert: Rutil und Anatas. Die Thermodynamik des Systems wird betrachtet und die Transportrate rechnerisch ($R_T = 0{,}18\,mg \cdot h^{-1}$) und experimentell ($R_T = 0{,}12\,mg \cdot h^{-1}$) bestimmt. – Der Rutil wurde an Hand seiner Farbe, Dichte und Röntgenwerte auf seine Stöchiometrie untersucht, und es wird geschlußfolgert, daß die Zusammensetzung zwischen $TiO_{1,95}$ und $TiO_{1,97}$ liegt.

A review is given on important physical properties of TiO_2 and the dependence of these properties on stoichiometry. With chemical transport reaction $TiO_2 + 4\,HCl = TiCl_4 + 2\,H_2O$ were synthetized two of the modifications of TiO_2: rutile and anatase. Experiments was carried out in closed quartz ampoules. The thermodynamics of the system is described and the rate of transport determined: theoretical $R_T = 0{,}18\,mg \cdot h^{-1}$, experimental $R_T = 0{,}12\,mg \cdot h^{-1}$. – The rutile was examined in regard of its stoichiometry. Colour, density and X-ray dates refer to compose of $TiO_{1,95}$ to $TiO_{1,97}$.

1. Einleitung

Von Titandioxid sind vier Modifikationen bekannt. Unter normalen Druck-Temperatur-Bedingungen treten Rutil (4/m m m), Brookit (m m m) und Anatas (4/m m m) auf. Von Dachille und Roy wurde 1962 erstmals eine Hochdruckmodifikation von TiO_2 synthetisiert, das TiO_2 II (m m m).

Die Stabilität der drei in der Natur vorkommenden Modifikationen nimmt in der oben angegebenen Reihenfolge ab. Es ist sicher, daß Rutil die von Raumtemperatur bis zum Schmelzpunkt stabile Phase des TiO_2 ist. Ob Anatas irgendein stabiles Existenzgebiet hat oder ob es im Hinblick auf Rutil bei allen Temperaturen metastabil ist, ist ungeklärt. Die Umwandlungstemperatur von Anatas in Rutil liegt zwischen 800 und 900 °C. Auch über den Stabilitätsbereich von Brookit gibt es nur Vermutungen. Wahrscheinlich liegt er zwischen dem von Anatas und Rutil. Aus dem kleinen Existenzbereich wird auch das seltene natürliche Vorkommen von Brookit erklärt. (vergl. Gmelin)

In den letzten Jahren ist das wissenschaftliche und technische Interesse am TiO_2 erheblich gestiegen. Die chemisch so widerstandsfähige Verbindung gibt bei hohen Temperaturen leicht einen Bruchteil ihres Sauerstoffs ab. Das führt zu einer außerordentlich hohen Variationsbreite der physikalischen Eigenschaften, insbesondere sind das veränderte elektrische, optische und magnetische Verhalten zu nennen.

Die Einkristallzüchtung dieser interessanten Substanz ist wegen des hohen Schmelzpunktes von Titandioxid – er liegt bei 1850 °C – nicht ohne Schwierigkeit. Bei diesen hohen Temperaturen führt die thermische Zersetzung stets zu einer Unstöchiometrie. Künstlich hat man bisher nur Rutil, eine der drei herkömmlichen Modifikationen, als Einkristall herstellen können.

In der vorliegenden Arbeit wurde versucht, die Methode der chemischen Transportreaktion auf die Einkristallsynthese von TiO_2 anzuwenden. Diese seit einiger Zeit als leistungsfähig bekannte Methode zur Darstellung einkristalliner Stoffe bietet bei Wahl einer geeigneten chemischen Reaktion die Möglichkeit, bei einer wesentlich niedrigeren Temperatur als der Schmelztemperatur gut ausgebildete Kristalle aus der Dampfphase abzuscheiden. Es sollte geprüft werden, ob es ein Transportsystem gibt, das es gestattet, stöchiometrische TiO_2-Einkristalle oder solche Einkristalle zu züchten, die einen definierten Sauerstoffunterschuß haben und damit die oben genannten interessanten physikalischen Eigenschaften besitzen.

2. Wichtige Eigenschaften von TiO_2

Wie die meisten nichtmetallischen Stoffe ist stöchiometrisch zusammengesetztes TiO_2 bei Raumtemperatur ein Isolator, die spezifische elektrische Leitfähigkeit $\varkappa$ beträgt 10^{-15} Ohm^{-1} cm^{-1}. Bei höheren Temperaturen ist TiO_2 Halbleiter:

T	30°	130°	260°	420°	500 °C
$\varkappa$	$5 \cdot 10^{-14}$	10^{-11}	$3{,}3 \cdot 10^{-9}$	$5 \cdot 10^{-7}$	$1{,}7 \cdot 10^{-6}$ Ohm^{-1} cm^{-1}

Wie extrem die spezifische elektrische Leitfähigkeit $\varkappa$ bei Sauerstoffunterschuß ansteigt, zeigt folgende Tabelle (Verwey)

Zusammensetzung	$TiO_{2,0000}$	$TiO_{1,9995}$	$TiO_{1,995}$	$TiO_{1,75}$
$\varkappa$	10^{-10}	10^{-1}	0,8	10^{2} Ohm^{-1} cm^{-1}

Bei Sauerstoffverlust wird TiO_2 (Rutil) zum n-Typ-Halbleiter. Bei der thermischen Zersetzung entstehen Donatorzentren in Form eines Überschusses von metallischen Bestandteilen des Oxides.

Reiner Rutil hat eine außerordentlich hohe Dielektrizitätskonstante ε, die sich durch geringe Empfindlichkeit gegenüber Temperatur- und Frequenzänderungen auszeichnet. Parallel zur tetragonalen Achse beträgt sie 172, senkrecht dazu 86 (vergl. Verwey, Bügel)

Auf Unterschiede in der Farbe von TiO_2-Kristallen wird später eingegangen. Hier sei nur angeführt, daß stöchiometrisch zusammengesetzte Rutilkristalle farblos sind. Die Farbe reagiert äußerst empfindlich gegen Abweichungen von der Ti-O-Zusammensetzung.

Rutilkristalle übertreffen durch ihre hohe Lichtbrechung und Dispersion sogar Diamanten an Glanz und Feuer. Bemerkenswert ist auch das hohe Reflexionsvermögen von TiO_2.

Bei Bestrahlung mit Licht wird TiO_2 leitend. Die Absorption von Lichtquanten setzt Elektronen frei.

Titandioxid ist schwach paramagnetisch. Die spezifische Suszeptibilität beträgt bei Raumtemperatur für Rutil $0{,}08 \cdot 10^{6}$. Durch quantitative Untersuchungen konnte Ehrlich nachweisen, wie sich die Suszeptibilitätswerte entscheidend bei Präparaten ändern, die von der stöchiometrischen Zusammensetzung auch nur wenig abweichen. In der Figur 1 ist die Suszeptibilitätskurve für verschiedene Sauerstoffgehalte dargestellt. Wird dem TiO_2 auch nur wenig Sauerstoff entzogen, ist ein starker Anstieg des Paramagnetismus zu beobachten. Es wird ein Maximum durchlaufen, der Paramagnetismus sinkt wieder und ändert sich nicht wesentlich bei weiterem O-Entzug.

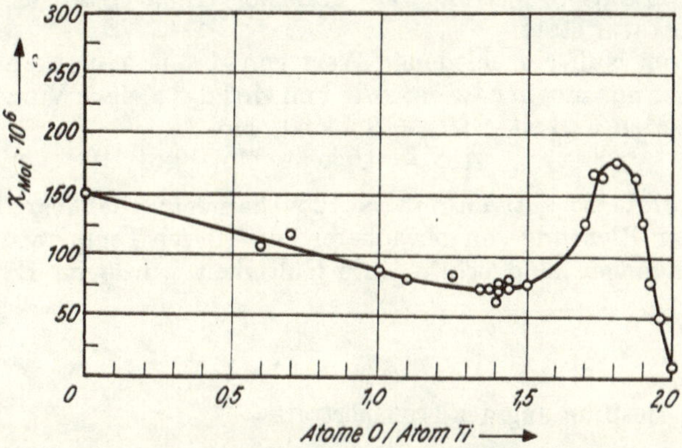

Fig. 1. Magnetische Suszeptibilität der Ti-O-Phasen (nach EHRLICH)

3. Versuchsdurchführung zur Einkristallsynthese

Es wurde in einem horizontalen, zylindrischen Zweizonenofen gearbeitet. Die chemische Transportreaktion fand in einer geschlossenen Quarzglasampulle, die mit der Ausgangssubstanz – einer synthetischen Anatasmodifikation von TiO_2 – und dem HCl-Transportgas gefüllt war, statt. Bei einer Vielzahl von Versuchen wurde festes NH_4Cl als Transportgas spendende Substanz eingewogen. Die Temperaturmessung erfolgte an den Enden der Ampulle mit Pt/PtRh-Thermoelementen.

4. Thermodynamik des untersuchten Systems

Die Reaktion zur Synthese von TiO_2-Einkristallen verläuft nach der Gleichung:

$$TiO_{2(f)} + 4\,HCl_{(g)} = TiCl_{4(g)} + 2\,H_2O_{(g)}\,. \qquad (I)$$

Die molaren Bildungsenthalpien der Reaktion betragen nach HAVEMANN:

TiO_2	HCl	$TiCl_4$	H_2O
−218,0	−21,0	−182,9	−57,8 kcal · Mol^{-1}

Die Bildungsenthalpie der obigen Reaktion ergibt sich zu

$$\Delta H_r^0 = 3{,}4\ \text{kcal} \cdot \text{Mol}^{-1}\,.$$

Die Angaben der Bildungsenthalpien für TiO_2 schwanken in den Tabellenwerken erheblich. FARBER, DARNELL geben für ΔH° TiO_2 −215 bis −225 kcal · Mol^{-1} an. Offensichtlich haben sie den Wert −225 kcal/Mol für die Berechnung der Bildungsenthalpie der Reaktion (I) zugrunde gelegt und bekommen so einen Wert

$$\Delta H_r^0 = 15{,}0\ \text{kcal} \cdot \text{Mol}^{-1}\,.$$

Schäfer hat einige thermodynamische Bedingungen für die Transportfähigkeit eines Systems gegeben. An Hand der bekannten thermodynamischen Daten für die Gleichung (I) sind folgende Punkte erfüllt:

Da, wie gefordert der Absolutwert von ΔH nicht zu groß ist, liegt keine extreme Gleichgewichtslage vor. ΔH_r ist auf jeden Fall >0, es handelt sich um einen endothermen Reaktionsverlauf. Der Transport findet von der höheren zur tieferen Temperatur statt.

Bei einem von Null verschiedenen Wert von $\Delta S°$ ist ein erheblicher Bodenkörpertransport nur möglich, wenn $\Delta H°$ und $\Delta S°$ das gleiche Vorzeichen haben.

Im vorliegenden Fall ist $\Delta H_r^0 = 3{,}4\ \text{kcal} \cdot \text{Mol}^{-1}$,

$$\Delta S_r^0 = 8{,}9\ \text{kcal} \cdot \text{Mol}^{-1}\,\text{grd}^{-1}\,.$$

Bei genügend großem positiven $\Delta S°$ ($\sim$ 10 cl) kann ein erheblicher Bodenkörpertransport nur in Richtung von der höheren zur tieferen Temperatur stattfinden.

Diese Bedingungen sind erfüllt, ihre Gültigkeit wurde im Experiment bestätigt.

5. Bestimmung der Transportrate

5.1. Theoretisch

$$TiO_2 + 4\,HCl = TiCl_4 + 2\,H_2O\,.$$

Mit Hilfe der von Hauptman, Schmidt und Banerjee angegebenen Formel wurde die Transportrate für obige Reaktion bestimmt. Es wird dabei vorausgesetzt, daß die Diffusion geschwindigkeitsbestimmender Schritt ist.

$$R_T = \frac{i}{j} \cdot M \cdot \frac{D_0}{\sum P} \cdot \left(\frac{T_M}{T_0}\right)^{1,8} \frac{(P_2 - P_1) \cdot q \cdot 3{,}6 \cdot 10^6}{s \cdot R \cdot T_M}$$

Darin ist:

R_T — Transportrate in mg h^{-1},

$\frac{i}{j}$ — stöchiometrischer Koeffizient der Reaktion, hier $\frac{1}{2}$

M — Molekulargewicht, hier 80,

D_0 — Diffusionskoeffizient unter Normalbedingungen. Er beträgt nach Schäfer 0,1 $cm^2 \cdot s^{-1}$,

T_M — Durchschnittstemperatur der Ampulle, hier 1120 °K,

$\sum P$ — Gesamtdruck in der Ampulle bei T_M, hier 4 at,

T_0 — 273 °K,

q — Querschnitt der Diffusionsstrecke, hier 0,8 cm^2,

$P_2 - P_1$ — Differenz der Partialdrucke von $TiCl_4$ zwischen Diffusions- und Wachstumszone, hier
bei 1200 °K p_{TiCl_4} $1{,}32 \cdot 10^{-3}$ at,
bei 1050 °K p_{TiCl_4} $0{,}97 \cdot 10^{-3}$ at,

s — Diffusionsstrecke, hier 10 cm,

R — Gaskonstante.

Die mit den angegebenen Werten berechnete Transportrate von TiO_2 ergab sich zu $R_T = 0{,}18\ \text{mg} \cdot h^{-1}$.

5.2. Experimentell

Aus dem Experiment wurde die Transportrate bestimmt, indem die abgeschiedenen Kristalle sorgfältig aus dem Transportrohr isoliert wurden. Sie wurden mit einem Skalpell vorsichtig von der Ampullenwand gelöst, auf ein Wägegläschen gebracht und gewägt. Als Grundlage für diese Bestimmung dienten die Versuche mit dem maximalen Transporteffekt, bei denen im Temperaturgefälle von 930° – 780 °C transportiert worden war. Der mittlere Wert für die Transportrate beträgt 0,12 mg · h^{-1}.

Für Anatas wurde die Transportrate nicht bestimmt, sie war geringer.

Die Übereinstimmung des beobachteten und des berechneten Wertes ist recht gut.

6. Beschreibung und Untersuchung der gezüchteten Kristalle

Während der Suche nach optimalen Bedingungen für die Züchtung möglichst großer Einkristalle wurden die Transportgasdrucke, die Transportzeiten und die Temperaturgradienten variiert. Im Laufe der Versuche hatte sich ein Transportgasdruck, der durch 10–15 mg NH_4Cl erzeugt wurde, als besonders günstig erwiesen. Bei dieser NH_4Cl-Menge liegt der Gasdruck in der Ampulle bei 4 at. Mit ihm wurde in der Hauptsache gearbeitet. Im untersuchten Abscheidungsgebiet zwischen 870°–600 °C konnten zwei der unter normalen p-T-Bedingungen existenten TiO_2-Modifikationen erhalten werden. Oberhalb 700 °C schieden sich stahlblaue Kristalle (Rutil) ab, unterhalb 700 °C kräftig grüne (Anatas).

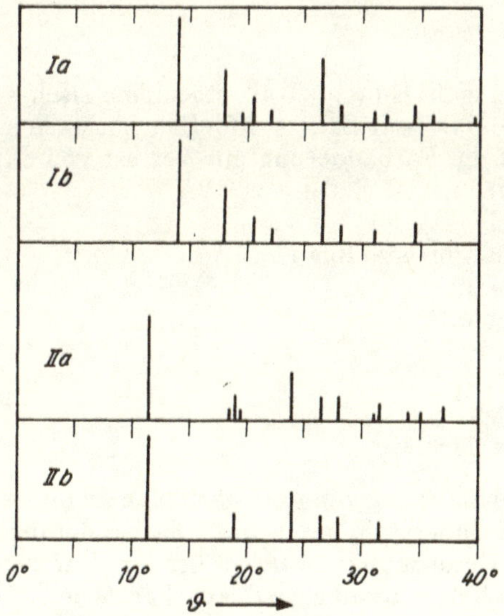

Fig. 2. Schematische Darstellung von TiO_2 Debyeogrammen
Ia – Rutil nach dem ASTM
Ib – stahlblaue Transportkristalle – Rutil
IIa – Anatas nach dem ASTM
IIb – grüne Transportkristalle – Anatas

6.1. Identifizierung durch Pulveraufnahmen

Die abgeschiedenen Kristalle wurden mit Hilfe von Debye-Scherrer-Aufnahmen identifiziert. Die stahlblauen Kristalle, die bei Abscheidungstemperaturen zwischen 850°—700 °C auftraten, konnten als Rutil diagnostiziert werden; die grünen Kristalle, die unterhalb 700 °C entstanden, sind Anatas. In der Literatur konnte keine Angabe über eine andere Art der erfolgreichen Einkristallsynthese von Anatas gefunden werden.

Aus der Umzeichnung der Pulveraufnahmen, verglichen mit den ASTM-Werten, kann entnommen werden, daß alle Linien mit starker Intensität sowohl für Rutil als auch für Anatas aufgefunden wurden.

6.2. Dichtebestimmung

Die Dichten von Titandioxidpräparaten sind schon mehrfach untersucht worden. Ehrlich gibt folgende Werte an:

Zusammensetzung	TiO_2	$TiO_{1,95}$	$TiO_{1,90}$	$TiO_{1,85}$	$TiO_{1,80}$	$TiO_{1,70}$
Dichte	4,197	4,183	4,170	4,150	4,171	4,207 $g \cdot cm^{-3}$

Wie man sieht, kann auch ein Abfall der Dichte den Sauerstoffverlust des TiO_2 anzeigen. Die Dichte des Rutils, der mit der Methode des chemischen Transports gezüchtet wurde, ist mit der Pyknometermethode bestimmt worden. Als Mittelwert ergab sich für die Dichte der blauen Rutilkristalle $\varrho = 4{,}176$ g $\times cm^{-3}$. Die Ti-O-Zusammensetzung muß demzufolge zwischen $TiO_{1,95}$ und $TiO_{1,90}$ liegen.

6.3. Farbe der Kristalle

Die zahlreichen Untersuchungen Ehrlichs an stöchiometrisch zusammengesetztem TiO_2 und an Titandioxid mit Sauerstoffdefizit beziehen sich auch auf die Farbe. Im Hinblick auf die Farbänderung mit Verlust von Sauerstoff gibt er folgenden Farbverlauf an:

TiO_2 — weiß
$TiO_{1,75}$—$TiO_{1,97}$ — dunkelblau bis stahlblau
$TiO_{1,60}$—$TiO_{1,73}$ — blauschwarz
$TiO_{1,50}$ — dunkelviolett
$TiO_{1,33}$ — dunkelbraun
$TiO_{1,25}$ — hellbraun
$TiO_{1,12}$ — bronzefarben
$TiO_{1,00}$ — hellbronzefarben

Auch Foex weist auf die Farbänderung von weiß nach blau schon bei quantitativ kaum mehr nachweisbarem Sauerstoffentzug hin. Farbänderungen von TiO_2 beim Erhitzen bis zum Schmelzpunkt werden in der Literatur mehrfach erwähnt. Die Endprodukte haben braune bis blaue Farbtöne.

Allerdings kann Blaufärbung auch durch Einbau von Kationen mit Wertigkeiten größer 4 hervorgerufen werden, wie Sb^{5+}, W^{6+}, Mo^{6+}. Dabei werden Ti^{3+}-Ionen gebildet, die eine höhere Lichtabsorption haben. In verschiedenen Arbeiten wird darauf hingewiesen, daß der Rutil weit empfindlicher auf die Abweichungen von der Stöchiometrie und auf Fremdbeimengungen reagiert als der Anatas. Damit kann der Unterschied in der Farbe zwischen beiden in dieser Arbeit dargestellten Modifikationen erklärt werden. Rutil hat in allen Fällen eine intensive stahlblaue Farbe, Anatas ist kräftig grün.

Die Ausgangssubstanz ist spektrographisch untersucht worden. In ihr sind Spuren von höherwertigen Kationen — Arsen und Wolfram — gefunden worden. Die Ausgangssubstanz ist jedoch rein weiß. Die Spurenelemente bewirken keine Färbung. So muß die Farbe der synthetisierten Kristalle wohl auf einen Sauerstoffunterschuß zurückgeführt werden; denn auch hierbei entstehen Ti^{3+}-Ionen, durch die die Blau- oder Grünfärbung hervorgerufen werden kann. Dabei ist zu bemerken, daß die Auflösungstemperaturen (950—750 °C) weit unter dem Schmelzpunkt (1850 °C) lagen, dennoch tritt eine Sauerstoffabgabe ein.

Ein Vergleich mit der oben angegebenen Farbtabelle zeigt, daß das Verhältnis von Ti–O zwischen $TiO_{1,75}$ und $TiO_{1,97}$ liegen muß.

Aus den Debye-Scherrer-Aufnahmen, die nur die TiO_2-Linien zeigen, deutet sich an, daß die Zusammensetzung mehr dem $TiO_{1,97}$ entsprechen muß. Denn nach den Arbeiten von Ehrlich und neueren Untersuchungen von Andersson werden bei Präparaten, deren Zusammensetzung bei $TiO_{1,90}$—$TiO_{1,95}$ liegt, die Debye-Scherrer-Linien bereits verwaschen und Spalten bei weiterem Sauerstoffverlust auf (vergl. Figur 3).

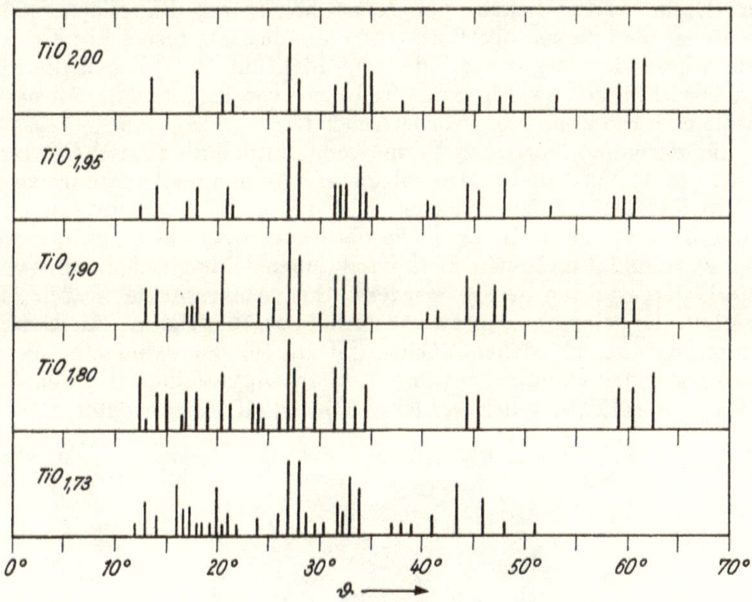

Fig. 3. Schematische Darstellung von Röntgenbildern verschiedener Titandioxide (nach Ehrlich)
Die Änderung im Röntgendiagramm bei Sauerstoffverlust des TiO_2 wird deutlich

Bei größerem O-Defizit entsteht dann ein Röntgendiagramm mit neuen Linien. Das trifft weder für den hier untersuchten Rutil noch für den Anatas zu. Die Wahrscheinlichkeit ist groß, Kristalle erhalten zu haben, deren stöchiometrische Zusammensetzung bei $TiO_{1,95}$ liegt, was durch die Dichtebestimmung erhärtet wird.

6.4. Größe, Tracht und Habitus der Kristalle

Eine deutliche Abhängigkeit der Kristallgröße von den variablen Versuchsparametern konnte nicht gefunden werden. Ein direkter Zusammenhang ergibt sich nur zwischen Kristallgröße und Versuchsdauer. Beste Versuchsergebnisse konnten bei Temperaturgefällen von 150 °C, das entspricht 15 grd $\times$ cm^{-1}, erreicht werden. Bei längerer Versuchsdauer wurden in diesem Temperaturgefälle bis zu 2 mm große Kristalle mit ausgezeichneter Flächenbeschaffenheit gezüchtet.

25 Stunden Versuchsdauer 0,5 mm max. Erstreckung,
50 Stunden Versuchsdauer 1,2 mm max. Erstreckung,
100 Stunden Versuchsdauer 2,0 mm max. Erstreckung.

Oft wird bei Arbeiten zur Kristallzüchtung mit chemischen Transportreaktionen auf eine interessante Änderung von Tracht und Habitus der auf diese Weise entstehenden Kristalle hingewiesen. Auch beim TiO_2, insbesondere beim Rutil, kann diese Erscheinung beobachtet werden. Die in der Natur auftretenden Rutilkristalle sind fast alle von kurzsäuligem bis nadligem Habitus. Das Dominieren von Dipyramiden wird selten beobachtet. Bei den hier künstlich hergestellten Kristallen tritt in der Regel kurzpyramidaler, teilweise sogar ein dicktafliger Habitus auf; verschiedene Dipyramiden bilden rundliche, nahezu „isometrische" Kriställchen, bei denen häufig die Basisfläche (001) erscheint. Seltener tritt die nadlige Form auf. Gerade diese gestreckte Form wäre energetisch besonders begünstigt; denn in Richtung [001] liegen die kürzesten Bindungsrichtungen zwischen den Ti-Ionen (vergl. Kleber). An natürlichen Kristallen ist die Zone [001] vorherrschend.

Bei der goniometrischen Vermessung wurden als typische Pyramidenflächen (101), (301), (231) und (331) beobachtet. An den nadligen Formen ist neben den (210) und (310) Prismenflächen und den genannten Dipyramiden wieder die (001)-Basisfläche zu finden. Die natürlichen Anataskristalle werden meist als spitzpyramidal und recht klein beschrieben. Das gleiche kann von den synthetisch dargestellten gesagt werden. Ihre Abmessungen waren in jedem Fall gering. Es gelang nie, ihre Größe erheblich zu steigern. Auch im Habitus war unter den verschiedenen Züchtungsbedingungen kaum eine Veränderung zu bemerken, es dominierten dünne, spitze Dipyramiden mit den Flächen (101), (112) und (111), seltener wurden (103) und (105) beobachtet.

Fig. 4. Rutilkristall

Fig. 5. Anataskristalle

Literatur

ANDERSSON, S. u. a.: Acta Chem. Scand. **11**, 1641–1652 (1957)
DACHILLE, F., ROY, R.: Geol. Soc. Amer. Annual Meeting, San Francisco 1966
EHRLICH, P.: Z. Elektrochem. **45**, 362–370 (1939)
FARBER, M., DARNELLE, A.: J. Chem. Phys. **23**, 1460 (1955)
FOEX, M.: Bl. Soc. Chim. **11**, 6–17 (1944)
Gmelins Handbuch der anorganischen Chemie. Titan. Weinheim 1951, S. 233
HAUPTMAN, Z., SCHMIDT, D., BANERJEE, S. K.: Coll. Czech. Chem. Commun. **32**, 2421–2430 (1967)
HAVEMANN, R.: Einführung in die chemische Thermodynamik. Berlin 1957
KLEBER, W.: N. Jb. Min. A Beilagebd. **75**, 72–89 (1940)
SCHÄFER, H.: Chemische Transportreaktionen. Weinheim 1962
VERWEY, E. J.: Phil. techn. Rdsch. **9**, 46–54 (1947)
VERWEY, E. J., BÜGEL, R. D.: Phil. techn. Rdsch. **10**, 232–239 (1947)

(Eingegangen am 20. Mai 1971)

Anschrift des Verfassers:

Dr. E. WÄSCH
Humboldt-Universität zu Berlin
Sektion Physik, Bereich Kristallographie
104 Berlin
Invalidenstraße 43

Kristall und Technik	7	1–3	1972	197–206

J. Doerschel

Institut für elektronische Bauelemente der VVB Bauelemente und Vakuumtechnik

Einkristallzüchtung und Versetzungsätzung von Antimon und antimonreichen Antimon-Wismut-Legierungen

Die Einkristallzüchtung von Antimon und antimonreichen Antimon-Wismut-Legierungen nach dem Zonenschmelzverfahren wird kurz beschrieben, speziell unter dem Aspekt der konstitutionellen Unterkühlung bei der Züchtung von Mischkristallen.

Zum Nachweis von Versetzungen in Antimon und Antimon-Wismut-Mischkristallen wird ein elektrolytisches Ätzverfahren angegeben. Aus den Ergebnissen verschiedener Untersuchungen kann geschlossen werden, daß Ätzgruben den Durchstoßpunkten von Versetzungen entsprechen.

Die Orientierungsabhängigkeit des Ätzverhaltens von Versetzungen bei Anwendung verschiedener Ätzmittel wurde untersucht.

The growth of single crystals of antimony and antimony-rich antimony-bismuth alloys by zone-melting is briefly described, especially with regard to constitutional supercooling at the growth of solid solutions.

An electrolytical method of etching dislocations in antimony and antimony-bismuth solid solutions is presented. It is concluded from the results of different investigations that etch pits correspond to dislocation sites.

The orientation dependence of etching behaviour of dislocations has been investigated using several etching reagents.

1. Einleitung

Das Verformungsverhalten von Antimonkristallen war wiederholt Gegenstand der Untersuchung verschiedener Autoren (u. a. Gough, Cox; Meyer, Brückner; Soifer, Ščegoleva). Welcher der möglichen Deformationsmechanismen (Bruchdeformation, Translations-, Zwillingsgleitung, Kinking) im Experiment tatsächlich beobachtet wird, hängt im wesentlichen vom Spannungszustand, von der Temperatur, der Verunreinigungskonzentration, der Kristallorientierung und der Verformungsgeschwindigkeit ab. In Weiterführung experimenteller Arbeiten (Meyer, Brückner) wurde der Einfluß dieser Parameter auf das Deformationsverhalten von Antimonkristallen und antimonreichen Antimon-Wismut-Mischkristallen untersucht.

Um einen schnellen Überblick hinsichtlich Versetzungsdichte und -anordnung vor und nach plastischer Verformung zu erhalten, sind Ätzverfahren gut geeignet, da sie relativ bequem und schnell die direkte Beobachtung der Durchstoßpunkte der Versetzungslinien an der Kristalloberfläche erlauben. Für quantitative Untersuchungen muß die Eignung des Ätzmittels zur Versetzungsanätzung nachgewiesen werden. Zur Prüfung der Korrelation zwischen Ätzgruben und Versetzungen wurden verschiedene Methoden vorgeschlagen, u. a. Vergleich des Ätzbildes korrespondierender Spaltflächen, Zuordnung der geätzten Ober- und Unterseite einer hinreichend dünnen Probe, Reproduzierbarkeit bei Ätzung mit verschiedenen Ätzmitteln bzw. aufeinanderfolgende Anwendung des gleichen Ätzmittels, Veränderung der linearen Ätzgrubendichte von Kleinwinkelkorngrenzen an Verzweigungen, Anordnung der Ätzgruben nach einer plastischen Deformation, Zusammenhang zwischen Desorientierungswinkel und linearer Ätzgrubendichte einer Subkorngrenze, Nachweis der Bewegung von Versetzungen unter Einwirkung von inneren und äußeren Spannungen, Vergleich mit Ergebnissen anderer Untersuchungsverfahren (z. B. Röntgentopographie, Transmissionselektronenmikroskopie, Versetzungsdekoration).

Bei der Durchführung bestimmter experimenteller Arbeiten besteht häufig die Notwendigkeit, Oberflächen unterschiedlicher Orientierung zu untersuchen. Hexagonale Metalle, beispielsweise Zink (Adams, Blish, Vreeland) und Cadmium (Rühlicke, Männel, Hötzsch), besitzen eine ausgeprägte Anisotropie des Ätzverhaltens von Versetzungen. Es ist daher erforderlich, für das benutzte Ätzmittel einen möglichst vollständigen Überblick bezüglich der Orientierungsabhängigkeit der Ätzgrubenausbildung zu erhalten.

2. Kristallzüchtung

Insbesondere für die Untersuchung des Deformationsverhaltens von Antimonkristallen bei hohen Verformungsgeschwindigkeiten mittels des Hopkinson'schen Spaltstabverfahrens (Kolsky) waren Proben mit größeren Durchmessern (bis 20 mm) erforderlich. Die Züchtung entsprechender Einkristalle erfolgte in Reinstgraphitschiffchen trapezförmigen Querschnitts (mittlere Abmessungen $22 \times 22 \times 285$ mm) nach dem horizontalen Zonenschmelzverfahren unter Argon bzw. Reinstickstoff (10 Torr Überdruck). Als Ausgangsmaterialien dienten Antimon und Wismut mit der Reinheit 99,999%. Die Geschwindigkeit des widerstandbeheizten und temperaturstabilisierten Ringofens ließ sich stufenweise von 1 mm/h auf 25 mm/h erhöhen. Impfkristalle wurden nicht verwendet. Mit Hilfe eines Keimselektionseinsatzes (Fig. 1). konnten über die gesamte Länge einkristalline Barren gezüchtet

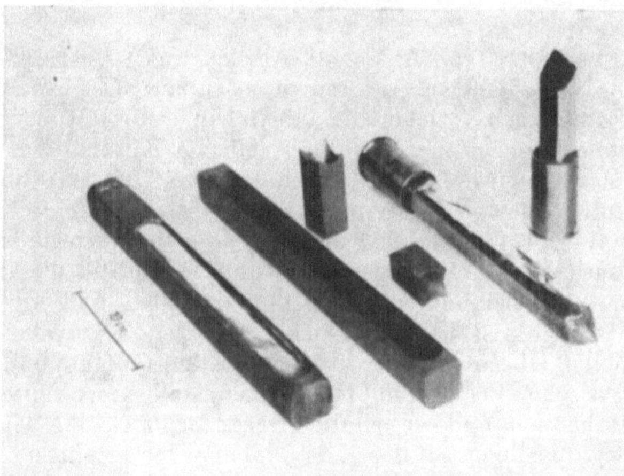

Fig. 1. Einkristalle und für die Züchtung benutzte Graphitschiffchen

werden. Der Winkel, den die Wachstumsrichtung mit der Basisfläche einschloß, bewegte sich im Bereich 0° bis 35°. Die Ätzgrubendichte auf (0001)-Spaltflächen[1]) lag zwischen 10^5 und 10^6 cm^{-2}. Die weitere Bearbeitung der Einkristalle ist in einer früheren Arbeit ausführlich beschrieben worden (DOERSCHEL, FISCHER).

Die Züchtung von Mischkristallen aus der Schmelze ist schwieriger als die Herstellung von Einkristallen der reinen Metalle, da mit steigender Konzentration der Zusätze sowohl die Keimbildungshäufigkeit als auch die Mikroinhomogenität zunehmen. Wesentlichste Ursache dafür ist die konstitutionelle Unterkühlung (TILLER, JACKSON, RUTTER, CHALMERS), verursacht durch unvollständige Durchmischung der Schmelzzone. In unmittelbarer Nähe der Kristallisationsfront kommt es innerhalb der Schmelzzone zu einer Anreicherung des Wismuts (Verteilungskoeffizient $k_0 < 1$). Die Liquidustemperatur der Schmelze und der Temperaturverlauf in der Schmelzzone überlagern sich derart, daß ein Bereich unterkühlter Flüssigkeit entsteht. Unter diesen Bedingungen ist eine ebene Phasengrenze instabil, es bildet sich eine zellulare oder dendritische Erstarrungsfläche aus. Das Wismut ist im Bereich der Zellgrenzen bzw. Dendritenzwischenräume angereichert. Um konstitutionelle Unterkühlung auszuschließen, muß der Quotient G/R (G Temperaturgradient, R Kristallisationsgeschwindigkeit) entsprechend gewählt werden. Nach BROWN und HEUMANN gilt unter der Voraussetzung, daß der Konzentrationsausgleich in der Schmelze nur durch Diffusion erfolgt, die Beziehung $G/R \geq T/D$, wobei T der Temperaturdifferenz zwischen Solidus- und Liquiduskurve bei der gegebenen Zusammensetzung und D dem Diffusionskoeffizienten des Wismuts in der Schmelze entspricht.

Analog den Beobachtungen an wismutreicheren Legierungen (BROWN, HEUMANN; YIM, DISMUKES) traten auch in den im Rahmen dieser Arbeit untersuchten antimonreichen Mischkristallen (2 bis 10% Bi), die bei einem ungünstigen G/R-Verhältnis gezüchtet wurden, Mikroinhomogenitäten auf. Figur 2 zeigt die elektrolytisch geätzte (0001)-Spaltfläche eines Mischkristalls mit 3% Bi,

Fig. 2. Elektrolytisch geätzte (0001)-Fläche eines Mischkristalls mit 3% Bi (gewachsen bei konstitutioneller Unterkühlung)

---

[1]) Der kristallographischen Indizierung dieser Arbeit liegt die kleine hexagonale Zelle mit $c/a = 2{,}617$ unter Verwendung Miller-Bravais'scher Indizes zugrunde.

gewachsen unter dem Einfluß einer größeren konstitutionellen Unterkühlung. Die als Folge der lokalen Schwankung der Zusammensetzung zu erwartenden elastischen Spannungen durch Änderung der Gitterkonstanten können bei hinreichend großen Konzentrationsunterschieden zur Entstehung von Versetzungen führen. Man beobachtet charakteristische Anordnung der Ätzgruben.

Die Wismutverteilung längs einiger der nach dem zone-leveling-Verfahren hergestellten Barren wurde komplexometrisch und röntgenfluoreszenzanalytisch verfolgt. Über die Hälfte der Gesamtlänge wies eine nur geringfügig schwankende Wismutkonzentration auf; bezogen auf eine Strecke von 100 mm betrug die Abweichung von der mittleren Konzentration $\pm 1{,}5\%$. Der Konzentrationsverlauf im letzten Teil entspricht einem normalen Erstarren.

3. Ätzuntersuchungen

Für Antimon sind von verschiedenen Autoren (Shigata, Hiramatsu; Wernick, Hobstetter, Lovell, Dorsi; Kosevič; Soifer, Starzev; Lange, Lange, Oganjan; Soifer; Cordes, Kruse; Meyer, Meier) chemische Ätzmittel mit teilweise sehr ähnlichen Zusammensetzungen zum Nachweis von Versetzungen angegeben worden. Für die eigenen Untersuchungen erwiesen sich die Ätzgemische von Soifer sowie Cordes und Kruse als vorteilhaft, da sie leichte Handhabung und gute Ätzeigenschaften vereinigen. Überwiegend wurde jedoch ein elektrolytisches Ätzverfahren angewandt. Der ursprünglich von Coons für das anodische Polieren von Antimon benutzte Elektrolyt hat die Zusammensetzung 30 T Methanol, 5 T konz. H_2SO_4 und 3 T konz. HCl. Geätzt wurde 8 bis 30 Sekunden bei einer Spannung von 0,5 bis 1,5 Volt. Da sich diese Lösung auch sehr gut zum elektrolytischen Polieren von Antimon und antimonreichen Mischkristallen eignet, lassen sich beispielsweise Abtragungsversuche zum Studium der räumlichen Versetzungsanordnung leicht ausführen, denn man kann die Probe nach dem Polieren einfach durch Herabsetzen der Spannung anätzen.

Zur Überprüfung der Korrelation zwischen Versetzungen und der durch elektrolytisches Ätzen erzeugten Ätzgruben wurden einige der einleitend aufgeführten Kriterien herangezogen. In Figur 3a—c ist eine (0001)-Spaltfläche gezeigt, die abwechselnd poliert (abgetragen) und geätzt wurde. Die Ätzgruben in Figur 3a und b sind elektrolytisch entwickelt worden, während Figur 3c den gleichen Kristallbereich nach Ätzen mit der von Soifer angegebenen Lösung wiedergibt. In den aufeinanderfolgenden Abbildungen haben nur einzelne Gruben keine entsprechende Fortsetzung. Diese Ausnahmeerscheinungen dürften auf Verzweigungen des Versetzungsnetzwerkes oder auch auf Bewegungen der Versetzungslinie während der Präparation zurückzuführen sein (Soifer). Die gleiche Korrelation beobachtet man auf korrespondierenden (0001)-Spaltflächen. In Figur 4a und b ist ein Spaltflächenpaar abgebildet; das Ätzen erfolgte mit dem Ätzmittel nach Cordes und Kruse (Fig. 4a) bzw. elektrolytisch (Fig. 4b).

Ein interessanter Aspekt ergibt sich an Verzweigungen von Kleinwinkelkorngrenzen, wenn es sich um etwa senkrecht zu (0001) liegende Kippgrenzen handelt (Wernick, Hobstetter, Lovell, Dorsi). Unter dieser Voraussetzung ist die Summe der Kippwinkel Θ_i gleich Null. Wenn n_i die lineare Versetzungsdichte des Korngrenzenzweiges i und φ_i der kleinste mögliche Winkel zwischen

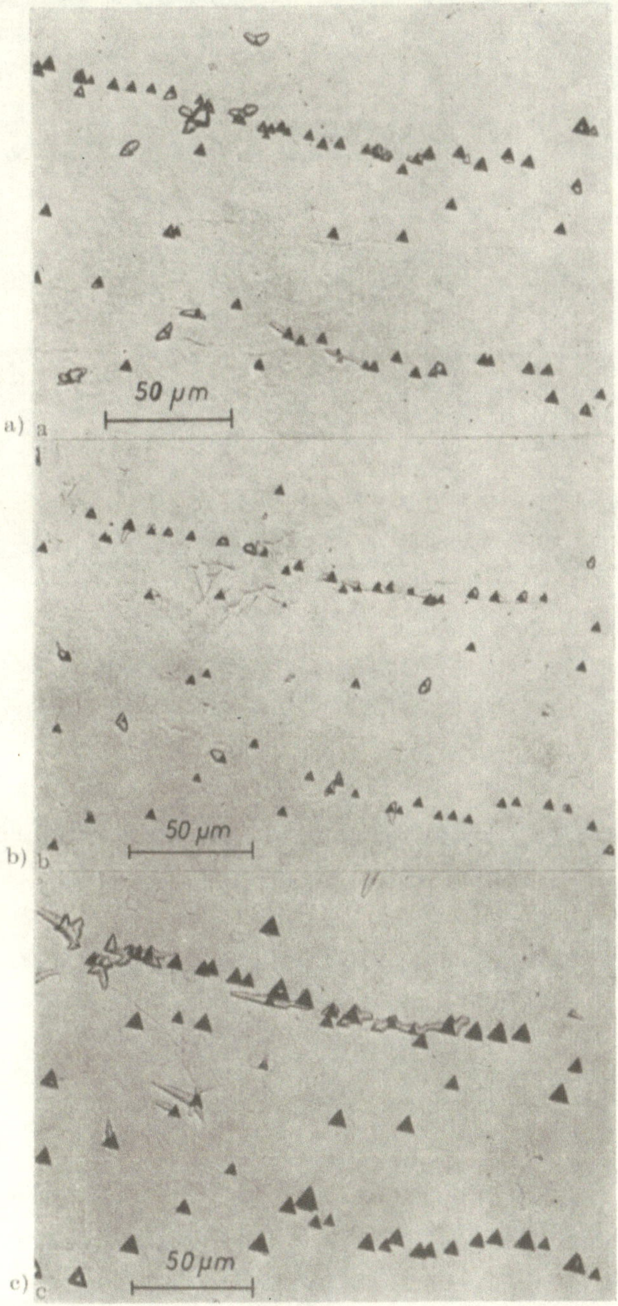

Fig. 3 a – c Schichtweise abgetragene und geätzte (0001)-Fläche eines Antimonkristalls
Fig. 3 a und b Elektrolytisch geätzt
Fg. 3 c. Ätzlösung nach Soifer

a)

b)

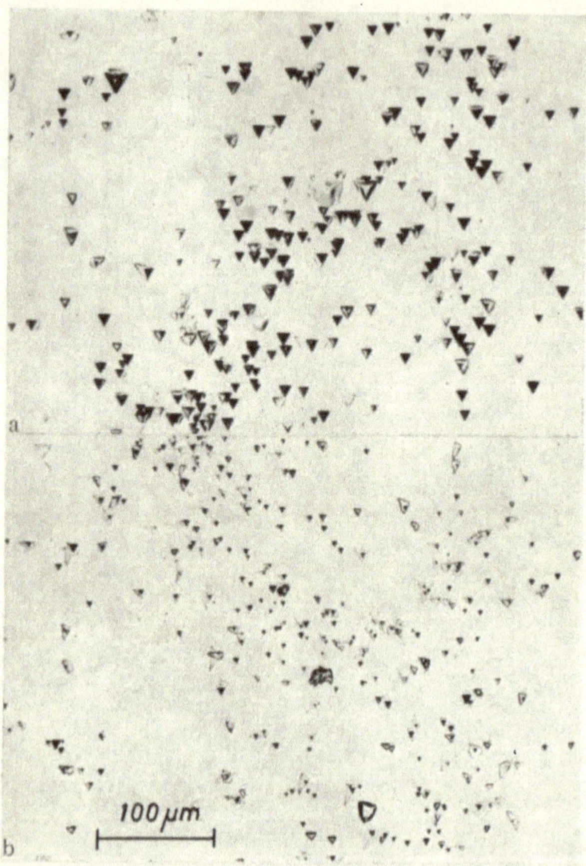

Fig. 4 a und b. Ätzgrubenanordnung auf korrespondierenden (0001)-Spaltflächen eines Antimonkristalls
Fig. 4 a. Ätzlösung nach Cordes und Kruse
Fig. 4 b. Elektrolytisch geätzt

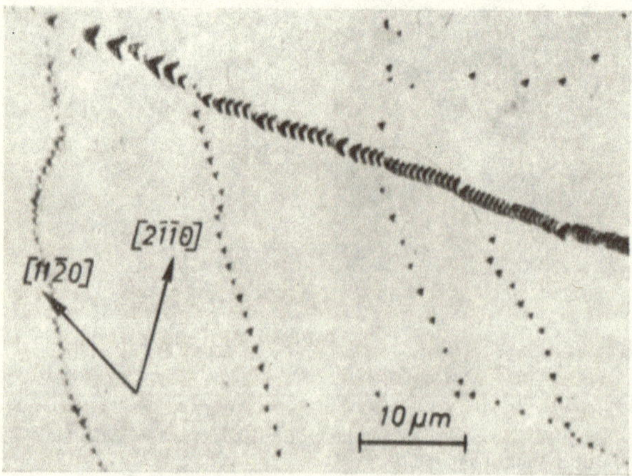

Fig. 5. Elektrolytisch geätzte (0001)-Fläche eines Mischkristalls mit 2,5 % Bi. Verzweigungen von Kleinwinkelkorngrenzen

einem der $\langle 11\bar{2}0\rangle$-Burgersvektoren und der Spur der Grenze sein soll, so gilt $n_A/\cos\varphi_A = n_B/\cos\varphi_B + n_C/\cos\varphi_C$. Im Rahmen der Meßgenauigkeit wurden diese Beziehung und damit die oben gemachten Voraussetzungen (Versetzungen mit dem Burgersvektor $\langle 11\bar{2}0\rangle$) bestätigt. Beispielsweise beträgt die maximal auftretende Abweichung bei den in Figur 5 sichtbaren Verzweigungen 9%.

4. Orientierungsabhängigkeit des Ätzverhaltens

Die von den verschiedenen Autoren durchgeführten Ätzuntersuchungen beziehen sich fast ausschließlich auf (0001)-Spaltflächen. Allgemein wurde festgestellt, daß sich Zwillingslamellen auf dieser Ebene nicht anätzen lassen. Cordes und Kruse beobachteten Ätzgruben auf wenig von der Basisorientierung abweichenden Oberflächen.

Um für die benutzten Ätzlösungen einen vollständigen Überblick über die Orientierungsabhängigkeit des Ätzverhaltens von Versetzungen zu erhalten, wurden Einkristallhalbkugeln benutzt. Mit Hilfe eines Zweikreisreflexionsgoniometers ließen sich die geätzten Bereiche ausmessen; ihre Lage bezüglich der Kristallachsen konnte durch Vermessen der Spuren von Zwillingen verschiedener Systeme bestimmt werden.

Nach Ätzen mit der von Soifer angegebenen Methanol-$FeCl_3$-Lösung wurde um den Flächenpol (0001) eine dreizählige Ätzrosette (Fig. 6a) beobachtet. Im Gegensatz zu den entsprechenden Angaben bei Soifer findet man also eine ausgeprägte Anisotropie des Ätzverhaltens auch hinsichtlich der trigonalen Symmetrieachse. Die schraffierten Bereiche in der stereographischen Projektion (Fig. 6d) stellen Gebiete mit unvollkommen ausgebildeten Ätzgruben dar.

Bei Anwendung der Ätzlösung nach Cordes und Kruse (Fig. 6b) ergaben sich nur bedingt reproduzierbare Ergebnisse. Besonders bei frisch angesetzten Ätzgemischen waren die Rosettenarme sehr schmal, wie in Figur 6e durch die gestrichelten Linien angedeutet. Die maximalen Winkel mit der [0001]-Richtung stimmten jedoch im wesentlichen überein. Das Übergangsgebiet war wenig ausgeprägt.

Im Gegensatz dazu zeigt eine durch elektrolytisches Ätzen entwickelte Rosette (Fig. 6c) relativ breite Zonen mit hinreichender bis schlechter Ausbildung der Ätzgruben, gekennzeichnet durch unterschiedliche Schraffur in der Figur 6f.

Außerhalb der angeätzten Bereiche konnte bei allen Ätzgemischen eine wechselnde Belegung mit Reaktionsprodukten beobachtet werden.

In Figur 7 ist die elektrolytisch geätzte Stirnfläche eines druckverformten Antimoneinkristalls ($\dot{a}_0 = 1{,}36 \cdot 10^{-4}\ s^{-1}$) nach einer Abgleitung von 3,56% wiedergegeben. Die Orientierung der Probe ($\Theta_0 = 40^1/_2{}^\circ$, $\lambda_0 = 55^1/_2{}^\circ$) lag nahe dem $(02\bar{2}7)$-Flächenpol. Die Ätzgruben sind parallel den Spuren der (0001)-Basisgleitebene angeordnet.

5. Zusammenfassung

Das horizontale Zonenschmelzverfahren ist für die Züchtung großer Antimoneinkristalle gut geeignet. Die Präparation homogener Antimon-Wismut-Mischkristalle erfordert eine geringe Kristallisationsgeschwindigkeit und einen steilen Temperaturgradienten im Bereich der Erstarrungsfront, um konstitutionelle Unterkühlung zu vermeiden. Mittels der zone-leveling-Technik lassen

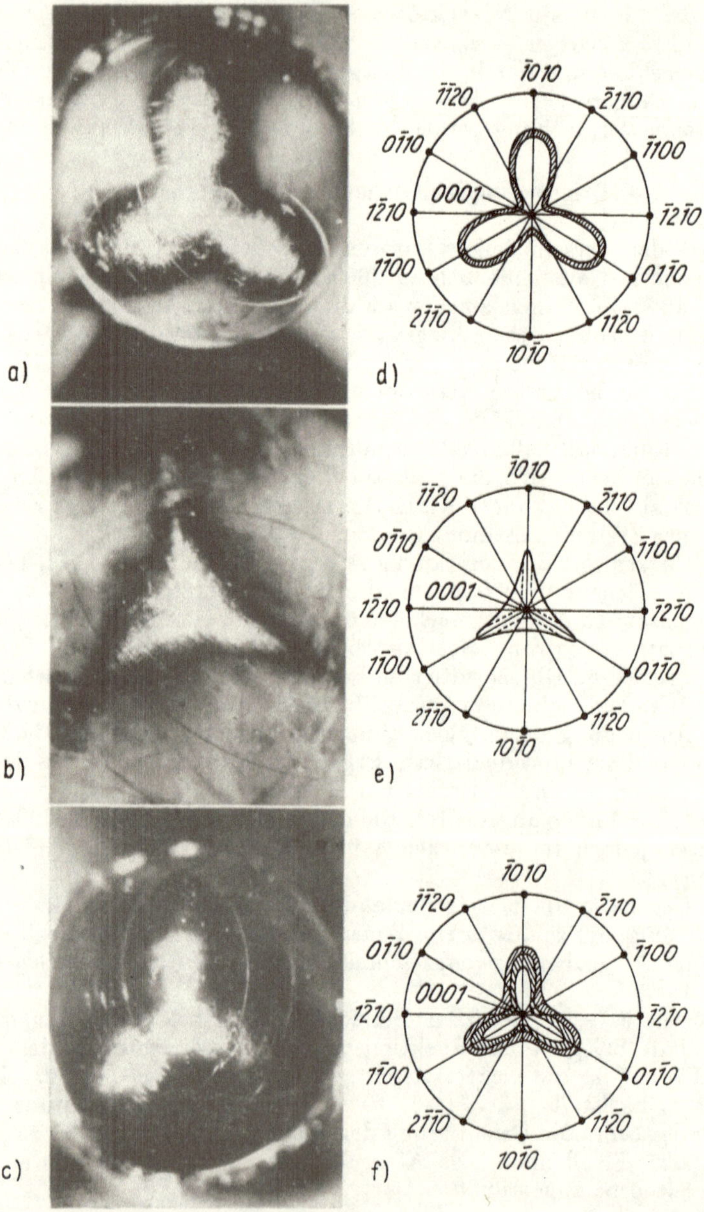

Fig. 6a–f. Orientierungsabhängigkeit des Ätzverhaltens bei Anwendung verschiedener Ätzmittel
Fig. 6a–c. Geätzte Antimoneinkristallhalbkugeln
Fig. 6d–f. Stereographische Projektionen der Orientierungsbereiche mit Ätzgrubenausbildung
Fig. 6a und d. Ätzlösung nach SOIFER
Fig. 6b und e. Ätzlösung nach CORDES und KRUSE
Fig. 6c und f. Elektrolytische Ätzung (Elektrolyt nach COONS)

Fig. 7. Elektrolytisch geätzte Stirnfläche eines druckverformten Antimonkristalls (Basisgleitung)
$a = 3{,}56\,\%$; $\dot{a}_0 = 1{,}36 \cdot 10^{-4}\,s^{-1}$; $\Theta_0 = 40^1/_2$, $\lambda_0 = 55^1/_2{}^\circ$
Der Winkel zwischen der [0001]-Richtung und der Flächennormalen beträgt $40^1/_2{}^\circ$

sich über einen relativ großen Bereich des Barrens Legierungen mit nur geringen Konzentrationsschwankungen herstellen.

Zum Nachweis von Versetzungen wurden chemische und ein elektrolytisches Ätzverfahren benutzt. Daß den Ätzgruben Durchstoßpunkte von Versetzungslinien entsprechen, ist durch folgende Untersuchungen sichergestellt worden; Reproduzierbarkeit der Ätzgruben nach schichtweisem Abtragen einer Kristalloberfläche, Entsprechung von Ätzgruben auf korrespondierenden Spaltflächen, Änderung der linearen Ätzgrubendichte an Verzweigungen von Kleinwinkelkorngrenzen, Anordnung von Ätzgruben parallel den Spuren der Gleitebenen nach plastischer Verformung.

Die Orientierungsabhängigkeit des Ätzverhaltens wurde mit einigen ausgewählten Ätzmitteln an Einkristallhalbkugeln untersucht. Man kann eine ausgeprägte Anisotropie beobachten; um den Flächenpol (0001) bildet sich eine Ätzrosette entsprechend der trigonalen Symmetrie der Richtung [0001].

Diese Arbeit wurde im Rahmen einer Aspirantur am Zentralinstitut für Physikalische Chemie der Deutschen Akademie der Wissenschaften zu Berlin durchgeführt. An dieser Stelle sei Herrn Prof. Dr. K. Meyer für sein stetes Interesse und die wertvolle Unterstützung gedankt.

Literatur

Adams, K. H., Blish, R. C., Vreeland, T.: J. appl. Physics **37**, 4291 (1966)
Brown, D. M., Heumann, F. K.: J. appl. Physics **35**, 1947 (1964)
Coons, W. C.: Metals Progress **84**, No. 6 (1963)
Cordes, H., Kruse, R.: Z. Metallkunde **57**, 563 (1966)
Doerschel, J., Fischer, G.: Kristall und Technik **6**, 153 (1971)
Gough, H. J., Cox, H. L.: Proc. Roy. Soc. [A] **127**, 431 (1930)
Kolsky, H.: Proc. Roy. Soc. [B] **62**, 676 (1949)
Kosevič, V. M.: Kristallografija **5**, 749 (1960)
Lange, V. N., Lange, T. I., Oganjan, S. G.: Kristallografija **10**, 260 (1965)
Meyer, K., Brückner, U.: Kristall und Technik **1**, 477 (1966)

Meyer, K., Meier, W.: Kristall und Technik 3, 399 (1968)
Rühlicke, D., Männel, P., Hötzsch, G.: Kristallisation, Leipzig 1969, S. 184
Shigata, J., Hiramatsu, M.: J. Physic. Soc. Japan 13, 1404 (1958)
Soifer, L. M.: Zaščita Metallov 1, 29 (1965)
Soifer, L. M., Ščegoleva, Z. A.: Fizika Metallov Metalloved. 24, 140 (1967)
Soifer, L. M., Starzev, V. I.: Dokl. Akad. Nauk SSSR 134, 795 (1960)
Tiller, W. A., Jackson, K. A., Rutter, J. W., Chalmers, B.: Acta Metallurgica 1, 428 (1953)
Wernick, H. J., Hobstetter, J. N., Lovell, L. C., Dorsi, D.: J. appl. Physics 29, 1013 (1958)
Yim, W. M., Dismukes, J. P.: J. Phys. Chem. Solids, Supplement 28, 187 (1967)

(Eingegangen am 20. 5. 1971)

Anschrift des Verfassers:

Dipl.-Min. J. Doerschel
VEB Werk für Fernsehelektronik, Bereich Forschung
116 Berlin
Ostendstraße 1–5

Kristall und Technik	7	1–3	1972	207–220

H.-J. Ullrich, G. E. R. Schulze

Technische Universität Dresden, Sektion Physik

Röntgenographische Mikrobeugungsuntersuchungen an kristallinen Festkörpern mittels Gitterquellen-interferenzen (Kossel-Linien) und Weitwinkel-interferenzen (Pseudo-Kossel-Linien)

Es wird eine Übersicht über Grundlagen und Methodik der an der Sektion Physik der TU Dresden verwendeten röntgenographischen Mikrobeugungsverfahren gegeben und über Gitterkonstantenmessungen, Ermittlung von Symmetrieerniedrigungen sowie Tief- und Hochtemperaturexperimente berichtet.

A view on the foundations and methods of the applied X-ray micro beam diffraction of the Department of Physic of the TU Dresden is given. It is reported upon the measure of lattice constants, the ascertainment of lowering of symmetry, and low- and high-temperature experiments.

W. Kleber hat stets dazu angeregt, neue Untersuchungsmethoden zur Lösung kristallphysikalischer Probleme heranzuziehen. Wir haben seine Aufgeschlossenheit gegenüber neueren Verfahren schon im Anfangsstadium unserer Arbeiten auf dem Gebiet der röntgenographischen Mikrobeugung in anregenden Diskussionen empfunden. Da er leider den Aufschwung, den diese Methoden erst in jüngster Zeit genommen haben, nicht mehr miterleben konnte, mag in diesem Gedächtnisheft eine Übersicht über unsere Arbeiten zu obiger Thematik gegeben werden.

1. Einleitung

Von den röntgenographischen Mikrobeugungsverfahren wie Laue- (mit 10 μm-Blende), Weitwinkel- (in Durchstrahlrichtung) und Kossel-Technik wird letztere Methode als die z. Zt. leistungsfähigste angesehen.

Sie benutzt die von Kossel 1924 vorhergesagten „Interferenzen aus Gitterquellen“, deren experimentellen Nachweis er mit seinen Mitarbeitern Loeck und Voges bei Elektronenstrahlanregung und Borrmann auch bei Röntgenstrahlanregung sicherstellen konnte.

Das Prinzip der Methode besteht in folgendem (s. Figur 1): Ein feinfokussierter Elektronenstrahl regt innerhalb des zu untersuchenden kleinen einkristallinen Volumens die charakteristische Röntgenstrahlung an, die mit den umgebenden Raumgitter interferiert. Die Interferenz-Strahlung liegt auf geraden Kreiskegeln, den sogenannten Kossel-Kegeln. Der halbe Öffnungswinkel der Kossel-Kegel beträgt $90° - \vartheta_{hkl}$ (Fig. 1b). Die Symmetrieachse des Kegels fällt mit der Normalen der reflektierenden Netzebenenschar zusammen. Verläßt die Interferenzstrahlung den Kristall in Rückstrahlrichtung, so werden auf

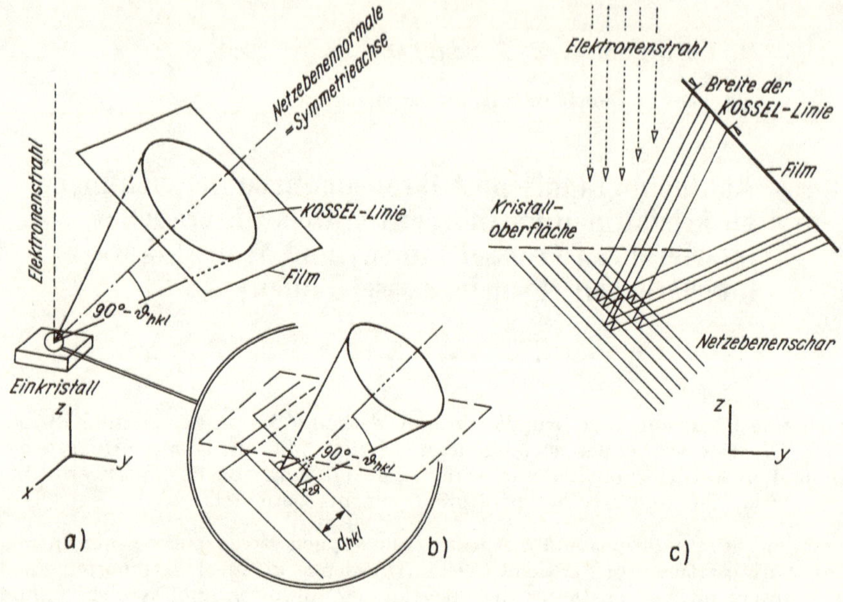

Fig. 1. Zur Entstehung der Kossel-Linien

einem ebenen Film Kegelschnittlinien registriert, die zu Ehren ihres Entdeckers auch Kossel-Linien genannt werden. Sie heben sich i. a. aus dem von der Bremsstrahlung und der nichtgebeugten charakteristischen Strahlung herrührenden Untergrund als dunkle Linien ab. Die Breite der Interferenzlinien auf dem Film entspricht bei vernachlässigter Dispersion dem Brennfleckdurchmesser der Elektronensonde, weil die Eindringtiefe der Elektronen bei den verwendeten Spannungen (3—5-fache der Anregungsspannung) in der Größenordnung Mikrometer liegt (Fig. 1c).

Die Fokussierungsmöglichkeit der Elektronen ermöglicht die Untersuchung mikroskopisch kleiner Gebiete (einige μm^3), also auch einzelner Körner polykristallinen Materials, die über diesen Abmessungen liegen.

Die aufgezählten Gesetzmäßigkeiten lassen folgende Anwendungsmöglichkeiten der Kossel-Technik erwarten:

1. Gitterkonstantenmessungen
2. Orientierungsbestimmungen
3. Ermittlung von Symmetrieelementen
4. Tief- und Hochtemperaturexperimente sollten neben den Ausdehnungskoeffizienten auch die Eigenschaften 1—3 bei Kristallgitterumwandlungen erfassen.

Bei den Anwendungen 1—4 wird die gegenseitige Lage der Kossel-Linien ausgewertet. Die Linien-Intensität und ihre Feinstruktur liefert Aussagen über

5. Gitterschwingungen
6. Phasen der gestreuten Wellen und die
7. Realstruktur des untersuchten Probengebietes.

Im folgenden wollen wir uns auf die Messungen der Gitterkonstanten und Ermittlung von Symmetrieelementen beschränken.

2. Bemerkungen zur Methodik

2.1. Röntgenstrahlanregung

Da diese Variante apparativ weniger Aufwand benötigte, haben wir ihren Anwendungsbereich untersucht (Ullrich, Küchler). Wir arbeiteten wie Borrmann mit einer zylindrischen Filmanordnung im Rückstrahlbereich, evakuierten aber zusätzlich die Kammer und konnten die Proben bis 600 °C aufheizen.

Obwohl im Vergleich zur Anregung mit Elektronen der Bremsstrahluntergrund fehlt, läßt sich bei Sekundäranregung nur in günstigen Einzelfällen eine ausreichende Anzahl von Kossel-Linien beobachten. Diese sind allenfalls als Hilfsmittel zur schnelleren Auswertung von Laue-Aufnahmen geeignet (Ullrich, Küchler; Brodowski). Wir haben von FeAl, Fe_3Al, Fe_3O_4, Fe, Cu, $MgCu_2$, $MgZn_2$, Cu_2O und $CuAl_2$ symmetrische Kossel-Aufnahmen angefertigt. Zur Präzisionsbestimmung von Gitterkonstanten (relativer Fehler kleiner als 10^{-4}) waren diese nicht geeignet. Denn, wie bei jedem röntgenographischen Verfahren zur Bestimmung der Gitterkonstanten, müssen Reflexe unter größten Bragg-Winkeln vermessen werden, um hohe Genauigkeiten zu erzielen. Gerade aber diese fehlen in Aufnahmen mit Sekundäranregung.

Die aufgezählten Nachteile veranlaßten uns, eine Apparatur mit Elektronenanregung aufzubauen (Ullrich 1966) und deren Anwendungsbereich bezüglich der untersuchbaren Substanzen zu studieren (Ullrich, Schulze).

2.2. Elektronenanregung

In Figur 2a ist eine Kossel-Aufnahme von Fe_3Al wiedergegeben. Die zugehörig Linienindizierung entnimmt man aus Figur 2b. Figur 2a enthält das vollständige Reflexsystem der Fe-Kα-Linien. Von den intensitätsreichsten Reflexen 101 und $\bar{1}01$ sind die β-Linien (im Original) deutlich zu erkennen. Nahezu alle Linien besitzen die dem Kossel-Effekt zugehörige Feinstruktur: An die dunkle Linie schließt sich auf der konvexen (kurzwelligen) Seite ein heller Saum an.

Der 022-Kreis geht auf der rechten Seite der Aufnahme in eine helle Linie über, weil dort der Kossel-Kegel unter niedrigem Abnahmewinkel aus der Kristalloberfläche austritt und dieser Teil bei seinem verlängerten Weg im Kristall dem Extinktionseffekt unterliegt.

Bezüglich des Anwendungsbereiches zeigte sich, daß die Kossel-Technik weit über den bisher verwendeten Wellenlängen-Bereich (Substanzbereich) von Fe—Kα bis As—Kα ausgedehnt werden kann. Wir konnten auswertbare Kossel-Aufnahmen mit Wellenlängen von S—Kα (!) bis Nb—Kα herstellen. Dadurch konnten nicht nur eine beträchtliche Anzahl weiterer Substanzen der Kossel-Technik zugänglich gemacht werden, sondern zusätzlich neue Problemstellungen wie Streuphasenbestimmungen (Blau et al.) oder Hochtemperaturuntersuchungen (Ullrich, Däbritz, Schreiber 1969) bearbeitet werden. Gleichzeitig traten auch die Grenzen der Kossel-Technik zutage. Dadurch, daß man auf die Eigenstrahlung der in der Probe vorliegenden Atome angewiesen ist, kann die Strahlung nicht dem Gitter „angepaßt" werden. Bei Legierungen kommt im Vergleich zu den Elementen noch erschwerend hinzu, daß die angeregten Atome in geringer Konzentration in der Substanz enthalten sind und damit ein Intensitätsverlust der Quellstrahlung einhergeht. Die Kossel-Technik versagt dann, wenn die im Gitter angeregte Wellenlänge die

a)

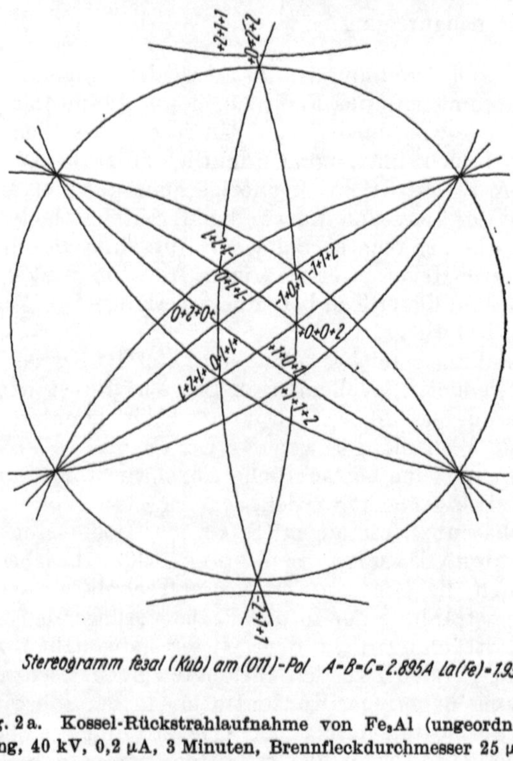

b)

Fig. 2a. Kossel-Rückstrahlaufnahme von Fe_3Al (ungeordnet) Elektronenstrahlanregung, 40 kV, 0,2 μA, 3 Minuten, Brennfleckdurchmesser 25 μm · λ(Fe) = 1.936 Å
Fig. 2b. Computergezeichnetes Stereogramm zur Indizierung von Abb. 2a

Interferenzbedingungen für die erste reflexionsfähige Netzebene nicht mehr erfüllt (z. B. Mg, Al, Si) oder wenn die Wellenlänge wie bei Mo so klein ist, daß das Reflexionsvermögen stark zurückgeht und von den 740 Kα-Reflexen im Halbraum nur einzelne registriert werden können.

Beim Weitwinkelverfahren im Rückstrahlbereich (Fig. 3a) wird durch geeignete Wahl des Antikatodenmaterials die Strahlung dem Gitter angepaßt. Der Kristall braucht nicht im Vakuum zu stehen. Die Aufnahmen zeichnen sich durch hohen Kontrast aus, da der intensitätsreiche Untergrund fehlt. Die Linien stellen bei ebenem Film Kurven 4. Ordnung dar (Maier). Wegen der Ähnlichkeit von Weitwinkelrückstrahlaufnahmen mit den Kossel-Aufnahmen hat Imura dafür die Bezeichnung „Pseude-Kossel-Pattern" eingeführt, obwohl diese Variante bereits auf Seemann zurückgeht und von Hess schon zur Bestimmung von Gitterkonstanten mit geringer Genauigkeit ausgenutzt wurde.

In der angelsächsischen Literatur faßt man häufig unberechtigterweise unter dem Begriff Kossel-Technik die Anwendung von Gitterquellen (Kossel)- und Weitwinkel (Pseudo-Kossel)-Interferenzen zusammen. Schülke und Brümmer

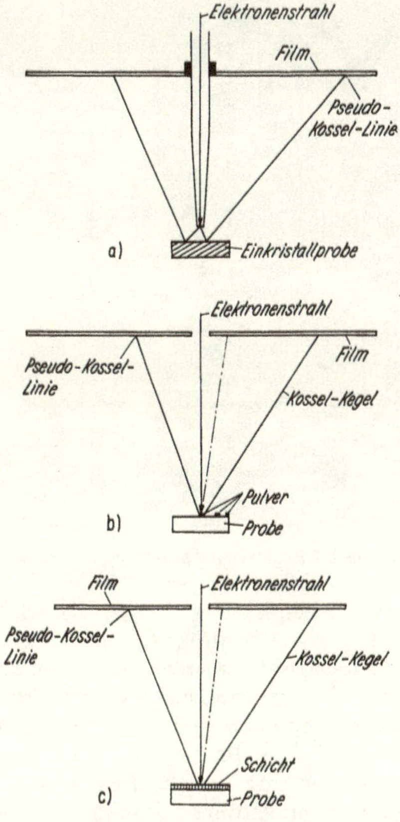

Fig. 3a. Weitwinkel-Rückstrahl (Pseudo-Kossel)-Anordnung
Fig. 3b. Punkt-Pseudo-Kossel-Anordnung mit Metall-Pulver
Fig. 3c. Punkt-Pseudo-Kossel-Anordnung mit Aufdampfschicht als Antikatodenmaterial auf der Probe

haben zeigen können, daß bei Lage der Gitterquellen in Realkristallen zumindest auf Durchstrahl-Aufnahmen Weitwinkel (Pseudo-Kossel)-Linien – also keine Kossel-Linien – beobachtet werden.

Der Nachteil der Weitwinkel-Rückstrahltechnik besteht darin, nur ebene größere Flächen von mindestens 4 mm² untersuchen zu können (ELLIS et al.). Um den nichtintegralen Charakter der Kossel-Technik experimentell möglichst wieder nahezukommen, haben wir zwei Wege beschritten: Erstens ist es uns durch Verbesserung des Auflösungsvermögens des Weitwinkelverfahrens gelungen, Flächen von weniger als 1 mm² zu untersuchen. Figur 4 zeigt eine solche Weitwinkelaufnahme von GaAs. Zweitens konnten wir die Vorteile der Kossel- und Weitwinkeltechnik, d. h. „punktförmiger" Untersuchungsbereich bzw. Wahl der günstigsten Wellenlänge in einem Verfahren vereinigen.

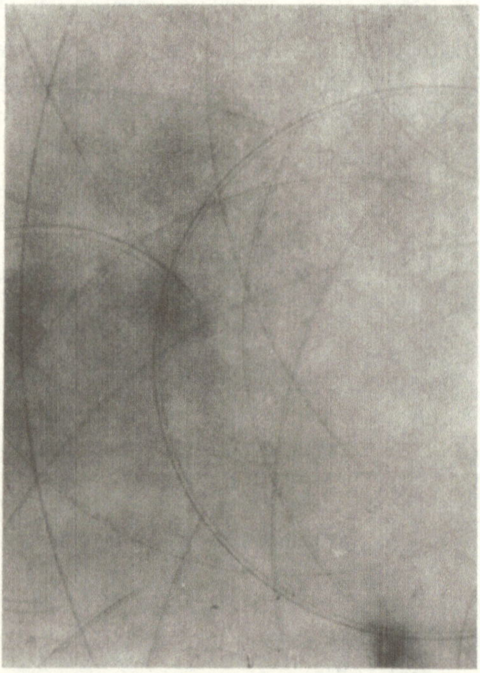

Fig. 4. Weitwinkel-(Pseudo-Kossel)-Rückstrahl-Aufnahme von GaAs mit Cu-Strahlung

Wir haben geeignetes Antikatodenmaterial (z. B. Cr, Ni, Cu) als Pulver (Fig. 3b) oder als Aufdampfschicht (Fig. 3c) direkt auf die Probenoberfläche aufgebracht und in diesem die Eigenstrahlung angeregt (ULLRICH 1966). Da die Röntgenquellen nun außerhalb des zu untersuchenden Gitters liegen, die Strahlung aber mit einem Probengebiet interferiert, welches nur wenig größer als das des Kossel-Verfahrens ist und deshalb die Methode ebenfalls zu den röntgenographischen Mikrobeugungsverfahren zählt, haben wir die Bezeichnung Punkt-Pseudo-Kossel-Verfahren eingeführt. Punkt-Pseudo-Kossel-Aufnahmen haben wir von Diamant und Al (mit Cu-Strahlung) an Si (mit Cr- und Cu-Strahlung), an Fe (mit Co-Strahlung) an Ni (mit Cu-Strahlung) an Cu (mit Cr-, Ni- und Zn-Strahlung), an Ge (mit Cr-, Ni-, Cu- und Zn-Strahlung) an Mo (mit Ni- und Cu-Strahlung), an Nb (mit Cu-Strahlung) sowie an W (mit Ni-Strahlung) hergestellt. Untersuchungen mit der Anordnung Figur 3c sind bei McCall, Strabel, Duerr) veröffentlicht.

Figur 5 zeigt eine Aufnahme von Nb mit Cu-Strahlung. Der Kontrast dieser Aufnahmen lag noch unter dem der Kossel-Aufnahmen, obwohl wir eingehend nach optimalen Herstellungsbedingungen gesucht haben.

Fig. 5. Punkt-Pseudo-Kossel-Aufnahme von Nb mit Cu-Strahlung

Wegen des Extinktionskontrastes bestehen bei Durchstrahlaufnahmen günstigere Voraussetzungen zur Kontraststeigerung. Hier können die Parameter Wellenlänge, Anregungsspannung und Kristalldicke einigermaßen aufeinander abgestimmt werden (Yakowitz; Brümmer, Dräger, Schülke).

Noch vor wenigen Jahren wurde die größte Zahl der Arbeiten mit der Pseudo-Kossel-Durchstrahl-Technik durchgeführt. Folgende Gründe sind dafür maßgebend, daß neuerdings immer mehr die Kossel-(Rückstrahl)Technik bevorzugt wird:

1. Aufwendige Propenpräparationen, wie sie für die Durchstrahltechnik oder andere Röntgenverfahren notwendig sind, entfallen i. a. Es können beliebig geformte Proben unterschiedlicher Abmessungen untersucht werden.
2. Die Belichtungszeit zur Herstellung der Kossel-Aufnahmen liegen zwischen einigen zehn Sekunden und wenigen Minuten.
3. Eine Kopplung Elektronenstrahl-Mikroanalyse und Kossel-Technik läßt sich vorteilhaft verwirklichen. Neben der chemischen Konzentrationsbestimmung erhält man zusätzlich vom gleichen Probenvolumen die Gitterkonstante, Orientierung, Aussagen über Gitterverzerrungen u. a. m.

Andere Mikrobeugungsverfahren weisen nicht diese Vorteile auf. Insbesondere besteht bei ihnen die Schwierigkeit, die Probengebiete schnell und treffsicher aufzusuchen. Kossel-Kameras, die in Elektronenstrahl-Mikroanalysatoren eingebaut sind, besitzen dafür die vorzüglich geeignete rasterelektronenmikroskopische Einrichtung (Scanningsystem).

3. Gitterkonstantenmessung

Die Kossel-Technik hat ihren Aufschwung Anfang der sechziger Jahre vor allem der Tatsache zu verdanken, daß mit keinem anderen Verfahren z. Zt. Gitterkonstantenmessungen mit solch großen relativen Genauigkeiten (z. B. $\Delta a/a = 10^{-5}$) in mikroskopisch kleinen Gebieten (Durchmesser einige μm) kompakter Proben möglich sind. Damit ist man z. B. in der Lage, die Gitterkonstanten in einzelnen Körnern eines Festkörpers miteinander zu vergleichen oder Gitterkonstantenänderungen in der Nähe von Korngrenzen auszumessen.

Zur Ermittlung der Gitterkonstanten aus Kossel-Aufnahmen verwendeten wir häufig die ursprüngliche Kosselsche Kompensationsmethode (Ullrich 1967a). Dazu mußte eine symmetrische Kossel-Aufnahme angefertigt und geeignete Begegnungspole der Kossel-Linien aufgesucht werden. Da sich der Film-Fokus-Abstand bei diesem Verfahren eliminieren läßt, wurde in einigen Fällen die relative Genauigkeit von 10^{-5} unterschritten.

Die Indizierung der einzelnen Linien erfolgte meist über stereographische Projektionen (vgl. Figur 2b), die wir vom Computer zeichnen lassen (Ullrich et al.).

Neuerdings werden die Linienkkordinaten von der Kossel-Aufnahme mit dem Meßgerät „Stecometer" vom VEB Carl Zeiß Jena auf einen Lochstreifen gestanzt, der dem Computer eingegeben wird. Der Rechenautomat übernimmt (bei Eingabe der ungefähren Gitterparameter) die Indizierung, berechnet die ϑ_{hkl} und druckt die Gitterparameter einzelner Netzebenen mit den Meßfehlern aus (Ullrich et al. 1971). Diese Auswertung benötigt keine symmetrische Kossel-Aufnahme, dafür muß aber der Fokus-Film-Abstand genau bekannt sein. Deshalb liegen die relativen Genauigkeiten noch über denen für die günstigsten Fälle der Kompensationsmethode. Zur Auswertung der Punkt-Pseudo-Kossel-Aufnahme wurden die eben beschriebenen Verfahren herangezogen. Die Weitwinkel-Aufnahmen werteten wir mit einem gesonderten Verfahren aus. Hinsichtlich weiterer Auswertungstechniken verweisen wir auf Tixier, Wache.

Unsere Erfahrungen bei der Ausmessung von Gitterkonstanten mit den oben beschriebenen Methoden lassen sich qualitativ wie folgt zusammenfassen:

Als Proben zur Untersuchung mit der Kossel-Technik sind die Elemente Ni und Cu besonders gut geeignet. Hier liegt das Verhältnis λ/a (Wellenlänge/Gitterkonstante), welches über die Reflexanzahl pauschal Auskunft gibt, um 0,4. Die Zahl der Reflexe ist hier so groß, daß sich mit hoher Wahrscheinlichkeit günstige Linienpaare unter großen Bragg-Winkeln finden lassen. Der Intensitätsüberschuß reicht unter den üblichen Beobachtungsbedingungen (30 kV) aus, um die Linien genau genug ausmessen zu können.

An Fe erweist sich das λ/a als ungünstiger. Kossel-Aufnahmen an Fe–Co-Verbindungen und Punkt-Pseudo-Kossel-Aufnahmen mit Co-Strahlung liefern Reflexe mit größerer Dispersion.

Obwohl wir an NaCl und KCl jeweils das vollständige Reflexsystem der Cl–Kα bzw. K–Kα-Strahlung erhielten, ließen sich Präzisionsgitterkonstantenbestimmungen an diesen Substanzen mit der Kossel-Technik nicht durchführen. Einmal werden durch die auftreffenden Elektronen Strahlenschäden erzeugt und damit die Gitterkonstante verändert. Andererseits besteht das vollständige Reflexsystem nur aus wenigen Linien, deren Dubletts kaum aufgelöst sind.

Wie Kossel-Aufnahmen von FeS_2 und ZnS beweisen, lassen sich Gitterkonstantenbestimmungen auch mit großen, bisher für solche Zwecke nicht verwendeten Wellenlängen durchführen. An FeS_2 und ZnS eignen sich die S—Kα-Reflexe wegen ihrer hohen Intensität (bei 15 kV) und Dispersion besser als die K-Reflexe von Fe bzw. Zn, obwohl S nur in annähernd gleicher bzw. geringerer Massenkonzentration als die Schwermetalle in diesen beiden Verbindungen enthalten ist.

Die durch den Elektronenstrahl verursachte Temperaturerhöhung im Brennfleck läßt sich experimentell bestimmen. Dazu wurden Gitterkonstantenmessungen vom gleichen Kristallort bei verschiedenen Strahlströmen vorgenommen, d. h., das Gitter im untersuchten Volumen aufgeheizt und die Gitterkonstante für Strahlstrom Null extrapoliert. Für schlecht wärmeleitendes Material wie Fe_3O_4 ergab sich eine max. Temperaturerhöhung von 13,5 °C während der Exposition der Aufnahmen. Daß die Strahlbelastungen der Probe nicht so von entscheidender Bedeutung sein kann wie bisher angenommen, beweisen Kossel-Aufnahmen von CaF_2 mit scharfen Linien und deutlichen Dublettenaufspaltungen.

Bei Fe_3O_4 liegt eine große Wahrscheinlichkeit für Reflexe mit hoher Dispersion vor, deren Intensität unter den üblichen Beobachtungsbedingungen (30 kV) zu gering ist. Hier konnte gezeigt werden, daß höhere Beschleunigungsspannungen einen besseren Kontrast hervorrufen. So ist es z. B. gelungen, am [111]-Pol mit 60 kV noch die 555-Kα_1-Linie unter 86,5° zu registrieren. Die Kα_2-Linie taucht gerade bei 90° als diffuser Punkt auf. Der Dublettabstand beträgt auf der Originalaufnahme 16 mm!

Trotz der erhöhten Reflexanzahl bei $ZrFe_2$, Cu_2O und $MgCu_2$ eigneten sich die Aufnahmen zur Präzisionsgitterkonstantenbestimmung. Die Kossel-Aufnahmen der nichtkubischen Substanzen Zn, $CaZn_5$, $MgZn_2$ und $CuAl_2$ belegen, daß alle Gitterparameter aus einer Aufnahme gewonnen werden können.

Für Elemente oberhalb Zn im Periodensystem überwiegt bei Kossel-Rückstrahluntersuchungen der Einfluß der kleinen Wellenlänge auf das Reflexionsvermögen. Deshalb mißlangen Präzisionsbestimmungen an Ge und Mo, obwohl das λ/a-Verhältnis noch günstiger als bei Fe_3O_4 und $MgCu_2$ liegt.

Genaue Gitterkonstantenmessungen mit der Lα_1-Strahlung sind bei Ta und W möglich.

Die Substanzen, die sich mit der Kossel-Technik nicht ausreichend genau untersuchen lassen, werden von der Punkt-Pseudo-Kossel-Technik (Fig. 3b, c) erfaßt. In vielen Fällen konnte das vollständige Reflexsystem der Pseudo-Kossel-Linien erzeugt und Präzisionsbestimmungen durchgeführt werden. Das gilt auch für leichte Elemente.

Tabelle 1 enthält in Ergänzung zu Daten von Ullrich, Schulze einige unserer Meßwerte.

Diskussion:

Bei den angegebenen Präzisionsbestimmungen handelt es sich (außer beim Weitwinkel-Verfahren) um Relativmessungen in Mikrobereichen der kompakten Proben, bei denen der Fehler der Wellenlängen (z. B. Einfluß der chemischen Bindung) unberücksichtigt blieb.

Tabelle 1
Ergebnisse von Gitterkonstantenmessungen mit verschiedenen Mikrobeugungsverfahren

Substanz	Struktur-typ	Methode	ausgenutzte Wellenlänge	Auswerte-verfahren	Ergebnisse ($a \pm \Delta a$) (10^{-10} m)	Bemerkung
Mn_2O_3	$D5_3$	K	Mn-$K\alpha_1$	A	9,4154 $\pm 3 \cdot 10^{-4}$	Kristalle, die mit Wachs-
$(Mn_{0.62}Fe_{0,38})_2O_3$	$D5_3$	K	Mn-$K\alpha_1$	A	9,4174 $\pm 3 \cdot 10^{-4}$	tumsflächen begrenzt waren
Fe_3O_4I	$H1_1$	K	Fe-$K\alpha_1$	A	8,3996 $\pm 2 \cdot 10^{-4}$	Verschiedene Kristallexem-
Fe_3O_4II	$H1_1$	K	Fe-$K\alpha_1$	A	8,3966 $\pm 5 \cdot 10^{-4}$	plare v. gleichen Fundort
Fe_3Al	(A2)	K	Fe-$K\alpha_1$	A	2,89322 $\pm 7 \cdot 10^{-5}$	(−49,0 °C)
X10CrAl18	(A2)	K	Fe-$K\alpha_1$	C	2,8749 $\pm 6 \cdot 10^{-4}$	
X5CrNiTi18,10	(A1)	K	Fe-$K\alpha_1$	C	3,5895 $\pm 5 \cdot 10^{-4}$	
X5CrNiTi18,10	(A1)	K	Fe-$K\alpha_1$	C	3,5906 $\pm 5 \cdot 10^{-4}$	abgeschreckte Probe
Ni	A1	K	Ni-$K\alpha_1$	A	3,51676 $\pm 3 \cdot 10^{-5}$	(−126,5 °C)
Ni	A1	K	Ni-$K\alpha_1$	C	3,5233 $\pm 1 \cdot 10^{-4}$	(20,0 °C)
Cu	A1	K	Cu-$K\alpha_1$	A	3,60900 $\pm 7 \cdot 10^{-5}$	(− 85,3 °C)
$MgZn_2$	C14	K	Zn-$K\alpha_1$	B	5,221 $\pm 2 \cdot 10^{-3}$	$c = 8{,}546 \pm 2 \cdot 10^{-3}$
$NbFe_2$	C15	K	Fe-$K\alpha_1$	B	4,811 $\pm 8 \cdot 10^{-3}$	
$CuAl_2$	C16	K	Cu-$K\alpha_1$	B	6,064 $\pm 1 \cdot 10^{-3}$	$c = 4{,}874 \pm 1 \cdot 10^{-3}$
C	A4	P	Cu-$K\alpha_1$	B	3,5664 $\pm 9 \cdot 10^{-4}$	
Si	A4	W	Cu-, Zn-$K\alpha_1$	D	5,4348 $\pm 7 \cdot 10^{-4}$	
Nb	A2	P	Cu-$K\alpha_1$	C	3,3003 $\pm 1 \cdot 10^{-4}$	

Erläuterungen der Abkürzungen

K = Kossel-Technik
P = Punkt-Pseudo-Kossel-Technik
W = Weitwinkel-Rückstrahl-Technik

Auswerteverfahren

A = Kompensationsmethode über Kristallwinkel
B = Kompensationsmethode über Wellenlängenvariationen (Schreiber)
C = Computerrechnung
D = Auswertung für Weitwinkelverfahren

Die mit B und C bezeichneten Gitterkonstanten wurden von Schreiber übernommen.

Die mit dem Auswerteverfahren A ermittelten Fehler entsprechend denen einer Einzelmessung. Bei Aufnahmen mit scharfen Interferenzlinien, die zehnmal mit dem Komparator vermessen wurden, lag die relative Genauigkeit bei $6 \cdot 10^{-6}$. Die Schwankungen des Absolutwertes der so genau ermittelten Gitterkonstanten über die gesamte Kristallfläche lagen aber weitaus höher. Daraus ergibt sich die Schlußfolgerung, daß mit der Kossel-Technik eine sehr feine Meßsonde zur Verfügung steht, die eine genaue Überprüfung der Homogenität zuläßt. Solche Homogenitätsuntersuchungen wurden an Ni (verunreinigt), $MgCu_2$, $MgZn_2$, FeAl, Mn_2O_3 und Fe_3O_4 vorgenommen.

3.1. Beispiel Ni

In Tabelle 2 sind unsere Meßwerte mit denen anderer Autoren verglichen. Die Gitterkonstante des von uns verwendeten Einkristalls war an verschiedenen „Punkten" der Oberfläche genauer als $\pm 1.10^{-14}$ m definiert, was wir auf geringere mechanische Spannungen an den betreffenden Stellen zurückführen.

Tabelle 2

Vergleich der Gitterkonstanten von Ni mit Literaturwerten

Methode	a (10^{-10} m)	Reinheit	T (°C)	Bemerkung
Kossel-Rückstrahl-Aufnahme	3,52036 ± 0,00011	99,5%	20	über Kristall gemittelt
	3,52030 ± 0,00003	99,5%	20	Messung für einen Punkt des gleichen Kristalls
Durchstrahl-	3,52606 ± 0,00002	–	–	HANNEMAN
Weitwinkel-	3,5286 ± 0,0001	–	–	HEISE
Autnahme	3,5238 ± 0,0001	–	–	MACKEY
Debye-Scherrer	3,5251	–	–	TAYLOR, KAGLE

Der Einkristall besaß eine Reinheit von 99,5% und hatte eine kleinere Gitterkonstante als die übrigen in Tabelle 2 angegebenen Vergleichswerte.

Tabelle 2 läßt den Schluß zu, daß die Kossel-Technik für Realkristalle zu größeren Genauigkeiten führen kann, als die „integralen" Methoden. So mitteln Pulveraufnahmen immer über die infolge der Inhomogenitäten in der Probe vorhandenen (und durch die Präparation zusätzlich hineingebrachten) unterschiedlichen Gitterkonstanten. Zu Kossel-Aufnahmen tragen nur mikroskopisch kleine Bereiche bei, in denen mit größerer Wahrscheinlichkeit ausreichende Homogenität vorliegt. Aus diesem Grunde sind auch beim Kossel-Verfahren hohe absolute Genauigkeitsangaben berechtigt, die aber nur für den untersuchten Proben-„Punkt" gelten. Besonders deutlich zeigte sich das auch an polykristallinem $MgCu_2$.

3.2. Beispiel $MgCu_2$

In der Literatur findet man die Gitterkonstanten $a = 7{,}080 \cdot 10^{-10}$ m (TAYLOR, KAGLE) bzw. $a = 7{,}048 \cdot 10^{-10}$ m (PEARSON). Bei 22 °C ermittelten wir $a = (7{,}0365 \pm 0{,}0002) \cdot 10^{-10}$ m. Solch hohe Genauigkeiten aus einer Einzelmessung erreichten andere Verfahren an dieser Substanz nicht. Unsere Auswertungen führten wir an den Linien 008—266 am Pol mit den Millerschen Indizes 7. 21. 40 durch. Bei den bei TAYLOR, KAGLE, PEARSON angegebenen Gitterkonstanten müßte der Schnitt der 008—$\bar{2}$66-Kegel bereits bei Zimmertemperatur erfolgt sein. Wir beobachteten ihn erst bei 70 °C.

Wie die angegebenen Gitterkonstanten beweisen, sind selbst für intermetallische Verbindungen Gitterkonstantenangaben nur mit genauen Konzentrationswerten berechtigt. An der untersuchten Kristallstelle bestimmten wir die Zusammensetzung mit der Elektronenstrahl-Mikroanalyse zu 65,5 at % Cu.

Innerhalb eines Korns der gleichen Substanz konnten wir Gitterkonstantenänderungen von $-5 \cdot 10^{-14}$ m beobachten. Unter der Annahme, daß nur Konzentrationsänderungen dafür verantwortlich sind, läßt das auf einen Mg-Verlust von 0,03 Masse-% schließen (ULLRICH 1967b).

3.3. Beispiel Fe_3O_4

Die in den Tabelle 1 enthaltenen Werte von verschiedenen Fe_3O_4-Exemplaren unterscheiden sich, wie selbstverständlich unterschiedliche Probenbehandlungen zu veränderten Gitterkonstanten (z. B. ferritischer Stahl X5CrNiTi18.10) führen.

4. Ermittlung von Symmetrieerniedrigungen

Kossel-Aufnahmen enthalten Punkte, in denen sich drei oder mehr Linien exakt schneiden. Diese Schnitte liegen unabhängig von der Wellenlänge (oder Gitterkonstanten) fest und sind nur durch die Kristallsymmetrie bedingt. Bei Punkt-Pseudo-Kossel-Aufnahmen muß man vorsichtig mit solch einer Deutung sein, da die exakte Linienüberdeckung eine Funktion der Schichtdicke des Antikatodenmaterials ist. In Weitwinkel-Rückstrahl-Aufnahmen trifft dieser Sachverhalt nicht mehr zu, wird aber sehr häufig vorgetäuscht.

Das Kossel-Diagramm für das kubischraumzentrierte Gitter (Fig. 2a) enthält z. B. je 4 Fünferschnitte und je 8 Dreierschnitte der $K\alpha_1$- und $K\alpha_2$-Reflexe. Sind diese Vielfachschnitte aufgelöst, dann liegt eine Abweichung vom kubischen Gitter vor. Diese Linienlagen bilden somit eine empfindliche Meßgröße zum Nachweis von Symmetrieerniedrigungen. Zur Bestimmung solcher Gitterverzerrungen versagt die Kompensationsmethode, da die aus dem Kristallgitter entnommenen Winkel undefiniert von denen des Idealgitters abweichen können. In einigen Fällen reichte sogar die Genauigkeit der Computerauswertung nicht aus, um die mit dem Auge erkennbaren Abweichungen wiederzugeben. Hier erarbeiteten wir eine besondere Methode.

4.1. Beispiel $(M_{1-x}nFe_x)_2O_3$

An einzelnen $(Mn_{1-x}Fe_x)_2O_3$-Exemplaren mit $x \leqq 0{,}38$, die nicht durch Wachstumsflächen begrenzt waren, konnte aus der Auflösung der 626-, 440- und 2$\bar{2}$6-Mn—$K\alpha_1$-Dreierschnitten auf eine tetragonale Verzerrung der Größe $8 \cdot 10^{-4}$ geschlossen werden.

4.2. Beispiel FeAl

Um eine statistische Aussage über die Gitterkonstante in Abhängigkeit vom Ort auf der Oberfläche eines 50 mm langen FeAl-Einkristalls zu treffen, wurden über 100 Kossel-Aufnahmen angefertigt. Nur in wenigen untersuchten Oberflächen„punkten“ lag ein exakt kubisches Gitter vor. Die Aufnahmen entsprachen dann der von Figur 2a nur mit veränderter Gitterkonstante.

Da in den meisten Aufnahmen alle Fünferschnitte und nur der Dreierschnitt vom Typ 002, 101, $\bar{1}01$ aufgelöst waren, konnte gezeigt werden, daß es sich um eine tetragonale Verzerrung handelte (Ullrich, Schreiber, Däbritz, Kleinstück; Ullrich, Däbritz, Schreiber 1968). Die Rechnung ergab für $c/a = 1 + (4{,}5 \pm 0{,}6) \cdot 10^{-3}$ mit $a = 2{,}9070 \cdot 10^{-10}$ m (Mittelwert). Nach 15-stündigem Tempern bei 900 °C im Wasserstoffstrom ging die Symmetrieabweichung zurück.

5. Tief- und Hochtemperaturexperimente

Tieftemperatur-Kossel-Aufnahmen wurden von Fe_3O_4, Cu (Ullrich, Schulze), Ni und Fe_3Al exponiert.

Die Hochtemperaturexperimente an Cu, Fe und Fe_3O_4 sind in (Ullrich, Däbritz, Schreiber 1969) beschrieben. Da wir an Fe oberhalb der $\alpha-\gamma$-Umwandlung keine Kossel-Aufnahmen infolge der Polygonisation erhalten haben, führten wir Hochtemperaturexperimente an polykristallinem Co aus. Hier konnten innerhalb eines Korns auch oberhalb der $\alpha-\beta$-Umwandlung gut auswertbare Kossel-Aufnahmen hergestellt werden. Die in der Literatur veröffentlichten kristallographischen Beziehungen des Umklappprozesses konnten wir bestätigen. Das untersuchte Korn zerfiel auch nach häufigen Umwandlungen nicht in kleinere Körner.

Während der Durchführung der Arbeiten hat uns Herr Prof. Dr. K. Kleinstück unterstützt, wofür wir ihm bestens danken. Den Herren Dipl.-Phys. L. Küchler, Dr. W. Blau, W. Damm, S. Däbritz, H. Schreiber, K. Mecke, U. Wegerer, F. Feldhofer, K. Götze, M. Hentsch, U. Weber, J. Steps und H. Scheiner danken wir für die Ausführung der Experimente.

Besonderer Dank gilt den Kollegen Dr. J. Spieß, Dipl.-Phys. U. Siegel und Dipl.-Ing. K. Schimmangk, mit denen wir gemeinsam die Stahluntersuchungen vornahmen.

Literatur

Blau, W., Schreiber, H., Schulze, G. E. R., Ullrich, H.-J.: Naturwissenschaften **54**, 535 (1967)

Borrmann, G.: Ann. Physik **27**, 686 (1936)

Brodowski, V.: Diplomarbeit TH Dresden 1960

Brümmer, O., Dräger, G., Schülke, W.: Exp. Techn. d. Physik **14**, 73 (1966)

Ellis, T., Nanni, L. F., Shrier, A., Weissmann, S., Padawer, G. E., Hosokowa, N.: J. appl. Physics **35**, 3364 (1964)

Hanneman, R. E., Ogilvie, R. E., Modrezejewski, A.: J. appl. Physics **33**, 1429 (1962)

Hase, W., Brückner, W., Tobisch, J., Ullrich, H.-J., Wegerer, U.: Z. Krist. **129**, 360 (1969)

Heise, B. H.: J. appl. Physics **33**, 938 (1962)

Hess, B.: Z. Krist. **104**, 294 (1942)

Imura, T.: Bull. Naniwa Univ. Ser. A **2**, 51 (1954)

Kossel, W.: Z. Physik **23**, 278 (1924)

Kossel, W., Loeck, V., Voges, H.: Z. Physik **94**, 139 (1935)

Mackay, K. J. H.: Vortrag 4th International Congress on X-Ray Optics and Microanalysis. Paris, September 1965

Maier, W.: Ann. Physik **40**, 85 (1941)

McCall, J. L., Strabel, G. L., Duerr, J. S.: J. appl. Physics **38**, 2695 (1967)

Pearson, W. B.: A handbook of lattice spacings and structures of metals and alloys. New York 1964

Schreiber, H.: Dissertation TU Dresden 1972

Schülke, W., Brümmer, O.: Z. Naturforsch. **17**a, 207 (1962)

Seemann, H.: Ann. Physik **6**, 793 (1930)

Taylor, A., Kagle, B. J.: Crystallographic Data on Metals and Alloy Structures. New York 1963

Tixier, R., Wache, C.: J. appl. Cryst. **3**, 466 (1970)

Ullrich, H.-J. (1966): Vortrag auf dem 7. Kongreß der Internationalen Union für Kristallographie, Moskau, 12. bis 21. Juli 1966

Ullrich, H.-J. (1967a): Vortrag gehalten auf der Arbeitstagung der Unterkommission Metallphysik der DAdW am 3. 3. 1967 in Dresden

Ullrich, H.-J. (1967b): phys. stat. sol. **20**, K 113 (1967)

Ullrich, H.-J., Däbritz, S., Schreiber, H.: Proceedings of the 5th International Congress on X-Ray Optics and Microanalysis, edited by G. Möllenstedt und K. H. Gaukler, Berlin-Heidelberg-New York 1969, p. 406

Ullrich, H.-J., Däbritz, S., Schreiber, H.: Exponat auf der Ausstellung anläßlich des 5th International Congress on X-ray Optics and Microanalysis, Tübingen, 9.—14. September 1968

Ullrich, H.-J., Küchler, L.: unveröffentlichte Untersuchungen (1965)

Ullrich, H.-J., Schreiber, H., Däbritz, S., Kleinstück, K.: Vortrag auf der Arbeitstagung Elektronenmikroskopie 24.—28. 2. 1969 in Dresden

Ullrich, H.-J., Schreiber, H., Feldhofer, F., Thiele, K., Kleinstück, K.: Vortrag auf der Arbeitstagung „Mikrosonde" 21.—23. 1. 1971 in Berlin

Ullrich, H.-J., Schulze, G. E. R.: Mikrochimica Acta, Suppl. III, 188 (1968)

Ullrich, H.-J., Thiele, K., Kleinstück, K., Däbritz, S., Schreiber, H., Götze, K., Feldhofer, F.: Computergesteuertes Zeichnen stereographischer Projektionen von Kossel-Linien (in Vorbereitung)

Yakowitz, H.: J. appl. Physics **37**, 4455 (1966)

(Eingegangen am 19. Juni 1971)

Anschrift der Verfasser:

Prof. Dr. G. E. R. Schulze
Doz. Dr. H.-J. Ullrich
Technische Universität Dresden
Sektion Physik
8027 Dresden
Münchnerplatz 3

Kristall und Technik	7	1–3	1972	221–227

F. Liebau and A. Amel-Zadeh

Mineralogisch-Petrographisches Institut der Universität Kiel

The Crystal Structure of $Ca[Zn_2(OH)_6] \cdot 2\,H_2O$ — a Retarder in the Setting of Portland Cement

Wird das Abbinden von Ca_3SiO_5 durch Zugabe von Zinkverbindungen verzögert, dann entsteht intermediär das Calciumhydroxozinkat-Hydrat $Ca[Zn_2(OH)_6] \cdot 2\,H_2O$. Die Gitterkonstanten dieser Phase sind:

$$a = 6{,}384\ \text{Å}, \quad b = 10{,}977\ \text{Å}, \quad c = 5{,}756\ \text{Å}, \quad \beta = 78{,}02°,$$
$$Z = 2, \quad P\,2_1/c.$$

Die Kristallstruktur wurde mit zweidimensionalen Röntgendaten bestimmt. $[Zn(OH)_4]$-Tetraeder sind über Ecken zu Ketten parallel [001] verknüpft. Die Ketten werden durch $[Ca(OH)_4(H_2O)_2]$-Oktaeder zusammengehalten.

When the setting of Ca_3SiO_5 is retarded with zinc compounds a calcium hydroxozincate hydrate $Ca[Zn_2(OH)_6] \cdot 2\,H_2O$ is formed intermediately. The lattice constants of this compound are

$$a = 6.384\ \text{Å}, \quad b = 10.977\ \text{Å}, \quad c = 5.756\ \text{Å}, \quad \beta = 78.02°,$$
$$Z = 2, \quad P\,2_1/c.$$

The crystal structure was determined with 2-dimensional X-ray data. $[Zn(OH)_4]$ tetrahedra are connected via corners to chains parallel [001]. These chains are bonded by $[Ca(OH)_4(H_2O)_6]$ octahedra.

1. Introduction

For special purposes retarded setting of cement is desired. For Portland cements especially carbohydrates, lignin sulfonic acid and lead and zinc compounds are used as retarders. The mode of action of zinc compounds during setting of Portland cement has been studied by Lieber. He discovered that calcium ions dissolved from the clinker mineral Ca_3SiO_5 by addition of water are bonded by zinc ions under formation of $Ca[Zn(OH)_3H_2O]_2$. Only after all zinc is fixed in this compound additional dissolved calcium ions initiate the setting process during which the calcium hydroxozincate hydrate is decomposed. In order to get an insight into this process a 2-dimensional crystal structure analysis of this compound has been performed.

2. Experimental data and structure determination

Single crystals of calcium hydroxozincate hydrate were prepared by adding a hydrous solution of $ZnCl_2$ to a saturated solution of $Ca(OH)_2$ at room temperature (Scholder, Breuning).

Table 1

Unit cell data of $Ca[Zn_2(OH)_6] \cdot 2\,H_2O$

a	6.384 ± 0.007 Å	density 2.62 g cm^{-3}
b	10.977 ± 0.011 Å	cell content $Ca_2[Zn_2(OH)_6]_2 \cdot 4\,H_2O$
c	5.756 ± 0.008 Å	space group $C_{2h}^5 - P\,2_1/c$
β	$78.02 \pm 0.13°$	

Table 2

X-ray powder diagram of $Ca[Zn_2(OH)_6] \cdot 2\,H_2O$ indexed with the lattice parameters given in table 1

d_{exp} [Å]	J	hkl	d_{exp} [Å]	J
6.237	8	100	1.964	5
5.735	1	001	1.957	1
5.488	1	020	1.946	4
5.008	9	011	1.882	3
4.116	8	120	1.828	6
3.926	8	021	1.820	7
3.560	7	121	1.772	4
3.123	9	200	1.754	3
		102	1.739	3
3.069	1	031	1.731	2
2.881	10	131	1.725	3
2.814	7	002	1.670	6
2.792	2	102	1.658	3
2.739	2	040	1.601	5
2.726	3	012	1.577	6
2.706	5	220	1.564	2
		112	1.549	2
2.647	5	221	1.545	2
		131	1.530	2
2.505	7	140	1.523	2
		022	1.452	2
2.454	6	$\bar{2}11$	1.440	3
2.388	5	$\bar{1}02$	1.372	3
2.368	3	141	1.346	2
2.322	1	231	1.339	3
2.295	2	212	1.332	2
2.224	5	032	1.290	2
		141		
		132		
2.157	4	222		
2.083	1	300		
2.079	2	$\bar{2}31$		
		150		
2.060	2	311		
		240		
2.043	4	310		
		051		

Lattice constants (Table 1) where determined from single crystal X-ray photographs and subsequent least squares refinement with 31 powder reflexions recorded with a GUINIER-JAGODZINSKI Camera. The indexed powder diagram is listed in Table 2. The density was measured pycnometrically under xylene at 18 °C. From systematic extinctions the space group symmetry $P\,2_1/c$ was deduced in accordance with the structure determination described. For intensity measurements with MoKα radiation a crystal of $0.3 \times 0.37 \times 0.004$ mm was used. Weissenberg photographs taken with the multiple film technique were measured with a microdensitometer from JENOPTIK GMBH, Jena. Visible intensities which were to weak to be measured with the microdensitometer were estimated visually.

Of 136 possible *hk*0 intensities 88 were measured and 26 visually estimated; of 86 possible 0*kl* intensities 48 were measured and 23 estimated. The 0*kl* intensities were corrected for absorption with the aid of the Program System for X-Ray Crystallography, The University of Maryland, Edition April 1965. The structure was determined by Patterson syntheses, minimum functions, structure factor calculations, electron density and difference Fourier syntheses, and least squares refinement of the atomic coordinates with individual isotropic temperature coefficients. The atomic scattering factors for Ca^+, Zn^+ and O^- were taken from MIRKIN. With the parameters given in Table 3 final values of $R = \frac{\sum ||F_0| - |F_c||}{\sum |F_0|}$ are $R\,(hk0) = 0.13$ and $R\,(0kl) = 0.20$ including the unobserved intensities with one third of the smallest observed value. Observed and calculated structure factors are listed in Table 4.

Table 3

Positional and thermal parameters for $Ca[Zn_2(OH)_6] \cdot 2\,H_2O$. Standard errors in parantheses, refer to last digit

	x	y	z	$B_{x.y}$	$B_{y,z}$
Ca	0	0	0	0.89 (17)	1.24 (40)
Zn	0.467 (1)	0.164 (0)	0.166 (2)	0.86 (9)	0.87 (14)
O(1)	0.170 (4)	0.120 (3)	0.259 (13)	1.27 (53)	2.14 (1.47)
O(2)	0.664 (4)	0.045 (2)	0.240 (12)	0.40 (36)	1.40 (1.19)
O(3)	0.503 (5)	0.182 (3)	0.818 (11)	0.62 (57)	1.57 (1.20)
O(4)	0.953 (5)	0.165 (3)	0.742 (10)	2.18 (71)	0.28 (91)

3. Discussion of the structure

The zinc ions are tetrahedrally coordinated by four oxygen atoms two of which — O(3) — are crystallographically equivalent (Fig. 1). Adjacent $[ZnO_4]$ tetrahedra are linked via these O(3) atoms to chains parallel [001]. These chains have two tetrahedra in the identity period of the chain. Each calcium ion is octahedrally coordinated by six oxygen atoms, four of which are non-bridging oxygen atoms O(1) and O(2) of $[ZnO_4]$ tetrahedra belonging to four different tetrahedral chains; the other two are O(4) atoms not coordinated to zinc.

The bond distances listed in Table 5 are in the range found in other phases with $[ZnO_4]$ tetrahedra and $[CaO_6]$ octahedra. Although the accuracy of bond

Table 4

Observed and calculated structure factors of $Ca[Zn_2[OH)_6] \cdot 2\ H_2O$

*hk*0	$\lvert F_0 \rvert$	F_c	*hk*0	$\lvert F_0 \rvert$	F_c
020	14.2	−11.90	640	3.96	+ 0.53
040	27.77	−24.32	740	4.38	+ 6.05
060	49.06	+58.26	840	6.68	+ 6.11
080	11.48	− 8.81	940	7.30	+ 5.16
0100	13.36	−11.30	1040	7.72	+ 5.90
0120	23.17	+21.82	150	15.86	−17.25
0140	7.52	− 5.30	250	17.12	−19.57
0160	3.13	− 2.77	350	15.66	+13.25
0180	12.94	+13.49	450	19.83	−20.22
100	40.71	−37.91	550	10.44	+10.96
200	68.27	−69.51	650	19.21	−19.37
300	31.11	−24.55	750	17.32	+18.46
400	44.26	−36.40	850	15.86	−15.86
500	11.06	−11.10	950	10.65	+10.61
600	24.22	+26.15	1050	7.72	− 7.26
700	4.59	− 2.30	160	25.05	−24.37
800	5.01	+ 2.85	260	48.64	+51.71
900	10.85	+ 5.74	360	16.91	−17.69
1000	6.05	− 2.32	460	30.27	+31.88
1100	6.26	+ 6.58	560	5.43	− 8.62
110	12.32	−10.87	660	14.20	+13.92
210	9.81	+11.51	760	2.50	− 0.20
310	12.73	− 3.29	170	5.22	+ 5.48
410	32.15	+31.26	270	10.85	+10.54
510	16.70	−16.05	370	9.60	−10.33
610	21.50	+21.26	470	17.54	+17.31
710	16.07	−16.23	570	18.37	−17.90
810	12.32	+13.08	670	16.91	+16.02
910	12.11	−10.34	180	27.97	+26.38
120	29.02	+37.10	280	9.18	− 8.72
220	28.18	−22.69	380	14.82	+11.95
320	40.50	+37.49	480	2.087	− 3.43
420	8.98	− 9.67	580	15.03	+15.43
520	15.24	+16.27	190	2.087	− 0.26
620	7.31	+ 4.29	290	2.087	− 1.94
720	7.93	+ 7.84	390	2.087	− 1.81
820	5.22	+ 3.09	490	8.14	+ 5.59
920	5.64	+ 6.26	590	2.50	− 3.21
1020	5.84	+ 3.70	690	2.71	+ 2.64
130	1.25	− 0.65	1100	19.83	+20.66
230	13.36	−12.19	2100	6.26	− 5.11
330	1.46	− 0.37	3100	11.69	+10.52
430	12.94	− 9.49	4100	2.50	− 0.45
530	2.087	− 2.00	5100	11.48	+11.05
630	2.087	+ 1.16	6100	2.71	+ 0.38
140	24.84	+26.53	2110	12.52	−10.73
240	13.78	−10.22	3110	9.60	+ 9.29
340	37.37	+35.27	4110	12.94	−11.49
440	4.59	− 3.78	5110	12.32	+12.87
540	14.2	+12.22	6110	13.99	−13.56

Table 4 (continued)

*hk*0	$\|F_0\|$	F_c	*hk*0	$\|F_0\|$	F_c
7110	8.35	+ 9.61	6140	3.13	− 1.55
8110	8.77	−11.16	1150	2.92	− 1.32
1120	12.32	−11.70	2150	2.92	+ 4.58
2120	25.47	+26.82	3150	2.92	− 2.02
3120	8.97	− 9.89	4150	3.13	− 0.63
4120	20.67	+18.92	5150	3.13	− 1.45
5120	2.71	− 3.7	6150	3.13	+ 2.47
6120	6.26	+ 4.96	1160	10.23	+10.46
1130	2.50	− 2.77	2160	3.13	− 3.03
2130	6.68	+ 7.25	3160	3.13	+ 6.20
3130	6.68	− 5.84	4160	3.13	− 0.56
4130	5.51	+ 6.68	5160	3.13	+ 5.60
5130	8.14	− 6.69	1170	3.13	− 0.27
6130	8.56	+ 8.48	2170	3.13	− 2.68
2140	2.71	− 0.92	3170	3.13	+ 5.07
3140	7.93	+ 8.22	4170	3.34	− 8.50
4140	2.92	+ 0.25	1180	3.34	− 5.17
5140	9.18	+ 7.64	2180	10.44	+10.23

Table 4 (continued)

0*kl*	$\|F_0\|$	F_c	0*kl*	$\|F_0\|$	F_c
020	17.299	−12.17	042	3.45	+31.59
040	31.49	−24.67	052	29.13	+24.83
060	50.64	+59.87	062	20.40	−15.87
080	12.46	−10.88	072	24.87	−18.31
0100	15.71	−12.80	082	21.40	+17.49
0120	27.91	+25.02	092	8.61	− 6.22
0140	4.69	− 6.57	0102	22.36	+16.63
0160	6.17	− 5.26	0112	21.55	+17.40
0180	14.78	+14.72	0122	5.32	− 4.99
011	28.09	+33.82	0132	12.90	− 9.13
021	30.24	−37.96	0142	13.05	+10.43
031	16.19	−12.74	0152	6.10	− 4.74
041	7.02	+ 2.40	0162	10.31	+ 7.12
051	29.24	+23.23	0172	10.64	+10.55
061	5.25	+ 5.45	013	13.75	−12.56
071	23.32	+17.73	023	3.58	+ 1.57
081	24.91	−19.83	033	49.05	+49.84
091	10.90	− 7.73	043	16.11	+14.51
0101	9.16	+ 5.32	053	11.97	− 9.77
0111	15.82	+10.21	063	4.14	+ 0.52
0121	8.35	+ 6.00	073	9.98	− 7.43
0131	13.16	+ 8.94	083	4.84	− 2.46
0141	10.31	− 8.21	093	33.41	+30.80
0151	5.91	− 3.44	0103	11.60	+ 6.86
012	24.65	−27.99	014	22.40	+19.42
022	22.47	+24.23	024	18.04	+18.32
032	4.99	− 4.89	034	6.80	+ 3.92

Table 4 (continued)

0*kl*	$\|F_o\|$	F_c
044	11.38	+11.43
054	16.78	−18.38
064	7.06	− 3.44
074	14.56	+14.51
084	15.30	+ 8.65
094	9.94	+ 6.00
0104	11.97	+ 6.49
0114	17.22	−14.80
015	17.07	+15.45
025	15.49	+17.34
035	6.65	− 6.58
045	14.67	−22.20
055	7.65	+13.21
065	3.80	− 4.26
075	7.91	+ 9.33
085	11.05	+16.50
095	4.47	− 5.59
0105	12.42	−13.78
0115	8.35	+ 7.24
026	7.65	− 7.43
066	11.16	+22.32
0126	11.20	+14.67
017	7.76	+ 8.62
027	7.35	−11.18
037	11.8	− 6.56
047	7.09	+13.17
057	6.95	+ 7.52
067	4.25	+ 3.85
077	6.94	+ 9.38
002	51.23	−51.21
004	13.93	+ 8.09
006	18.15	+22.01

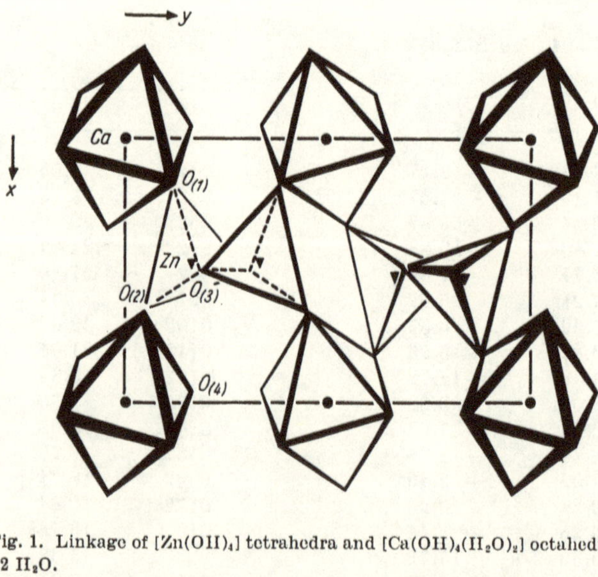

Fig. 1. Linkage of $[Zn(OH)_4]$ tetrahedra and $[Ca(OH)_4(H_2O)_2]$ octahedra in $Ca[Zn_2(OH)_6]$ $\cdot$ 2 H_2O.
Projection parallel [001], the direction of the tetrahedral chains

distances from this 2-dimensional structure analysis is rather low a decision between hydroxyl groups and water molecules with the DONNAY-ALLMANN method was successful. From this calculation it follows that the zinc ions are coordinated by hydroxyl groups O(1), O(2) and O(3) only, while the calcium ions are surrounded by four hydroxyl groups O(1) and O(2) and by two water molecules O(4). The formula of this calcium hydroxozincate hydrate should therefore be written

$$Ca^{[6]}[Zn_2^{[4]}(OH)_6] \cdot 2\,H_2O \quad \text{instead of} \quad Ca[Zn(OH)_3H_2O]_2$$

given by SCHOLDER and BREUNING and by LIEBER.

The authors are indebted to Dr. G. Bissert for assistance in using computer programs.

Table 5

Bond lengths in Å and bond angles in degrees.
Standard errors in parentheses, refer to last digit

$[Zn(OH)_4]$	tetrahedron	$[Ca(OH)_4(H_2O)_2]$	octahedron
Zn-O(1)	1.91 (3)	Ca-O(1)	2.39 (5)
Zn-O(2)	1.88 (3)	Ca-O(2)	2.31 (3)
Zn-O(3)	1.84 (4)	Ca-O(4)	2.32 (5)
Zn-O(3)	1.96 (7)	average	2.34
average	1.90	O(1)-O(2)	3.28 (7)
O(1)-O(2)	3.25 (4)	O(1)-O(2)	3.38 (4)
O(1)-O(3)	2.96 (7)	O(1)-O(4)	2.83 (8)
O(1)-O(3)	3.03 (5)	O(1)-O(4)	3.03 (4)
O(2)-O(3)	2.98 (4)	O(2)-O(4)	3.25 (7)
O(2)-O(3)	3.18 (8)	O(2)-O(4)	3.30 (4)
O(3)-O(3)	3.19 (2)	average	3.18
average	3.10	O(1)-Ca-O(2)	91.7 (1.5)
O(1)-Zn-O(2)	117.9 (1.5)	O(1)-Ca-O(2)	88.3 (1.8)
O(1)-Zn-O(3)	99.8 (2.3)	O(1)-Ca-O(4)	99.8 (1.6)
O(1)-Zn-O(3)	107.6 (1.7)	O(1)-Ca-O(4)	80.2 (1.6)
O(2)-Zn-O(3)	106.2 (2.0)	O(2)-Ca-O(4)	89.3 (1.4)
O(2)-Zn-O(3)	111.8 (2.0)	O(2)-Ca-O(4)	90.7 (1.4)
O(3)-Zn-O(3)	113.7 (1.7)		
average	109.5		
Zn-O(3)-Zn	124.3 (2.3)		

Literatur

DONNAY, G., ALLMANN, R.: Amer. Mineralogist **55**, 1003 (1970)

LIEBER, W.: Zement, Kalk, Gips **22**, 161 (1969)

MIRKIN, J. L.: Handbook of X-ray Analysis of Polycrystalline Materials, New York 1964

SCHOLDER, R., BREUNING, H.: Diplomarbeit von H. BREUNING, TH Karlsruhe 1956

(Received June 4, 1971)

Anschrift der Verfasser:

Prof. Dr. F. LIEBAU
Dipl.-Min. A. AMEL-ZADEH
Mineralogisch-Petrogr. Institut
der Universität Kiel
23 Kiel
Olshausenstr. 40—60, B 1

Kristall und Technik | 7 | 1–3 | 1972 | 229–233

H. Saalfeld, W. Depmeier

Mineralogisch-Petrographisches Institut, Universität Hamburg

Silicon-Free Compounds with Sodalite Structure

Mit Hilfe von Sinterreaktionen wurden sieben chemisch unterschiedliche Verbindungen mit Sodalithstruktur hergestellt. Für $Ca_4Al_6O_{12} \cdot SO_4$ sind Vorschläge über die Atomanordnung in einer Subzelle (a = 9,20 Å) mit der Symmetrie I $\bar{4}$3 m sowie in der echten Zelle (a = 18,41 Å) mit der Symmetrie I $4_1$32 gemacht worden.

With solid state reactions seven compounds of different chemical composition have been synthesized. A proposal of the atomic structure in a subcell (a = 9.20 Å) with the symmetry I $\bar{4}$3 m as well as in the real unit cell (a = 18.41 Å) with the symmetry I $4_1$32 is discussed.

1. Introduction

Sodalite, $Na_8[Cl_2Al_6Si_6O_{24}]$ and the related minerals nosean, haüyne and lasurite are alumosilicates with a 3-dimensional framework of SiO_4- and AlO_4-tetrahedra. Cl^-, SO_4^{2-} and other atoms or atomic groups are located in large vacancies, the negative charges being balanced by Na^+, K^+ or Ca^{2+}. In these minerals the ratio Si : Al is almost exactly 1 : 1.

During their investigations of the system $CaO-Al_2O_3-SO_3$ Halstead and Moore found the ternary compound 3 CaO · 3 Al_2O_3 · $CaSO_4$ (= $Ca_4Al_6O_{12} \cdot SO_4$). By X-ray powder photographs they confirmed the strong structural relationship of this silicon-free compound to the minerals of the sodalite-group. With solid state reactions Kondo succeeded in substituting other cations and anions for Ca^{2+} and SO_4^{2-}. Löns suggested a solid solution ranging from haüyne, $(Na_5KCa_2)Al_6Si_6O_{24} \cdot (SO_4)_2$, to the silicon-free compound $Ca_4Al_6O_{12} \cdot SO_4$. This transition is accompanied by the replacement of (Na, Si) by (Ca, Al).

In the following investigation new silicon-free compounds with sodalite structure will be described and more details about the crystal structure are presented.

2. Experimental

Stoichiometric amounts of the different source materials have been well mixed and pressed into pellets, the carbonates and hydrates being prefired to eliminate CO_2 and H_2O. The pellets were heated for 12–14 hours at temperatures between 800 to 1400 °C, crushed, mixed and heated again. The reaction products were examined by microscopy and X-ray diffractometry. In no case was it possible to synthesize crystals large enough for single-crystal X-ray investigations.

3. Experimental results

In most cases the synthesis of phases with sodalite structure was successful. Table 1 gives the data for these phases with some physical properties. The chemical composition of these phases is noted with the abbreviations used in the cement industry (C = CaO, A = Al_2O_3).

Table 1

Compounds with sodalite structure

substance	temperature °C	refractive index	lattice constant Å	calculated density
3 CA · $CaSO_4$	1350	1.57	18.41	2.61
3 CA · $CaWO_4$	1350	1.62	18.62	3.14
3 CA · $CaMoO_4$	1200	1.62	18.62	2.77
3 CA · $CaCrO_4$	1300	1.66	18.43	2.68
(3 SrO · Al_2O_3) · $SrSO_4$	1300	1.58	18.89	3.15
(3 SrO · Al_2O_3) · $SrMoO_4$	1300	1.62	18.90	3.41
(3 SrO · Al_2O_3) · $SrWO_4$	1300	1.64	18.96	3.71

It was impossible to synthesize sodalite-like substances with the compounds $BaSO_4$, $ZnSO_4$, $MgWO_4$, $PbSO_4$, $CdWO_4$, $CdMo_4$. KONDO reports the compounds 3 CA · $CdSO_4$ and 3 CA · $CaMoO_4$, but under our conditions of crystal synthesis we only found α-Al_2O_3, $CdWO_4$ and $CdMoO_4$. The formation of a spinel $CdAl_2O_4$ could not be established with certainty. An attempt was also made to substitute Ga_2O_3 for Al_2O_3. A partial substitution is possible up to 20 mole % Ga_2O_3.

The microscopic investigation of all specimens shows that the crystal size ranges between 10–30 μ. The high refractive index of 3 CA · $CaCrO_4$ is remarkable, but cannot yet be explained. The lattice constant of all compounds given in table I is twice that of the sodalite minerals. Unlike sodalite, all phases reveal a body-centred unit cell. Neglecting most of the weak reflections, a body-centred subcell with a demi-lattice-constant can be assumed.

4. Proposal for the structure of 3 CA · $CaSO_4$ with the symmetry I $\bar{4}$3 m

Among the sodalite minerals only the reflections of haüyne can be attributed to a nearly body-centred unit cell. LÖHN and SCHULZ explain the deviation from the exact body-centred cell by different Si—O and Al—O distances in the tetrahedral framework. A different tetrahedral size, however, does not exist for 3 CA · $CaSO_4$; thus an exact body centred unit cell results. Haüyne is one end-member of the above mentioned solid solution series. Therefore the determination of the atomic parameters of 3 CA · $CaSO_4$ should begin with the haüyne parameters.

Based on the measured intensities of 18 powder lines, a structure refinement of the subcell with the symmetry I $\bar{4}$3m was calculated. A special computer program was used. The starting parameters were those reported by LÖNS for a solid solution haüyne crystal. After some cycles of refinement the R-factor was finally 0.162. Due to the small number of reflections further refinement was not possible. Table 2 shows the refined atomic parameters.

Table 2

Refined parameters of the subcell of 3 CA · $CaSO_4$

atom and number of atoms	Wyckoff notation and equipoint	parameter
12 Al	12d 1/4, 1/2, 0	
8 Ca_I	8c x, x, x	$x = 0.170$
8 Ca_{II}	8c x, x, x	$x = 0.241$
24 $O_{framework}$	24g x, x, z	$x = 0.163$
		$z = 0{,}444$
2 S	2a 0, 0, 0	
8 $O_{sulfate}$	8c x, x, x	$x = -0.079$

The Al—O-distance of 1.78 Å calculated from these values was found to be in good agreement with the data of similiar substances. The S—O-distance of 1.26 Å proved to be too low, so that the proposed positions of the SO_4-groups are surely incorrect. Moreover, it can be seen that the Ca-ions are located about two equipoints, probably due to the SO_4distribution.

The refinement of the subcell confirms the tetrahedral framework but the problems concerning the position of the SO_4-groups as well as the Ca-distribution are still unsolved. An answer may be given by a proposal for the structure of 3 CA · $CaSO_4$ with doubled lattice constant and symmetry I 4_132.

5. Proposal for the structure of 3 CA · $CaSO_4$ with the symmetry I 4_132

The systematically absent reflections indicate that the structure of 3 CA · $CaSO_4$ as well as that of the other isotypic compounds must be based on the space group I 4_132. With the lattice constant of 18.41 Å the volume of the cubic unit cell is 8-times larger than that of the subcell. One can assume that the framework structure remains unchanged. Only the distribution of the SO_4-tetrahedra and the Ca^{2+}-ions may be different. The transition from space group I $\bar{4}3$m to I 4_132 affects only the positions of atoms distributed among the equipoint xxx. In this equipoint SO_4-groups and calcium-ions are located. Figure 1 shows diagrammatically the arrangement of the SO_4-tetrahedra in both space groups. In the space group I $\bar{4}3$m all tetrahedra are oriented parallel whereas in the space group I 4_132 these tetrahedra occupy alternating positions being rotated about 90° with respect to each other. This alternation results in doubling the lattice constant as compared to the subcell. A statistically alternating arrangement of the SO_4-tetrahedra was found by Löhn and Schulz in the cause of the structure refinement of a disordered haüyne (space group P $\bar{4}3$m).

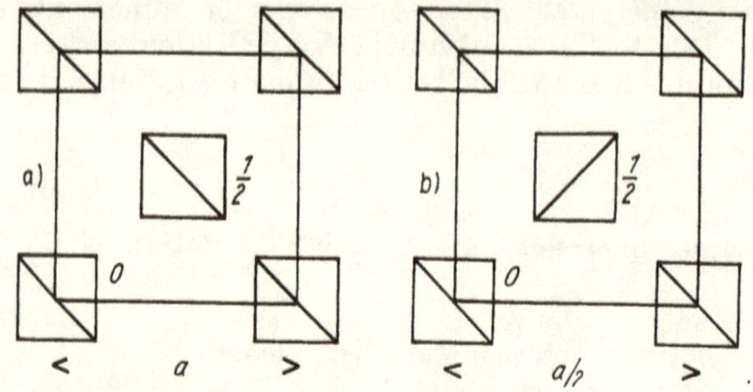

Fig. 1. Arrangement of the SO_4-tetrahedra
a) in the subcell with symmetry I $\bar{4}3$ m
b) in the unit cell with symmetry I 4_132

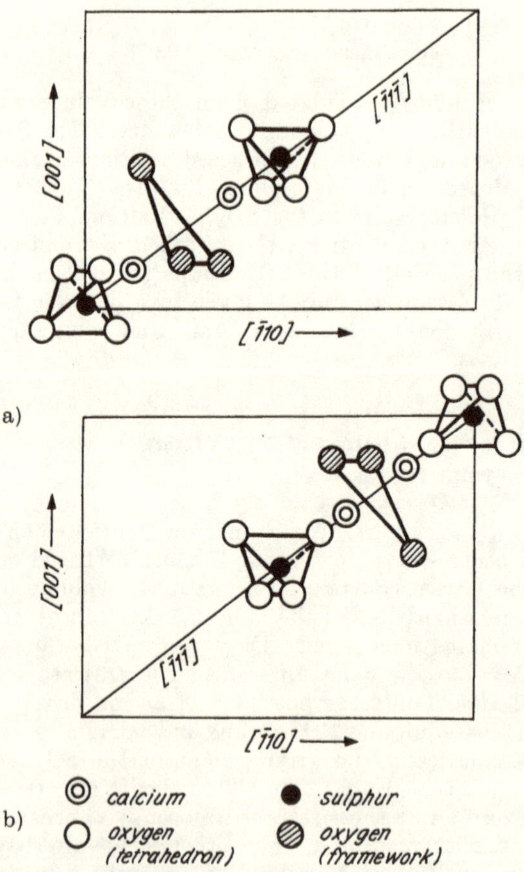

Fig. 2. Coordination calcium-oxygen:
a) calcium in x, x, x
b) calcium in $\bar{x}, \bar{x}, \bar{x}$

The distribution of the 64 Ca^{2+}-ions poses some difficulties. Owing to the alternating distribution of the SO_4-tetrahedra, there may exist two different positions of the Ca^{2+}-ions among the equipoints xxx and $\bar{x}\bar{x}\bar{x}$. Calcium ions located in xxx are surrounded by 3 oxygens of the framework and 3 oxygens of the SO_4-group, thus resulting in a coordination number 6. If, however, the calcium ions occupy the equipoint $\bar{x}\bar{x}\bar{x}$ then they are locatedbetween 1 oxygen on of the SO_4-group and 3 oxygens of the framework. Figure 2 demonstrates both possibilities. The fourfold coordination of calcium relative to oxygen is uncommon. The distribution of all calcium ions about two equipoints xxx, however, has the consequence that half of these cations exhibit distances which cannot exist. Therefore a statistical distribution of all cations is assumed. Half of them occupy xxx with $x = 0.085$ and the other half is distributed $\bar{x}\bar{x}\bar{x}$ among with $x = 0.120$. Such a distribution has the following advantages:

a) uniform charge distribution within a high symmetrical space
b) distribution of the calcium ions among two equipoints just as it was found for the subcell
c) compatible atomic distances.

In Table 2 all atomic parameters are listed on the basis of the space group I 4_132 as derivated from the subcell with the symmetry I $\bar{4}3$m.

With this model a structure refinement by LSQ-calculations was tried. Due to the low number of reflections and the high number of parameters the necessary convergence could not be attained. A final confirmation or a possible modification of the model will only be possible with single crystal methods.

The authors whish to express their thanks to Dr. K. KATO for cooperation in programming.

References

HALSTEAD, P. E., MOORE, A. E.: J. appl. Chem. **12**, 413 (1962)
KONDO, R.: Yogyo Kyokai Shi **71**, 1 (1965)
LÖHN, J., SCHULZ, H.: N. Jb. Miner. Abh. **109**, 201 (1968)
LÖNS, J.: Dissertation Hamburg 1969

(Received June 7, 1971)

Author's address:

Prof. Dr. H. SAALFELD, Dipl.-Min. W. DEPMEIER
Mineralog.-Petrogr. Institut der Universität Hamburg
2 Hamburg 13, Grindelallee 48

Kristall und Technik	7	1–3	1972	235–246

G. Schubert

Humboldt-Universität zu Berlin
Sektion Physik, Bereich Kristallographie

Die Bestimmung der intrakristallinen Magnesium-Eisen-Verteilung in Orthopyroxenen mit röntgen-strukturanalytischen Mitteln

Von metamorphen und vulkanischen Orthopyroxenen wurden die Intensitätsdaten ($hk0$) und ($h0l$) aus integrierten Weißenbergaufnahmen mit MoK-Strahlung gewonnen. Die Gitterkonstanten wurden aus geeichten Weißenbergaufnahmen mit CuK-Strahlung ermittelt. Der Chemismus für die metamorphen Proben ergab sich aus Elektronenstrahlmikroanalysen. Mit Hilfe von Elektronendichteprojektionen $\varrho(xy)$ und $\varrho(xz)$ konnten die Literaturangaben der Atomkoordinaten bestätigt, bzw. zum Teil neue bestimmt werden. Die intrakristalline Mg—Fe-Verteilung zwischen den Kationenlagen M1 und M2 ist einmal aus Elektronendichteprojektionen und zum anderen aus Strukturfaktorrechnungen gewonnen werden. Allgemein ist eine Preferenz des Eisens gegenüber der verzerrten M2-Lage festzustellen. Die metamorphen Orthopyroxene weisen einen hohen Ordnungsgrad der Mg—Fe-Verteilung auf. Der vulkanische Orthopyroxen zeigt dagegen eine weniger geordnete Verteilung. Die Gründe für die Unterschiede werden diskutiert.

The intensity data ($hk0$ und $h0l$) of metamorphic and volcanic orthopyroxenes have been determined by integrated Weissenberg photographs, using MoK radiation. The cell dimension were measured by calibrated Weissenberg photographs, using CuK radiation. The chemical formulae have been determined by electron probe microanalysis. The reference data of atomic coordinates have been confirmed and partly new coordinates have been determined by electron density projections $\varrho(xy)$ and $\varrho(xz)$. The intracrystalline Mg, Fe distribution over the octahedrally coordinated sites M1 and M2 have been determined by electron density projections and by structur factor computations. Generally an Fe-preference for the distorted M2 site can be observed. The metamorphic orthopyroxenes show a high degree of order in the Mg, Fe distribution. The volcanic orthopyroxene however shows a less ordered distribution. The reasons for this difference are the subject of discussion.

1. Einleitung

Nachdem schon um 1930 das Bauprinzip der Pyroxenstruktur geklärt worden war (Warren, Bragg; Warren, Modell), gelangen in den sechziger Jahren auf Grund der stark verbesserten Röntgenbeugungsaufnahmetechnik und der Anwendung neuer Methoden in der Strukturaufklärung die Strukturverfeinerungen fast aller Vertreter der Pyroxengruppe (u. a. Ghose; Morimoto, Koto). Diese neue Qualität der zur Verfügung stehenden Strukturdaten bot im Zusammenhang mit neuen Ergebnissen spektroskopischer Untersuchungen (u. a. Bancroft, Burns) die Möglichkeit, Zusammenhänge zwischen dem realisierten Strukturzustand und den physikalisch-chemischen Parametern der genetischen und postgenetischen Bedingungen zu betrachten (Virgo, Hafner). Die allgemeine Bedeutung der Kenntnis der kristallchemisch-strukturellen Phasenzustände für die Deutung der Minero- und Petrogenese und damit die Wichtigkeit für die Petrologie wurde u. a. von Kleber und Buerger postuliert. Die erste Mineralgruppe, bei der diese Fragen komplex untersucht und überwiegend gelöst wurden, waren die Feldspäte. Strukturelle Untersuchungen an Amphibolen unter genetischen Aspekten folgten. All diesen Betrachtungen liegt die Tatsache zu Grunde, daß die Strukturbesonderheiten von der physikalisch-chemischen Entwicklungsgeschichte (besonders der Temperaturgeschichte) mehr oder weniger stark beeinflußt werden, und sich daraus die Möglichkeit ergibt, aus dem realisierten Strukturzustand, der sich in bestimmten physikalischen (z. B. röntgenographische und optische) Eigenschaften offenbart, auf die physikalisch-chemischen Parameter der genetischen und postgenetischen Bedingungen (insbesondere die Temperaturgeschichte) zu schließen (Bautsch).

Ziel der vorliegenden Arbeit war es, die intrakristalline Me—Fe-Verteilung einmal von metamorphen Orthopyroxenen (Proben 9, 28,31) aus Metapyroxeniten des Granulitgebirges (Steinbruch Diethendorf/Sachsen) und zum anderen eines vulkanischen Orthopyroxens (Probe V1) der Eifel (Dreiser Weiher) zu bestimmen, und die Ergebnisse unter dem Aspekt der unterschiedlichen Genese bzw. Temperaturgeschichte zu diskutieren.

2. Beschreibung der Orthopyroxenstruktur

Mineralogisch versteht man unter Orthopyroxenen die Komponenten der Mischungsreihe $Mg_2Si_2O_6$—$Fe_2Si_2O_6$ mit der orthorhombischen Raumgruppe Pbca. Detaillierte Strukturkenntnisse über diese Mineralgruppe basieren auf den dreidimensionalen röntgenographischen Strukturverfeinerungen von Orthoenstatit—$Mg_2Si_2O_6$ — (Morimoto, Koto), Orthohypersthen—$MgFeSi_2O_6$ — (Ghose) und Orthoferrosilit—$Fe_2Si_2O_6$ — (Burnham). Danach ist in diesen Strukturen das allgemeine Bauprinzip der Pyroxene verwirklicht, das in der Existenz von SiO_4-Tetraedern besteht, die miteinander über die Tetraederecken zu sogenannten Zweiereinfachketten, deren Translationsperiode in Kettenrichtung zwei Tetraeder (5,2 Å) beträgt, verbunden sind. Die drei verfeinerten Orthopyroxenstrukturen zeigen dieselben Strukturprinzipien und auch dieselben Strukturdetails, die nur durch den Ersatz von Magnesium durch Eisen modifiziert werden. Von besonderem Interesse sind im Zusammenhang mit dieser Arbeit die Kationenplätze M1 und M2 (Nomenklatur nach Burnham, Clark, Papike, Prewitt), die von Magnesium bzw. Eisen besetzt werden und von denen es in der Struktur kristallographisch jeweils nur eine Art gibt (siehe Fig. 1). Der M1-Kationenplatz ist relativ regelmäßig oktaedrisch von Sauerstoffionen koordiniert. Beispielsweise beträgt der mittlere M1—O-Bindungsabstand bei Orthohypersthen 2,09 Å mit einer maximalen Abweichung von 0,07 Å. Die M2-Position ist stark verzerrt sechsfach von Sauerstoffionen koordiniert. Die starke Unregelmäßigkeit des M2-Oktaeders wird durch folgende Daten belegt: Im Orthohypersthen beträgt die mittlere M2—O-Bindungsdistanz für vier Sauerstoffionen 2,10 Å mit einer maximalen Abweichung von 0,07 Å, während die beiden anderen Sauerstoffionen 2,41 Å bzw. 2,52 Å

(O6 bzw. O3 in Fig. 1) vom Zentralion entfernt sind. Die zwei anderen Orthopyroxenstrukturen zeigen ähnliche Verzerrungen des M2-Kationenpolyeders.

Von besonderer Bedeutung ist die von GHOSE röntgenstrukturanalytisch ermittelte stark geordnete Mg—Fe-Verteilung zwischen den Plätzen M1 und M2. Für den von ihm untersuchten Orthohypersthen mit der vereinfachten che-

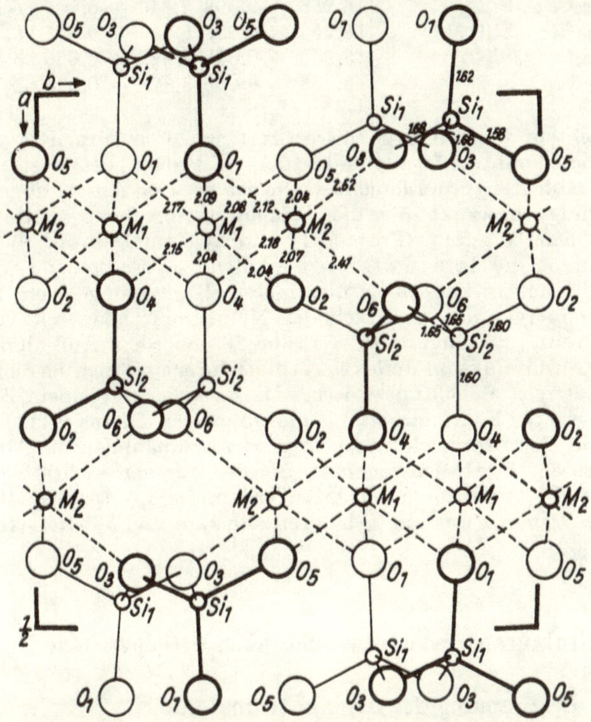

Fig. 1. Struktur von Orthohypersthen in $x-y$-Projektion (nach GHOSE)

mischen Formel $Mg_{0,93}Fe_{1,07}Si_2O_6$ fand er eine Besetzung der M1-Position mit 85% Mg^{2+} und 15% Fe^{2+} und der M2-Position mit 10% Mg^{2+} und 90% Fe^{2+}. Eisen zeigt also gegenüber der stark verzerrten M2-Position eine eindeutige Preferenz. Diese hochgeordnete intrakristalline Kationenverteilung, die von VIRGO, HAFNER auf spektroskopischem Wege bestätigt wurde, gab zu verschiedenen Betrachtungen und Untersuchungen Anlaß. Unter anderem vermuteten GHOSE und ZUSSMAN eine Temperaturabhängigkeit der intrakristallinen Mg—Fe-Verteilung und damit einen Einfluß der genetischen Bedingungen auf die Mg—Fe-Verteilung.

3. Experimentelle Arbeiten

Die vereinfachten chemischen Formeln der untersuchten Proben, abgeleitet für die Proben 9, 28 und 31 aus Elektronenstrahlmikroanalysen und für die Probe V1 aus Literaturangaben (RAMMELSBERG), sind mit den ermittelten Translationsperioden in Tabelle 1 angegeben. Die Translationsperioden wurden aus Weißenbergaufnahmen bestimmt, die mit CuK-Strahlung angefertigt, mit Siliziumpulver-

Tabelle 1

Chemismus und Translationsperioden der untersuchten Proben

Pro	vereinfachte Formel	a (Å)	b (Å)	c (Å)
28	$Mg_{1,56}Fe_{0,39}Ca_{0,05}Si_2O_6$	18,258 ± 0,008	8,840 ± 0,005	5,192 ± 0,005
31	$Mg_{1,58}Fe_{0,37}Ca_{0,05}Si_2O_6$	18,253 ± 0,008	8,834 ± 0,005	5,191 ± 0,005
9	$Mg_{1,54}Fe_{0,41}Ca_{0,05}Si_2O_5$	18,262 ± 0,008	8,842 ± 0,005	5,193 ± 0,005
V1	$Mg_{1,78}Fe_{0,22}Si_2O_6$	18,238 ± 0,008	8,828 ± 0,005	5,184 ± 0,005

aufnahmen geeicht und deren Ergebnisse für $\Theta = 90°$ extrapoliert wurden. Die Raumgruppe Pbca konnte für alle untersuchten Proben bestätigt werden. Die Zahl der Formeleinheiten pro Elementarzelle ist 8. Der Absorptionskoeffizient μ für MoK-Strahlung schwankt für die Proben entsprechend dem angegebenen Chemismus zwischen 16,9 cm^{-1} (Probe V1) und 23,8 cm^{-1} (Probe 9).

Für die Sammlung der Intensitätsdaten wurden von weitgehend würfligen einkristallinen Spaltstücken mit einer maximalen Kantenlänge von $2{,}2 \cdot 10^{-2}$ cm integrierte Weißenbergaufnahmen nach der Mehrfilmmethode mit verschiedenen Belichtungszeiten um [001] und von der Probe 28 außerdem Aufnahmen um [010] hergestellt. Zur Gewinnung von Reflexen geringer Intensität dienten nichtintegrierte Aufnahmen mit langen Belichtungszeiten. Die aus den mit einem Zeiss-Schnellphotometer bestimmten Filmschwärzungen gewonnenen Intensitäten wurden mit dem Lorentz- und Polarisationsfaktor korrigiert und aus ihnen die Absolutbeträge der Strukturfaktoren $|F_0(hk0)|$ berechnet. Bei den aus den nichtintegrierten Aufnahmen gewonnenen Intensitäten wurde außerdem die $\alpha_1 - \alpha_2$-Aufspaltungskorrektur durchgeführt. Zur Auswertung gelangten alle $hk0$- und $h0l$-Reflexe im Bereich von $\sin \Theta/\lambda = 1{,}00$.

4. Strukturanalyse der metamorphen Orthopyroxene

4.1. Bestimmung der Atomkoordinaten

Für die Bestimmung der Atomkoordinaten wurden die Intensitäten der Probe 28 benutzt. Mit den von Ghose angegebenen Koordinaten und B-Werten eines Orthohypersthens (siehe Tab. 2) sind die Strukturfaktoren $F_c(hk0)$ und $F_c(h0l)$ berechnet worden. Für die Atomformfaktoren konnten die Werte der neutralen Atome aus den International tables for x-ray crystallography entnommen werden. Die Berechnung der Atomformfaktoren für die Lagen M2 und M2 erfolgte entsprechend folgender Kationenverteilungsannahme: $M1 = Mg_{1,00}$, $M2 = Mg_{0,56} + Fe_{0,39} + Ca_{0,05}$. Die Bestimmung des R-Wertes $R = \frac{||F_0| - |F_c||}{|F_0|}$ von 200 symmetrieunabhängigen, von Null verschiedenen Reflexen $F(hk0)$ ergab $R = 0{,}1248$. Für 154 symmetrieunabhängige, von Null verschiedene Reflexe $F(h0l)$ berechnete sich ein R-Wert von $R = 0{,}1445$. Die mit den Koordinaten und B-Werten von Ghose ermittelten Strukturfaktoren $F_c(hk0)$ bzw. $F_C(h0l)$ zeigen also eine relativ gute Übereinstimmung mit den beobachteten Beträgen der Strukturfaktoren $|F_0(hk0)|$ bzw. $|F_0(h0l)|$. Mit den Vorzeichen der berechneten Strukturfaktoren und den beobachteten Beträgen der Strukturfaktoren konnten die Elektronendichteprojektionen $\varrho(x\,y)$ und $\varrho(xz)$ nach den für beide Projektionen in Frage kommenden Elektronendichteformeln

Tabelle 2

Atomkoordinaten und B-Werte eines Orthohypersthens nach Ghose (1) und die neu ermittelten z-Koordinaten eines metamorphen Orthopyroxens (2)

	1				2
	x	y	z	B (Å^{-2})	z
M1	$0{,}12484 \pm 12 \cdot 10^{-5}$	$0{,}1547 \pm 3 \cdot 10^{-4}$	$0{,}8740 \pm 7 \cdot 10^{-4}$	0,45	$0{,}8694 \pm 1 \cdot 10^{-3}$
M2	$0{,}37845 \pm 7 \cdot 10^{-5}$	$0{,}0167 \pm 1 \cdot 10^{-4}$	$0{,}8666 \pm 4 \cdot 10^{-4}$	0,43	$0{,}8632 \pm 1 \cdot 10^{-3}$
Si1	$0{,}47389 \pm 12 \cdot 10^{-5}$	$0{,}1635 \pm 2 \cdot 10^{-4}$	$0{,}2973 \pm 8 \cdot 10^{-4}$	0,29	$0{,}2982 \pm 1 \cdot 10^{-3}$
Si2	$0{,}27121 \pm 13 \cdot 10^{-5}$	$0{,}1593 \pm 2 \cdot 10^{-4}$	$0{,}5512 \pm 8 \cdot 10^{-4}$	0,30	$0{,}5501 \pm 1 \cdot 10^{-3}$
O1	$0{,}06210 \pm 34 \cdot 10^{-5}$	$0{,}1643 \pm 6 \cdot 10^{-4}$	$0{,}2069 \pm 21 \cdot 10^{-4}$	0,51	$0{,}2007 \pm 3 \cdot 10^{-3}$
O2	$0{,}18948 \pm 32 \cdot 10^{-5}$	$0{,}0009 \pm 6 \cdot 10^{-4}$	$0{,}0524 \pm 20 \cdot 10^{-4}$	0,33	$0{,}0468 \pm 3 \cdot 10^{-3}$
O3	$0{,}44653 \pm 34 \cdot 10^{-5}$	$0{,}2023 \pm 7 \cdot 10^{-4}$	$0{,}5924 \pm 19 \cdot 10^{-4}$	0,65	$0{,}5974 \pm 3 \cdot 10^{-3}$
O4	$0{,}18402 \pm 33 \cdot 10^{-5}$	$0{,}1625 \pm 6 \cdot 10^{-4}$	$0{,}5444 \pm 20 \cdot 10^{-4}$	0,46	$0{,}5387 \pm 3 \cdot 10^{-3}$
O5	$0{,}43445 \pm 36 \cdot 10^{-5}$	$0{,}0163 \pm 7 \cdot 10^{-4}$	$0{,}2034 \pm 20 \cdot 10^{-4}$	0,64	$0{,}1942 \pm 3 \cdot 10^{-3}$
O6	$0{,}30236 \pm 31 \cdot 10^{-5}$	$0{,}2344 \pm 7 \cdot 10^{-4}$	$0{,}8176 \pm 21 \cdot 10^{-4}$	0,67	$0{,}8248 \pm 3 \cdot 10^{-3}$

der Ebenengruppe *pmg* berechnet werden (siehe Fig. 2 und 3). Mit den aus diesen Elektronendichteprojektionen gewonnenen Koordinaten, die sich nur in den z-Werten merklich von den Ghose-Koordinaten unterscheiden, sind wiederum mit den B-Werten von Ghose Strukturfaktoren berechnet worden, deren Vergleich mit den beobachteten Beträgen der Strukturfaktoren für die x—y-Projektion einen schlechteren R-Wert von 0,2003, dagegen für die x—z-

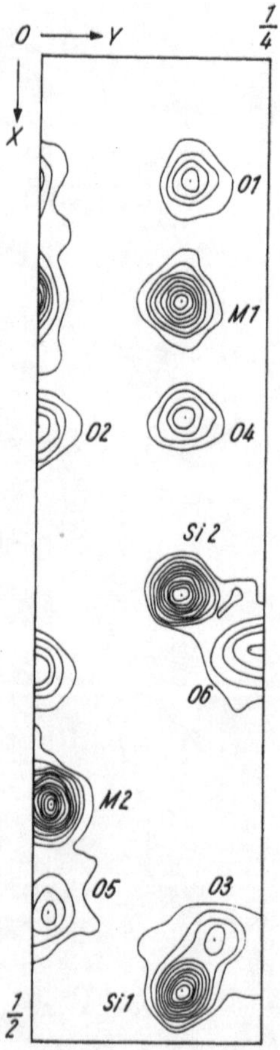

◀ Fig. 2. Elektronendichteprojektion $\varrho(x\,y)$ eines metamorphen Orthopyroxens (Probe 28)

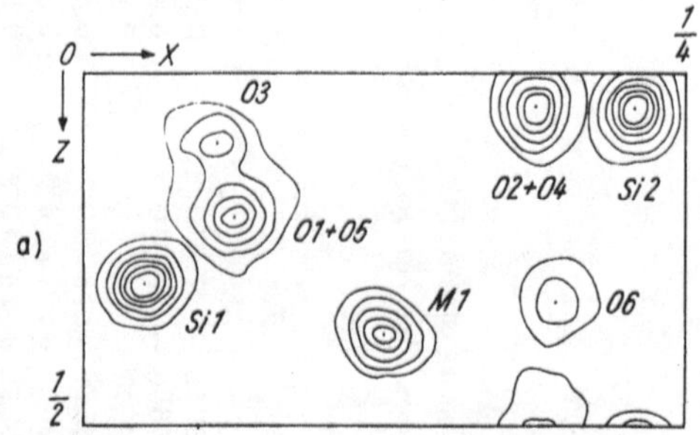

a) unter Abzug von M2,

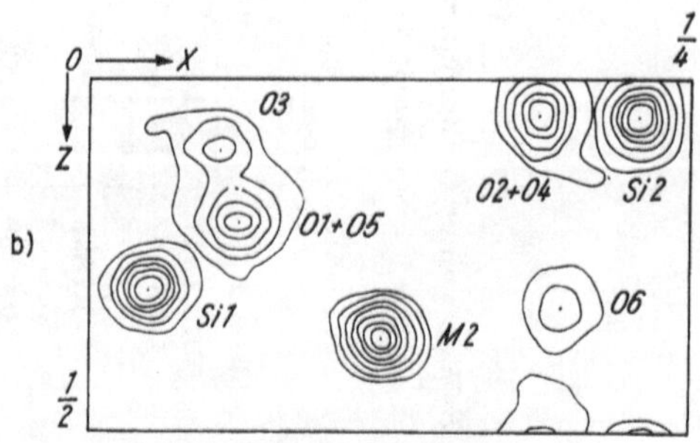

b) unter Abzug von M 1

Fig. 3. Elektronendichteprojektion $\varrho(x\,z)$ eines metamorphen Orthopyroxens (Probe)

Projektion einen verbesserten R-Wert von 0,1268 ergab. Weitere Versuche zur Koordinatenverfeinerung mittels Differenzelektronendichteprojektionen oder unter Verwendung der für Orthoenstatit von Morimoto, Koto und auch von Lindemann angegebenen Koordinaten führten zu keiner weiteren Verbesserung des R-Wertes, so daß als Grundlage für die weiteren Arbeiten die x- und y-Koordinaten von Ghose und die selbst ermittelten z-Koordinaten dienten. Die x—y—z-Koordinaten von Ghose, die von ihm ermittelten B-Werte und die neuen z-Koordinaten sind in Tabelle 2 angegeben.

4.2. Bestimmung der intrakristallinen Mg—Fe-Verteilung

In den Elektronendichteprojektionen $\varrho(xy)$ und $\varrho(xz)$ (Fig. 2 und 3) sind deutliche Unterschiede in der Höhe der Elektronendichtemaxima der Lagen M1 und M2 offensichtlich. Da die Elektronendichte an der Stelle eines Atoms direkt proportional der Ordnungszahl des betreffenden Elements ist und sich die Ordnungszahlen von Magnesium und Eisen relativ stark unterscheiden (für Fe $Z = 26$, für Mg $Z = 12$), kann aus den unterschiedlichen Höhen der Elektronendichtemaxima auf eine stark geordnete Mg—Fe-Verteilung zwischen den Lagen M1 und M2 geschlossen werden. Eine quantitative Berechnung ergibt für die Elektronendichteprojektion $\varrho(xy)$ ein Verhältnis der Maxima $\varrho(\mathrm{M2}):\varrho(\mathrm{M1}) = 1{,}57$, das einer Platzbesetzung $\mathrm{M1} = \mathrm{Mg}^{2+}_{0,995} + \mathrm{Fe}^{2+}_{0,005}$ und $\mathrm{M2} = \mathrm{Mg}^{2+}_{0,565} + \mathrm{Fe}^{2+}_{0,385} + \mathrm{Ca}^{2+}_{0,005}$ mit dem Verhältnis der Ordnungszahlen $Z(\mathrm{M2}):Z(\mathrm{M1}) = 1{,}57$ entspricht. Aus der Elektronendichteprojektion $\varrho(xz)$ ergibt sich ein Verhältnis der Maxima $\varrho(\mathrm{M2}):\varrho(\mathrm{M1}) = 1{,}48$, das folgender Platzbesetzung entspricht: $\mathrm{M1} = \mathrm{Mg}^{2+}_{0,97} + \mathrm{Fe}^{2+}_{0,03}$; $\mathrm{M2} = \mathrm{Mg}^{2+}_{0,59} + \mathrm{Fe}^{2+}_{0,36} + \mathrm{Ca}^{2+}_{0,05}$.

Eine exaktere Möglichkeit zur Bestimmung der intrakristallinen Mg—Fe-Verteilung besteht in der Benutzung des schon in 4.1. definierten R-Wertes als Führer zur realisierten Verteilung. Da der R-Wert ein Gradmesser für die Richtigkeit einer Strukturbestimmung ist, kann er auch für die hier speziell zu lösende Strukturverfeinerung herangezogen werden. Bei der Berechnung der Strukturfaktoren $F_c(hkl) = \sum_{j=1}^{N} f_j \cdot \exp(-B \sin^2\Theta/\lambda^2) \cdot \exp 2\pi i (h x_j + k y_j + l z_j)$ werden die Atomformfaktoren für die Lagen M1 und M2 f_{M1} und f_{M2} als Variable behandelt und entsprechend der angenommenen Verteilung variiert. Die Verteilung, deren Strukturfaktoren den niedrigsten R-Wert ergeben, kann als die realisierte intrakristalline Mg—Fe-Verteilung in der Struktur angesehen werden.

Die gerundeten x—y-Koordinaten der Lagen M1 und M2 (M1: $x \approx 1/8$, $y \approx 1/8$; M2: $x \approx 3/8$, $y \approx 0$) legen die Vermutung nahe, daß in den Beiträgen, die die Lagen M1 und M2 besetzenden Atome zu den Strukturfaktoren leisten, eine Systematik liegt. Die Berechnung der Beiträge von M1 und M2 zu den $F(hk0)$ nach den für die Ebenengruppe pgm gültigen Formeln

$$F(hk0) = 4\cos 2\pi h x \cdot \cos 2 k y \qquad \text{für } k = 2n$$
$$F(hk0) = 4\sin 2\pi h x \cdot \sin 2\pi k y \qquad \text{für } k = 2n + 1$$

mit den oben angegebenen gerundeten Koordinaten ergibt, daß die Lage M1 zu den $F(hk0)$ mit $k = 4n + 2$ und $k = 2n + 1$ und die Lagen M1 und M2 zu den $F(hk0)$ mit $h = 4n$ und $k = 2n$ maximale Beiträge liefern. Prüft man die Beträge der für verschiedene Mg—Fe-Verteilungen berechneten Strukturfaktoren $F_c(hk0)$, so bestätigt sich diese Systematik, und es ergeben sich 35 Strukturfaktoren $F_c(hk0)$, deren Beträge sich mit der Variierung der Mg—Fe-Verteilung stark ändern, die also auf die Verteilung signifikant ansprechen. Die Ergebnisse von Übersichtsrechnungen und ein Vergleich der ermittelten R-Werte für alle Strukturfaktoren $F(hk0)$ und für 35 ausgewählte Strukturfaktoren zeigt Tabelle 3. Der Vergleich bestätigt die Übereinstimmung der Ergebnisse der Rechnungen mit allen Strukturfaktoren $F(hk0)$ und mit den ausgewählten Strukturfaktoren. Die Vorteile der Verwendung ausgewählter Strukturfaktoren sind offensichtlich und bestehen unter anderem in einer schärferen Herausarbeitung des minimalen R-Wertes und damit einer Erhöhung der Genauigkeit der Verteilungsbestimmung, in einer Verringerung

Tabelle 3

Ermittelte *R*-Werte für verschiedene Platzbesetzungen (Probe 28)

Platzbesetzung					*R*-Wert	
M1		M2				
Mg	Fe	Mg	Fe	Ca	alle *F*(*hk*0)	ausgewählte *F*(*hk*0)
1,00	0,00	0,56	0,39	0,05	12,48	9,53
0,95	0,05	0,61	0,34	0,05	12,39	8,95
0,90	0,10	0,66	0,29	0,05	13,36	12,16
0,80	0,20	0,76	0,19	0,05	17,85	21,85
0,70	0,30	0,86	0,09	0,05	23,34	33,27

der Zahl der zu vermessenden Reflexe bei Routineuntersuchungen und in einer Verkürzung der Rechenzeiten auf ein Fünftel.

Eine weitere Verfeinerung der Platzbesetzung unter Verringerung der Schrittweite der Platzbesetzungsvariation erschien auf Grund der zu erzielenden Genauigkeit sinnvoll. Dazu wurden auch die *F*(*h*0*l*) und die ausgewählten *F*(*hk*0) weiterer Kristalle derselben Gesteinsprobe herangezogen. Die Ergebnisse für die Probe 28 sind in Tabelle 4 zusammengefaßt. Aus allen durchgeführten Bestimmungen der intrakristallinen Mg—Fe-Verteilung ergibt sich

Tabelle 4

Ermittelte *R*-Werte für verfeinerte Platzbesetzungen (Probe 28)

Platzbesetzung					*R*-Werte					
M1		M2			28.14	28.14	28.0	28.0	28.5	28.8
Mg	Fe	Mg	Fe	Ca	*hk*0[1])	*hk*0[2])	*h*0*l*[1])	*hk*0[2])	*hk*0[2])	*hk*0[2])
1,00	0,00	0,56	0,39	0,05	12,48	9,53	12,68	9,85	11,12	16,17
0,99	0,01	0,57	0,38	0,05		9,27	12,60	9,66	10,78	16,06
0,98	0,02	0,58	0,37	0,05		9,24	12,59	9,55	10,52	15,89
0,97	0,03	0,59	0,36	0,05		9,13	12,68	9,70	10,18	16,01
0,96	0,04	0,60	0,35	0,05		9,10		9,91	10,04	16,19
0,95	0,05	0,61	0,34	0,05	12,39	8,95		10,12	9,95	
0,94	0,06	0,62	0,33	0,05		9,08			10,02	
0,93	0,07	0,63	0,32	0,05		9,57			10,25	

[1]) Verwendung aller *F*(*hk*0) bzw. *F*(*h*0*l*).
[2]) Verwendung ausgewählter *F*(*hk*0).

für die Probe 28 folgende mittlere Platzbesetzung: M1 = $Mg_{0,972} + Fe_{0,028}$, M2 = $Mg_{0,588} + Fe_{0,362} + Ca_{0,050}$ mit einer Standardabweichung für die Platzbesetzung von $\sigma = \pm\ 0{,}017$, für Probe 31: M1 = $Mg_{0,96}\ Fe_{0,04}$, M2 = $Mg_{0,62}\ Fe_{0,33}Ca_{0,05}$, für Probe 9: M1 = $Mg_{0,93}Fe_{0,07}$, M2 = $Mg_{0,61}\ Fe_{0,34}Ca_{0,05}$ mit ähnlichen Fehlergrößen.

5. Strukturanalyse eines vulkanischen Orthopyroxens

5.1. Bestimmung der $x-y$-Koordinaten

Da sich in der Bestimmung der intrakristallinen Mg—Fe-Verteilung der Probe 28 eine gute Übereinstimmung der Ergebnisse aus der $x-y$-Projektion und aus der $x-z$-Projektion ergab, wurde für die Probe V1 auf eine Bearbeitung der $x-z$-Projektion verzichtet. Mit den Koordinaten und B-Werten von Ghose wurden die Strukturfaktoren $F_c(hk0)$ unter Annahme folgender Platzbesetzung $M1 = Mg_{1,00}$, $M2 = Mg_{0,78}\ Fe_{0,22}$ berechnet. Die Bestimmung des R-Wertes für 204 symmetrieunabhängige, von Null verschiedene Strukturfaktoren $F(hk0)$ ergab $R = 0{,}1588$. Mit den Vorzeichen der berechneten Strukturfaktoren und den beobachteten Beträgen der Strukturfaktoren wurde die Elektronendichteprojektion $\varrho(xy)$ berechnet (Fig. 4). Mit den aus dieser Projektion gewonnenen $x-y$-Koordinaten konnten die Strukturfaktoren $F_c(hk0)$ berechnet werden, die eine bessere Übereinstimmung mit den beobachteten

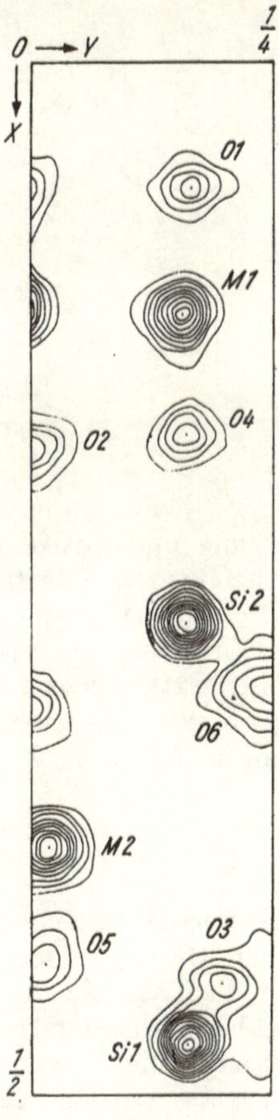

Fig. 4. Elektronendichteprojektion $\varrho(x\,y)$ eines vulkanischen Orthopyroxens (Probe V 1)

Tabelle 5

$x-y$-Koordinaten eines vulkanischen Orthopyroxens (Probe V1)

	x	y
M1	$0{,}1238 \pm 4 \cdot 10^{-4}$	$0{,}1543 \pm 6 \cdot 10^{-4}$
M2	$0{,}3785 \pm 4 \cdot 10^{-4}$	$0{,}0178 \pm 6 \cdot 10^{-4}$
Si1	$0{,}4740 \pm 4 \cdot 10^{-4}$	$0{,}1636 \pm 6 \cdot 10^{-4}$
Si2	$0{,}2703 \pm 4 \cdot 10^{-4}$	$0{,}1577 \pm 6 \cdot 10^{-4}$
01	$0{,}0623 \pm 1 \cdot 10^{-3}$	$0{,}1639 \pm 2 \cdot 10^{-3}$
02	$0{,}1890 \pm 1 \cdot 10^{-3}$	$0{,}0008 \pm 2 \cdot 10^{-3}$
03	$0{,}4465 \pm 1 \cdot 10^{-3}$	$0{,}2020 \pm 2 \cdot 10^{-3}$
04	$0{,}1810 \pm 1 \cdot 10^{-3}$	$0{,}1591 \pm 2 \cdot 10^{-3}$
05	$0{,}4345 \pm 1 \cdot 10^{-3}$	$0{,}0133 \pm 2 \cdot 10^{-3}$
06	$0{,}3025 \pm 1 \cdot 10^{-3}$	$0{,}2345 \pm 1 \cdot 10^{-3}$

$|F_0(hk0)|$ aufweisen (R-Wert 0,1472). Eine sukzessive Untersuchung der neu gewonnenen Koordinaten zeigte, daß die Ghose-Koordinaten für Si1 einen besseren R- Wert ergeben. Die endgültig gewonnenen $x-y$-Koordinaten, die auch Grundlage der weiteren Untersuchungen waren, sind in Tabelle 5 angegeben.

5.2. Bestimmung der intrakristallinen Mg—Fe-Verteilung

Die Elektronendichtemaxima der Lagen M1 und M2 zeigen in der Elektronendichteverteilung $\varrho(xy)$ (Fig. 4) nur geringe Unterschiede, die auf eine ungeordnete Mg—Fe-Verteilung zwischen den beiden Lagen hinweisen. Das Verhältnis der Elektronendichtemaxima $\varrho(M2):\varrho(M1) = 1{,}09$ entspricht dem Verhältnis der Ordnungszahlen bei einer Platzbesetzung von $M1 = Mg^{2+}_{0,925} + Fe^{2+}_{0,075}$, $M2 = Mg^{2+}_{0,855} + Fe^{2+}_{0,145}$. Die Ergebnisse der Verfeinerung der Platzbesetzung mit Hilfe des R-Wertes unter Benutzung der 35 ausgewählten Strukturfaktoren sind in Tabelle 6 angegeben. Daraus kann die realisierte

Tabelle 6

Ermittelte R-Werte für verfeinerte Platzbesetzungen (Probe VI)

Platzbesetzung				R-Werte
M1		M2		
Mg	Fe	Mg	Fe	
1,00	0,00	0,78	0,22	0,1510
0,98	0,02	0,80	0,20	0,1303
0,95	0,05	0,83	0,17	0,1186
0,93	0,07	0,85	0,15	0,1123
0,92	0,08	0,86	0,14	0,1119
0,91	0,09	0,87	0,13	0,1099
0,90	0,10	0,88	0,12	0,1111
0,89	0,11	0,89	0,11	0,1125

Mg—Fe-Verteilung mit $M1 = Mg_{0,91} + Fe_{0,09}$, $M2 = Mg_{0,87} + Fe_{0,13}$ entnommen werden. Ein Vergleich der mit dieser Platzbesetzung und mit den in Tabelle 5 angegebenen Koordinaten berechneten Strukturfaktoren $F_c(hk0)$ mit den beobachteten $|F_0(hk0)|$ ergab einen R-Wert von $R = 0{,}1384$.

Versuche zur weiteren Koordinatenverfeinerung unter Benutzung der ermittelten Platzbesetzungen führten weder für die metamorphen Orthopyroxene noch für den vulkanischen Orthopyroxen zu positiven Ergebnissen. Vielmehr wurden die mit den Ausgangsplatzbesetzungen ermittelten Koordinaten bestätigt.

6. Diskussion

Die Ergebnisse der Bestimmung der intrakristallinen Mg—Fe-Verteilung zeigen sowohl für die metamorphen als auch für die vulkanischen Orthopyroxene eine mehr oder weniger eindeutige Preferenz des Eisens für die Lage M2 und befinden sich damit qualitativ in Übereinstimmung mit den über andere Orthopyroxene veröffentlichten Daten (Ghose und Virgo, Hafner). Die Ursache dieser Erscheinung kann nicht auf geometrischem Wege erklärt werden, da die Ionenradien für Mg^{2+} und Fe^{2+} (0,78 Å bzw. 0,82 Å) nicht wesentlich verschieden sind, so daß die Kationenplätze M1 und M2 für die Besetzung mit Mg^{2+} als auch mit Fe^{2+} gleichermaßen in Frage kommen. Dazu steht das Verhalten von Ca^{2+} im Gegensatz, das auf Grund seines größeren Ionenradius (1,06 Å) und in Analogie zum Diopsid von vornherein die gestörtere und größere Lage M2 besetzen wird. Die bevorzugte Besetzung der Lage M2 durch Eisen kann mit Hilfe der Kristallfeldstabilisierungsenergie des Fe^{2+} in beiden Lagen erklärt werden. Das *d*-Orbital der Elektronenhülle des Fe^{2+} spaltet im elektrostatischen Feld der umgebenden Ionen auf, und es ergibt sich eine zusätzliche Kristallfeldstabilisierungsenergie. Die Aufspaltung der Niveaus und damit die Größe der Kristallfeldstabilisierungsenergie ist um so größer, je stärker die Verzerrung des umgebenden oktaedrischen Feldes ist. Burns bestimmte aus den Positionen der Absorptionsbanden im Elektronenspektrum von Orthopyroxenen die Kristallfeldstabilisierungsenergie für Fe^{2+} in beiden Kationenlagen. Für M1 ergaben sich etwa 12 kcal · Mol^{-1} und für M2 20 kcal · Mol^{-1}. Der Tendenz der Mg—Fe-Ordnung infolge der unterschiedlichen Kristallfeldstabilsierungsenergien des Fe^{2+} steht das Bestreben nach statistischer Verteilung der Mg- und Fe-Ionen zwischen beiden Gitterplätzen auf Grund des damit verbundenen Ansteigens der Entropie gegenüber. Der letztere Effekt wird sich mit steigender Temperatur verstärken, so daß der als Kompromiß zwischen beiden gegenläufigen Tendenzen realisierte Verteilungszustand gewisse Rückschlüsse auf die Temperaturverhältnisse der genetischen bzw. postgenetischen Bedingungen ermöglichen sollte. Kinetische Untersuchungen der Ordnungs- und Unordnungsvorgänge von Virgo, Hafner an Orthopyroxenen zeigten, daß der Mg—Fe-Austausch zwischen den Lagen M1 und M2 als ein schnell verlaufender Prozeß mit niedriger Energiebarriere charakterisiert werden muß. Virgo, Hafner ziehen daraus die Schlußfolgerung, daß Orthopyroxene eine Mg—Fe-Verteilung zeigen müssen, die dem letzten Gleichgewichtszustand entspricht. Langsam abgekühlte Orthopyroxene (z. B. aus metamorphen Gesteinskomplexen) müßten also eine Gleichgewichtsverteilung aufweisen, die der vollkommen geordneten Verteilung nahe kommt, schnell abgekühlte Orthopyroxene (z. B. aus vulkanischen Gesteinskomplexen) müßten jedoch metastabile Verteilungen zeigen, die höheren Temperaturen entsprechen und damit eine höhere Unordnung realisiert haben. Die Arbeiten von Virgo, Hafner ergaben außerdem das Vorliegen einer maximalen Ordnung, die bei 480 °C und einer maximalen Unordnung, die bei 1000 °C erreicht wird. Diese beiden erreichten Zustände entsprechen jedoch nicht der idealen Ordnung bzw. Unordnung.

Benutzt man den von Virgo, Hafner definierten Unordnungsparameter $p = X_1(1 - X_2)/X_2(1 - X_1)$ mit X_1 und X_2 als Platzbesetzungszahlen von Fe^{2+} in Mol in M1 bzw. M2, so kann man den Grad der Eisenunordnung in einem Orthopyroxen charakterisieren und Orthopyroxene mit verschiedenem Chemismus miteinander vergleichen. In Tabelle 7 sind die Eisenplatzbesetzungszahlen und die daraus berechneten Unordnungsparameter angegeben. Während die Werte der metamorphen Proben einen relativ niedrigen Unordnungsgrad der Magnesium-Eisenverteilung charakterisieren, ist der vulkanische Orthopyroxen durch einen hohen Unordnungsgrad gekennzeichnet. Dafür ist in Übereinstimmung mit den oben angeführten Gründen die unterschiedliche postgenetische Temperaturgeschichte verantwortlich. Die Unordnungsparameter der metamorphen Orthopyroxene stimmen mit gewissen Abweichungen mit den von Virgo, Hafner angegebenen überein (die sich aus den Unordnungs-

Tabelle 7
Eisenplatzbesetzungszahlen, Unordnungsparameter und Gleichgewichtstemperaturen von Orthopyroxenen

Probe	X_1 (Mol)	X_2 (Mol)	p	T (°C)
28	0,028	0,362	0,0508	495
31	0,040	0,330	0,0813	530
9	0,070	0,340	0,1358	590
V1	0,090	0,130	0,6620	–

parametern nach Virgo, Hafner ergebenden Gleichgewichtstemperaturen sind in Tabelle 7 mit angegeben). Dagegen übertrifft der Wert des Unordnungsparameters p für den vulkanischen Orthopyroxen den von Virgo, Hafner bestimmten maximalen Unordnungsparameter von $p = 0{,}25$ bei 1000 °C erheblich. Eine extrem unterschiedliche Vorgeschichte der untersuchten Proben könnte dafür der Grund sein.

Zu Dank verpflichtet bin ich Herrn Prof. Brümmer und Herrn Dipl.-Phys. Röder von der Martin-Luther-Universität Halle, Sektion Physik für die gegebene Möglichkeit, Elektronenstrahlmikroanalysen durchzuführen, weiterhin dem Zentralinstitut für Physikalische Chemie, Abteilung Strukturforschung für die Überlassung von Rechenprogrammen und dem Rechenzentrum der Humboldt-Universität zu Berlin für die am ZRA-1 zur Verfügung gestellten Rechenzeiten.

Es soll bemerkt werden, daß zur Zeit die Bestimmung der Kationenverteilung unter Benutzung der Intensitätsdaten höherer Schichten des reziproken Gitters mit Hilfe des „least-square“-Verfahrens präzisiert wird.

Literatur

Bancroft, G. M., Burns, R. G.: Amer. Mineralogist **52**, 1278 (1967)
Bautsch, H.-J.: Habilitationsarbeit, Humboldt-Univ. Berlin 1965
Buerger, M. J.: Amer. Mineralogist **33**, 101 (1948)
Burnham, C. W.: Carnegie Inst. Wash. Year Book **65**, 285 (1966)
Burnham, C. W., Clark, J. R., Papike, J. J., Prewitt, C. T.: Z. Kristallogr. **125**, 1 (1967)
Burns, R. G.: Origin and distribution of the elements (Herausgeber L. H. Ahrens), Oxford 1968, S. 1151
Ghose, S.: Z. Kristallogr. **122**, 81 (1965)
International tables for x-ray crystallography, Bd. 3, Birmingham 1962, S. 202
Kleber, W.: Forsch. u. Fortschr. **25/27**, 203 (1944)
Lindemann, W.: N. Jb. Miner. Mh **10**, 226 (1961)
Morimoto, N., Koto, K.: Z. Kristallogr. **129**, 65 (1969)
Rammelsberg, C. F.: Pogg. Annal. **141**, 514 (1870)
Virgo, D., Hafner, S. S.: Mineral. Soc. Amer. Spec. Paper **2**, 67 (1969)
Virgo, D., Hafner, S. S.: Amer. Mineralogist **55**, 201 (1970)
Warren, B. E., Bragg W. L.: Z. Kristallogr. **69**, 168 (1928)
Warren, B. E., Modell, D. J.: Z. Kristallog. **75**, 1 (1930)
Zussman, J.: Earth-Sci. Rev. **4**, 39 (1969)

(Eingegangen am 20. Mai 1971)

Anschrift des Verfassers:
Dipl.-Krist. Gernot Schubert
Humboldt-Universität zu Berlin
Sektion Physik, Bereich 11 Kristallographie
104 Berlin
Invalidenstr. 43

Kristall und Technik | 7 | 1–3 | 1972 | 247–251

F. Frölich

Central Earth Physics Institute, Potsdam
German Academy of Sciences Berlin

Spin Pairing in Spinel-Type Minerals of Iron (II) and Corresponding Phases

For interpretations of the results of high-temperature high-pressure research which nowadays is used to clarify problems of the upper and deeper Earth's interior, the application of modern solid-state physics is necessary. Informations from such studies are the better, the better the approximation of laboratory technique to the correspondent conditions of the respective depth is realized and the higher the approach in the description of the real lattice situation in detail can be made which must be analysed. In the latter case, the crystal field approach is to be used, because a lower approximation is completely insufficient. That is discussed in the following for two different situations of the same (spinel) symmetry.

Für die Deutung der Ergebnisse von Hochtemperatur-Hochdruck-Untersuchungen, die heute zur Klärung von Problemen des oberen und tieferen Erdinnern verwendet werden, ist die Anwendung der modernen Festkörperphysik erforderlich. Die Informationen aus solchen Untersuchungen sind um so zutreffender, je besser die Annäherung der Laboratoriumstechnik an die korrespondierenden Bedingungen der entsprechenden Tiefe realisiert ist und je genauer die Beschreibung der wirklichen Situation im Gitter, die zu untersuchen ist, im einzelnen gelingt. Im letzteren Fall wird die Kristallfeldnäherung benutzt, weil eine schwächere Approximation völlig unzureichend ist. Dieser Zusammenhang wird im folgenden für zwei verschiedene Situationen derselben (Spinell-) Symmetrie diskutiert.

The spinel structure O_h^7-Fd3m, belonging to the relative rare cases of the volume-distance paradoxon, is an instructive type. As is well known, the tetrahedral coordination of the A-sites cations and the octahedral coordination of the cations at B-sites are coupled together by common edges. A widening of the A-sublattice leads therefore to simultaneous shrinkages of B and consequently to a diminution of the interstitial radius (r_B) of the B-sites:

$$r_B = 0.375\, a - (1 + 1/\sqrt{3})\, r_0 - r_{A/\sqrt{3}} .$$

This means that the anionic perturbation field increases within B-sublattices and the splitted 3 d^6 state of ferrous cations (Fe^{II+}) in these sites can be turned from the high-spin state $\begin{Bmatrix} e_g^2 \\ t_{2g}^4 \end{Bmatrix}$

to the low-spin state $\begin{Bmatrix} e_g^0 \\ t_{2g}^6 \end{Bmatrix}$

if only the intensity of the crystal field is sufficient high.

On the other hand, a widening of the tetrahedral anion groups (A) is accompanied by an increasing lattice constant. A transition into the low-spin state, as mentioned above, therefore especially can be expected in the direction to the upper boundary of spinel-type system.

So much to the conception used here.

1st problem:

The experimental results on titanomagnetites, gained by Kawai in 1957, could not be interpreted up to now, i.e. without such a conception. Figure 1 demonstrates the statements under increasing pressure and temperature in that system.

On the base of the hard-sphere ionic model, such a behaviour could not be understood. This the more, since high-pressure research meantimes has shown that the bonding character will be modified from the ionic to the covalent type under increasing load.

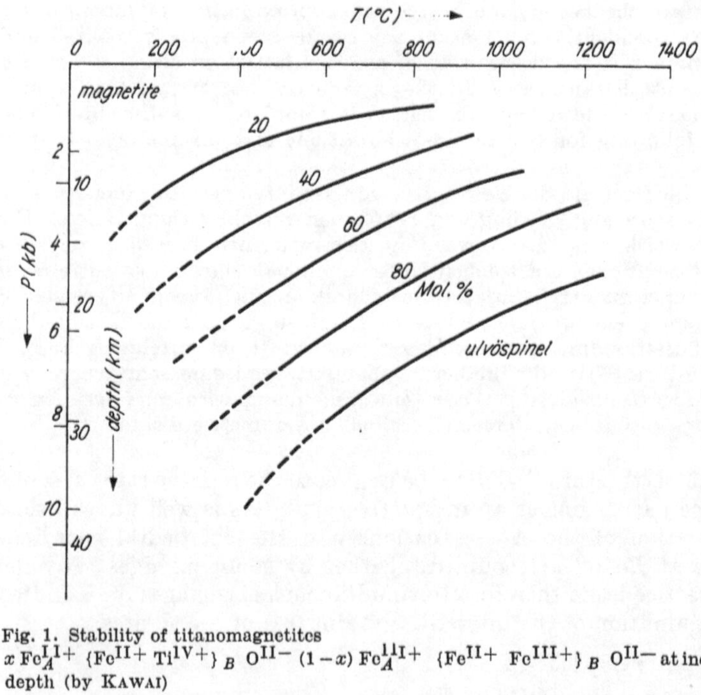

Fig. 1. Stability of titanomagnetites
$x\, Fe_A^{II+}\, \{Fe^{II+}\, Ti^{IV+}\}_B\, O^{II-}\ (1-x)\, Fe_A^{III+}\, \{Fe^{II+}\, Fe^{III+}\}_B\, O^{II-}$ at increasing depth (by Kawai)

Thus from this point of view, the differences of stability within the Ti-magnetite system at increasing depth is intelligible as consequence of additional covalent bonds in the titanium-rich components.

Finally on the basis of solid-state physics, transitions from the high- to the low-spin state enable an arising of covalent bonds, e.g. of d^2sp^3-type for an octahedral perturbation field, because of unoccupying of the high-energy level (e_g) in the latter case.

Detailed studies (Frölich 1964/70) show the following:
The Ti-rich components of the system concerned favours the low-spin state.

1. in consequence of the increasing lattice parameter

   from $a = 8.39$ Å ($x = 0$) magnetite
   to $a = 8.53$ Å ($x = 1$) ulvöspinel (Nagata)

2. since, hereby released, the crystal field within B-sites will be strengthened
3. because of the growing portion of iron (II) and the simultaneous diminution of iron (III) portions (particularly the latter hinder the low-spin state in magnetite).

This $3\,d^6$ low-spin state of iron (II) at B-sites with no unpaired electrons is necessary a diamagnetic one, just as the $3\,d^0$-state of titanium (IV), the participating cation of ulvöspinel, which prefers the same (B-) sites and favours for its part (by the small cation radius) the above alternative situation, as opposed to magnetite where such a possibility not exists.

2nd-problem:

The high-pressure phase change of olivine from the neso-silicate type to the spinel-type has been frequently discussed in the past. The experimental results, however, meantimes gained by Ringwood in 1969, show a surprising barrier at 80 Mol-% forsterite (Fig. 2) which likewise can not be understood on the base of a hard-sphere ionic model of spinel-type structure.

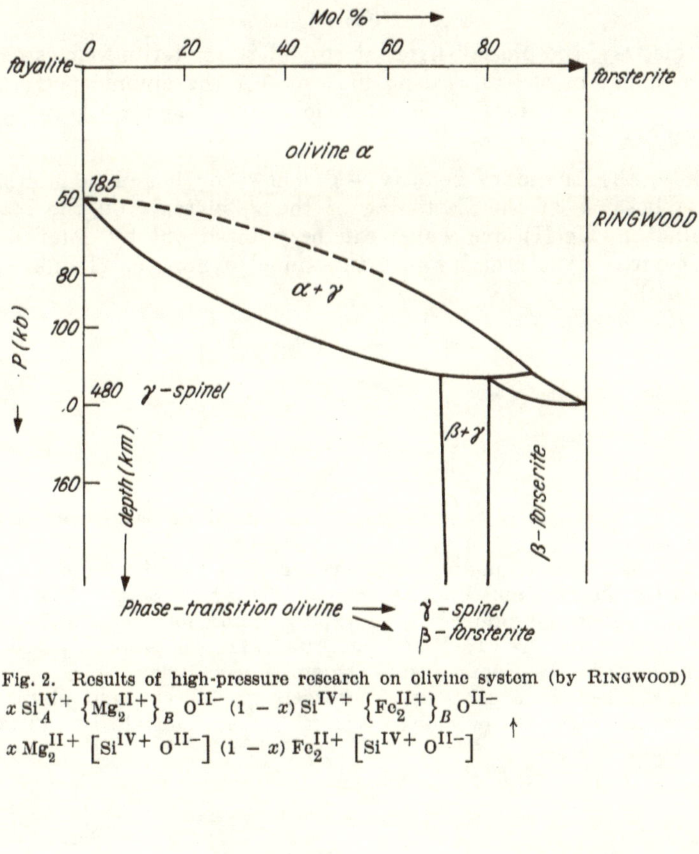

Fig. 2. Results of high-pressure research on olivine system (by Ringwood)

$$x\,\mathrm{Si}_A^{IV+}\,\{\mathrm{Mg}_2^{II+}\}_B\,\mathrm{O}^{II-}\;(1-x)\,\mathrm{Si}^{IV+}\,\{\mathrm{Fe}_2^{II+}\}_B\,\mathrm{O}^{II-}$$

$$\uparrow\qquad\qquad\qquad\qquad\qquad\qquad\qquad\uparrow$$

$$x\,\mathrm{Mg}_2^{II+}\,[\mathrm{Si}^{IV+}\,\mathrm{O}^{II-}]\;(1-x)\,\mathrm{Fe}_2^{II+}\,[\mathrm{Si}^{IV+}\,\mathrm{O}^{II-}]$$

In this case also, the higher approach with co-regarding of crystal-field effects enables the interpretation of high-pressure spinel-type phase in detail:

1. Transitions into the low-spin state are eliminated by nature from the first in forsterite in consequence of the unfit electron configuration of Mg (II)-cations.

2. On the other hand, pure γ-fayalite shows the maximum value of the spinel-lattice parameter ($a = 8.23$ Å). This value decreases with increasing forsterite content (to $a = 8.11$ Å at 80 Mol-% $SiMg_2O_4$). That means, as derived at the beginning, that the highest crystal field within B-sites exists in pure γ-fayalite. Here therefore the lowest pressure energy for the phase transition ($\alpha \to \gamma$) must be expected.

3. Such a phase change, however, results in a decrease of volume: The unit-cell of α-olivine with the same content of ions as the spinel-cell is marked by the following lattice constants: $a = 10.21$ Å, $2\,b = 2 \cdot 4.775$ Å, $c = 5.985$ Å which corresponds with a spinel parameter $a = 8.345$ Å
4. That diminution of volume during the phase transition ($\alpha \to \gamma$) must be compensated by a corresponding increase of internal energy in order to guarantee the equilibrium condition ($dG = 0$) of the thermodynamical potential (G). Hence one must expect an increasing value of bond-energy, i.e. an arising of additional covalent bonds in the γ-phase of fayalite which on the other side could only be realized by a transition, e.g. of B-sites iron (II), into the spin-paired state $\begin{Bmatrix} e_g^0 \\ t_{2g}^6 \end{Bmatrix}$.

5. On this way, the phase barrier at 80 Mol-% forsterite follows necessarily in consequence of the decreasing intensity of the anionic perturbation field with increasing values of the intersticial radius r_B for a growing content of $Si\{Mg_2\}O_4$

6. Finally, the boundary zone ($\gamma + \beta$) between 70 and 80 Mol-% forsterite arises because of the flattening of the r_B-increase on the reason of the dominating Mg(II)-ions which can be pointed out by detailed studies on the typical structural laws of the spinel symmetry (FRÖLICH, LÖFFLER, STILLER).

References

FRÖLICH, F., LÖFFLER, H., STILLER, H. in: Abhandlung 34, Geom. Inst. Potsdam, Berlin 1964

FRÖLICH, F.: Elektrische und magnetische Eigenschaften der festen Körper. Handbuch Geomagnetismus und Aeronomie, Band I/3. Berlin 1964

FRÖLICH, F.: Beiträge zum Erkundungsprogramm: Materieparameter im Bereich der Erdkruste. Teil I: Verfahrensgrundlagen, Erprobungsergebnisse, Testtechnik. Abh. 41, Geom. Inst. Potsdam, Berlin 1969

FRÖLICH, F.: Teil II: Ergänzende festkörperphysikalische und physikochemische Untersuchungen, Auswertung. Veröff. Nr. 4, Zentral-Inst. Phys. d. Erde, Potsdam 1970

KAWAI, N.: J. Geomagn. Geoelectr. 9, 144 (1957)
NAGATA, T.: Rock magnetism, 2nd ed., Maruzen, Tokyo 1961
RINGWOOD, A. E.: Earth a. plan. letters 5, 401 (1969)

(Received May 20, 1971)

Authors' address:

F. FRÖLICH
Central Earth Physics Institute
15 Potsdam
Telegraphenberg

Kristall und Technik	7	1–3	1972	253–376

A. Messerschmidt

Humboldt-Universität zu Berlin
Sektion Physik, Bereich Kristallographie

Reflexionselektronenbeugungsuntersuchungen an HV-aufgedampften CdSe-Schichten auf CaF_2

Es wurden Reflexionselektronenbeugungsuntersuchungen an HV-aufgedampften CdSe-Schichten auf an Luft gespaltenen (111)-Flächen von CaF_2 durchgeführt. Es treten nebeneinander die kubische (111)-Orientierung und die hexagonale (0001)-Orientierung auf. Bei niedrigen Substrattemperaturen (440 °C) dominiert die kubische Phase; bei höheren Substrattemperaturen (600 °C) herrscht die hexagonale Phase vor.

In allen Schichten werden Zwillinge und Mehrfachzwillinge beobachtet. Entlang den ⟨111⟩-Richtungen der Zwillinge bzw. Mehrfachzwillinge treten ebenfalls hexagonale Schichtenfolgen auf. Bei Zunahme der Schichtdicke geht die Anzahl der Mehrfachzwillinge zurück. Starke Erhöhung der Aufdampfrate ($80\,Å \cdot s^{-1}$) erhöht den Defektgehalt der Schichten (polykristalline Anteile). Es wird das Auftreten von Satellitenreflexen bei der Durchstrahlungselektronenbeugung und der Reflexionselektronenbeugung diskutiert.

Reflection electron diffraction investigations were performed with high vacuum deposited films of CdSe on to air cleaved (111) planes of CaF_2. Together there occur the cubic (111) orientation and the hexagonal (0001) orientation. At lower substrate temperatures (440 °C) the cubic phase dominates. At higher substrate temperatures (600 °C) the hexagonal phase is predominant.

Twins and multiple twins are observed in all films. Along the ⟨111⟩ directions of the twins and multiple twins, respectively, additionally there occur hexagonal stacking sequences. With increasing film thickness multiple twins become less frequently. A strong increasing of the deposition rate ($80\,Å \cdot s^{-1}$) increases the content of defects in the films (polycrystalline fractions). The appearance of satellite spots in transmission electron diffraction patterns and in reflection electron diffraction patterns will be discussed.

1. Einleitung

Die möglichen Anwendungen von dünnen Schichten der $A^{II}B^{VI}$-Verbindungen in Elektronik und Mikroelektronik haben intensive Untersuchungen auf diesem Gebiet ausgelöst. Besonders die Miniaturisierung der Bauelemente und Schaltkreise erfordert die Einführung der Dünnschichttechnik. Zahlreiche Arbeiten zur Herstellung dünner Schichten von $A^{II}B^{IV}$-Verbindungen sind in den letzten Jahren durchgeführt worden. Besonders hervorzuheben sind Arbeiten von Holloway, Wilkes (CdS), Kalinkin, Sergeyeva et al. (CdS, CdSe, ZnS, ZnSe, ZnTe), Wilcox, Holt (CdS), Dhere, Goswami (CdSe, CdTe),

Shalimova et al. (CdSe), Shallcross (CdSe), Yasuda (CdSe), Novik et al. (CdSe), Holt, Woodcock (ZnS), Woodcock, Holt 1969 (ZnS) und Holt (ZnTe). Bei diesen zitierten Arbeiten, die nur eine kleine Auswahl darstellen und besonders kristallographische Probleme behandeln, wurden als Substrate entweder Glas oder ionogene Substrate verwendet. Zwei neue interessante Veröffentlichungen von Rawlins und Lilley et al. behandeln die Herstellung epitaxialer ZnS-Schichten auf Silizium. Die Autoren weisen auf die Wichtigkeit der Verwendung technisch interessanter Substrate wie z. B. Saphir (Isolator) oder Silizium (Halbleiter) hin.

Von besonderer Wichtigkeit für die Herstellung dünner Schichten sind geeignete Untersuchungsmethoden, die zur physikalischen und strukturellen Charakterisierung der Schichten herangezogen werden können. Breite Anwendung zur strukturellen Charakterisierung haben die konventionelle Durchstrahlungselektronenmikroskopie und Durchstrahlungselektronenbeugung gefunden. Grenzen sind diesen Methoden in der Schichtdicke (einige 1000 Å) und in der Zerstörung der Probe durch die Präparation gestellt. Dadurch ist eine physikalische Charakterisierung der Schichten im Anschluß an die Untersuchungen nicht mehr möglich. Eine Untersuchungsmethode zur strukturellen Kennzeichnung, wobei die Probe nicht zerstört wird, ist die Reflexionselektronenbeugung. Leider ist die Reflexionselektronenbeugung bis jetzt noch nicht so zum Einsatz gekommen wegen der etwas schwierigen Deutung der resultierenden Beugungsdiagramme.

In dieser Arbeit werden Ergebnisse aus Reflexionselektronenbeugungsuntersuchungen an CdSe-Schichten auf (111)-Spalfflächen von CaF_2 berichtet. Es wird versucht zu zeigen, wie in Verbindung mit anderen Überlegungen die Reflexionselektronenbeugung effektiv eingesetzt werden kann.

2. Durchführung der Untersuchungen

Die Untersuchungen wurden an Schichten vorgenommen, die in der von Däweritz beschriebenen Anordnung hergestellt wurden. Als Substrate dienten an Luft gespaltene (111)-Flächen von CaF_2. Das Arbeitsvakuum war besser als 10^{-5} Torr. Die Aufdampfrate wurde von 5—80 Å/sec, die Substrattemperatur von 440—600 °C und die Schichtdicke von 100—10000 Å variiert. Alle Schichten wurden im Elektronendiffraktograph KD3 der Firma Trüb & Täuber bei Beschleunigungsspannungen von 50 kV untersucht.

3. Ergebnisse der Untersuchungen

In Übereinstimmung mit der Literatur (z. B. Kleber, Mietz, Elsasser; Däweritz) wurden die für Kristalle mit kubisch-flächenzentriertem Gitter oder hexagonal-dichtgepackter Struktur üblichen Verwachsungsgesetze auf Flächen mit dreizähliger Symmetrie festgestellt. Sie lauten:

$$(111)\ CdSe_{kub.} \parallel (111)\ CaF_2$$

$$[1\bar{1}0]\ CdSe_{kub.} \parallel [\bar{1}10]\ CaF_2$$

$$(0001)\ CdSe_{hex.} \parallel (111)\ CaF_2$$

$$[11.0]\ CdSe_{hex.} \parallel [\bar{1}10]\ CaF_2$$

Wie bereits von STREHLOW, COOK für CdS berechnet und von DÄWERITZ an CdSe diskutiert, wachsen die Doppelschichten des Deposits mit den Nichtmetallatomen zuerst auf der CaF_2-(111)-Fläche auf. Daher tritt nur eine azimutale Stellung des Deposits auf. Die Stellung

$$(111)\ \mathrm{CdSe}_{\mathrm{kub.}} \parallel (111)\ \mathrm{CaF_2}$$
$$[\bar{1}10]\ \mathrm{CdSe}_{\mathrm{kub.}} \parallel [\bar{1}10]\ \mathrm{CaF_2}$$

ist energetisch nicht gleichberechtigt. Dennoch kann in den Beugungsdiagrammen eine zweite azimutale Orientierung vorgetäuscht werden, die aber durch Zwillingsbildung nach $[\bar{1}\bar{1}\bar{1}]$ entsteht.

Es ist allgemein bekannt, daß CdSe in der Zinkblende- und Wurtzitstruktur kristallisiert. Beide lassen sich durch Stapelung dichtgepackter Doppelschichten, z. B. $(A\alpha)$, entlang [111] bzw. [00.1] erklären, wobei die kubische Struktur eine Stapelfolge $(A\alpha)$, $(B\beta)$, $(C\gamma)$, $(A\alpha)$, $(B\beta)$, $(C\gamma)$, ... und die hexagonale Struktur eine Stapelfolge $(A\alpha)$, $(B\beta)$, $(A\alpha)$, $(B\beta)$, ... aufweisen. Durch Änderung der Stapelfolge treten meist beide Phasen gemeinsam auf. Dabei gilt folgende Orientierungsbeziehung:

$$(0001)_{\mathrm{hex.}} \parallel \{111\}_{\mathrm{kub.}}$$
$$[11.0]_{\mathrm{hex.}} \parallel \langle 110 \rangle_{\mathrm{kub.}}$$

Außerdem kann bei der kubischen Phase Zwillingsbildung nach $\langle 111 \rangle$ und Mehrfachverzwillingung (HOLLOWAY 1966) auftreten. Es wird in dieser Arbeit bei der Erklärung der Zeichnungen und bei der Beschreibung der Beugungsdiagramme die von HOLLOWAY 1966 vorgeschlagene Bezeichnungsweise für Zwillinge und Mehrfachzwillinge bei Substanzen mit Zinkblendestruktur verwendet. Ein Mehrfachzwilling, der z. B. erst nach $[\bar{1}\bar{1}\bar{1}]$, dann nach $[11\bar{1}]$ und anschließend nach $[1\bar{1}1]$ verzwillingt ist, erhält folgendes Symbol: $T_{\bar{1}\bar{1}\bar{1}/11\bar{1}/1\bar{1}1}$. Damit ist eine kurze und exakte Kennzeichnung der verschiedenen Zwillinge in übersichtlicher Form möglich. Da im Anfangsstadium des Schichtwachstums die kubische Phase in der Form von kleinen Tetraedern auf der Substratfläche aufwächst, ist es möglich, daß die hexagonale Phase auch auf den anderen schrägen $\{111\}$-Flächen des Wachstumstetraeders oder der Mehrfachzwillinge aufwächst. Neben diesen regelmäßigen Stapelfolgen sind natürlich auch Stapelfehler entlang allen $\langle 111 \rangle$-Richtungen möglich.

Nach PASHLEY, STOWELL, LAW ergibt Zwillingsbildung bei kubisch-dichtgepackten Strukturen zusätzliche reziproke Gittervektoren, die von der Form $R = r_a \pm 1/3\,(hkl)$ sind, wobei r_a ein erlaubter reziproker Matrixvektor ist und (hkl) die Zwillingsebene ($\{111\}$) darstellt. Dabei treten nur solche Zusatzreflexe durch Zwillingsbildung auf, die die Bedingung $h\,u + k\,v + l\,w = (3\,N + 1)$ erfüllen, wobei u, v, w die Indizes eines Matrixvektors und h, k, l wieder die Indizes der entsprechenden Zwillingsebene sind. Durch die hexagonal-dichtgepackte Phase, die auf den $\{111\}$-Ebenen aufwächst, entstehen zusätzliche reziproke Gittervektoren, die in drei Klassen eingeteilt werden können:

1. Reziproke Gittervektoren, die mit reziproken Gittervektoren der kubischen Phase zusammenfallen.
2. Reziproke Gittervektoren der Form $R = r_x \pm 1/3\,(hkl)$, wobei h, k, l die Indizes der $\{111\}$-Zwillingsebene sind.
3. Reziproke Gittervektoren der Form $R = r_a \pm 1/6\,(hkl)$.

Die 3. Klasse von Vektoren, die nicht mit reziproken Vektoren der kubischen Phase zusammenfallen, sind zur eindeutigen Identifizierung der hexagonalen Phase geeignet. Sind diese reziproken Gittervektoren in hexagonalen Indizes *HKIL* gegeben, so ist *L* immer eine ungerade ganze Zahl. Holloway 1968 hat ausgeführt, daß diese Bedingung eindeutig ist, da durch Mehrfachverzwillingung in der Phase mit Zinkblendestruktur und durch Doppelbeugung an Zwillingsgrenzen nur reziproke Gittervektoren entstehen, die in hexagonalen Indizes ausgedrückt ein geradzahliges *L* haben.

a) b) c) d) e)

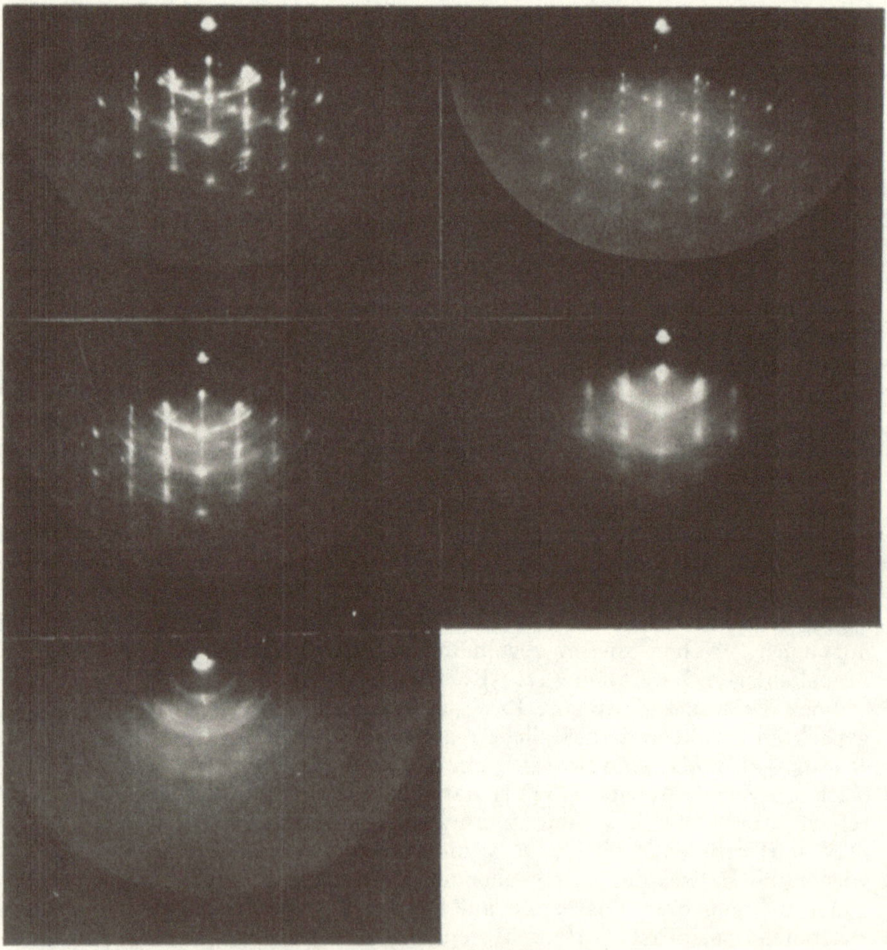

Fig. 1. RHEED[1])-Aufnahmen von CdSe-Schichten auf CaF_2-(111)-Spaltflächen a) T_S[2]) = 440 °C, R[3]) = 20 Å · s^{-1}, D[4]) = 420 Å; b) T_S = 440 °C, R = 20 Å · s^{-1}, D = 2000 Å; c) T_S = 440 °C, R = 40 Å · s^{-1}, D = 450 Å; d) T_S = 440 °C, R = 80 Å · s^{-1}, D = 250 Å; e) T_S = 440 °C, R = 80 Å · s^{-1}, D = 800 Å; [1$\bar{1}$0]-Einstrahlazimut

[1]) „RHEED" bedeutet: reflection high energy electron diffraction

[2]) T_S – Substrattemperatur

[3]) R – Aufdampfrate

[4]) D – Schichtdicke

Sämtliche Zusatzreflexe sind also entlang den ⟨111⟩-Richtungen des reziproken Gitters angeordnet, die mit den ⟨111⟩-Richtungen des direkten Gitters übereinstimmen.

Aus den oben genannten Gründen wurden die Elektronenbeugungsdiagramme unter ⟨110⟩-Einstrahlazimut aufgenommen, da in diesem Fall zwei ⟨111⟩-Richtungen direkt mit der Ewald-Fläche zusammenfallen. Bei diesem Einstrahlazimut können kubische und hexagonale Phase gut unterschieden werden.

Die Figuren 1a–c und 2a–c zeigen die erhaltenen Beugungsdiagramme von Schichten mit unterschiedlichen Herstellungsparametern. Figur 3a–c enthält die entsprechenden theoretischen Muster. Die theoretischen Beugungsdiagramme wurden an der BESM 6 mit einem nach OLSEN, JESSER modifizierten

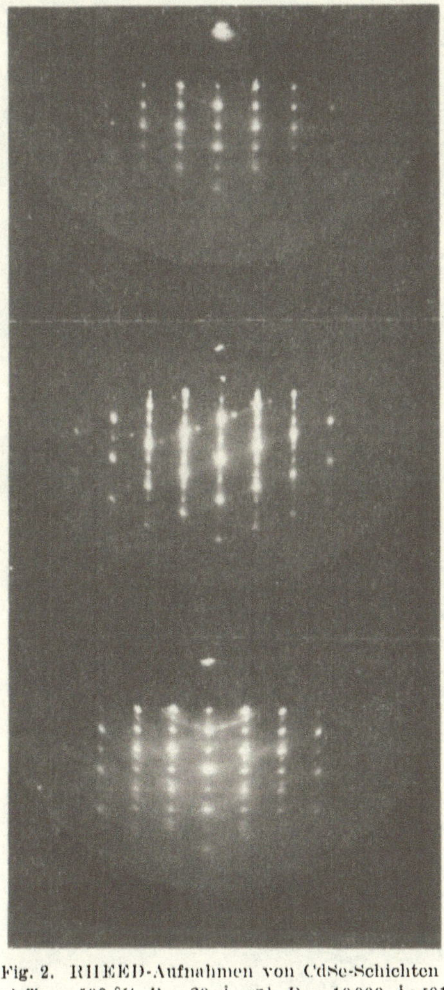

a)

b)

c)

Fig. 2. RHEED-Aufnahmen von CdSe-Schichten auf CaF_2-(111)-Spaltflächen.
a) $T_S = 500$ °C, $R = 20$ Å · s^{-1}, $D = 10000$ Å; [01.0]-Einstrahlazimut; b) $T_S = 580$ °C, $R = 20$ Å · s^{-1}, $D = 300$ Å; [$\bar{1}$10]-Einstrahlazimut; c) $T_S = 600$ °C, $R = 20$ Å · s^{-1}, $D = 500$ Å; [1$\bar{1}$0]-Einstrahlazimut

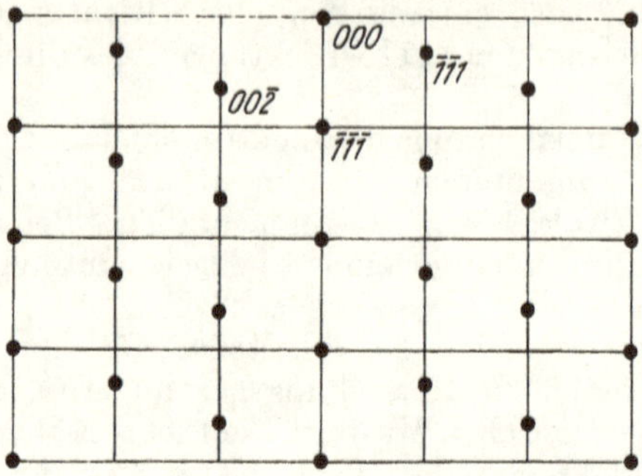

a)

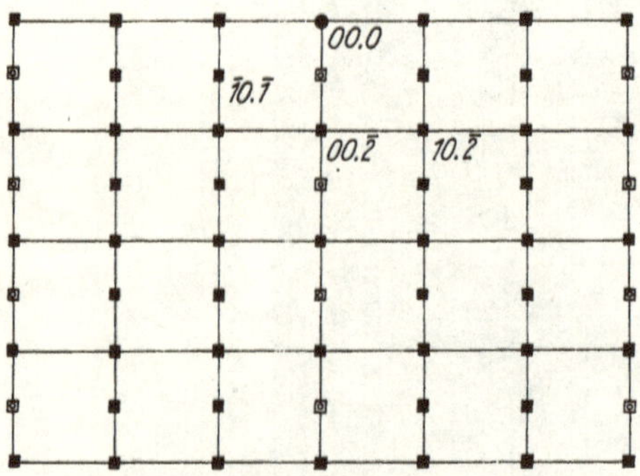

b)

■ erlaubter Reflex, hexagonale Phase, 01.0 - Azimut
▣ durch Doppelbeugung entstandener Reflex

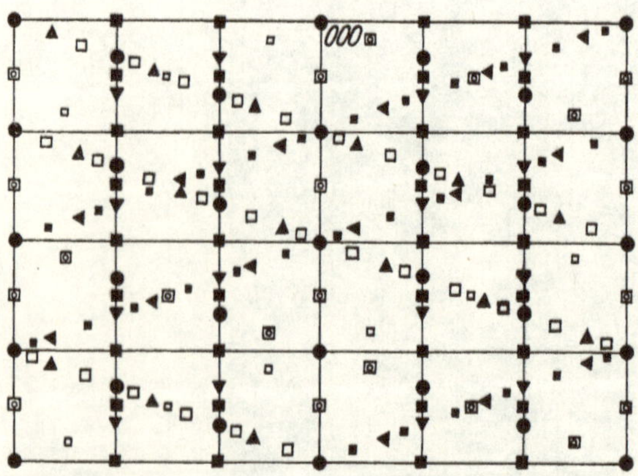

● kub. Matrix, [$\bar{1}10$]-Azimut ▼ Zwilling $T_{\bar{1}\bar{1}\bar{1}}$
▲ Zwilling $T_{\bar{1}\bar{1}1}$ ◀ Zwilling $T_{\bar{1}\bar{1}\bar{1}/\bar{1}\bar{1}1}$
■ hex. Phase II ($\bar{1}\bar{1}\bar{1}$)-Matrix □ hex. Phase II ($\bar{1}\bar{1}1$)-Matrix
▫ hex. Phase II ($\bar{1}\bar{1}1$) – $T_{\bar{1}\bar{1}\bar{1}}$

c)

(lies [$1\bar{1}0$]-Azimut in c), d), e))

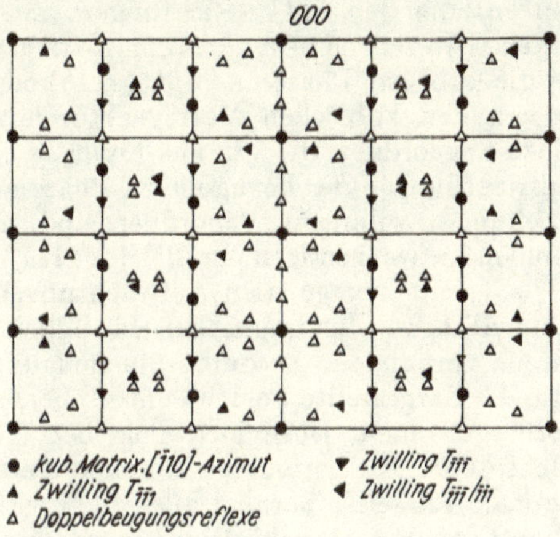

d)

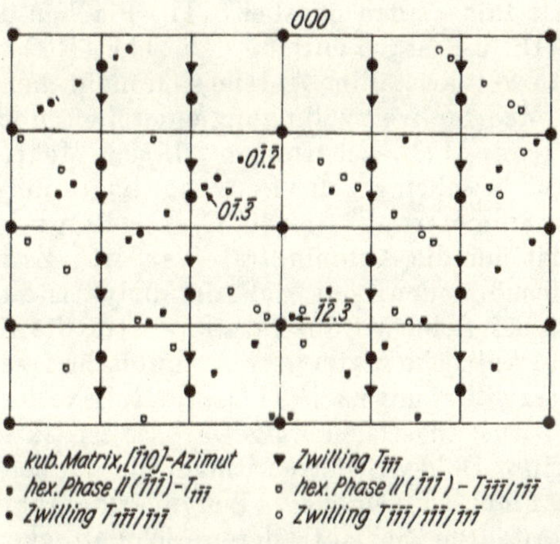

e)

Fig. 3. Theoretische Beugungsdiagramme für RHEED-Aufnahmen von (111)- oder (0001)-orientierten Proben mit Zinkblende- oder Wurtzitstruktur unter $[1\bar{1}0]$- oder [01.0]-Einstrahlazimut. a) (111)-orientierte kubische Matrix, $[1\bar{1}0]$-Azimut; b) (0001)-orientierte hexagonale Phase, [01.0]-Azimut; c) Kubische (111)-Orientierung, hexagonale (0001)-Orientierung, kubische Zwillinge $T_{\langle 111\rangle}$ und $T_{\bar{1}1\bar{1}/\langle 111\rangle}$ sowie hexagonale Schichtenfolgen entlang den $\langle 111\rangle$-Richtungen der Matrix und des Zwillings $T_{\bar{1}1\bar{1}}$; d) kubische (111)-Orientierung, kubische Zwillinge $T_{\langle 111\rangle}$ und $T_{\bar{1}1\bar{1}/\langle 111\rangle}$ sowie Doppelbeugungsreflexe entstanden durch Doppelbeugung an Zwillingsgrenzen; e) kubische (111)-Orientierung, kubische Zwillinge $T_{\bar{1}1\bar{1}}$, hexagonale Schichtenfolgen parallel der $\{111\}$-Flächen der Zwillinge $T_{\langle 111\rangle}$ und $T_{\bar{1}1\bar{1}/\langle 111\rangle}$ und Zwillinge der Form $T_{\langle 111\rangle/\langle 111\rangle}$ und $T_{\bar{1}1\bar{1}/\langle 111\rangle/\langle 111\rangle}$.

Rechenprogramm berechnet, das es gestattet, bei vorgegebenen kristallographischen Orientierungsbeziehungen zwischen verschiedenen kristallinen Phasen unter Berücksichtigung des Einstrahlazimuts die Geometrie des resultierenden Beugungsdiagramms zu berechnen. Figur 1a zeigt ein typisches Beugungsdiagramm von Schichten, die bei einer Substrattemperatur von 440 °C mit einer Aufdampfrate von 20 Å · s^{-1} aufgedampft wurden. Neben den starken Matrixreflexen der kubischen Phase treten viele zusätzliche Reflexe auf. Einmal die Reflexe, die entstehen durch Zwillingsbildung nach $\langle 111\rangle$ und durch hexagonale Schichtenfolgen entlang den $\langle 111 \rangle$-Richtungen. Sie ordnen sich entlang den in der reziproken Gitterebene liegenden $\langle 111 \rangle$-Richtungen in Abständen von 1/3 ($\langle 111 \rangle$) für die kubische Phase und 1/3 ($\langle 111 \rangle$) oder 1/6 ($\langle 111 \rangle$) für die hexagonale Phase von den kubischen Matrixvektoren an. Entlang, $[00\bar{1}]$ sind zusätzliche Reflexe angeordnet, die wie aus Figur 3e ersichtlich ist, Sekundärzwillingen und Aufwachsungen der hexagonalen Phase auf Tetraederflächen von Primär- und Sekundärzwillingen zuzuordnen sind. Die Zwillinge der Form $T_{\langle 111 \rangle}$ mit hexagonalen Aufwachsungen parallel der $\{111\}$-Flächen und die Zwillinge der Form $T_{\bar{1}\bar{1}\bar{1}/\langle 111 \rangle}$ mit hexagonalen Aufwachsungen parallel der $\{111\}$-Flächen der Sekundärzwillinge treten in gleicher Häufigkeit auf. Zusätzlich sind poykristalline Anteile vorhanden, die durch die Beugungsringe angedeutet werden. Die in Figur 1b dargestellte Schicht unterscheidet sich nur in der Schichtdicke von der vorhergehenden. Die Schichtdicke beträgt hier 2000 Å. Es herrscht die kubische Phase vor. Mit geringeren Anteilen treten Zwillingen nach $\langle 111 \rangle$ auf. Die hexagonale Phase ist parallel aller $\{111\}$-Flächen aufgewachsen. Wie der Vergleich mit dem theoretischen Beugungsdiagramm Figur 3e zeigt, wächst die hexagonale Phase auch parallel $\{111\}$-Flächen der kubischen Zwillinge 1. Ordnung auf. Diffuse Stäbe entlang den $\langle 111 \rangle$-Richtungen zeigen vorhandene Stapelfehler an. Polykristalline Anteile sind nicht mehr vorhanden. Orientierungen, die durch Mehrfachverzwillingung entstehen und hexagonale Schichtenfolgen entlang den $\langle 111 \rangle$-Richtungen dieser Mehrfachzwillinge, treten stark zurück. In Figur 1c sehen wir das Beugungsdiagramm einer Schicht mit folgenden Herstellungsparametern, $T_S = 440$ °C, $R = 40$ Å s^{-1}, $D = 450$ Å. Zur Schicht aus Figur 1a ist nur die Aufdampfrate variiert. Es ist keine entscheidende Veränderung festzustellen, nur daß die polykristallinen Anteile zunehmen. In Figur 1d ist bei gleicher Substrattemperatur die Aufdampfrate auf 80 Å s^{-1} erhöht. Es sind kubische Matrixreflexe, kubische Zwillingsreflexe, die entstanden sind durch Verzwillingung nach $[\bar{1}\bar{1}\bar{1}]$, und Reflexe der hexagonalen Phase, die parallel der $(\bar{1}\bar{1}\bar{1})$-Ausgangsfläche aufgewachsen ist, zu sehen. Dominierend sind aber stark diffuse Debey-Scherrer-Ringe, die auf sehr kleine Kristallite schließen lassen. Figur 1e zeigt ein Beugungsdiagramm einer bei gleichen Parametern hergestellten Schicht, deren Schichtdicke aber 750 Å beträgt. Die Einkristallreflexe sind nur noch ganz undeutlich zu sehen. Die Schicht ist fast polykristallin. Bei der in Figur 2a untersuchten Schicht wurde die Substrattemperatur auf 500 °C erhöht. Die Aufdampfrate betrug 20 Å · s^{-1}, die Schichtdicke lag bei 10000 Å. Es liegt die hexagonale Phase vor. Ganz schwach sind kubische Pulveranteile und verzwillingte Kristallite zu erkennen. Die Schicht, die durch Figur 2b repräsentiert wird, hat folgende Herstellungsparameter: $T_S = 580$ °C, $R = 20$ Å s^{-1}, $D = 300$ Å. Kubische und hexagonale Phase sind ungefähr zu gleichen Teilen vertreten. Es treten Zwillinge nach allen $\langle 111 \rangle$-Richtungen auf. Die entsprechenden Sekundärzwillinge der Form $T_{\bar{1}\bar{1}\bar{1}\langle 111 \rangle}$ (sie täuschen die um 180° gedrehte zweite azimutale Stellung vor) sind nur in geringer Anzahl vorhanden. Die hexagonale Phase ist parallel aller $\{111\}$-Flächen aufgewachsen. Ebenfalls wie in Figur 1b wächst die hexagonale Phase auch parallel der $\{111\}$-Flächen der kubischen Zwillinge 1. Ordnung auf. Stapelfehler sind zu beobachten. Eine letzte ausgewählte Schichtaufnahme zeigt Figur 2c. Die Herstellungsparameter der Schicht sind: $T_S = 600$ °C, $R = 20$ Å · s^{-1}, $D = 500$ Å. Die hexagonale und kubische Phase sind vorhanden, wobei die hexagonale Phase überwiegt. Zwillinge nach $\langle 111 \rangle$ treten auf. Starke polykristalline Anteile werden beobachtet. Stapelfehler sind nicht angedeutet.

3. Diskussion

3.1. Bemerkungen zur Deutung der Reflexionselektronenbeugungsdiagramme

Die Figur 3a—e zeigt die berechneten theoretischen Beugungsdiagramme. Nicht alle eingezeichneten Reflexe fallen mit der Ewald-Fläche direkt zusammen. Bei RHEED-Aufnahmen beträgt der Einfallswinkel des Elektronenstrahls zur Probenoberfläche nicht 0° wie hier angenommen, sondern einige Winkelgrade, in den meisten Fällen $\approx 1°$. Trotzdem stellt diese Betrachtungsweise eine gute Näherung dar. Die relativen Fehler bei der Bestimmung der Position der Reflexe im Beugungsdiagramm liegen unter 0,1%. Die Reflexe der Figuren 3a—d liegen genau in der Ewald-Fläche. Tritt eine Phase in der dazu entsprechenden Orientierung auf, dann werden diese Reflexe erscheinen. Das zeigen auch alle hier angeführten Aufnahmen. Reflexe, die Abweichungen von der Ewald-Fläche aufweisen, können in der Durchstrahlungselektronenbeugung durch die immer mehr oder weniger vorhandene Folienverbiegung noch bei Abweichungen von 10° von der Ewald-Fläche auftreten. Untersucht man kubisch-flächenzentrierte Metalle oder Verbindungen mit Zinkblende- und Wurtzitstruktur in Transmission unter [111]-Einstrahlazimut, dann treten neben den Matrixreflexen zahlreiche Satellitenreflexe auf. Diese Satellitenreflexe entstehen durch Zwillingsbildung, durch Doppelbeugung an Zwillingsgrenzen und durch hexagonale Schichtenfolgen entlang den ⟨111⟩-Richtungen. Sie haben Abweichungen bis zu 10° von der Ewald-Fläche. Sind die Schichten verhältnismäßig dünn, so wird die Folienverbiegung groß sein; die Satellitenreflexe werden auftreten und so die richtige Schichtstruktur widerspiegeln. Nimmt die Schichtdicke zu, so wird die Folienverbiegung geringer und es könnte vorkommen, daß kein Satellitenreflex zur Abbildung kommt. Es würde also eine idealere Struktur vorgetäuscht. Zusätzlich liegen bei Aufnahmen, die im [111]-Einstrahlazimut angefertigt wurden, Schwierigkeiten bei der Unterscheidung von hexagonaler und kubischer Phase vor. Reflexe der hexagonalen Phase, die keine Abweichungen von der Ewald-Fläche aufweisen, haben sämtlich Indizes $HKIL$ mit $L = 2N$, wobei N eine ganze Zahl ist. Ein Teil der Reflexe fällt mit den kubischen Matrixreflexen zusammen, der andere Teil kann durch Zwillinge oder durch Doppelbeugung an Zwillingsgrenzen entstehen. Man kann diese Schwierigkeit umgehen, wenn man die Folie zum Elektronenstrahl kippen und somit eine ⟨110⟩-Einstrahlrichtung einstellen kann. Hier liegen, wie Figur 3b zeigt, Reflexe der hexagonalen Phase mit dem Index $L = 2N - 1$ in der Ewald-Fläche. Ebenfalls befinden sich, wie schon ausgeführt, jeweils zwei ⟨111⟩-Richtungen in der Ewald-Fläche, wobei sämtliche Zusatzreflexe entlang diesen ⟨111⟩-Richtungen angeordnet sind. Leider ist in konventionellen Elektronenmikroskopen eine Kippung der Folie im Gerät um so große Winkel nicht möglich.

In der Reflexionselektronenbeugung läßt sich bei (111)-orientierten Proben das ⟨110⟩-Azimut ohne weiteres einstellen. Eine Folienverbiegung liegt nicht vor. Sämtliche Zusatzreflexe, die mit der Ewald-Fläche zusammenfallen, werden zur Abbildung gelangen. Reflexe, die Abweichungen von der Ewald-Fläche aufweisen, können nur bei ganz geringen Abweichungen (0—2°) im Beugungsdiagramm beobachtet werden. Entschärfung der Bedingung kann nur auftreten, wenn die Kristallite, die parallel zur Substratoberfläche angeordnet sind, gewisse azimutale Winkelabweichungen aufweisen. Das Auftreten solcher Reflexe läßt dann ohne weiteres auf die azimutale Winkelverteilung der einzelnen Kristallite um die ideale Lage schließen. Die Satellitenreflexe in Figur 3e zeigen alle mehr oder weniger starke Abweichungen von der Ewald-Fläche. Abweichungen von 0,7° haben die mit $0\bar{1}.\bar{3}$ und $\bar{1}2.\bar{3}$ indizierten Reflexe. Sie werden fast immer auftreten, wenn diese Orientierung vorhanden ist, denn azimutale Abweichungen der Kristallite von 1° sind in epitaxialen Schichten fast immer vorhanden. Der nächste hexagonale Reflex, vom $0\bar{1}.\bar{3}$ Reflex in Richtung $\bar{1}\bar{1}\bar{1}$ Matrixreflex gesehen, hat eine Abweichung von 6°. Er wird erst bei größerer Unperfektion der Schichten zu sehen sein, was Figur 1a beweist. In dieser Aufnahme treten fast alle in Figur 3e angedeuteten Zusatzreflexe auf.

Die Ausarbeitungen dieses Abschnitts zeigen, welche wichtigen Aussagen zur strukturellen Charakterisierung von dünnen Schichten vorgenommen werden können, wenn auch die zusätzlichen Reflexe zur Diagnostik herangezogen werden. Es muß bemerkt werden, daß eine solche detaillierte Deutung ohne das erwähnte Rechenprogramm nicht möglich wäre. Weiterhin ist eine genaue Beachtung der Besonderheiten der verwendeten Methode notwendig.

3.2. Korrelation der Aussagen aus den RHEED-Aufnahmen mit den Herstellungsparametern

Bei niedrigen Substrattemperaturen (440 °C) wird vorwiegend die kubische Phase ausgebildet. Wie in Abschnitt 2 beschrieben treten Zwillinge der Form $T_{\langle 111\rangle}$, $T_{\bar{1}\bar{1}\bar{1}/\langle 111\rangle}$, $T_{\langle 111\rangle/\langle 111\rangle}$ und $T_{\bar{1}\bar{1}\bar{1}/\langle 111\rangle/\langle 111\rangle}$ mit hexagonalen Schichtenfolgen parallel der {111}-Flächen der Zwillinge und Mehrfachzwillinge auf. Die Zwillinge $T_{\langle 111\rangle}$ und $T_{\bar{1}\bar{1}\bar{1}/\langle 111\rangle}$ sind in gleicher Häufigkeit vorhanden. Nimmt die Schichtdicke zu, so wird die Anzahl der Mehrfachzwillinge stark reduziert. Dieser Effekt tritt bei allen untersuchten Schichten auf. Er ist darauf zurückzuführen, daß nach dem Schließen der Schicht zweidimensionales Wachstum vorherrscht.

Wird die Aufdampfrate stark erhöht (80 Å · s^{-1}), so nehmen die Verzwillingung und die polykristallinen Anteile erheblich zu. Aufgrund der höheren Übersättigung findet in größerem Maße dreidimensionales Wachstum statt, was zur Ausbildung von kleinen stark verzwillingten Wachstumstetraedern führt. Bei solchen komplizierten geometrischen Gebilden ist eine parallele Ausrichtung der einzelnen Kristallite kaum möglich.

Erhöhung der Substrattemperatur auf 500—600 °C führt zur immer intensiveren Ausbildung der hexagonalen Phase. Das deckt sich mit der Tatsache, daß wie von Pashinkin und Saposhnikov gezeigt die hexagonale Wurtzitphase von CdSe die Hochtemperaturmodifikation darstellt. Erhöhung der Schichtdicke führt wieder zu einer Verbesserung der Schichtstruktur. Es muß jedoch bemerkt werden, daß alle Schichten sowohl die hexagonale als auch die kubische Phase enthielten. In allen Schichten waren Zwillinge nach ⟨111⟩ und hexagonale Aufwachsungen parallel der {111}-Flächen des kubischen Matrixtetraeders zu beobachten. Es muß noch hinzugefügt werden, daß bei ansteigenden Substrattemperaturen (bis 600 °C) die hexagonale Phase immer stärker auftritt, gleichzeitig aber ein größerer Defektgehalt (polykristalline Anteile) zu verzeichnen ist.

Danken möchte ich Herrn Dr. L. Däweritz, der freundlicherweise die Schichten für diese Untersuchungen zur Verfügung stellte und durch wertvolle Diskussionen den Fortgang der Arbeit förderte. Ebenfalls zu Dank verpflichtet bin ich Frau Dr. I. Rechenberg für wertvolle Hinweise auf dem Gebiet der Elektronenbeugung und für die Unterstützung bei der experimentellen Einführung der RHEED-Technik.

Literatur

Däweritz, L.: Kristall u. Technik **6**, 281 (1971)
Dhere, N. G., Goswami, A.: Indian J. pure appl. Physics **7**, 398 (1969)
Holloway, H.: The Use of Thin Films in Physical Investigations (Herausgeber J. C. Anderson), London 1966, S. 111
Holloway, H.: J. Physics Chem. Solids **29**, 1977 (1968)
Holloway, H., Wilkes, E.: J. appl. Physics **39**, 5807 (1968)
Holt, D. B.: J. Mater. Sci. **4**, 935 (1969)
Holt, D. B., Woodcock, J. M.: J. Mater. Sci. **5**, 275 (1970)
Kalinkin, I. P., Muravyeva, K. K., Sergeyeva, L. A., Aleskovski, V. B., Bogomolov, N. S.: Kristall u. Technik **5**, 51 (1970)
Kleber, W., Mietz, I., Elsasser, U.: Kristall u. Technik **2**, 327 (1967)
Lilley, P., Jones, P. L., Litting, C. N. W.: J. Mater. Sci. **5**, 891 (1970)
Novik, F. T., Rumsch, M. A., Simkina, T. M.: Kristallografiya **8**, 378 (1963)
Olsen, G. H., Jesser, W. A.: Mater. Sci. Eng. **5**, 135 (1969/70)
Pashinkin, A. S., Saposhnikov, R. A.: Kristallografiya **7**, 623 (1962)
Pashley, D. W., Stowell, M. J., Law, T. J.: phys. stat. sol. **10**, 153 (1965)
Rawlins, T. G. R.: J. Mater. Sci. **5**, 881 (1970)
Shalimova, K. V., Dimitriyev, V. A., Rogge, K., Botnev, A. F.: Kristallografiya **15**, 342 (1970)
Shallcross, F. V.: RCA Review 676 (1963)
Strehlow, W. H., Cook, E. L.: Physic. Rev. **188**, 1256 (1969)
Wilcox, D. M., Holt, D. B.: J. Mater. Sci. **4**, 672 (1969)
Woodcock, J. M., Holt, D. B.: Brit. J. appl. Physics **2**, 775 (1969)
Yasuda, Y.: Japan. J. appl. Physics **7**, 1171 (1968)

(Eingegangen am 20. Mai 1971)

Anschrift des Autors:
Dr. Albrecht Messerschmidt
Humboldt-Universität zu Berlin
Sektion Physik/Bereich Kristallographie
104 Berlin
Invalidenstr. 43

| Kristall und Technik | 7 | 1–3 | 1972 | 265–270 |

M. FRENZEL

Arbeitsstelle für Molekularelektronik, Dresden
(Leiter: Prof. Dr.-Ing. habil. W. Hartmann)

Ein Beitrag zur Lokalisierung von Maskierungs- und Passivierungsfehlern von Deckschichten auf Silizium

Bei Diffusionsprozessen können Defekte in Deckschichten im oberflächennahen Silizium parasitäre pn-Übergänge erzeugen, deren Konzentration, Verteilung und Größe Aussagen über die Qualität von Maskierungs- und Passivierungsschichten von Halbleiterbauelementen geben. Die Markierung der durch Diffusantendurchbrüche erzeugten pn-Übergänge erfolgt mit einer Chrom-Ätzlösung, die Aussagen werden mit Ergebnissen der Kupfer- und Silberdekoration verglichen.

Defects in masking oxide layers on Silicon will produce parasite p—n junctions in the Silicon surface during diffusion processes. Area, density and distribution of these junctions will provide information about the quality of masking and passivating layers on semiconductor devices. The junctions were made visible by Sirtl etching, the results of this technique are compared with those obtained by copper and silver plating.

1. Einleitung

Von besonderem Interesse bei der Herstellung von Halbleiterbauelementen sind die Maskierungs- und Passivierungsfähigkeit von Deckschichten auf Silizium, speziell von Oxid-, Silikat- und Mischsilikatschichten.

Fragen dieser Art sind mit einer direkten Untersuchung der Topologie, Morphologie und Struktur der einzelnen in Frage kommenden Schichten zu beantworten. Derartige Untersuchungen erfordern Verfahren mit hohem Auflösungsvermögen, hauptsächlich elektronenmikroskopische Methoden.

Eine andere Untersuchungsmöglichkeit bieten die parasitären pn-Übergänge bzw. Sprünge in der Dotierungskonzentration des Siliziums, die nach Diffusionsprozessen entstehen. Sie lassen indirekte Schlüsse auf die Qualität von Maskierungs- und Passivierungsschichten zu.

Die Markierung der pn-Übergänge muß möglichst in ihrer gesamten Größe und Gestalt erfolgen. Infolge der geringen Ausdehnung sind Widerstands- (z. B. Vierspitzenmethode) und Leitungstypmessungen mit ihrem schlechten Auflösungsvermögen nicht brauchbar. Besser geeignet sind die allgemein üblichen Lokalisierungsverfahren der stromlosen Metallabscheidung (WHORISKEY; ILLES, COPPEN; TURNER). Zum anderen kann man Staining-Methoden (BOGENSCHÜTZ) und chemische Ätzer verwenden. Je nach Konzentration und Art der Dotierungsatome werden die entsprechenden Bereiche des Siliziums unterschiedlich angeätzt. An Dotierungsgrenzen entstehen scharfe Stufen, die mikroskopisch zu beobachten sind. Zur Auswahl steht eine Vielzahl von Ätzlösungen der unterschiedlichsten Zusammensetzung und Qualität.

Im folgenden soll hauptsächlich über die Anwendung der Chromätzlösung nach SIRTL und ADLER (nachfolgend Chrom-Ätzer genannt) berichtet werden, über deren Verwendung zur Lokalisierung großflächiger pn-Übergänge schon von CHU und GAVALER berichtet wurde. Die Aussagen werden mit Ergebnissen der Silber- und Kupferdekoration verglichen. Staining-Ätzlösungen zur Lokalisierung der durch Diffusantendurchbrüche erzeugten pn-Übergänge sind ungeeignet und finden daher keine spezielle Betrachtung.

2. Probenmaterial und -behandlung

Die Siliziumproben wurden aus nach dem Czochralski-Verfahren gezogenen Kristallen verschiedener Widerstandswerte parallel {111} geschnitten, geläppt und mechanisch mit 0,25 µm-Diamantpaste poliert.

Anschließend erfolgte bei 1200 °C eine thermische Oxydation in Sauerstoff, der mit Wasser bei 90 °C gesättigt wurde. Die entstandenen Oxidschichten hatten eine Dicke von 0,2···0,5 µm. Die nach der Bor- und Phosphordiffusion maximal erreichten Oberflächenkonzentrationen an der freien Siliziumoberfläche betrugen $5 \cdot 10^{19}$ Atome/cm^3 (Bor) und $1 \cdot 10^{21}$ Atome/cm^3 (Phosphor).

3. Lokalisierungsverfahren

Nach Entfernung des Oxid- und Silikataufbaues mit konz. Flußsäure wurden die Diffusionsdurchbrüche auf der Siliziumoberfläche mit Silber- und Kupferdekoration bzw. durch Behandlung mit Chrom-Äther lokalisiert.

Die Silberdekoration erfolgte in einer verdünnten AgF-Lösung (unter 2%) bei starker Beleuchtung.

Für die Kupferdekoration wurde ein Lösungsgemisch aus 20 g $CuSO_4 \cdot 5\,H_2O$, 100 ml H_2O und 1 ml Flußsäure (48%) verwendet.

Der Chrom-Ätzer bestand aus 1 Volumenteil 48%iger Flußsäure und 2 Volumenteilen Chromoxidlösung (50 g CrO_3 in 100 g H_2O); die Ätzzeiten betrugen bei Zimmertemperatur 15 s bis 5 min.

Bei allen drei Verfahren wurden die Proben anschließend sofort in destilliertem Wasser gespült und im heißen Luftstrom getrocknet.

Die Beobachtung der Dekorationseffekte erfolgte meist im Auflicht-Dunkelfeld. Die im Silizium entstandenen Ätzkalotten beim Chrom-Ätzverfahren lassen sich im Auflicht-Hellfeld oder -Dunkelfeld nachweisen, bei sehr geringem Ätzabtrag ist eine Beobachtung im Auflicht-Phasenkontrast erforderlich.

4. Experimentelle Ergebnisse

4.1. Silberdekoration

Bei der Lokalisierung mit AgF scheidet sich im allgemeinen das Silber schneller auf dem n-Gebiet ab. Mitunter ist auch der umgekehrte Fall zu beobachten.

Neben den mehr oder weniger typisch dekorierten, in den meisten Fällen kreisrunden bis nierenförmigen Diffusantendurchbrüchen, erfolgt zusätzlich eine erhöhte Silberabscheidung an Polierspuren, Schmutzpartikeln und Strukturfehlern des Siliziums (z. B. Mikrobrüche, Ausscheidungen). Sind keine Oberflächenstörungen in der Art von Polierspuren vorhanden, so besitzt der Niederschlag eine homogene, körnige Struktur. Oft ist auch eine stark ausgeprägte Bildung von dendritischen bis skelettartigen Silberkristallisaten zu beobachten. Diese können sich beim Spülvorgang, der an die Dekoration angeschlossen wird, ablösen und so einen pn-Übergang vortäuschen. Benetzungsunterschiede dürften den gleichen Effekt hervorrufen.

4.2. Kupferdekoration

Bei der Lokalisierung der Diffusantendurchbrüche mit Kupfer sind etwa die gleichen Verhältnisse wie bei der Silberdekoration zu beobachten. Der Dekorationseffekt ist aber nicht so ausgeprägt (Fig. 1). Die Neigung zum dendritischen Wachstum ist geringer.

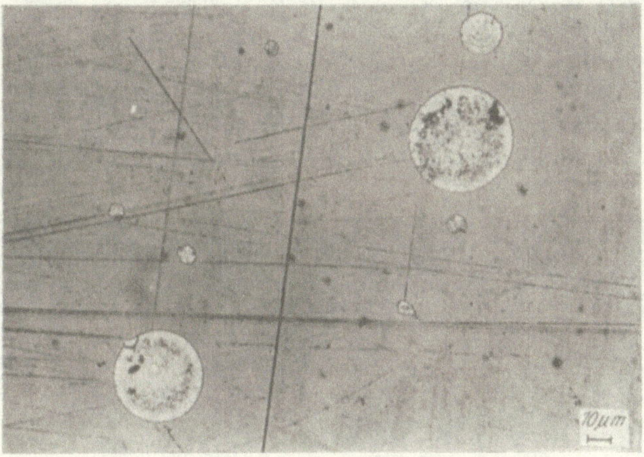

Fig. 1. Kupferdekoration von n-Gebieten. Auflicht-Hellfeld

4.3. Lokalisierung mit dem Chrom-Ätzer

Mit dem Chrom-Ätzer sind bereits nach kurzen Ätzzeiten (unter 15 s) die scharf begrenzten kreis- und nierenförmigen Ätzkalotten, die die Diffusantendurchbrüche lokalisieren, nachzuweisen (Fig. 2). Ein wichtiges Kriterium zu ihrer Beurteilung ist die stets gerundete, scharfe, stufenartige Begrenzung, die etwa mit dem pn-Übergang identisch ist. Die lokalisierten Diffusantendurchbrüche sind daher in jedem Fall von anderen Strukturfehlern des Siliziums, die bei längeren Ätzzeiten mit dem Chrom-Ätzer gleichfalls „entwickelt" werden, zu unterscheiden. Meist sind innerhalb der Ätzkalotten Ausscheidungen zu erkennen.

Entsprechend der Größe der Dotantenkonzentration sind unterschiedliche Lösungsraten zu beobachten. Naturgemäß sind sie größer bei hohen Phosphor- und Borkonzentrationen. Anderseits sind mit dem gleichen Ätzer auch die geringen wachstumsbedingten Unterschiede der Dotierungskonzentration im Ausgangsmaterial noch nachzuweisen (Fricke).

Auf Grund der typischen Ätzerscheinungen lassen sich Diffusionsdurchbrüche von minimal 2···3 μm Durchmesser noch erkennen.

Bei der Silber- und Kupferdekoration Phosphor-diffundierter Proben ist eine visuelle Trennung der lokalisierten Diffusantendurchbrüche von den zahlreichen formenreichen zusätzlichen Dekorationseffekten nicht möglich. Hier deckt erst der Chromätzer die wahren Verhältnisse auf (Fig. 3a und 3b).

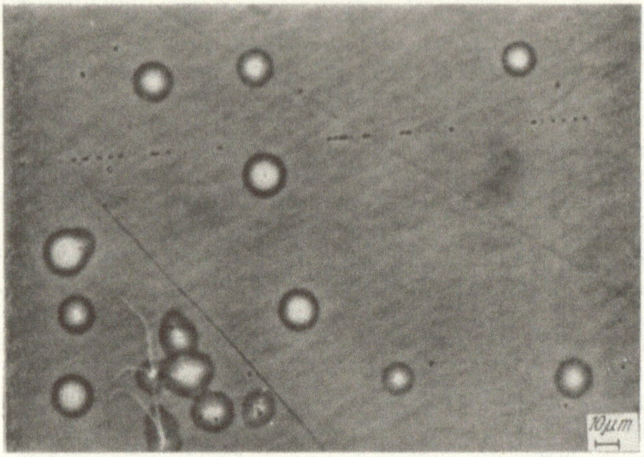

Fig. 2. Lokalisierung von n-Gebieten mit dem Chrom-Ätzer, Ätzzeit 15 s. Auflicht-Phasenkontrast

a)

b)

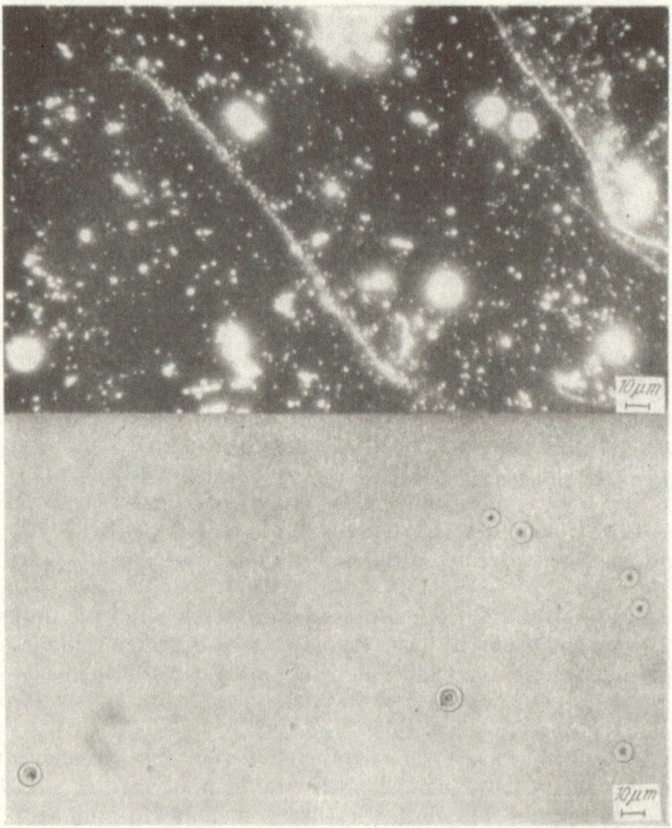

Vergleich von Kupferdekoration und Chrom-Ätzer: Lokalisierung von n-Gebieten
Fig. 3a. Kupferdekoration. Auflicht-Dunkelfeld
Fig. 3b. Lokalisierung mit dem Chrom-Ätzer, Ätzzeit 30 s. Auflicht-Hellfeld

Die Fig. 4a und 4b ermöglichen einen Vergleich der Lokalisierung der Diffusantendurchbrüche mit AgF und mit Chrom-Ätzer. Obwohl durch die Silberabscheidung eine geringe Menge des Siliziums abgetragen wurde (die erzeugte Stufe ist nachzuweisen), ist nach der Behandlung mit dem Chrom-Ätzer infolge

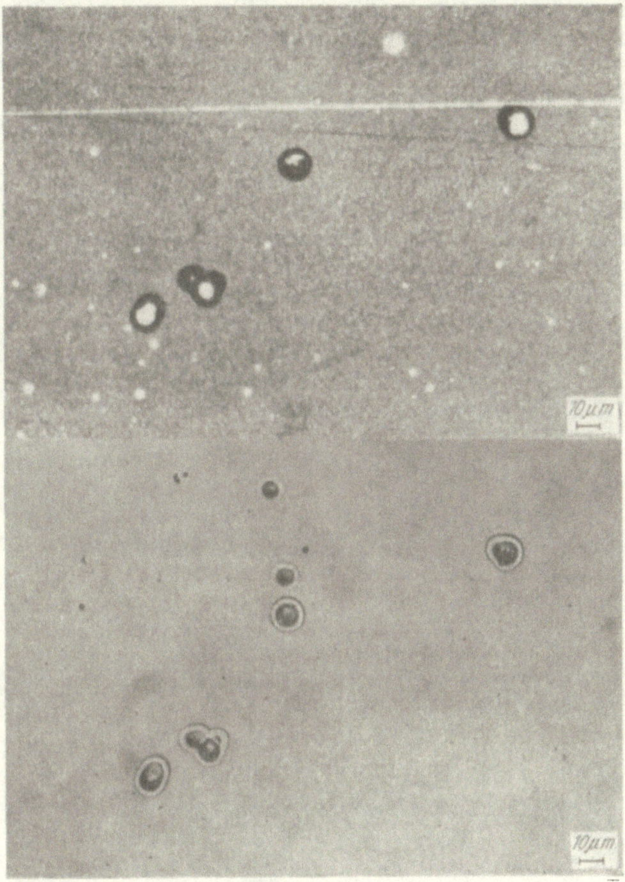

a)

b)

Vergleich von Silberdekoration und Chrom-Ätzer: Lokalisierung von p-Gebieten

Fig. 4a. Silberdekoration. Auflicht-Dunkelfeld

Fig. 4b. Lokalisierung mit dem Chrom-Ätzer, Ätzzeit 1 min, Auflicht-Hellfeld

der unterschiedlichen Dotierungskonzentration ein relativ großer Unterschied der Lösungsrate zu beobachten. Zusätzlich erfolgt die Bildung von Ätzkalotten an Stellen, die von der Silberdekoration nicht lokalisiert werden. Bei diesen Ätzkalotten ist die Ätzrate geringer.

Im allgemeinen liegt bei einer vergleichsweisen zahlenmäßigen Erfassung der Diffusantendurchbrüche der Wert nach der Lokalisierung mit dem Chrom-Ätzer bis zu einer Größenordnung höher als bei der Silber- und Kupferdekoration.

5. Diskussion

Bei einem Vergleich der drei Verfahren zum Nachweis von Diffusantendurchbrüchen erweisen sich die stromlosen Metallabscheidungsmethoden als weniger brauchbar. Bei der geringen Ausdehnung und unterschiedlichen Form der einzelnen Diffusantendurchbrüche und infolge zusätzlicher Metallabscheidung an wachstumsprädestinierten Stellen des Siliziumsubstrates ist ein exaktes Erkennen der Verhältnisse oft sehr schwierig. Dies wirkt sich besonders nachteilig bei der zahlenmäßigen Erfassung der Diffusantendurchbrüche im Zusammenhang mit Ausbeuteuntersuchungen aus.

Die Silber- und Kupferdekoration zeigen im Gegensatz zum Chrom-Ätzverfahren nur Diffusantendurchbrüche an, die einen Leitungstypumschlag mit einem z. Z. noch unbekannten Kompensationsgrad erzeugen. Der Chrom-Ätzer ermöglicht eine quantitative Erfassung der Diffusantendurchbrüche und — bei einiger Erfahrung — bis zu einem gewissen Grade auch die qualitative Beurteilung der Konzentrationsverhältnisse. Dagegen ist die Bestimmung der Eindringtiefe mit dem Chrom-Ätzer im Gegensatz zu den stromlosen Metallabscheidungsmethoden nicht möglich. Schon bei geringen Ätzzeiten ist ein gewisser Silizium-Abtrag zu beobachten, der die wahre Eindringtiefe verfälscht.

Den Mitarbeitern der Arbeitsstelle für Molekularelektronik Dresden bin ich für die Bereitstellung des Untersuchungsmaterials zu Dank verpflichtet.

Literatur

Bogenschütz, A. F.: Ätzpraxis für Halbleiter, München 1967
Chu, T. L., Gavaler, J. R.: J. electrochem. Soc. **110**, 54 C (1963)
Fricke, P.: priv. Mitteilung
Illes, P. A., Coppen, P. J.: J. appl. Physics **29**, 1514 (1958)
Sirtl, E., Adler, A.: Z. Metallkunde **53**, 529 (1961)
Turner, D. R.: J. electrochem. Soc. **106**, 701 (1959)
Whoriskey, P. J.: J. appl. Physics **29**, 867 (1958)

(Eingegangen am 20. Mai 1971)

Anschrift des Verfassers:
Dipl.-Min. M. Frenzel
Arbeitsstelle für Molekularelektronik
808 Dresden
Königsbrücker Landstr. 159

Kristall und Technik | 7 | 1–3 | 1972 | 271–283

W. Muschner, Chr. Männel

Kombinat VEB Halbleiterwerk Frankfurt/Oder
Betrieb Spurenmetalle Freiberg (WIB), Muldenhütten

Ein Beitrag zur lichtoptischen Orientierungsbestimmung von Galliumphosphid-Einkristallen

Es wird über die lichtoptische Orientierungsbestimmung von niedrigindizierten GaP-Flächen berichtet. Nach einer kurzen Erläuterung des apparativen Aufbaues wird auf die Präparation der Kristallflächen zur Erzeugung der Lichtfiguren eingegangen. Die benutzte Methode gestattet die Unterscheidung der polaren {111}-Flächen und die Orientierungsbestimmung von *A* (111)- und erstmals auch von *B* (111)-, {100}-, {110}-, {112}- und {113}-GaP-Flächen mittels Lichtfiguren. Die erhaltenen Lichtfiguren werden anhand der kristallographischen Symmetrie diskutiert und Lichtreflex-Polfiguren in Abhängigkeit von der Probenpräparation angegeben.

Von {111}-, {100}- und {110}-Probenflächen kann die Fehlorientierung mit einer Genauigkeit von $\leq \pm 20'$ bestimmt werden, bei {112}- und {113}-Probenflächen beträgt die Genauigkeit etwa ± 1 bis $\pm 2°$.

This paper is concerned with light-optical determination of the orientation of low-indexed GaP faces. After a short description of the apparatus preparation of crystal faces for generating light figures is discussed. The method applied allows the polar faces {111} to be distinguished and the orientations of the GaP faces *A* (111) and, for the first time, also *B* (111), {100}, {110}, {112}, and {113} to be determined by means of light figures. The light figures obtained are discussed with regard to crystallographic symmetry, and light-reflection pole figures are given as a function of specimen preparation.

Disorientation of the specimen faces {111}, {100} and {110} can be determined with an accuracy of $\leq \pm 20'$. For the faces {112} and {113} the accuracy is about $\pm 1°$ to $\pm 2°$.

1. Einleitung

Für die Herstellung von Bauelementen aus der halbleitenden Verbindung Galliumphosphid werden Scheiben bestimmter kristallographischer Orientierung benötigt. Dazu ist es notwendig, kompakte GaP-Einkristalle orientiert zu schneiden. Desweiteren besteht mitunter die Forderung nach möglichst genauer Bestimmung der Fehlorientierung, d. h. des Desorientierungswinkels einzelner GaP-Scheiben.

Zur Orientierungsbestimmung können grundsätzlich röntgenographische und lichtoptische Verfahren angewendet werden. Zur Ermittlung vor allen Dingen unbekannter kristallographischer Orientierungen von Einkristallproben dient die klassische Laue-Rückstrahl-Methode und für Kontroll- bzw. Eichzwecke und zur genauen Fehlorientierungsbestimmung im Bereich $<5'$ erweist sich das Röntgen-Zählrohrgoniometer als nützlich. Für Routineuntersuchungen der Orientierung erscheint die Verwendung der lichtoptischen Methode durch Erzeugung von Lichtfiguren als geeignet, zumal sie keinen großen Geräteaufwand benötigt und relativ leicht — auch im betriebsmäßigen Einsatz — durchführbar ist.

2. Vorbemerkungen

Das Prinzip der lichtoptischen Orientierung von Kristallflächen besteht darin, daß von speziell präparierten Oberflächen durch Reflexion eines Lichtbündels unter definiertem Winkel bezogen auf diese Flächen Licht- bzw. Reflexfiguren erzeugt werden. Aus deren Lage — im Hinblick auf die optische Achse der Orientierungsvorrichtung — und symmetrischer Form kann die kristallographische Orientierung bestimmt werden. Bei fehlorientierten Kristallproben erscheint die Lichtfigur vom Mittelpunkt des Reflexschirmes verschoben, durch Zentrierung derselben auf dem Schirm wird der sog. Fehl- bzw. Desorientierungswinkel erhalten. Er gibt die Abweichung der Probenoberfläche von einer kristallographischen Vorzugsorientierung an.

In der Literatur nehmen die Arbeiten über die lichtoptische Orientierungsbestimmung von Einkristallen der $A^{III}B^{V}$- und z. B. der $A^{II}B^{VI}$-Verbindungen relativ breiten Raum ein. Speziell auf GaP angewandt, sind jedoch nur sehr wenige Publikationen erschienen, die sich — ohne Angabe der erhaltenen Lichtfiguren — mit den Lichtreflex-Mustern von $\{111\}$-Flächen beschäftigen. MARIANO und WOLFF erhielten von chemisch geätzten $\{111\}$-GaP-Flächen nur von den $A(111)$ eine Lichtfigur, deren Strahlen in die $\langle\bar{1}\bar{1}2\rangle$-Richtungen zeigten; die geätzten $B(111)$ ergaben auf Grund der Polarität im vorliegenden Gittertyp ($F\bar{4}3m$) runde Ätzgruben, die keine Lichtfigur erzeugten. Zu einem analogen Ergebnis kamen VASILEV und VJATKIN bei der Untersuchung von $\{111\}$-GaAs-Oberflächen, wobei die Orientierungsgenauigkeit der $A(111)$ etwa 1° betrug. WOLFF, FRAWLEY und HIETANEN geben einige allgemeine Lichtreflex-Muster für $\{111\}$-GaP-Flächen nach dem Ätzen in verschiedenen wäßrigen Lösungen von Halogenwasserstoffen an, die allerdings nicht zur Orientierungsbebestimmung benutzt werden. WOLFF, BRODER und WOLFF bearbeiteten die Probenoberflächen von $A^{III}B^{V}$-Verbindungen nach der Methode des „rollenden Korns" mit SiC (150 bzw. 250 Maschen) und erzeugten an den erhaltenen Mikrospaltflächen Lichtfiguren. Dabei wurde die ausgezeichnete Spaltbarkeit von $A^{III}B^{V}$-Verbindungen nach (110) ausgenutzt. BURDUKOV und SEDOV benutzten zur Orientierungsbestimmung von $\{111\}$-, $\{100\}$- und $\{110\}$-Probenflächen von $A^{III}B^{V}$-Verbindungen (z. B. InAs) mit SiC 50 (0,5 mm Korndurchmesser!) grob geläppte Flächen; hierbei betrug die Meßgenauigkeit $\pm 1°$. AKASAKI und KOBAYASI erzeugten Lichtfiguren durch Reflexion an geläppten und geätzten $A(111)$- und $B(111)$-GaAs-Oberflächen. Sie stellten fest, daß sich durch den Ätzangriff die Orientierung der dreizähligen Lichtfigur der $A(111)$ ändert, während bei der $B(111)$ keine Richtungsänderung gefunden wurde.

Die vorliegende Arbeit verfolgt das Ziel, mittels der Lichtfiguren-Methode natürlich gewachsene niedrigindizierte Flächen von GaP-Kristallen, hergestellt nach dem Verfahren der chemischen Transportreaktion (JACOBI, MUSCHNER), zu identifizieren, kompakte Kristalle orientiert zu schneiden und von Kristallscheiben die Fehlorientierung zu bestimmen.

3. Experimentelles

3.1. Apparativer Aufbau

Figur 1 zeigt den Aufbau der Orientierungseinrichtung zur Ermittlung der Fehlorientierung einer Kristallscheibe. Durch eine Blende wird ein nahezu punktförmiges Lichtbündel einer Hg-Höchstdrucklampe HBO 50 ausgeblendet und mit einem Kondensor gesammelt. Ein fehlerkorrigiertes Objektiv 180/3,5 dient dazu, die durch das Lichtbündel auf der am Theodoliten (Theo 6.1 vom VEB Freiberger Präzisionsmechanik) befestigten präparierten Einkristallscheibe erzeugte Lichtfigur auf einem Reflexschirm abzubilden. Die einzelnen Teile der Einrichtung sind auf einer optischen Band angeordnet, die eine einwandfreie Justierung der optischen Achse gestattet.

Für die Identifizierung der natürlich gewachsenen Flächen eines GaP-Kristalls dient anstelle des Theodolits ein Kugelgelenk, an welchem der Kristall mit Plastelina gehaltert wird.

Zum orientierten Schneiden von kompaktem Kristallmaterial wird die gesamte Kristallhalterung des Schneidautomaten mit dem aufgeklebten Kristall auf die optische Bank anstelle des Theodolits gesetzt.

Fig. 1. Orientierungseinrichtung zur Ermittlung des Fehlorientierungswinkels einkristalliner scheibenförmiger Proben

3.2. Probenpräparation

Zur Erzeugung von für die Orientierungsbestimmung geeigneter Lichtfiguren ist eine spezielle Oberflächenbehandlung notwendig, deren Ziel die Ausbildung von Mikrosspaltflächen ist. Die in einer früheren Arbeit (Muschner) erwähnte Vorbehandlung der $\{111\}$-GaP-Oberflächen durch Läppen mit SiC F40 und anschließendem Ätzen mit alkalischer Kaliumhexacyanoferrat(III)-Lösung gestattet nur die Orientierung der $A(111)$-Flächen, da die Ätzlösung auf den $B(111)$-Flächen polierend wirkt und somit keine Lichtfigur erzeugt wird.

Die Untersuchungen zur Probenvorbehandlung konzentrierten sich auf

a) Mechanische Bearbeitung der Oberflächen mit Läppulver SiC F40 bzw. nach der Methode des „rollenden Korns“ (Korndurchmesser $\geq 80\,\mu m$) mit dem Ziel, durch Ausspalten Mikrospaltflächen freizulegen. Um GaP-Scheiben mit Läppulver der Körnung $80 \cdots 100\,\mu m$ bearbeiten zu können, müssen diese etwa 1 mm dick sein. Es ist empfehlenswert, bei GaP nicht mit Korndurchmessern über 100 μm zu arbeiten, da die Gefahr des Kristallbruchs zunimmt und keine Verbesserung des Ausspaltens — im Hinblick auf die Güte der erzeugten Lichtfigur — erreicht wird;

b) Läppen der Oberflächen mit Läppulver SiC F40 und anschließendes Ätzen mit der von Richards und Crocker sowie Saul angegebenen Lösung, bestehend aus 6 ml HNO_3 + 4 ml HF + 10 ml $AgNO_3$ + 8 ml H_2O, die die polaren {111}-Flächen anätzt. Dies konnte auch durch eigene Untersuchungen bestätigt werden. Die Ätzdauer betrug 8···10 min in siedender Lösung.

Die präparierte Probenfläche muß mindestens so groß sein, daß auf ihr ein Lichtstrahl von etwa 1,5 mm Durchmesser reflektiert werden kann.

Die Oberflächenbehandlung zur Identifizierung von natürlich gewachsenen Kristallflächen, zum orientierten Schneiden und zur Bestimmung der Fehlorientierung von scheibenförmigen Proben ist gleichartig.

3.3. Versuchsdurchführung

Die Identifizierung der natürlich gewachsenen Flächen eines GaP-Kristalls erfolgt dadurch, daß die entsprechend Abschnitt 3.2 präparierten Flächen nacheinander in den Strahlengang der Orientierungseinrichtung gebracht werden. Anhand der auf dem Schirm sichtbar gemachten Lichtfiguren wird auf die vorliegenden kristallographischen Flächen geschlossen.

Für das orientierte Schneiden wird die präparierte Fläche des am Kristallhalter aufgeklebten Kristalls in den Strahlengang gebracht und die erzeugte Lichtfigur auf dem Reflexschirm durch Bewegung des Kristallhalters zentriert. Nach dem Umsetzen des Halters auf den Schneidautomaten erfolgt — ohne nochmalige Winkeleinstellung — das Schneiden parallel zu der orientierten Probenoberfläche.

Zur Bestimmung des Desorientierungswinkels einkristalliner Scheiben ist es notwendig, die gesamte Anordnung zu eichen. Dies geschieht mit Hilfe eines planparallelen Spiegels, der auf die PVC-Platte des Theodolits geklebt und in Reflexstellung gebracht wird. Der zu dieser Theodolitenstellung gehörige Winkelwert W_0 wird abgelesen. Danach wird die zu messende, nach Abschnitt 3.2 präparierte Einkristallscheibe auf den Spiegel aufgeklebt und durch Horizontal- bzw. Vertikalverstellung des Theodolits der Mittelpunkt der Lichtfigur ins Zentrum des Reflexschirmes gebracht. Nach dieser Einstellung kann am Theodolit der Winkelwert W_K abgelesen werden. Die Orientierungsabweichung, d. h. der sog. Fehl- oder Desorientierungswinkel ΔW beträgt demnach

$$\Delta W = W_K - W_0 .$$

3.4. Meßgenauigkeit des Verfahrens

Zur Prüfung der Genauigkeit der Methode wurden u. a. an {111}-GaP-Scheiben mit unterschiedlichem Desorientierungswinkel je 10 voneinander unabhängige Messungen durchgeführt. Die Flächen waren durch Läppen mit SiC F40 und anschließendem Ätzen mit der angegebenen Lösung vorbehandelt worden. Die Prüfung ergab, daß der absolute Orientierungsfehler für $A(111)$- und $B(111)$-Flächen bei einer statistischen Sicherheit von 95% $\leqq \pm 20'$ beträgt und unabhängig vom Orientierungswinkel — zumindestens bis 10° — ist. Das Ergebnis wurde unter Berücksichtigung der Studentschen t-Verteilung ermittelt. Analoge Ergebnisse sind für {100}- und {110}-Flächen zu erwarten, da die Symmetrie der Lichtfiguren und deren Schärfe die angegebene Meßgenauigkeit gestattet. Bei {112}- und {113}-Flächen ist das genaue Einstellen des Mittelpunktes der Lichtfigur auf das Zentrum des Schirmes wegen der komplizierten symmetrischen Form derselben und der teilweise diffusen Lichtreflexe schwieriger. Aus diesen Gründen kann eine Orientierungsgenauigkeit von nur ± 1 bis $\pm 2°$ angegeben werden.

Für die Einhaltung der Meßgenauigkeit ist die Planparallelität der Scheiben und die Güte der Präparation wichtig; desweiteren ist eine definierte Auflage der Scheibe auf dem Spiegel zu gewährleisten.

Beim Schneiden von ⟨111⟩-orientierten Scheiben beträgt die Orientierungsgenauigkeit gleichfalls $\leqq \pm 20'$, d. h. sie wird durch das Umsetzen des Kristallhalters von der Orientierungsvorrichtung auf den Schneidautomaten nicht beeinträchtigt.

4. Ergebnisse und Diskussion

Von {111}- und {110}-Flächen wurden sowohl nach der mechanischen Bearbeitung durch grobes Läppen (Korndurchmesser 80···100 µm) als auch nach kombiniertem Läppen mit SiC F40 und chemischen Ätzen zur Orientierungsbestimmung verwendbare Lichtfiguren erhalten. Von {100}-, {112}- und {113}-Flächen konnten nur nach der kombinierten Oberflächenbehandlung auswertbare Lichtfiguren erzeugt werden.

Durch die zusätzliche Ätzbehandlung der Mikrospaltflächen enthaltenen Probenoberflächen konnten einmal verschiedentlich die Lichtfiguren verbessert werden und zum anderen traten, durch den Ätzangriff der polaren {111}-Netzebenen bedingt, in ihrer äußeren Form veränderte Lichtfiguren auf.

Nachfolgend sollen die nach unterschiedlicher Oberflächenbehandlung erzeugten Lichtfiguren beschrieben und diskutiert werden.

4.1. {111}-Flächen

Infolge Fehlens des Symmetriezentrums im vorliegenden Gittertyp sind die ⟨111⟩-Richtungen polar. Daraus folgt u. a., daß sich Unterschiede in den Lichtfiguren — bezogen auf Richtung und Gestalt — und den damit in Zusammenhang stehenden Mikrospaltflächen und dem Ätzverhalten bemerkbar machen, die neben der Orientierung dieser Flächen gleichzeitig ihre Identifizierung gestatten.

Die zur Mikrospaltung beitragenden (011)-, (110)- und (101)-Flächen des positiven Tetraeders $A(111)$ und die $(\bar{1}0\bar{1})$, $(0\bar{1}\bar{1})$- und $(\bar{1}\bar{1}0)$-Flächen des negativen Tetraeders $B(111)$ konnten mit einem Zweikreis-Reflexionsgoniometer indiziert werden.

4.1.1. Grobgeläppte {111}-Flächen

Mit Läppulver SiC 10 (Korndurchmesser 100 µm) bearbeitete {111}-Flächen ergaben für $A(111)$ und $B(111)$ gleichartige, scharfausgebildete dreizählige Sternfiguren („Mercedesstern"), wie sie die Figuren 2a und 2b zeigen. Sie unterschieden sich nur in der Richtung ihrer Sternspitzen. Während auf $A(111)$-

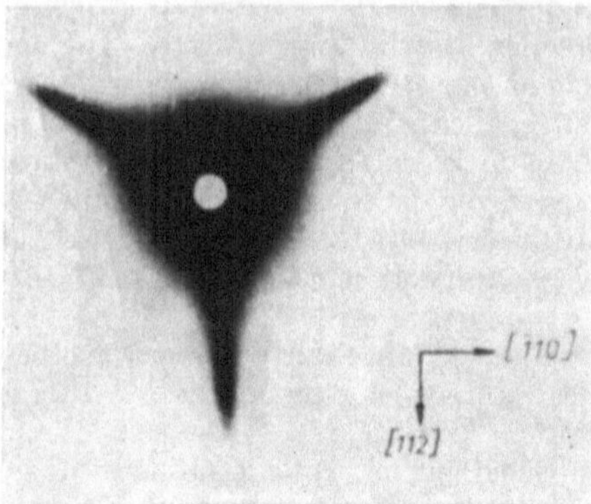

Fig. 2. a $A(111)$-Fläche

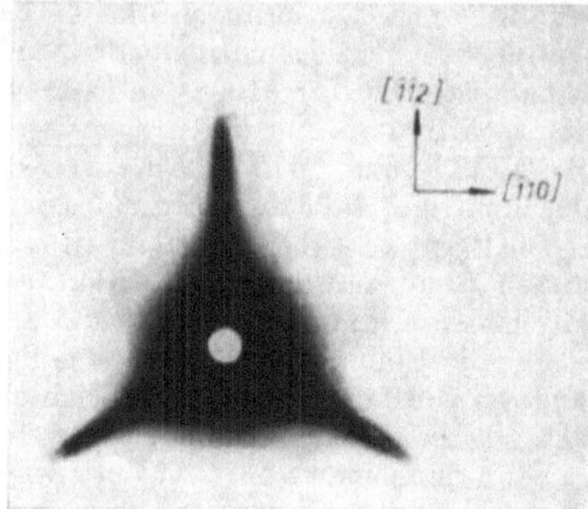

Fig. 2 b $B(111)$-Fläche

Lichtfiguren von mit SiC 10 geläppten polaren {111}-Probenflächen

Flächen die Strahlen in die 3 $\langle 11\bar{2}\rangle$-Richtungen zeigten, wiesen sie auf $B(111)$-Flächen in die 3 $\langle \bar{1}\bar{1}2\rangle$-Richtungen. Die Lichtfiguren eignen sich ausgezeichnet zur Orientierungsbestimmung, allerdings gestatten sie keine Identifizierung der polaren {111}-Flächen.

Bei der mechanischen Bearbeitung mit Läppulver SiC F40 wurden ebenfalls Mikrospaltflächen freigelegt, die eine durchaus noch auswertbare Lichtfigur erzeugten, obwohl dieselbe an Helligkeit und Konturenschärfe gegenüber der mit SiC 10 erzeugten abgenommen hat. Das Resultat verschlechterte sich wesentlich nach Behandlung mit Läppulver SiC F20 bzw. SiC F10.

4.1.2. Geätzte {111}-Flächen

Die mit Läppulver SiC F40 und der in Abschnitt 3.2 angegebenen Ätzlösung behandelten A- und B-Flächen ergaben unterschiedliche Lichtfiguren, die sich zur Orientierungsbestimmung und zur Identifizierung der polaren Probenoberflächen eigneten.

Für die Ausbildung der Lichtfiguren waren die Ätzdauer und -temperatur nicht kritisch. Wurde allerdings die angegebene Ätzdauer stark überschritten, waren die erhaltenen Reflexe sehr diffus.

4.1.2.1. $A(111)$-Flächen

Die in die 3$\langle 11\bar{2}\rangle$-Richtungen weisenden Strahlen der Lichtfigur waren relativ breit und an den Enden aufgespalten (Fig. 3a). Ihre Form ist auf das Anätzen von positiven Deltoiddodekaederflächen $\{hhl\}$ zurückzuführen. Dieser Figur überlagert traten drei mehr oder weniger gut ausgebildete, zur Grundfigur um 180° gedrehte, in $\langle \bar{1}\bar{1}2\rangle$ weisende Strahlspitzen auf (Fig. 3a), die durch die Ätzbehandlung beeinflußbar waren. Für die Entstehung dieser Zusatzreflexe wird angenommen, daß die in den $\langle \bar{1}\bar{1}2\rangle$-Richtungen liegenden positiven Tristetraederflächen $\{hll\}$ angeätzt wurden, wie auch orientierte Ätzversuche mit $\{hll\}$-Probenflächen gezeigt haben. Eine durch den Ätzangriff bedingte grundsätzliche Richtungsumkehrung der Reflexfigur, wie sie von Akasaki und Kobayasi an GaAs festgestellt worden ist, konnte bei GaP nicht beobachtet werden.

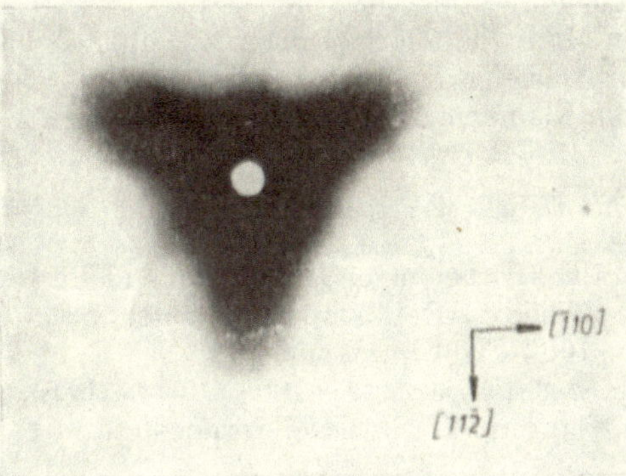

Fig. 3a. Lichtfigur einer geätzten A(111)-Probenfläche

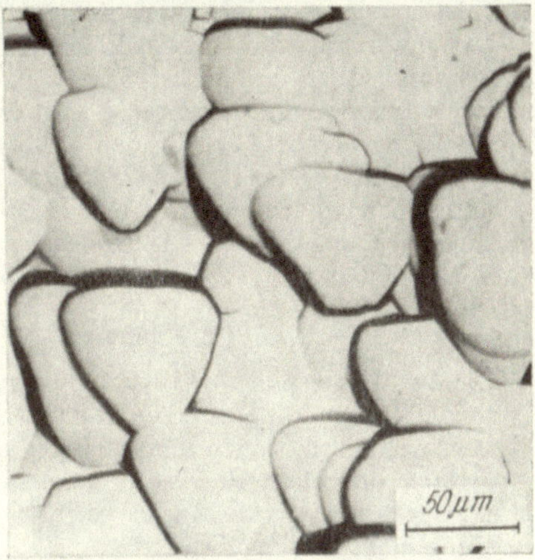

Fig. 3b. Angeätzte Mikrospaltflächen auf A(111)

Die zur Erzeugung der Lichtfiguren benutzten geätzten Mikrospaltflächen zeigten ausnahmslos gut ausgebildete abgerundete Dreiecke (Fig. 3b).

4.1.2.2. B(111)-Flächen

Die Lichtfigur der B-Fläche zeigte eine gut ausgebildete dreizählige Symmetrie mit ⟨112⟩-Strahlrichtungen (Fig. 4a). Die Reflexe waren schmaler als die der A-Fläche und sind durch das Anätzen von negativen Deltoiddodekaederflächen $\{\bar{h}\bar{h}\bar{l}\}$ bedingt. Der kleine Lichthof um die B(111)-Polfläche (auf der Figur schwarz) war größtenteils in Richtung der Strahlen verzerrt, was wahrscheinlich auf einen verstärkten Ätzangriff der negativen Deltoiddodekaederflächen in der Nähe des B(111)-Poles zurückzuführen ist.

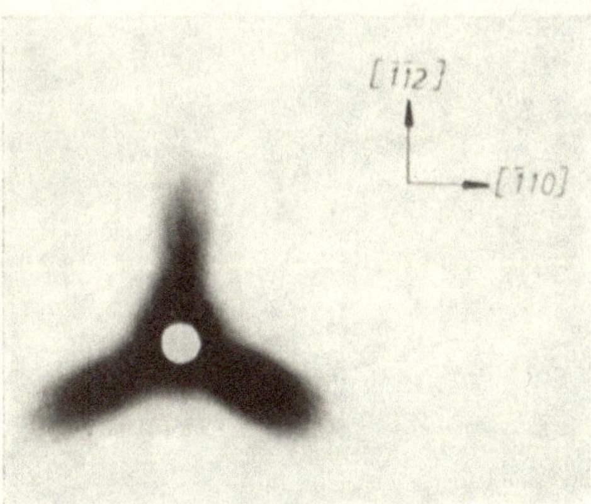

Fig. 4a. Lichtfigur einer geätzten B(111-Probenfläche

Fig. 4b. Angeätzte Mikrospaltflächen auf B(111)

In den uns zur Verfügung stehenden Veröffentlichungen ist bisher noch keine Lichtfigur für die $B(111)$-Flächen beschrieben worden. Die entsprechend Abschnitt 3.2b präparierten $B(111)$-Flächen erzeugten Lichtfiguren, welche zur Orientierungsbestimmung gut geeignet waren und die eindeutige Unterscheidung der $B(111)$- von den $A(111)$-Flächen gestatteten.

Wie Figur 4b zeigt, wiesen die geätzten Mikrospaltflächen der $B(111)$ gleichseitige bzw. gleichschenkelige Dreiecke mit scharfen Kanten entlang den $\langle 110 \rangle$-Gitterrichtungen auf. Die Dreiecke waren kleiner ausgebildet als die auf den $A(111)$-Flächen.

4.2. $\{100\}$-Flächen

Die zur Mikrospaltung auf $\{100\}$-Probenflächen beitragenden Rhombendodekaederflächen konnten ebenfalls mit dem Zweikreis-Reflexionsgoniometer indiziert werden; sie lagen um 45° gegen den $\{100\}$-Flächenpol geneigt.

Nach der Methode des „rollenden Korns" präparierte $\{100\}$-Oberflächen erzeugten nur unvollständige diffuse Lichtfiguren, die zur Orientierungsbestimmung nicht geeignet waren.

Von nach dem Läppen geätzten Probenoberflächen wurden dagegen Lichtfiguren erzeugt, deren Symmetrie zur Orientierung des $\{100\}$-Flächenpoles benutzt werden konnten (Fig. 5a). Anhand der Ausbildung der Reflexe in den $\langle 100 \rangle$-Richtungen wird auf ein Anätzen der Tetrakishexaederflächen $\{hk0\}$ geschlossen. Zusätzlich wurde ein merklicher Ätzangriff der negativen und positiven Tristetraederflächen $\{hll\}$ beobachtet, wobei bei letzteren der Reflex strahlenförmiger ausgebildet war und auf eine der $A(111)$-Ebenen hindeutet (auf Fig. 5a kaum erkennbar). Die Lichtfigur der (001) und $(00\bar{1})$ unterschieden sich nur in ihrer Lage zueinander, indem beide durch Drehung um 90° in der Ebene zur Deckung gebracht werden konnten.

Die die Lichtfiguren erzeugenden angeätzten Mikrospaltflächen waren dachförmig angeordnet, wobei auf einer (001)-Probenfläche die Kanten parallel zur [110]-Richtung lagen (Fig. 5b).

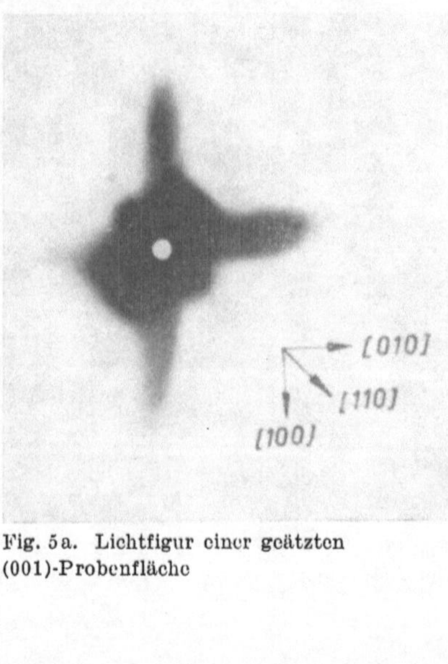

Fig. 5a. Lichtfigur einer geätzten (001)-Probenfläche

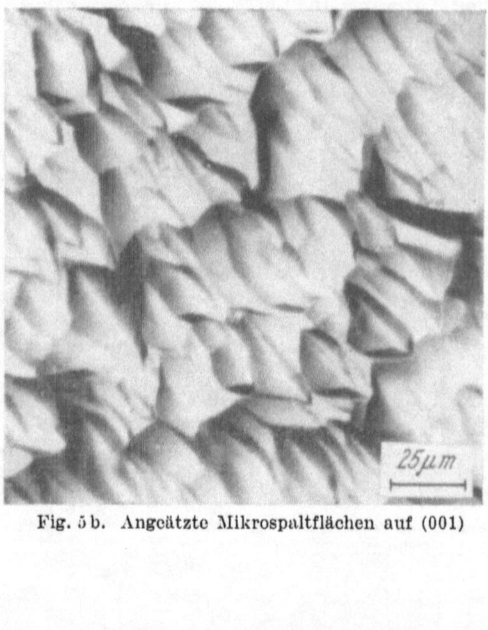

Fig. 5b. Angeätzte Mikrospaltflächen auf (001)

4.3. {110}-Flächen

Die Mikrospaltflächen auf einer {110}-Probenoberfläche lagen gegen den Flächenpol um 60° geneigt und ließen sich mit dem Goniometer gut bestimmen.

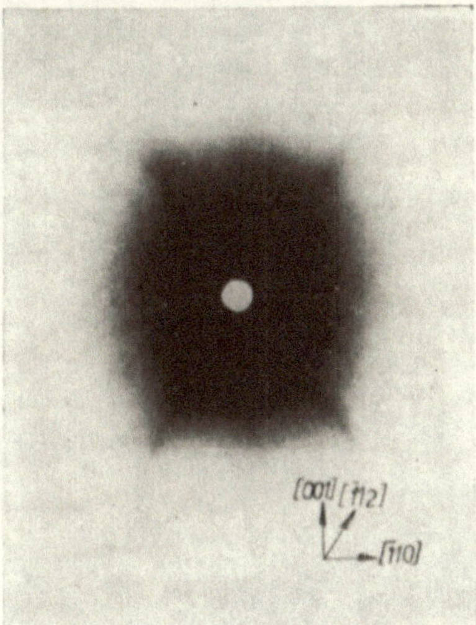

Fig. 6. Lichtfigur einer mit SiC 10 geläppten (110)-Probenfläche

4.3.1. Grobgeläppte {110}-Flächen

Mit Läppulver SiC 10 bearbeitete Probenflächen ergaben Lichtfiguren, die neben einem relativ großen Lichthof (auf der Figur 6 schwarz dargestellt) vier schmale lange Reflexe in ⟨112⟩ enthielten (Fig. 6). Ein verschiedentlich schwach ausgebildeter Reflex in ⟨001⟩ kann vermutlich mit einem Ausspalten von Deltoiddodekaederflächen in Verbindung gebracht werden.

4.3.2. Geätzte {110}-Flächen

Eine merklich andere Form der Lichtfigur erzeugten nach dem Läppen geätzte {110}-Oberflächen, da die klaren Symmetrieverhältnisse der Lichtfigur auf Figur 6 nicht ohne weiteres zu erkennen waren (Fig. 7a).

Faust jr. und Sagar sowie Gatos und Lavine zeigten an anderen $A^{III}B^{V}$-Verbindungen, daß u. a. auf {110}-Flächen durch spezifische Ätzmittel {111}-Gitterebenen angeätzt werden. Da mit der in Abschnitt 3.2b angegebenen Ätzlösung sowohl A(111)- als auch B(111)-Flächen angeätzt wurden, lassen sich die von {110}-Probenflächen erzeugten Lichtfiguren dahingehend interpretieren, daß der Ätzangriff bevorzugt an der positiven und negativen Tetraederfläche erfolgte. Diese Tetraederflächen liegen dem {110}-Flächenpol am nächsten. Wie Figur 7a zeigt, ist der Ätzangriff der A(111) wesentlich stärker als der der B(111). Desweiteren ist anzunehmen, daß ein zusätzliches Anätzen von Deltoiddodekaederflächen erfolgte.

Die angeätzten Mikrospaltflächen auf {110} zeigten ein fischschuppenähnliches Aussehen (Fig. 7b).

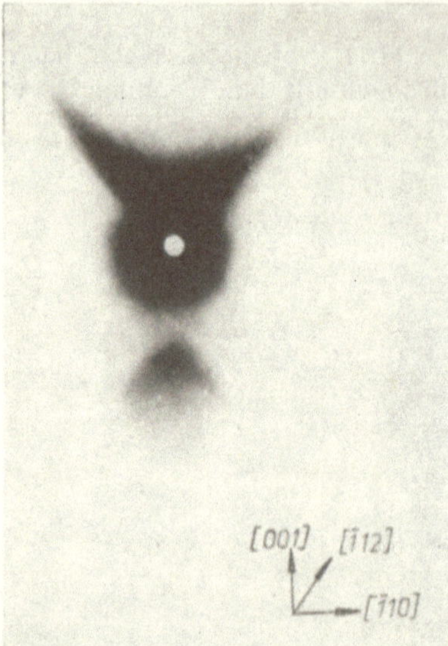

Fig. 7a. Lichtfigur einer geätzten (110)-Probenfläche

Fig. 7b. Angeätzte Mikrospaltflächen auf (110)

4.4. {112}-Flächen

Eine exakte Zuordnung von Rhombendodekaederflächen, die zur Mikrospaltung bei {112}-Probenflächen beitragen, kann nicht gegeben werden, da sich die Flächen goniometrisch nicht vermessen ließen.

Die von geläppten {112}-Flächen erzeugte Lichtfigur stellte einen breiten diffusen Reflex parallel zur ⟨110⟩-Richtung dar, der zur optischen Orientierungsbestimmung nicht geeignet war.

Auswertbare Lichtfiguren erzeugten dagegen nach dem Läppen geätzte Probenoberflächen (Fig. 8). Hierbei wurden besonders stark die Flächenlagen zwischen der (011) und (101) und die *A*(111) angeätzt.

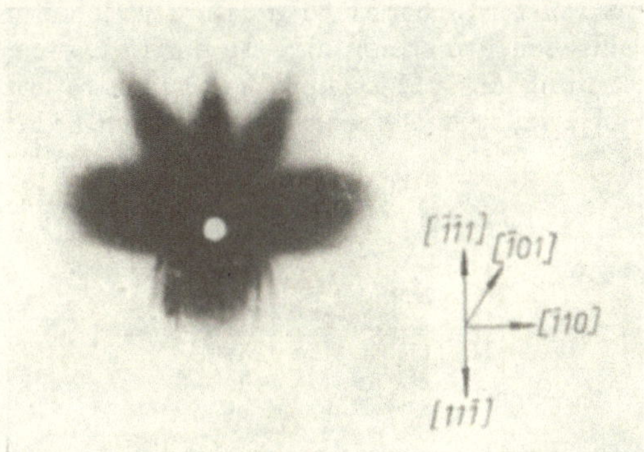

Fig. 8. Lichtfigur einer geätzten (112)-Probenfläche

4.5. { 113 }-Flächen

Aus der Winkeldifferenz der { 112 }- und { 113 }-Flächenlagen von 10° 1′ resultiert, daß sich die Symmetrieverhältnisse nicht wesentlich voneinander unterscheiden, und demzufolge die von beiden Probenflächen erzeugten Lichtfiguren Ähnlichkeiten aufweisen. Bei der { 113 }-Lichtfigur, die von nach dem Läppen geätzten Probenoberflächen erhalten wurde, ist der *A*(111)-Reflex wesentlich länger als bei der { 112 }-Lichtfigur und die Reflexe der angeätzten Flächenlagen zwischen (113) und (011) sowie (113) und (101) sind ca. 10° geneigt (Fig. 9). Die { 113 }-Lichtfigur eignet sich relativ gut zur Orientierungsbestimmung.

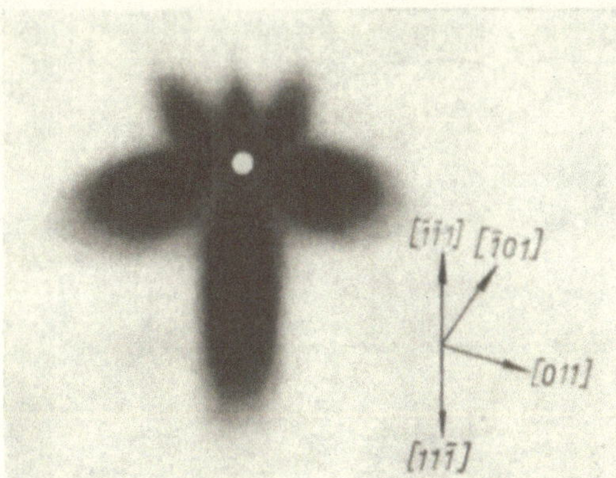

Fig. 9. Lichtfigur einer geätzten (113)-Probenfläche

4.6. Darstellung von Lichtreflex-Polfiguren

Lichtreflex-Polfiguren stellen Stereogramme dar, in denen die Strahlen der von niedrigindizierten, speziell präparierten Kristallflächen erzeugten Reflex- bzw. Lichtfiguren untereinander verbunden sind. Grundlage dafür bildeten die stereographischen Projektionen der A(111)-, B(111)-, (001)- und (110)-Flächen in Verbindung mit den jeweiligen Lichtfiguren. Zur Darstellung der Lichtreflex-Projektionen ist die [001]-Polfigur gewählt worden. Figur 10 zeigt die Lichtreflex-Polfigur für geläppte GaP-Einkristallflächen. Die ausgezogenen Linien auf der Figur stellen die Spuren der Lichtreflexe und die gestrichelten die Grundsymmetrie der stereographischen Projektion dar. In der Polfigur ist der Übersicht wegen auf die Darstellung der {112}- und {113}-Lichtreflexe verzichtet worden; sie liegen auf den jeweiligen Zonenkreisen zwischen {111} und (001).

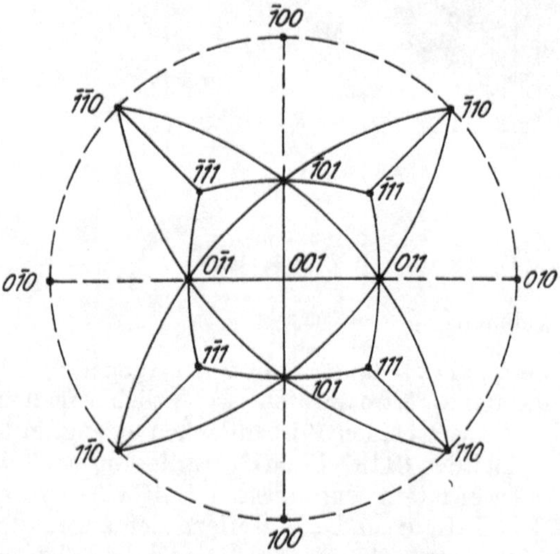

Fig. 10. Lichtreflex-Polfigur bezogen auf [001] für geläppte {111}-, {100}- und {110}-Probenflächen

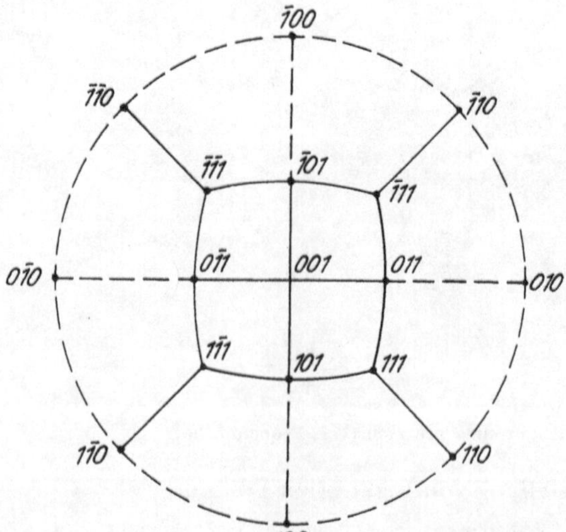

Fig. 11. Lichtreflex-Polfigur bezogen auf [001] für nach dem Läppen geätzte {111}-, {100}- und {110}-Probenflächen

Für nach dem Läppen geätzte {111}-, {100}- und {110}-Probenflächen ergibt sich die in Figur 11 wiedergegebene Lichtreflex-Polfigur. Im Vergleich zu Figur 10 ist erkennbar, daß bei Lichtfiguren, die von chemisch geätzten Flächen erzeugt wurden, der Einfluß der Anätzbarkeit der {111}-Gitterebenen deutlich hervortritt.

Aufgrund der Symmetrieähnlichkeit zwischen kristallographischer Projektion, bezogen auf einen der Orientierung zugrundeliegenden Flächenpol, und der Lichtfigur, lassen sich Winkelschablonen auf dem Reflexschirm der Orientierungsvorrichtung anbringen, die der Ermittlung des Fehlorientierungswinkels bzw. der Identifizierung der Kristallfläche dienen.

Die angegebenen Lichtreflex-Polfiguren erleichtern die Übersicht bei einer Orientierungsbestimmung von relativ stark fehlorientierten Kristallflächen, da durch Verschwenken der Probe die Form der Lichtfigur verändert werden kann.

Literatur

Akasaki, J., Kobayasi, H.: J. electrochem. Soc. **112**, 757 (1965)

Burdukov, Ju. M., Sedov, V. E.: Kristallografija **13**, 556 (1968)

Faust, jr., J. W., Sagar, A.: J. appl. Physics **31**, 331 (1960)

Gatos, H. C., Lavine, M. C.: J. electrochem. Soc. **107**, 433 (1960)

Geist, D., Preuss, E.: Z. angew. Physik **9**, 526 (1957)

Jacobi, G., Muschner, W.: Wiss. Z. Karl-Marx-Univ. Leipzig, Math.-Naturwiss. R. **20**, 63 (1971)

Mariano, A. N., Wolff, G. A.: Z. Kristallographie **126**, 244 (1968)

Muschner, W.: Kristall und Technik **5**, 577 (1970)

Richards, J. U., Crocker, A. J.: J. appl. Physics **31**, 611 (1960)

Saul, R. H.: J. electrochem. Soc. **115**, 1184 (1968)

Vasilev, A. P., Vjatkin, A. P.: Kristallografija **8**, 248 (1963)

Wolff, G. A.: Compound semiconductors **1**, 34 (1962)

Wolff, G. A., Broder, J. D.: Acta Cryst. **12**, 313 (1959)

Wolff, G. A., Frawley, I. J., Hietanen, J. R.: J. electrochem. Soc. **111**, 22 (1964)

(Eingegangen am 19. Juni 1971)

Anschrift der Verfasser:

Dipl. Min. Wolfgang Muschner
Ing. Christa Männel
Kombinat VEB Halbleiterwerk Frankfurt/Oder
Betrieb Spurenmetalle Freiberg (WIB)
9206 Hilbersdorf

Kristall und Technik	7	1–3	1972	285–302

P. Fricke

Arbeitsstelle für Molekularelektronik, Dresden
(Leiter: Prof. Dr.-Ing. habil. W. Hartmann)

Bestimmung der Scheibenorientierung und Schichtdicke an epitaktischen Siliziumschichten auf Silizium

Durch fehlerhaftes epitaktisches Wachstum entstehen in der epitaktischen Schicht Stapelfehler. Beim Wachstum an gestörten Substratstellen durchsetzen die Stapelfehler die gesamte Epitaxieschicht. Form und Größe der Stapelfehler sind abhängig von Schichtdicke und Scheibenorientierung. Für verschiedene Scheibenorientierungen (*hk*0), (*hll*) und (*hhl*) für *h*, *k*, *l* von 1 bis 10 bezogen auf die Ebenen (100), (110) und (111) werden die Stapelfehlerwinkel und die Seitenlängen der Stapelfehler für eine normierte Schichtdicke ermittelt und in Diagrammen aufgetragen. An Hand einiger Beispiele wird dieses Verfahren geprüft und mit herkömmlichen Methoden zur Bestimmung der Epitaxieschichtdicke und Scheibenorientierung verglichen.

Stacking faults in the epitaxial layer are formed by defective growth. While growing in defective points of the substrates the stacking faults are penetrating the epitaxial layer. Form and size of stacking faults are dependent on the thickness of epitaxial silicon films and on slice orientation. For different slice orientation (*hk*0), (*hll*) and (*hhl*) with *h*, *k*, *l* from 1 to 10 related to the planes (100), (110) and (111) the angles and the side lengths of stacking faults for a normalized layer thickness are calculated and registered in diagrams. The method is tested and compared with usual methods of determination of the epitaxial layer thickness and slice orientation.

1. Einleitung

Durch epitaktisches Wachstum von Siliziumschichten auf gestörten Siliziumsubstraten bilden sich beim Wachstum an gestörten Substratstellen oder durch lokale Differenzen in den Abscheidungsbedingungen Stapelfehler, deren Form und geometrische Abmessungen von der Scheibenorientierung und der Schichtdicke abhängen. Die an gestörten Substratstellen entstehenden Stapelfehler werden zur Ermittlung der Dicke der abgeschiedenen Schicht verwendet. Dabei ist die Kenntnis der Substratorientierung notwendig. Für einige niedrig indizierte Wachstumsebenen wurden die Faktoren zur Ermittlung der Schichtdicke berechnet (Chu und Gavaler, Dash, Mendelson).

Im folgenden wird eine Berechnungs- und Bestimmungsmethode beschrieben, die unter Verwendung der Seitenlängen und der Winkel der auf der Schichtoberfläche sichtbaren Stapelfehlergebilde die Bestimmung der Schichtdicke und der Scheibenorientierung für eine große Zahl von Orientierungen ermöglicht.

Für die Bestimmung geringer Orientierungsabweichungen von der (111)-Ebene aus der Seitenlänge der Stapelfehler wurde bereits von Diment ein Nomogramm angegeben.

An 16 Scheiben verschiedener Orientierung wurde nach Abscheidung der epitaktischen Schicht die Stapelfehlergeometrie (Stapelfehlerwinkel und Seitenlängen) gemessen und ihre Scheibenorientierung und Epitaxieschichtdicke bestimmt. Diese Werte werden mit Messungen verglichen, die mit bereits bekannten Verfahren zur Bestimmung der Scheibenorientierung und der Schichtdicke erzielt wurden.

2. Berechnung der Schichtdicke und Scheibenorientierung

Die möglichen Stapelfehlerebenen bilden als geometrische Form einen Oktaeder, aus dem je nach Scheibenorientierung verschiedenartige Schnittgebilde entstehen können. Die Berechnung der geometrischen Beziehungen dieser Körper erfolgt zur Vereinfachung für die Ebenen $\{hk0\}$ $(h > k)$, $\{hll\}$ $(h > l)$ und $\{hhl\}$ $(h > l)$ für h, k, l von $1 \cdots 10$ bezogen auf die Ebenen (100), (110) und (111). Als Beispiel soll in diesem Rahmen die Berechnung für die Bezugsebene (111) durchgeführt werden. Die Ergebnisse für (110) und (100) sind den entsprechenden Kurven und Tabellen zu entnehmen.

Der Rechengang für die einzelnen Bestimmungsmöglichkeiten ist folgender:

1. Für die Bestimmung der Scheibenorientierung aus den Stapelfehlerwinkeln werden zuerst die Winkel zwischen den Ebenen $(hk0)$, (hll), (hhl) und (100), (110), (111) berechnet. Danach erfolgt die Ermittlung der sogenannten Stapelfehlerwinkel aus den Richtungen der Strecken, die das Stapelfehlerdrei- oder -viereck auf der Scheibenoberfläche begrenzen.
2. Zur Ermittlung der Schichtdicken der abgeschiedenen epitaktischen Schichten aus den Seitenlängen der Stapelfehlergebilde bei bekannter Scheibenorientierung, d. h. vorgegebener (hkl)-Ebene, wird der Zusammenhang zwischen Seitenlängen der Stapelfehlergebilde und der Höhe der zugehörigen Pyramide ermittelt. Die Proportionalitätsfaktoren (k-Werte) sind eine Funktion der Scheibenorientierung. Durch Multiplikation mit den gemessenen Stapelfehlerlängen liefern sie die Schichtdicken.
3. Als Folgerung aus den Punkten 1 und 2 kann die Schichtdicke aus den Stapelfehlerwinkeln und den Seitenlängen bestimmt werden. Dazu werden die k-Werte in Abhängigkeit von der Summe der Absolutbeträge der Differenzen zwischen den Stapelfehlerwinkeln für (hkl) und (100), (110) und (111) aufgetragen und zur Ablesung der k-Werte verwendet.

Den gesamten Berechnungen liegen die Gleichungen für den Winkel zwischen zwei Ebenen $(h_1k_1l_1)$ und $(h_2k_2l_2)$

$$\cos \delta_{(h_1k_1l_1)(h_2k_2l_2)} = \frac{h_1 h_2 + k_1 k_2 + l_1 l_2}{\sqrt{h_1^2 + k_1^2 + l_1^2} \cdot \sqrt{h_2^2 + k_2^2 + l_2^2}} \tag{1}$$

und zwischen zwei Richtungen $u_1v_1w_1$ und $u_2v_2w_2$

$$\cos \delta_{[u_1v_1w_1][u_2v_2w_2]} = \frac{u_1 u_2 + v_1 v_2 + w_1 w_2}{\sqrt{u_1^2 + v_1^2 + w_1^2} \cdot \sqrt{u_2^2 + v_2^2 + w_2^2}} \tag{2}$$

zugrunde.

2.1. Bestimmung der Scheibenorientierung aus den Stapelfehlerwinkeln

Als erster Schritt werden aus Gleichung (1) die Winkel zwischen den speziellen Ebenen $(hk0)$, (hll), (hhl) und (100), (110), (111) berechnet. In Tabelle 1 sind die entsprechenden $\cos\delta$-Werte zur Berechnung der Winkel zwischen den Ebenen zusammengestellt. Die Zeilen geben die Bezugsebenen an und die Spalten die zu berechnenden speziellen Ebenen.

Tabelle 1
Cos δ-Werte zur Berechnung der Winkel zwischen den Ebenen $(hk0)$, (hll), (hhl) und (100), (110), (111)

	$(hk0)$	(hll)	(hhl)
(100)	$\sqrt{\frac{h^2}{h^2 + k^2}}$	$\sqrt{\frac{h^2}{h^2 + 2l^2}}$	$\sqrt{\frac{h^2}{2h^2 + l^2}}$
(110)	$\sqrt{\frac{(h+k)^2}{2(h^2 + k^2)}}$	$\sqrt{\frac{(h+l)^2}{2(h^2 + 2l^2)}}$	$\sqrt{\frac{2h^2}{2h^2 + l^2}}$
(111)	$\sqrt{\frac{(h+k)^2}{3(h^2 + k^2)}}$	$\sqrt{\frac{(h+2l)^2}{3(h^2 + 2l^2)}}$	$\sqrt{\frac{(h+2l)^2}{3(2h^2 + l^2)}}$

Aus den für die (111)-Bezugsorientierung möglichen Stapelfehlerebenen $(1\bar{1}\bar{1})$, $(\bar{1}1\bar{1})$ und $(\bar{1}\bar{1}1)$ und den Ebenen $(hk0)$, (hll) und (hhl) werden die gemeinsamen Zonen ermittelt, um daraus den Winkel zwischen diesen Richtungen bestimmen zu können. Die Kombination der Ebenen $(hk0)$ mit den Stapelfehlerebenen $(1\bar{1}\bar{1})$, $(\bar{1}1\bar{1})$ und $(\bar{1}\bar{1}1)$ ergeben die Zonen

	Zonen	Zonennummer
$(hk0)$ $(1\bar{1}\bar{1})$	$\bar{k}\,h\,(-h-k)$	A_1
$(hk0)$ $(\bar{1}1\bar{1})$	$\bar{k}\,h\,(h+k)$	A_2
$(hk0)$ $(\bar{1}\bar{1}1)$	$\bar{k}\,h\,(-h+k)$	A_3

und ihre Gegenrichtungen mit

$k\,\bar{h}\,(h+k)$	$\bar{A}_1$
$k\,\bar{h}\,(-h-k)$	$\bar{A}_2$
$k\,\bar{h}\,(h-k)$	$\bar{A}_3$.

Durch Einsetzen der beiden Richtungen, die einen Stapelfehlerwinkel aufspannen — Ursprung der Vektoren liegt im Scheitelpunkt des Winkels — in Gleichung (2) werden drei Gleichungen für die Stapelfehlerwinkel einer $(hk0)$-Ebene bezogen auf (111) abgeleitet. Durch Einsetzen der Zonen A_1 und $\bar{A}_2$ in Gleichung (2) ergibt sich

$$\cos\alpha_{(hk0)(111)_1} = \frac{2hk}{\sqrt{2h^2 + 2hk + 2k^2} \cdot \sqrt{2h^2 + 2hk + 2k^2}} = \frac{hk}{h^2 + hk + k^2}, \quad (3)$$

aus den Zonen A_2 und $\bar{A}_3$ erhält man:

$$\cos\alpha_{(hk0)(111)_2} = \frac{2h^2}{\sqrt{2h^2+2hk+2k^2}\cdot\sqrt{2h^2-2hk+2k^2}} = \sqrt{\frac{h^2}{h^4+h^2k^2+k^4}} \tag{4}$$

und schließlich aus A_3 und $\bar{A}_1$:

$$\cos\alpha_{(hk0)(111)_3} = \frac{2k^2}{\sqrt{2h^2-2hk+2k^2}\cdot\sqrt{2h^2+2hk+2k^2}} = \sqrt{\frac{k^2}{h^4+h^2k^2+k^4}}\,. \tag{5}$$

Die Ableitung der Beziehungen der Stapelfehlerwinkel für die (hll)-Ebenen bezogen auf (111) ergibt bei Kombination von (hll) mit $(\bar{1}1\bar{1})$, $(1\bar{1}\bar{1})$ und $(\bar{1}\bar{1}1)$ die Zonen

	Zonen	Zonennummer
(hll) $(\bar{1}1\bar{1})$	$-2l\,(-l+h)\,(h+l)$	B_1
(hll) $(1\bar{1}\bar{1})$	$0\,(l+h)\,(-h-l)$	B_2
(hll) $(\bar{1}\bar{1}1)$	$2l\,(-l-h)\,(-h+l)$	B_3

und ihre Gegenrichtungen

$2l\,(l-h)\,(-h-l)$	$\bar{B}_1$
$0\,(-l-h)\,(h+1)$	$\bar{B}_2$
$-2l\,(l+h)\,(h-l)$	$\bar{B}_3$.

Aus B_1 und $\bar{B}_3$ erhält man durch Einsetzen in Gleichung (2)

$$\cos\alpha_{(hll)(111)_1} = \frac{2l^2+2h^2}{\sqrt{6l^2+2h^2}\cdot\sqrt{6l^2+2h^2}} = \frac{h^2+l^2}{3l^2+h^2}\,, \tag{6}$$

aus B_2 und $\bar{B}_1$

$$\cos\alpha_{(hll)(111)_2} = \frac{2l^2+2hl}{\sqrt{2h^2+4hl+2l^2}\cdot\sqrt{6l^2+2h^2}} = \sqrt{\frac{l^2}{3l^2+h^2}} \tag{7}$$

und aus B_3 und $\bar{B}_2$

$$\cos\alpha_{(hll)(111)_3} = \frac{2l^2+2hl}{\sqrt{6l^2+2h^2}\cdot\sqrt{2h^2+4hl+2l^2}} = \sqrt{\frac{l^2}{3l^2+h^2}}\,. \tag{8}$$

Für die Ableitung der Ausdrücke für (hhl)-Ebenen bezogen auf (111) werden in den Gleichungen (6) bis (8) nur die Indizes von h mit l vertauscht, da man die Ebenen (hhl) mit $h > l$ auch als (hll)-Ebenen mit $l > h$ auffassen kann.

Die Gleichungen lauten dann entsprechend:

$$\cos \alpha_{(hhl)(111)_1} = \frac{h^2 + l^2}{3\,h^2 + l^2}\,, \tag{9}$$

$$\cos \alpha_{(hhl)(111)_{2/3}} = \sqrt{\frac{h^2}{3\,h^2 + l^2}}\,. \tag{10}$$

Der Gang der Ableitung für die verschiedenen Ebenen (hkl) bezogen auf (110) und (100) ist der eben beschriebenen Herleitung für (111) ähnlich. Zur Übersicht werden die Kosinus der Stapelfehlerwinkel für alle hier erörterten Orientierungen in Tabelle 2 zusammengestellt.

Tabelle 2
Kosinus der Stapelfehlerwinkel $\cos\alpha$ und $\cos\beta$ für die Ebenen $(hk0)$, (hll) und (hhl) bezogen auf die Ebenen (100), (110) und (111).

bezogen auf (100)

	$(hk0)$	(hll)	(hhl)
$\cos\beta_1$	$\frac{hk}{h^2 - hk + k^2}$	$\sqrt{\frac{l^2}{h^2 + 3\,l^2}}$	$-\sqrt{\frac{h^2}{3\,h^2 + l^2}}$
$\cos\beta_2$	$-\sqrt{\frac{k^4}{h^4 + h^2 k^2 + k^4}}$	$-\sqrt{\frac{l^2}{h^2 + 3\,l^2}}$	$\frac{h^2 + l^2}{3\,h^2 + l^2}$
$\cos\beta_3$	$-\frac{h\,k}{h^2 + h\,k + k^2}$	$-\sqrt{\frac{l^2}{h^2 + 3\,l^2}}$	$-\sqrt{\frac{h^2}{3\,h^2 + l^2}}$
$\cos\beta_4$	$-\sqrt{\frac{k^4}{h^4 + h^2 k^2 + k^4}}$	$\sqrt{\frac{l^2}{h^2 + 3\,l^2}}$	1

bezogen auf (110), (111)

$\cos\alpha_1$	$\frac{h\,k}{h^2 + h\,k + k^2}$	$\frac{h^2 + l^2}{3\,l^2 + h^2}$	$\frac{h^2 + l^2}{3\,h^2 + l^2}$
$\cos\alpha_2$	$\sqrt{\frac{h^2}{h^4 + h^2 k^2 + k^4}}$	$\sqrt{\frac{l^2}{3\,l^2 + h^2}}$	$\sqrt{\frac{h^2}{3\,h^2 + l^2}}$
$\cos\alpha_3$	$\sqrt{\frac{k^2}{h^4 + h^2 k^2 + k^4}}$	$\sqrt{\frac{l^2}{3\,l^2 + h^2}}$	$\sqrt{\frac{h^2}{3\,h^2 + l^2}}$

Aus der Tabelle 2 lassen sich einige Folgerungen über die Form der Stapelfehlergebilde auf den (hkl)-Schnittflächen ziehen. Bei $(hk0)$ bezogen auf (100) bilden die Stapelfehler Drachen, bei (hll) Trapeze und für (hll) und (hhl) bezogen auf (110) und (111) sind es gleichschenklige Dreiecke, wobei es sich bei (hll)-Ebenen um Dreiecke handelt, deren Scheitelwinkel kleiner als 60° sind, da $\frac{h^2 + l^2}{h^2 + 3\,l^2} > \frac{1}{2}$ (s. Gleichung (6)) ist. Für (hhl)-Ebenen gilt $\frac{h^2 + l^2}{3\,h^2 + l^2} < \frac{1}{2}$ (s. Gleichung (9)); die Dreiecke haben daher einen Scheitelwinkel von größer als 60°.

Zur Anwendung der mit den Gleichungen der Tabelle 2 berechneten numerischen Werte der Stapelfehlerwinkel für die Orientierungsbestimmung von epitaktischen Siliziumschichten werden die gewonnenen Winkel nach Umformung gegen die Flächennormalenwinkel zwischen (*hkl*) und (100), (110) und (111) aufgetragen. Die Umformung erfolgt durch Bildung der Summe der Absolutbeträge der Differenzen zwischen den Stapelfehlerwinkeln der Bezugsebenen (100), (110) und (111) und den Stapelfehlerwinkeln der (*hkl*)-Ebenen. Für die drei Bezugsebenen gelten dann die Beziehungen:

$$S_{(hkl)(100)} = \sum_{i=1}^{4} /(90° - \beta_i)/ , \tag{11}$$

$$S_{(hkl)(110)} = /(70°31'44'' - \alpha_1)/ + /(54°44'8'' - \alpha_2)/ + /(54°44'8'' - \alpha_3)/ \tag{12}$$

$$S_{(hkl)(111)} = \sum_{i=1}^{3} /(60° - \alpha_i)/ . \tag{13}$$

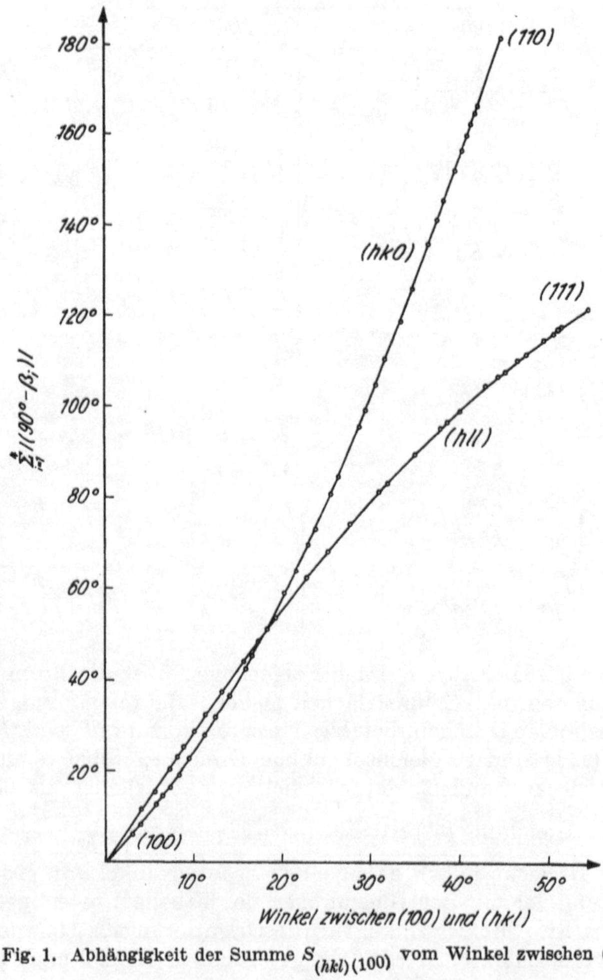

Fig. 1. Abhängigkeit der Summe $S_{(hkl)(100)}$ vom Winkel zwischen (100) und (*hkl*)

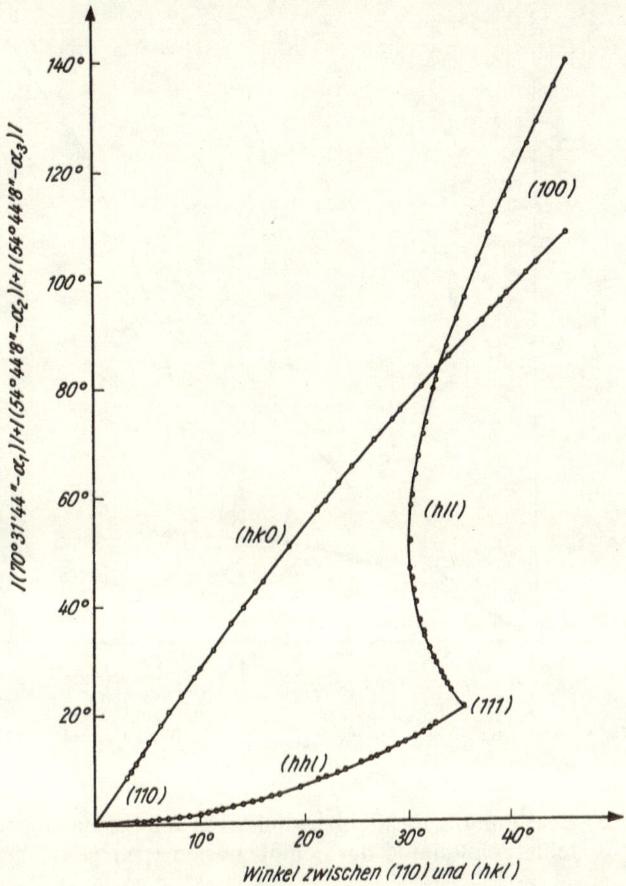

Fig. 2. Abhängigkeit der Summe $S_{(hkl)(110)}$ vom Winkel zwischen (110) und (*hkl*)

In den Abbildungen 1 bis 3 sind die Summen aus den Gleichungen (11) bis (13) in Abhängigkeit von den Winkeln zwischen den entsprechenden (*hkl*)-Ebenen und den Bezugsebenen (100), (110) und (111) aufgetragen. Aus den gemessenen Stapelfehlerwinkeln und ihren Größenverhältnissen läßt sich damit die Scheibenorientierung bestimmen. Die Genauigkeit kann nicht allgemein angegeben werden. Sie setzt sich zusammen aus dem Meßfehler des Winkels und dem Ablesefehler im Diagramm. Für verschiedene Ebenen kann die Genauigkeit des Verfahrens je nach Kurvenverlauf zwischen einigen zehntel Grad bzw. einigen Grad liegen. Bei allgemeiner Flächenlage und geringer Abweichung zu den Bezugsebenen (100), (110) und (111) beträgt die Genauigkeit einige zehntel Grad. Mit zunehmendem Fehlorientierungswinkel zur Bezugsebene nimmt die Genauigkeit ab. Nur wenn keine allgemeine Flächenlage (*hkl*), sondern eine spezielle Orientierung (z. B. (*hll*) oder (*hhl*)) vorliegt, bleibt auch bei größerem Fehlorientierungswinkel die Bestimmungsgenauigkeit von einigen zehntel Grad erhalten, da dann der gesuchte Diagrammwert auf einer der speziellen Kurven liegt.

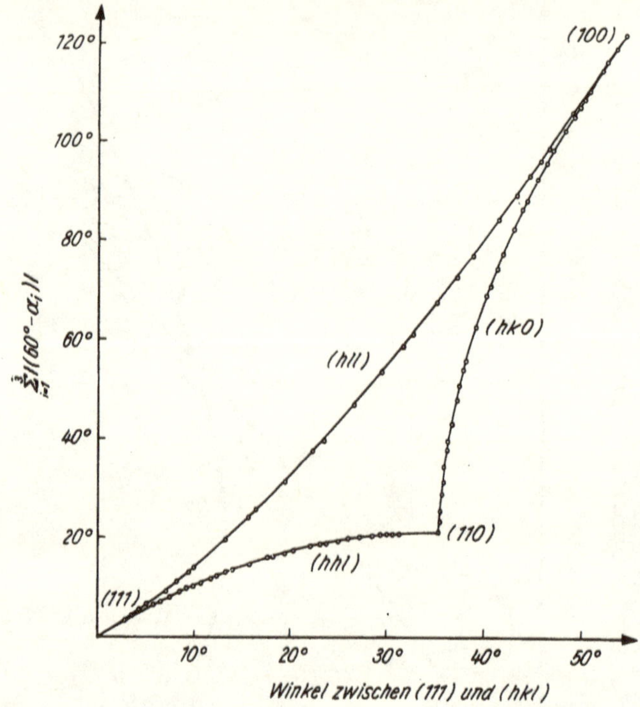

Fig. 3. Abhängigkeit der Summe $S_{(hkl)(111)}$ vom Winkel zwischen (111) und (hkl)

2.2. Bestimmung der Schichtdicke aus den Seitenlängen der Stapelfehlergebilde und der Scheibenorientierung

Gesucht ist bei dieser Bestimmung ein Faktor k, der mit der Seitenlänge der Stapelfehler (gemessen z. B. in μm) multipliziert die Schichtdicke der abgeschiedenen Epitaxieschicht ergibt.

Berechnet werden diese k-Werte für $(hk0)$-, (hll)- und (hhl)-Ebenen bezogen auf die Grundorientierungen (100), (110) und (111). Diese Untergliederung ist notwendig, da z. B. bei (100) das Stapelfehlergebilde ein Quadrat ist und bei (111) ein gleichseitiges Dreieck.

Für die Gesamtheit der Rechnungen soll hier eine Kombination als Beispiel durchgerechnet werden. Vorgegeben ist eine (hll)-Ebene bezogen auf $(\bar{1}11)$, d. h. das Schnittgebilde ist ein gleichschenkliges Dreieck (s. Fig. 4).

Die Höhe des gesamten Stapelfehlerkörpers $W_{(hll)}$ wird 1 gesetzt. Dann sind:

$$k_{(hll)(\bar{1}\bar{1}1)} = \frac{1}{l_{(hll)(\bar{1}\bar{1}1)}}, \tag{14}$$

$$k_{(hll)(\bar{1}1\bar{1})} = \frac{1}{l_{(hll)(\bar{1}1\bar{1})}} \tag{15}$$

und

$$k_{(hll)(1\bar{1}\bar{1})} = \frac{1}{l_{(hll)(1\bar{1}\bar{1})}}. \tag{16}$$

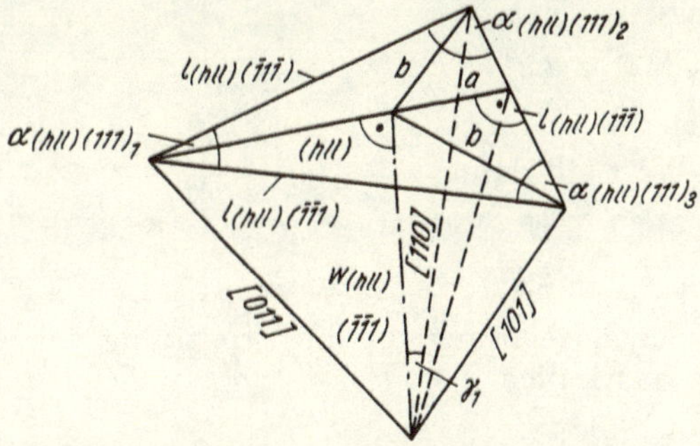

Fig. 4. Stapelfehlertetraeder mit Schnittfläche (*hll*)

Da es sich bei dem Stapelfehlergebilde gemäß Abschnitt 2.1. um ein gleichschenkliges Dreieck handelt, ist:

$$k_{(hll)(\bar{1}\bar{1}1)} = k_{(hll)(\bar{1}1\bar{1})} \,. \tag{17}$$

Für die Berechnung von $k_{(hll)(1\bar{1}\bar{1})}$ wird aus der Fig. 4

$$l_{(hll)(1\bar{1}\bar{1})} = 2\sqrt{b^2 - a^2} \tag{18}$$

entnommen.

Da

$$b = \tan \gamma_1 \tag{19}$$

und der Winkel γ_1 zwischen $[hll]$ und $[110]$ aus

$$\cos \gamma_1 = \frac{h + l}{\sqrt{2} \cdot \sqrt{h^2 + 2\,l^2}} \tag{20}$$

berechnet werden kann, ergibt sich für

$$b = \sqrt{\frac{(h - l)^2 + 2\,l^2}{(h + l)^2}} \,. \tag{21}$$

Für a gilt dann:

$$a = \frac{1}{\tan \gamma_2} \,. \tag{22}$$

Durch Einsetzen von γ_2 (Winkel zwischen (hll) und $(1\bar{1}\bar{1})$) in Form von

$$\cos \gamma_2 = \frac{-\,2\,l + h}{\sqrt{3} \cdot \sqrt{h^2 + 2\,l^2}} \tag{23}$$

in Gleichung (22) errechnet sich

$$a = \sqrt{\frac{(2\,l - h)^2}{2\,(h + l)^2}} \,. \tag{24}$$

Werden die beiden Ausdrücke für a bzw. b aus den Gleichungen (24) und (21) in Gleichung (18) eingesetzt, so erhält man:

$$l_{(hll)(1\bar{1}\bar{1})} = 2\sqrt{\frac{2\,(h - l)^2 + 4\,l^2 - (2\,l - h)^2}{2\,(h + l)^2}} \tag{25}$$

und nach Vereinfachung

$$l_{(hll)(1\bar{1}\bar{1})} = \sqrt{\frac{4\,l^2 + 2\,h^2}{(h + l)^2}}\,. \tag{26}$$

Für den zugehörigen k-Wert gilt dann

$$k_{(hll)(1\bar{1}\bar{1})} = \sqrt{\frac{(h + l)^2}{4\,l^2 + 2\,h^2}}\,. \tag{27}$$

Aus Figur 4 ist auch der Rechengang für die Ermittlung von $k_{(hll)(\bar{1}\bar{1}1)}$ und $k_{(hll)(\bar{1}1\bar{1})}$ ersichtlich. Es ist nämlich

$$l_{(hll)(\bar{1}1\bar{1})} = \frac{\frac{1}{2}\,l_{(hll)(1\bar{1}\bar{1})}}{\cos\alpha_{(hll)(111)_2}}\,. \tag{28}$$

Durch Einsetzen der Ausdrücke für $l_{(hll)(1\bar{1}\bar{1})}$ aus Gleichung (26) und für $\cos\alpha_{(hll)(111)_2}$ aus Gleichung (7) in die Beziehung (28) berechnet sich der k-Wert zu:

$$k_{(hll)(\bar{1}1\bar{1})} = \sqrt{\frac{2\,l^2\,(h + l)^2}{(2\,l^2 + h^2)\,(3\,l^2 + h^2)}}\,. \tag{29}$$

Da die für die anderen Kombinationen berechneten Ausdrücke sehr umfangreich sind, werden sie nicht aufgeführt. Das gezeigte Beispiel soll für alle stehen.

In den Figuren 5 bis 7 sind die k-Werte in Abhängigkeit von der Ebenenorientierung (hkl), d. h. in Abhängigkeit vom Winkel zwischen (100), (110), (111) und (hkl) aufgetragen.

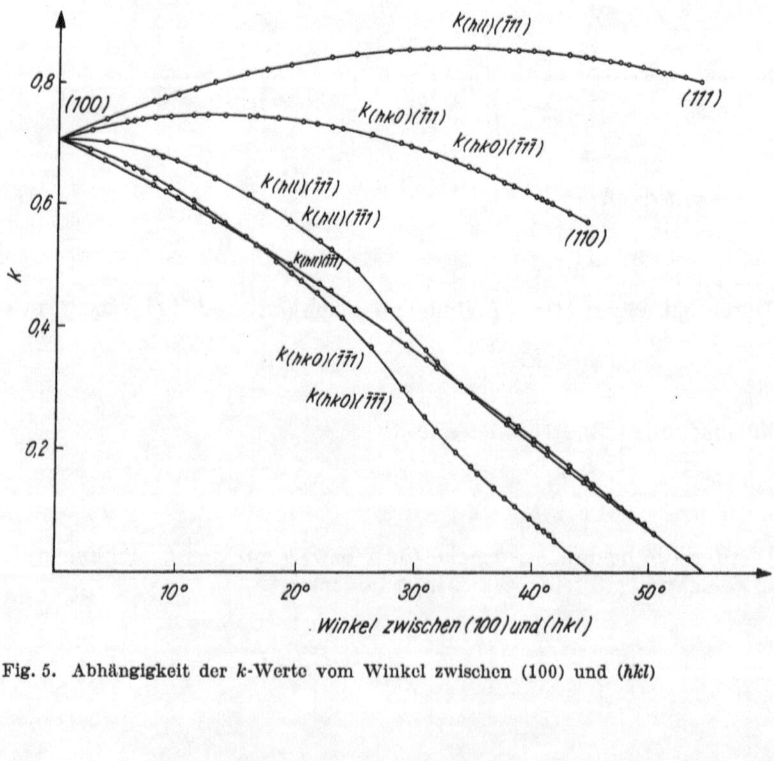

Fig. 5. Abhängigkeit der k-Werte vom Winkel zwischen (100) und (hkl)

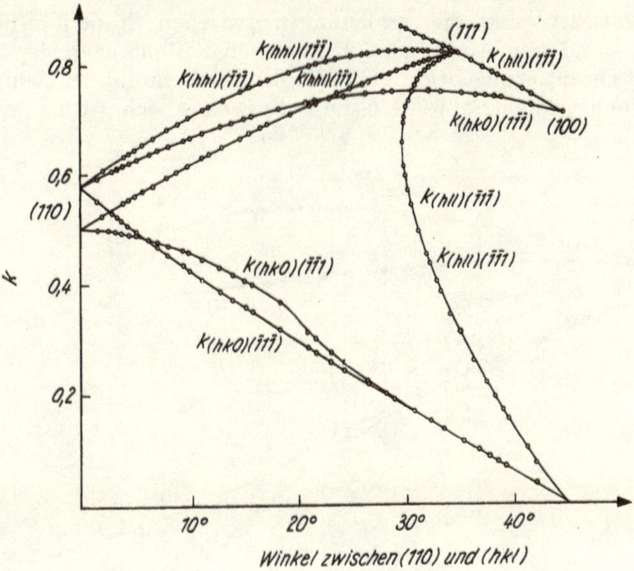

Fig. 6. Abhängigkeit der *k*-Werte vom Winkel zwischen (110) und (*hkl*)

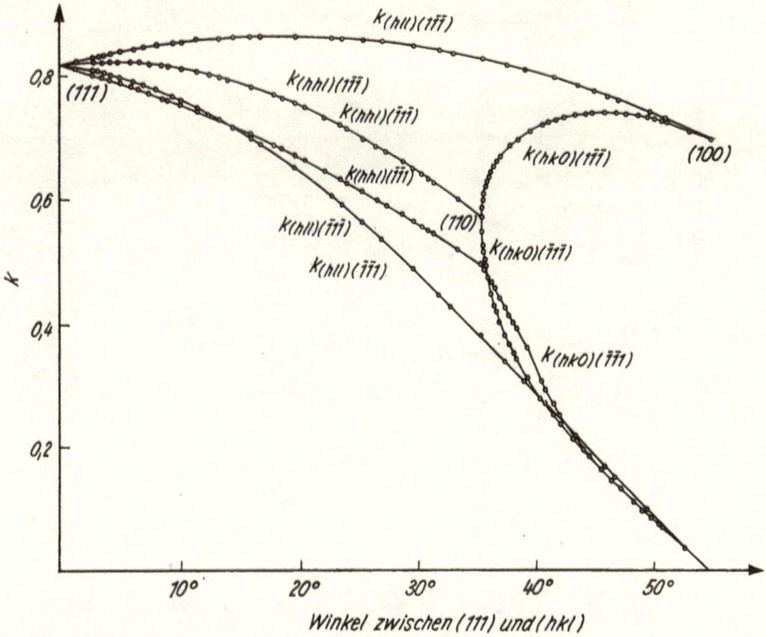

Fig. 7. Abhängigkeit der *k*-Werte vom Winkel zwischen (111) und (*hkl*)

2.3. Bestimmung der Schichtdicke aus der Stapelfehlergeometrie

Aus Abschnitt 2.1.' sind die Beziehungen zwischen Stapelfehlerwinkeln und Orientierung und aus Abschnitt 2.2. zwischen Seitenlängen der Stapelfehlergebilde und Orientierung bekannt. Für die Bestimmung der Schichtdicke aus der Stapelfehlergeometrie werden nun die aus Abschnitt 2.2. erhaltenen k-

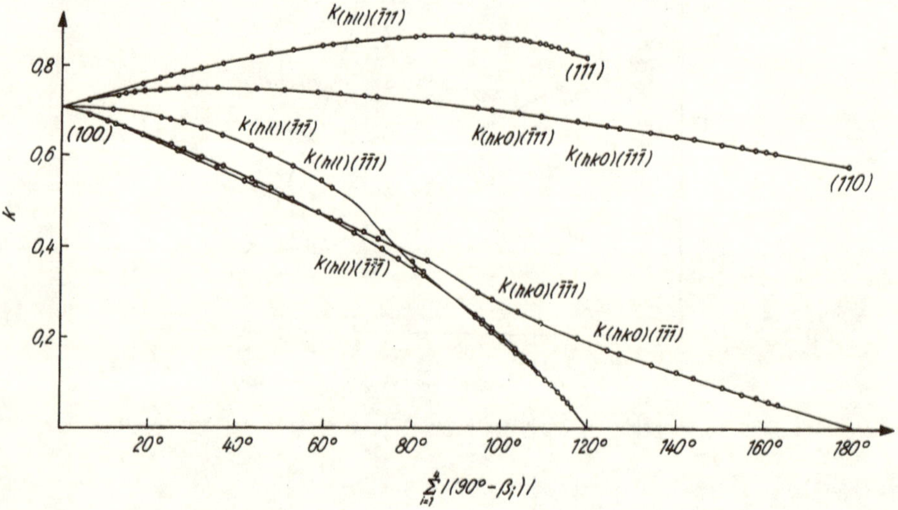

Fig. 8. Abhängigkeit der k-Werte von der Summe $S_{(hkl)(100)}$

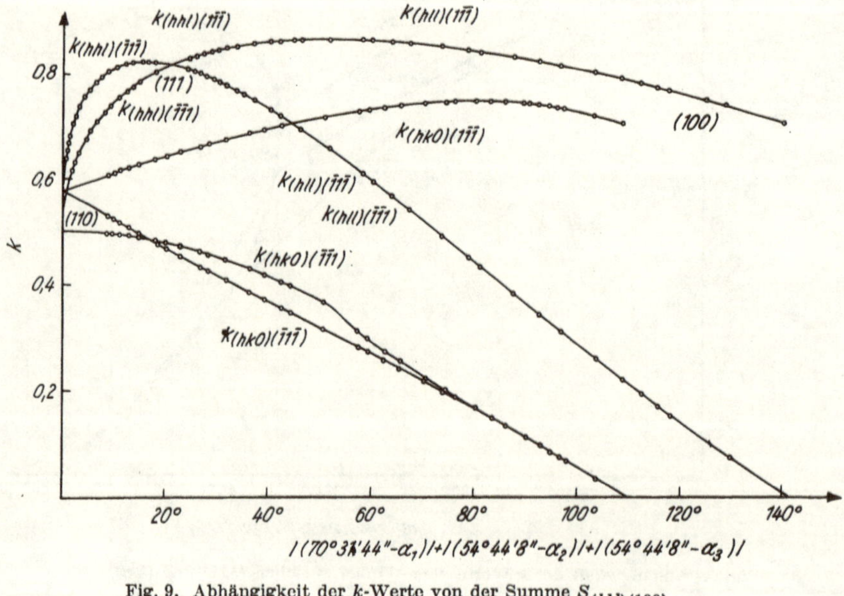

Fig. 9. Abhängigkeit der k-Werte von der Summe $S_{(hkl)(100)}$

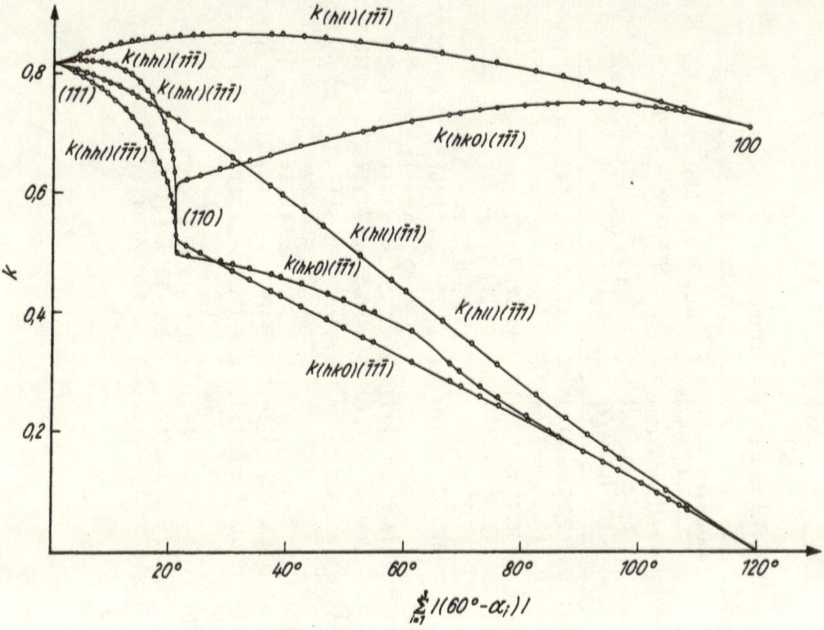

Fig. 10. Abhängigkeit der k-Werte von der Summe $S_{(hkl)(110)}$

Werte gegen die Stapelfehlerwinkel aufgetragen (Fig. 8 bis 10). Damit ist es möglich, die Schichtdicke der epitaktischen Schicht für Proben unterschiedlicher Orientierung nur aus der Geometrie der Stapelfehler – aus Winkeln, Seitenlängen und ihren Größenverhältnissen – zu ermitteln.

3. Beispiele zur Bestimmung der Scheibenorientierung und der Schichtdicke aus der Stapelfehlergeometrie

An 16 Scheiben verschiedener Orientierung wurden epitaktische Schichten durch Wasserstoffreduktion von Siliziumtetrachlorid bei 1250 °C abgeschieden. Figur 11 zeigt eine Zusammenstellung von acht Aufnahmen von Stapelfehlern an Scheiben mit verschiedenen Orientierungen.

An den verschieden orientierten Proben wurden zur Bestimmung der Schichtdicke und der Scheibenorientierung die Seitenlängen und die Winkel der Stapelfehlergebilde gemessen.

Die Längenmessung erfolgte lichtmikroskopisch mit einem geeichten Okularmikrometer (5:100) mit einer Genauigkeit von $\pm 1/2$ Skt.; das entspricht bei einer Vergrößerung von 800:1 einer Genauigkeit von $\pm 0{,}4$ µm.

Für die Winkelmessung wurde das Winkelmeßokular des Universalmeßmikroskops UP 1 vom VEB Carl Zeiss JENA mit einem besonderen Tubus an das Universalforschungsmikroskop Nu angepaßt. Die Meßgenauigkeit beträgt etwa $\pm 0{,}1°$.

Die Stapelfehlergebilde sind im Mikroskop entweder direkt im Auflicht-Phasenkontrast (Ogino) oder Auflicht-Interferenzkontrast (Dudley) sichtbar oder sie müssen durch spezielle Ätzgemische sichtbar gemacht werden. Am besten

Tabelle 3
Zusammenstellung der nach verschiedenen Verfahren ermittelten Scheibenorientierungen

Proben-nummer	röntgenografisch ermittelte Werte				aus Lichtfigur + U-Tisch ermittelten Werte bezogen auf Ebene (*hkl*)	aus Stapelfehlerwinkeln ermittelte Werte	
	Winkel δ zur Ebene (*hkl*)	Lage der Ebene (*hkl*) auf der Richtung	in der Nähe gelegene Ebene (*hkl*)	δ		δ zu (*hkl*)	(*hkl*)
1	17,6° zu (111)	$[2\bar{1}\bar{1}]$	(955)	16,6° zu (111)	17,3° zu (111)	17,3° zu (111)	(955)
2	12,7° zu (111)	$[2\bar{1}\bar{1}]$	(855)	13,3° zu (111)	13,1° zu (111)	12,9° zu (111)	(855)
3	35° zu (111)	$[10\bar{1}]$	(210)	39,2° zu (111)	35,3° zu (111)	ein Winkel 106,6°	—
4	11,5° zu (111)	$[\bar{2}11]$	(885)	11,4° zu (111)	11,8° zu (111)	10,5…11° zu (111)	—
5	9° zu (111)	$[10\bar{1}]$	(654)	9,3° zu (111)	7,7° zu (111)	7,2…9,2° zu (111)	—
6	2° zu (100)	—	(100)	—	1,5…2° zu (100)	2° zu (100)	(100)
7	27,6° zu (111)	$[10\bar{1}]$	(531)	28,6° zu (111)	—	22…34° zu (111)	—
8	14,3° zu (111)	$[2\bar{1}\bar{1}]$	(533)	14,4° zu (111)	14,8° zu (111)	14,5° zu (111)	(533)
9	1,2° zu (110)	—	(110)	—	0…0,5° zu (110)	35,5° zu (111)	(110)
10	38° zu (111)	$[2\bar{1}\bar{1}]$	(922)	37,3° zu (111)	—	37° zu (111)	(922)
						17,5° zu (100)	
11	30,4° zu (111)	$[\bar{2}11]$	(881)	30,2° zu (111)		29,5° zu (111)	
	4,9° zu (110)			5,0° zu (110)	4,0° zu (110)	10° zu (110)	(771)
12	20,2° zu (111)	$[2\bar{1}\bar{1}]$	(211)	19,5° zu (111)	20,5° zu (111)	19,2° zu (111)	(211)
13	17,5° zu (100)	—	(310)	18,4° zu (100)	—	17° zu (100)	(310)
14	12° zu (100)	—	(920)	12,5° zu (100)	8…9° zu (100)	9,9…11,4° zu (100)	—
15	28° zu (111)	$[\bar{2}11]$	(661)	28,5° zu (111)	26° zu (111)	19…33° zu (111)	—
16	3° zu (111)	—	—	—	3° zu (111)	2,9° zu (111)	—

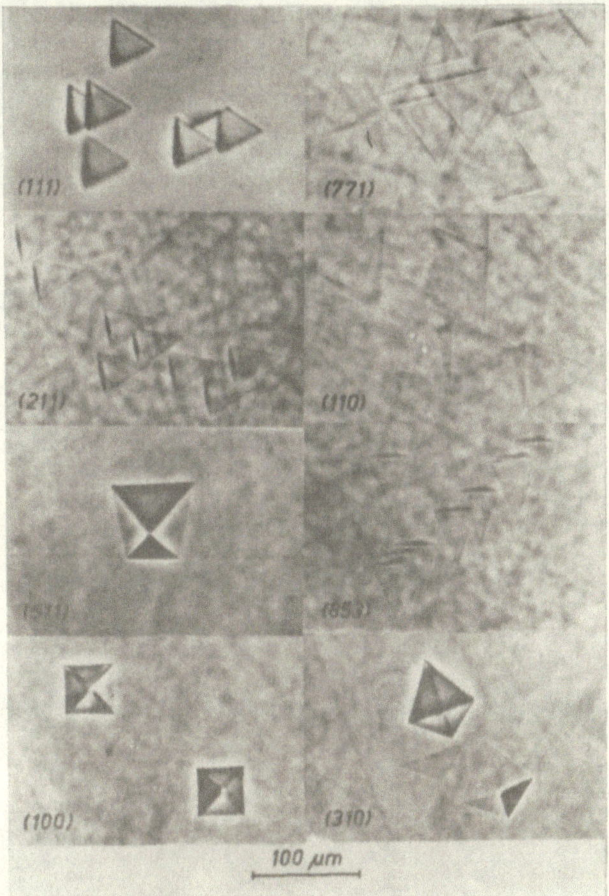

Fig. 11. Stapelfehler an epitaktischen Siliziumscheiben mit verschiedenen Scheibenorientierungen.
Lichtmikroskopische Aufnahmen im Auflicht-Phasenkontrast

ist für diesen Zweck Chromätzer, ein Gemisch aus Flußsäure und Chromoxidlösung in Wasser, geeignet (Sirtl, Adler).

Unter Anwendung der in den Abschnitten 2.1. bis 2.3. abgeleiteten Beziehungen und gezeigten Diagrammen wurden aus den Stapelfehlerwinkeln und Seitenlängen Scheibenorientierung und Dicke der abgeschiedenen Epitaxieschicht bestimmt.

Zur Gegenüberstellung der nach verschiedenen Verfahren ermittelten Scheibenorientierungen wurden die entsprechenden Meßwerte für die einzelnen Verfahren in Tabelle 3 zusammengestellt. In Spalte 2 ist der röntgenografisch mit Laueaufnahmen bestimmte Winkel δ bezogen auf eine der niedrig indizierten Ebenen (100), (110) oder (111) angegeben. Spalte 3 zeigt die azimutale Lage der Ebene, und die Spalten 4 und 5 geben die der gemessenen Orientierung am nächsten gelegenen Ebenen niedriger Indizierung mit dem zugehörigen Winkel an. Eine weitere Messung des Fehlorientierungswinkels — bezogen auf eine niedrig indizierte Ebene — wurde mit Hilfe der Lichtfiguren-Methode vorgenommen (Fricke). Die Ausmessung erfolgte nicht am Zweikreis-Reflex-Goniometer, da die Intensität der Lichtfiguren zu gering war. Für den hier vorliegenden Zweck wurde eine einfache Goniometereinrichtung durch Kombination eines Nf-Mikroskops vom VEB Carl Zeiss JENA mit einem Universaldrehtisch aufgebaut. Der U-Tisch diente als Winkelmeßgerät. Die Beobachtung der Lichtfigur erfolgte bei möglichst kleiner Aperturblende des Mikroskops mit einem Objektiv 25×/0,50, eingeschalteter Bertrandscher Linse und Fadenkreuz. Spalte 6 gibt die erhaltenen Meßwerte mit einer Genauigkeit von $\pm 0{,}5°$ an. Aus den Winkelmessungen an den Stapelfehlergebilden der epitaktischen Schichten lassen sich mit Hilfe der Gleichungen (11) bis (13) und der Diagramme 1 bis 3 die in den Spalten 7 und 8 angegebenen Fehlorientierungswinkel mit den entsprechenden Ebenen ableiten.

Die Schichtdicken wurden bestimmt (Tabelle 4) aus den Summen $S_{(hkl)(100)}$, $S_{(hkl)(110)}$ und $S_{(hkl)(111)}$ (Gleichungen (11) bis (13)) und den Winkeln $\delta_{(hkl)}$ (s. Spalte 7 der Tabelle 3). Die Unterschiede in den Schichtdicken (Tabelle 4), die aus $S_{(hkl)}$ und $\delta_{(hkl)}$ mit Hilfe der Diagramme der Figuren 5 bis 10 ermittelt wurden, sind auf Ablesefehler zurückzuführen. Zum Vergleich der so bestimmten Schichtdicken wurden Schrägschliffe angefertigt, die nach Silberdekoration

Tabelle 4

Zusammenstellung der nach verschiedenen Verfahren ermittelten Schichtdicken

Probennummer	Schichtdicke berechnet aus $S_{(hkl)}$		Schichtdicke berechnet aus $\delta_{(hkl)}$		Schichtdicken gemessen an Schrägschliffen		
	(hkl)	(μm)	(hkl)	(μm)	aus Dekoration (μm)	nach Chromätzung (μm)	Tiefe der Entstehungsfront der Stapelfehler (μm)
1	(111)	35,1	(111)	35,1	34,6	35,6	35,6
2	(111)	33,0	(111)	32,9	33,5	33,5	33,5
3	—	—	—	—	—	42,3···44,3	—
4	(111)	29,9···31,3	(111)	29,9···31,4	(25,5)	30,7	30,3
5	(111)	32,4···33,6	(111)	32,2···33,4	(30,4)	32,2···34,4	32,2···34,4
6	(100)	32,8	(100)	32,6	—	34,3···35,7	—
7	(111)	25,4···35,7	(111)	23,2···35,2	—	27,7···29,1	—
8	(111)	32,8	(111)	32,9	—	32,5	—
9	(111)	34,8	(111)	33,5	(27,3)	31,4	31,4
	(110)	34,3					
10	(100)	38,9	(100)	39,1	33,8	33,3···36,5	33,3
11	(111)	40,0	(111)	42,3	36,8	40,3	36,1
12	(111)	28,0	(111)	28,2	28,8	26,9···28,0	28,8
13	(100)	33,3	(100)	32,9	(38,6)	34,8	31···35
14	(100)	31,2···33,5	(100)	31,0···33,6	28,9···34,7	32,1···32,7	31,0
15	(111)	26,2···36,7	(111)	24,7···37,8	28,5	27,7	31,4···32,6
16	(111)	35,0	(111)	35,0	34,9	—	34,9

und Chromätzung die Bestimmung der Schichtdicken ermöglichten. Die Silberdekoration war ungenügend, da die abgeschiedenen epitaktischen Schichten *p*-leitend waren und das Substrat meistens den gleichen Leitungstyp hatte. Sehr fragwürdige Meßwerte sind in Tabelle 4 Spalte 6 in Klammer gesetzt. In Spalte 7 der Tabelle 4 sind die Werte angegeben, die sich nach Behandlung mit Chromätzer durch die unterschiedliche Abtragung verschieden dotierter Gebiete ergeben. Spalte 8 gibt die Tiefe der Entstehungsfront (Grenzfläche zwischen Schicht und Substrat) der Stapelfehler am Schrägschliff an. Aus den Werten der Tabelle 4 ist zu ersehen, daß die Schichtdicken, die aus der Stapelfehlergeometrie (Stapelfehlerwinkel oder Seitenlängen) bestimmt wurden, gut mit den nach anderen Verfahren gemessenen Dicken übereinstimmen und daß diese Meßmethode der Schrägschlifftechnik mit nachfolgender Dekoration überlegen ist, wenn durch die epitaktische Abscheidung kein *pn*-Übergang gebildet wird.

4. Zusammenfassung

1. Aus den Stapelfehlerwinkeln, den Winkeln zwischen den Seiten der Stapelfehlergebilde, ist auf Grund der vorliegenden Berechnungen die Bestimmung der Scheibenorientierung bezogen auf eine der Ebenen (100), (110) oder (111) möglich.
2. Aus den Seitenlängen ist die Bestimmung der Schichtdicke bei bekannter Scheibenorientierung möglich.
3. Bei unbekannter Scheibenorientierung kann durch Ermittlung der Stapelfehlergeometrie (Stapelfehlerwinkel und Seitenlängen) die Schichtdicke der abgeschiedenen epitaktischen Schicht bestimmt werden.
4. Das Verfahren zur Bestimmung der Scheibenorientierung und der Schichtdicke aus der Stapelfehlergeometrie ist schnell und bei Verwendung von lichtmikroskopischer Ausmessung im Auflicht-Phasenkontrast oder Auflicht-Interferenzkontrast zerstörungsfrei.
5. Wenn Scheiben mit Orientierungen in spezieller Lage vorliegen, beträgt die Genauigkeit bei der Bestimmung der Scheibenorientierung einige zehntel Grad.
6. Bei Scheibenorientierung in allgemeiner Lage ist der geringe Fehler von einigen zehntel Grad nur für kleinere Orientierungsabweichungen von den speziellen Lagen (100) und (111) gewährleistet; bei großen Abweichungen beträgt der Fehler einige Grad.
7. Das Verfahren von Diment hat den Nachteil, daß sich bei geringen Abweichungen von der Bezugsebene (z. B. (111)) die Seitenlängen wenig unterscheiden und durch den Ätzprozeß auch noch verfälscht werden können. Dagegen sind die Stapelfehlerwinkel meßtechnisch besser zu erfassen. Sogar Orientierungsabweichungen von einigen zehntel Grad zur Bezugsebene können mit Hilfe dieser Methode nachgewiesen werden. Dazu braucht man nur ausreichend große Stapelfehlergebilde (hohe Vergrößerungen bzw. große Schichtdicken).

Literatur

Chu, T. L., Gavaler, I. R.: J. electrochem. Soc. **110**, 388 (1963)
Dash, W. C.: J. appl. Physics **33**, 2395 (1962)
Diment, M. P.: Solid-State Electronics **11**, 1177 (1968)

Dudley, R. H.: J. appl. Physics **35**, 1360 (1964)
Fricke, P.: Kristall und Technik **3**, K 15 (1968)
Mendelson, S.: J. appl. Physics **35**, 1570 (1964)
Ogino, N.: Japan. J. appl. Physics **3**, 797 (1964)
Sirtl, E., Adler, A.: Z. Metallkunde **52**, 529 (1961)

(Eingegangen am 14. Juni 1971)

Anschrift des Verfassers:

Dr. P. Fricke
Arbeitsstelle für Molekularelektronik
808 Dresden
Königsbrücker Landstr. 159

Kristall und Technik	7	1–3	1972	303–312

K. Jegerlehner, U. Mohr

Kombinat VEB Halbleiterwerk Frankfurt/Oder
Gleichrichterwerk Stahnsdorf

Röntgentopographie von gekrümmten Halbleiterscheiben

In der vorliegenden Arbeit wird ein Überblick über röntgentopographische Methoden zur Untersuchung von Halbleitermaterialien gegeben. Speziell wird auf die SO-Technik eingegangen und die Apparatur einer Anwendungsvariante vorgestellt. An Hand von Beispielen werden Anwendungsmöglichkeiten der SO-Technik erläutert.

A survey is given of X-ray topographical methods for investigating semiconductor materials. Particulary scanning oscillator technique (SOT) is explained and a camera as practical example given. Application possibilities of SOT are explained.

1. Einleitung

Unter Röntgentopographie sollen alle die Verfahren verstanden werden, welche darauf beruhen, daß Röntgenstrahlen von einer kristallinen Probe nach dem Bragg'schen Gesetz abgebeugt werden und auf einer fotoempfindlichen Schicht eine Übereinanderprojektion der auf dem Weg durch den Kristall passierten Störungen aufzeichnen. Dadurch entsteht eine Abbildung der im durchstrahlten Volumen einer Probe enthaltenen Störungen unter einer bestimmten Strahlenrichtung in einer Ebene. Das Verfahren beinhaltet als Voraussetzung, daß monochromatische Strahlung ausgenutzt wird. Es handelt sich dabei also nicht um eine bloße Durchstrahlung, bei der Einschlüsse und Hohlräume, wie bei einer Grobstrukturaufnahme, sichtbar werden, sondern um die Abbildung von Störungen des periodischen Gitteraufbaus.

Die Röntgentopographie erreichte die ihr heute zukommende Bedeutung als selbständiges Untersuchungsgebiet erst durch die Anwendung in der Halbleitertechnik und auf Grund einer apparativen und verfahrensmäßigen Verfeinerung, welche sie durch die grundlegenden Arbeiten von Lang in der Jahren 1958/59 erfuhr (Anwendung auf Siliziumeinkristalle). Es ist der Zeitpunkt des Überganges von der „section-topograph-technique“ zur „projection-topograph-technique“. Von dort her stammt auch die im Deutschen übernommene Bezeichnung Topographie. Es sind auch die Bezeichnungen Röntgenbeugungsmikroskopie oder einfach Mikroskopie mit Röntgenstrahlen oder auch Röntgenbeugungsmikrographie verwendet worden.

Ursprünglich war die Röntgentopographie auf dem Halbleitergebiet allein ein Mittel der Erkundungsforschung. Heute überwiegen die Fälle der Anwendung für technologische Grundlagenuntersuchungen und Verfahrenskontrollzwecke. Die Ausweitung der Anwendungsgebiete der Topographie läuft konform mit der enormen Entwicklung der Halbleitertechnik in den letzten 10 Jahren. Es stiegen und steigen die Anforderungen hinsichtlich Miniaturisierung und Zuverlässigkeit sowie im Hinblick auf die Erreichung der theoretisch möglichen Bauelementparameter. Dadurch ist man in immer stärkerem Maß gezwungen, auf Störungen des Kristallgitters nicht nur im Ausgangsmaterial sondern auch während der einzelnen Bearbeitungsverfahren des Halbleitermaterials Rücksicht zu nehmen.

Ein weiterer Grund für die breite Anwendung der Röntgentopographie ist die apparative Weiterentwicklung für die Untersuchung von gekrümmten Einkristallproben, wie sie in der Technik häufig vorliegen, in der Art der „scanning-oszillator-technique“ (SOT) nach Schwuttke, die in dieser Arbeit in einer Ausführungsvariante und Anwendung näher beschrieben werden soll.

2. Die Anwendung der Röntgentopographie in der Halbleitertechnik

Es soll verfahrensmäßig zwischen 3 Gruppen unterschieden werden, welche Grundvarianten darstellen, eine historische Reihenfolge bilden, aber nebeneinander angewendet werden. Es sind gebräuchlich:

1. Parallelstrahlverfahren nach Berg und Barrett,
2. Verfahren nach Lang,
3. SO-Technik nach Schwuttke.

Alle Verfahren sind grundsätzlich in Reflexion oder in Transmission anwendbar. Die meisten Anwendungsfälle sind in Durchstrahlung bekannt. Zur Reflexionsmethode wird man immer dann greifen müssen, wenn die Kristallprobe einseitig mit wenig transparenten Schichten versehen ist, die eine Abbildung von Defekten verhindern.

Die Parallelstrahlmethode wird in den Fällen angewendet, wo ein rascher Überblick über den Zustand des Gitteraufbaus erhalten werden soll, bzw. wenn an die Auflösung im Topogramm nicht so hohe Ansprüche gestellt werden. Zur Lösung einer Reihe von Problemen in der Halbleitertechnik fand die Methode schon Anwendung (Newkirk; Schwuttke, Queisser; Mohr, Jegerlehner; Juleff, Lapierre). In der einfachen Strahlenanordnung sind $K_{\alpha 1}$- und $K_{\alpha 2}$-Reflex nicht voneinander getrennt. Die Belichtungszeiten sind im Verhältnis zur Lang-Technik kurz. Mit Hilfe der Lang-Technik können nicht nur Übereinanderprojektionen von Störungen im durchstrahlten Gesamtvolumen erhalten werden (Jenkinson, Lang; Jenkinson; Yoshiamaitsu), sondern auch Schichtaufnahmen von unterschiedlich tief unter der Probenoberfläche liegenden Schichten, was durch ein geeignetes Blendensystem, das in bezug auf die Strahlenquelle hinter der Probe liegt, erreicht wird (Lang). Die dabei notwendige Präzision schließt Probenkrümmungen aus. Bei einer ausschließlich translatorischen Bewegung der Probe, wie sie für das Lang-Verfahren üblich ist, würde ein benachbarter Bereich einer verkrümmten Probe nicht mehr der Reflexionsbedingung genügen. Soweit der Krümmungswinkel nicht größer als der Differenzbetrag der beiden Bragg'schen Winkel für die Reflexion der $K_{\alpha 1}$- und $K_{\alpha 2}$-Wellenlängen ist, wird ein monochromatisches Bild der gesamten Fläche dadurch erreicht, daß die Probe um den entsprechenden Reflexionswinkel oszilliert wird. Das gerade ist die technische Maßnahme, welche zur SO-Technik führte, die ein breites Gebiet der Anwendung in der Halbleitertechnik erschloß. Die Verfahrensunterschiede der 3 Grundvarianten sind grob der Tabelle 1 zu entnehmen. Die SO-Technik soll im folgenden noch ausführlicher beschrieben werden.

3. Eine Kamera zur Anwendung der SO-Technik

Das Prinzip der SO-Technik wurde von Schwuttke in der Literatur beschrieben. In etwas modifizierter Form liegt es dieser Arbeit zugrunde (Figur 1). Ausgehend von einem vertikal stehenden Strichbrennfleck wird ein Röntgenstrahlenbündel von Blende 1 und 2 so ausgeblendet, daß die Divergenz im Röntgenstrahl $\Delta\Theta \leqq \vartheta_2 - \vartheta_1$ ist. Das erfolgt aus dem Grunde, um das Auftreten von Doppelreflexen im Röntgentopogramm zu vermeiden. Der so ausgeblendete Röntgenstrahl trifft im ϑ-Winkel auf die einkristalline Probe und verläßt sie im Win-

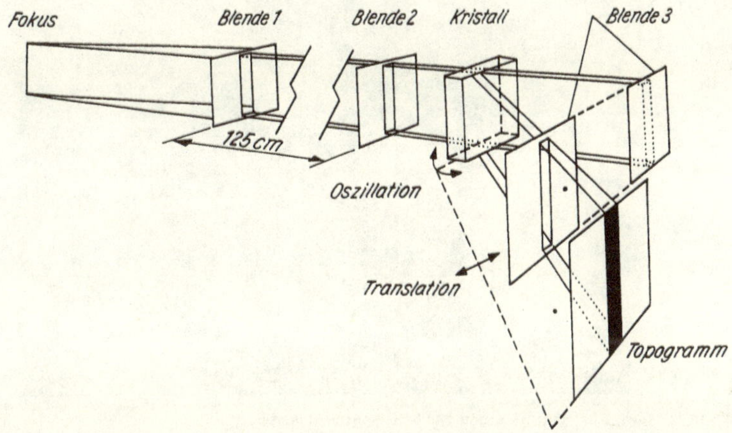

Fig. 1. Perspektivische Darstellung des Strahlenganges bei der „scanning oscillator technique“ (SOT)

kel 2ϑ. Da meist die Proben nicht so dick sind, daß der Primärstrahl vollständig absorbiert wird bzw. da der die Probe verlassende Röntgenstrahl in zwei Teilbündel aufspaltet, wobei der eine in Richtung des Primärstrahles weiterläuft, muß der zur Herstellung des Topogramms benutzte abgebeugte Röntgenstrahl ausgeblendet werden. Diesen Zweck erfüllt Blende 3. Kristall und Fotoplatte sind starr gekoppelt. Durch eine Hin- und Herbewegung dieser beiden erhält man dann ein Topogramm auf der Fotoplatte. Von großflächigen Einkristallproben erhält man aber nur dann ein Röntgentopogramm, wenn die zur Herstellung des Topogramms ausgenutzten Netzebenen über die gesamte Probe streng parallel verlaufen. Gibt es Abweichungen von dieser Ordnung, so findet man im Topogramm Bereiche, die keine Röntgenkontraste zeigen, weil für diese Bereiche die Bragg'sche Bedingung nicht mehr erfüllt ist. In diesem Fall kommt dann der eigentliche Vorteil der SO-Technik zum Tragen, denn zum Ausgleich dieser Fehlorientierung der Netzebenen in den oben genannten Bereichen wird die Probe im ϑ-Winkel um einen kleinen Winkelbetrag (bis $\sim 3{,}5'$) oszilliert. So erreicht man dann ein vollständiges Topogramm von Kristallproben, die gekrümmte Bereiche aufweisen. Die Figuren 2 und 3 zeigen Details der dem oben beschriebenen Prinzip zugrunde liegenden Kamera, die zur Herstellung von Röntgentopogrammen nach der SO-Technik benutzt wurde. Die Kamera ist eine für diese Zwecke umgebaute Weißenbergkamera (s. MOHR). Zur Hin- und Herbewegung wurde der vorhandene Wagen ausgenutzt. Der Wagen wurde so umgebaut, daß er zur Aufnahme des vom Goniometerkopf getragenen Probenhalters und der Plattenkamera (1 und 2 Figur 2) dienen kann. Zwischen Kamera und Probenhalter kann über eine Brücke die Blende 3 eingefahren werden (3 Figur 2). Im Hintergrund von Figur 2 (4) erkennt man Blende 2 am Ende des Kollimatorrohres. Die Grobjustierung einer Probe kann über ein Fernrohr (5 in Figur 2) optisch und die Feinjustierung mit Hilfe eines Feintriebes (6 in Figur 3) über ein Zählrohr röntgenographisch vorgenommen werden. Zur Durchführung der Hin- und Herbewegung konnte zwar die Origi-

Fig. 2. Detail einer SOT-Kameravariante
(1) Goniometerkopf mit Probenhalter, (2) Plattenkassette, (3) Blende 3, (4) Blende 2, (5) Fernrohr zur optischen Orientierung der Probe

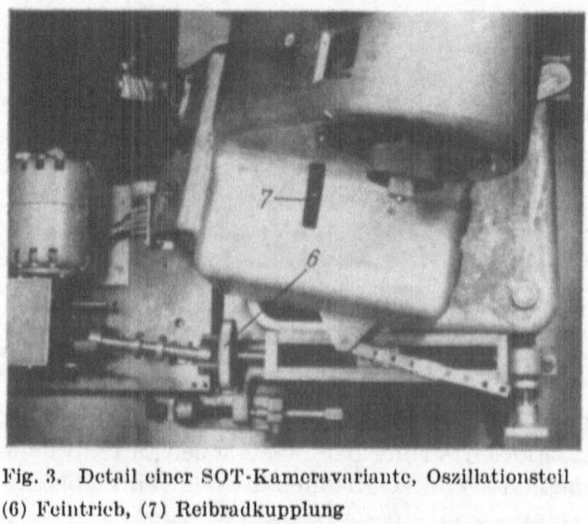

Fig. 3. Detail einer SOT-Kameravariante, Oszillationsteil
(6) Feintrieb, (7) Reibradkupplung

nalspindel aber nicht das Getriebe der Weißenbergkamera verwendet werden. Das Getriebe wurde durch Umbau untersetzt und eine Gummireibradkupplung (7 in Figur 3) eingebaut. Das Umschalten des Motors nimmt ein mit den Endlagen kombiniertes Relais vor. Zur Erzeugung der Oszillationsbewegung wurde eine Vorrichtung entwickelt (Figur 3), die aus einem umschaltbaren Motor, einem Getriebe und einer Kupplung besteht. Die Vorrichtung kann über die Kupplung fest über das der Feinjustierung dienende Handrad mit dem Feintrieb verbunden werden. Der Feintrieb und das Getriebe sind so konstruiert, daß ein Oszillieren bis zu einem Maximalwert von $\pm 6{,}5'$ möglich ist.

4. Beispiele der Anwendung der SO-Technik

Die Wirkung einer SO-Aufnahme im Verhältnis zu einer Aufnahme mit der Lang-Technik soll an einigen Beispielen demonstriert werden. Diese Beispiele unterstreichen weiterhin die Vielfalt der Anwendungsmöglichkeiten. Es handelt sich in allen Fällen um Siliziumproben. Grundsätzlich lassen sich all diese

a)

b)

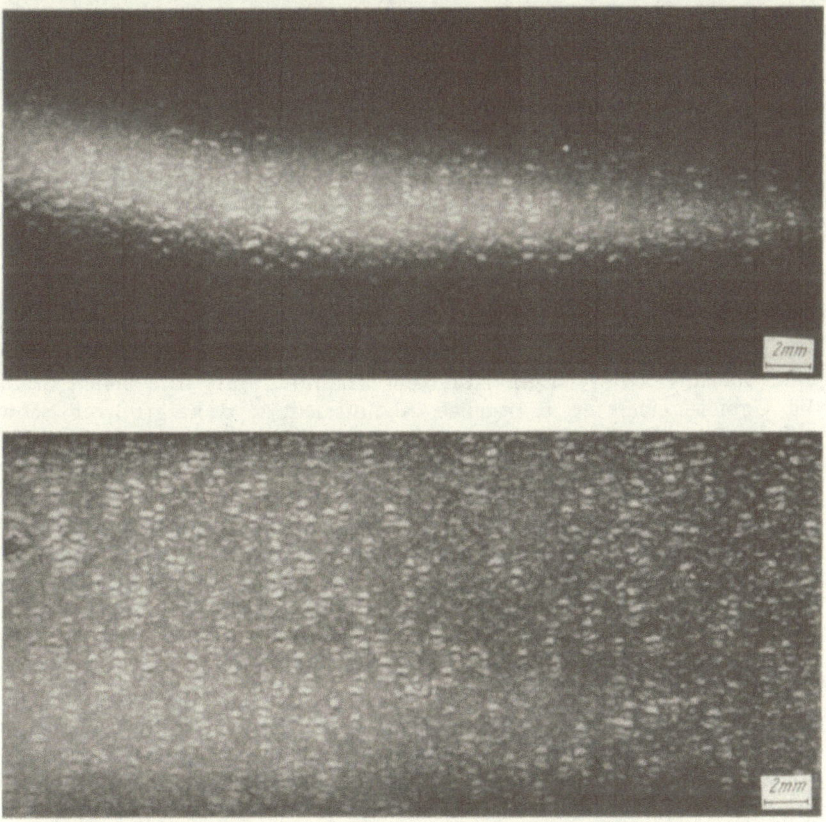

Fig. 4. Phosphordotierte Si-Probe, {220}-Reflexion, $MoK\alpha_1$.
(a) Lang-Topogramm
(b) SOT-Topogramm

Tabelle 1

Verfahren	Fokus	Mono-chromasie	Proben bewegung	bestrahlter Kristallbereich	Genauigkeit Auflösungs vermögen (μm)	Beugungsfall
Parallelstrahl-verfahren nach Berg/Barrett	Strich	$K\alpha_1$ und $K\alpha_2$ Reflex nicht getrennt	nein	flächenhaft (meist recht-eckige Fläche)	100 – 200	Transmission Reflexion
Lang-Technik	Punkt oder Strich	Mono-chromasie durch Aus-blendung	translatorisch (Einhaltung des Bragg-Winkels)	strichförmig (Bild wird aus Strichbildern kontinuierlich zusammengesetzt	3 – 5	Transmission (Reflexion im Prinzip möglich)
SO-Technik nach Schwuttke	Punkt oder Strich	Mono-chromasie durch Aus-blendung	translatorisch und Oszillations-bewegung (Änderung des Einfallwinkels) nur Schwenk-bewegung	strichförmig wie bei Lang flächenhaft	5 – 15	Transmission (Reflexion im Prinzip möglich)

Techniken auch für die anderen Halbleitereinkristalle anwenden, wie Ge, GaAs, GaP etc. Jedoch müssen für diese Materialien die Kontrastbedingungen in stärkerem Maße beachtet werden wie bei Silizium. Alle Aufnahmen sind unter den Bedingungen des Extinktionskontrastes aufgenommen worden und stellen Vergrößerungen der Original-Plattenaufnahmen dar. Die hellen Stellen sind die Gebiete erhöhter Röntgenstrahlenintensität, d. h. größerer Schwärzung auf der Fotoplatte.

In Figur 4 ist eine Aufnahme einer mit Phosphor dotierten Probe zu sehen. Die Ausgangsprobe war feingeläppt, sie wurde von allen Seiten diffundiert (Eindringtiefe $\sim$50 μm). Vor der Röntgenaufnahme wurde eine Phosphor-dotierte Schicht durch Ätzpolitur entfernt. Der im oberen Teilbild erkennbare helle Streifen (Lang-Aufnahme) kann im Hinblick auf die Störungsverteilung zu Fehlschlüssen führen. Die Aufnahme im unteren Teilbild mit der SO-Technik beweist, daß der Streifen-Effekt ausschließlich durch die Probenkrümmung hervorgerufen wird.

Figur 5 ist das Topogramm einer Si-Probe, welche einseitig mit einer amorphen Deckschicht versehen ist. Die Aufnahme im Teilbild b) beweist wieder, daß der Intensitätseffekt ausschließlich auf eine Verkrümmung der Probe zurückgeht. Bei einer Zunahme der Deckschichtdicke wird schließlich ein Stadium erreicht, wo die Krümmung der Si-Scheibe so groß wird, daß es zur Versetzungsbildung kommt. In Figur 6 ist links wieder die Lang-Aufnahme und rechts die SOT-Aufnahme zu sehen. Man erkennt hier an der Breite des aufgehellten Streifens, der über die Aufnahme geht, daß diese Probe stärker

Anwendungsmöglichkeiten	Einsatzgebiete in der Halbleitertechnik
Mosaik-Kristalle Zwillings-Kristalle Einkristalle: Grobe mechan. Störungen, Makroskop. Ausscheidungen; Segregationen Grobe elastische Verspannungen	Kristallzüchtung (kompakt, Epitaxieschichten) Mechan. Kristallbearbeitung Untersuchung von Oberflächenschichten in Reflexion
Schichtenaufnahmen Nachweis und Identifizierung von Baufehlern in nahezu idealen Einkristallen: Nachweis von Einzelversetzungen Burgers-Vektor-Bestimmungen einzelner Versetzungen Stapelfehler Segregationen Ausscheidungen Nachweis von mechan. Zerstörungen; Risse Deformationstiefe Nachweis elastischer Verspannungen, z. B. von unterbrochenen Deckschichten ausgehend	Einkristallzüchtung (kompakt, Dendriten, Epitaxie) Grundlagenforschung: z. B. Vergleichsmethode zur Eichung von anderen Nachweisverfahren für Baufehler, z. B. „etch-pits"-Ätzung Kristallbearbeitung: Prozeßkontrolle bei den mechan. Bearbeitungsverfahren der Kristallscheiben (Störungsnachweis beim Läppen, Polieren, Schleifen, Läppstrahlen, Trennschleifen) Gütekontrolle: Einkristallkontrolle Prozeßkontrolle und Verfahrensentwicklung für Verfahrensschritte zur Herstellung spezieller Bauelemente, wie z. B. Diffusion, Fotolithographie, Erzeugung von Deckschichten, Metallisierung, Kontaktierung, Vereinzelung
wie beim Lang-Verfahren insbesondere bei verkrümmten Proben	Gleiche Einsatzgebiete wie bei der Lang-Technik, jedoch insbesondere bei der Verfahrensentwicklung und Prozeßkontrolle von Verfahrensschritten zur Herstellung spezieller Bauelemente bei Einseitenprozessen bezügl. der Kristallscheibe bzw. des Kristallplättchens.

verkrümmt ist als die in Figur 5. Die in Figur 6 abgebildeten streifenförmigen Muster sind Mikrorisse, welche durch elastische Spannungen infolge der Deckschicht entstanden sind. Die Risse gehen durch die amorphe Deckschicht hindurch, was aus ätztechnischen Untersuchungen sicher festgestellt wurde. Offenbar enden sie in einer dünnen Oberflächenschicht des Silizium-Substrates, was durch das Röntgentopogramm belegt wird. Interessant ist die Tatsache, daß die Spalteigenschaften des einkristallinen Substrates sich in der Weise ausge-

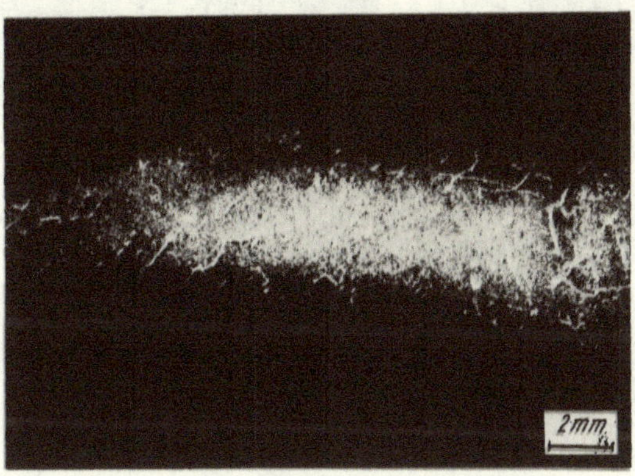

Fig. 5a)

b)

Fig. 5. Si-Scheibe mit einseitiger amorpher Deckschicht, {220}-Reflexion, MoK_{α_1}.
(a) Lang-Topogramm
(b) SOT-Topogramm

a)

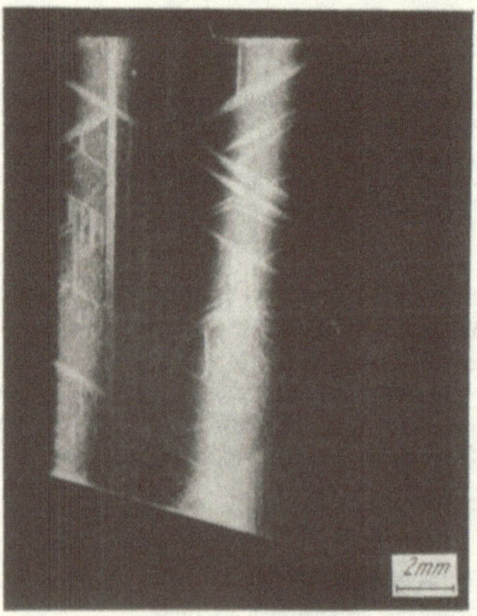

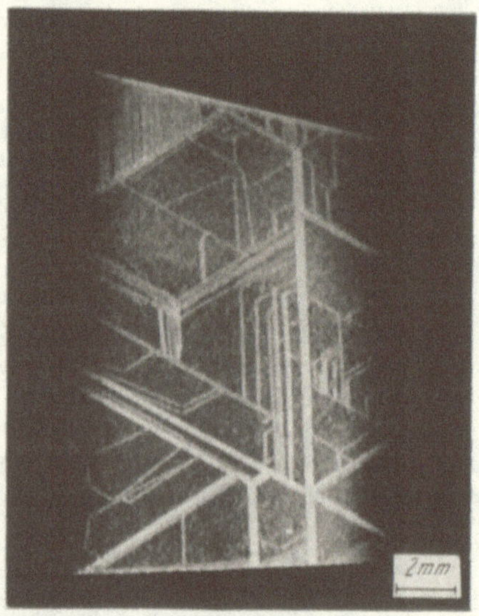

Fig. 6. Si-Scheibe mit einseitiger amorphen Deckschicht, {220}-Reflexion, MoK_{α_1}.
(a) Lang-Topogramm (K_{α_1} und K_{α_2}-Reflex)
(b) SOT-Topogramm

wirkt haben, daß ein Haar-Rißsystem im Silizium entstanden ist, bei dem annähernd alle Teilrisse dem {110}-Spaltsystem angehören. Mit dem Prozeß des Aufbringens der Deckschicht ist ein Temperprozeß verbunden, durch welchen eine Versetzungsbildung ermöglicht wird. In Figur 7 erkennt man Transistoranordnungen. Durch die einseitige Behandlung bei den einzelnen Prozeßschritten wurde die Si-Scheibe leicht verbogen. Wie in Figur 7a im Lang-Topogramm zu erkennen ist, läßt sich dadurch nur ein Teil der Si-Scheibe im Topogramm abbilden. Ganz anders in Figur 7b, hier ist von der gleichen Probe nach der SO-Technik ein Topogramm angefertigt worden. Alle auf dem Probenausschnitt vorhandenen Transistoren sind im Topogramm abgebildet. Zusätzlich sind in den Bildern 7 a und 7 b Versetzungen zu erkennen, die sich infolge der Diffusion über Gleitprozesse gebildet haben.

a)

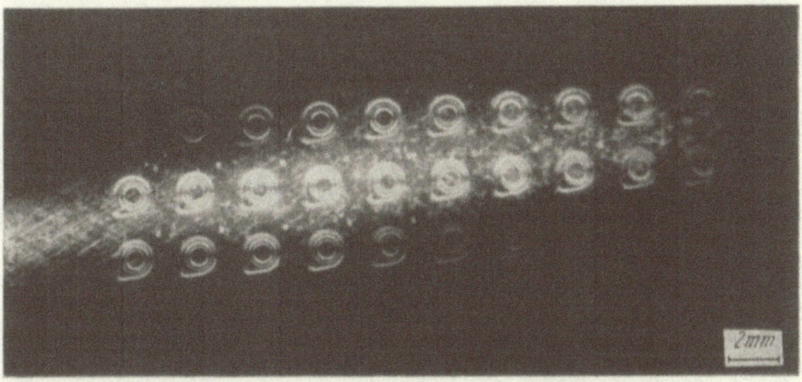

b)

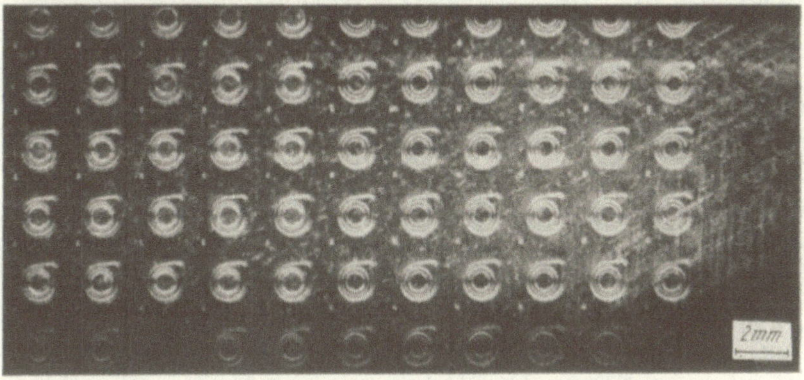

Fig. 7. Si-Scheibe mit Transistoranordnungen, {220}-Reflexion, $MoK\alpha_1$.
(a) Lang-Topogramm
(b) SOT-Topogramm

5. Zusammenfassung

Es wurde eine Einschätzung der Anwendbarkeit der bekannten röntgentopographischen Verfahren auf dem Gebiet der Halbleitertechnik gegeben. Eine wesentliche Eigenschaft, die für die Breite der Anwendung einer Methode ausschlaggebend ist, ist das Auflösungsvermögen. Während die Lang-Technik das beste Auflösungsvermögen (3—5 μm) zu realisieren gestattet, ist die „scanning-oszillator-technique" nach SCHWUTTKE mit einer etwas geringeren Auflösung (5—15 μm) zur Topographie von bearbeiteten Halbleiterproben (vorzugsweise bei ungleicher Behandlung der beiden Scheibenflächen) besonders geeignet, da Krümmungen der Netzebenen in bestimmten Grenzen nicht zu „blinden Bereichen“ durch Verletzung der Bragg’schen Reflexionsbedingungen führen. Die große Einsatzmöglichkeit der SO-Technik wurde an Hand von typischen Beispielen demonstriert.

Literatur

BARRETT, C. S.: Trans. AIME **161**, 15 (1945)
BERG, W. F.: Naturwissenschaften **19**, 391 (1931)
JENKINSON, A. E., LANG, A. R.: Direct Observation of Imperfections in Crystals, Newkirk und Wernick, Eds., New York (1962), S. 471
JENKINSON, A. E.: Philipps „Technische Rundschau“ 23. Jahrgang 198 (1961/62)
JULEFF, E. M., LAPIERRE, A. G.: Int. J. Electronics **20**, 273 (1966)
LANG, A. R.: J. appl. Physics **29**, 597 (1958); J. appl. Physics **30**, 1748 (1959); Acta Cryst. **12**, 249 (1959)
LANG, A. R.: Brit. J. appl. Physics **14**, 904 (1963)
MOHR, U.: Z. physik. Chemie **227**, 235 (1964)
MOHR, U., JEGERLEHNER, K.: Z. physik. Chemie **231**, 42 (1966)
NEWKIRK, J. B.: Physic. Rev. **110**, 1465 (1958); Trans. AIME **215**, 483 (1959)
SCHWUTTKE, G. H.: J. appl. Physics **36**, 2712 (1965)
SCHWUTTKE, G. H., QUEISSER, H. J.: J. appl. Physics **33**, 1540 (1962)
YOSHIMATSU, M.: J. Physic. Soc. Japan **18**, II 335 (1963)

(Eingegangen am 24. Juni 1971)

Anschrift der Verfasser:

Dipl.-Min. K. JEGERLEHNER, Dr. U. MOHR
Kombinat VEB Halbleiterwerk Frankfurt/Oder
Gleichrichterwerk Stahnsdorf
DDR 1533 Stahnsdorf,
Ruhlsdorfer-Weg

K. H. Jost

Institut für Anorganische Chemie
of the Deutsche Akademie der Wissenschaften zu Berlin
Berlin-Adlershof, DDR

Setting of Crystals on the Precession Camera

The general case of the setting of crystals on the procession camera is dealt with.

Es wird der allgemeine Fall der Justierung von Kristallen auf der Precession-Kamera behandelt.

Setting of crystals on the precession camera is discussed in literature (Buerger), so far known, only under simplifying assumptions on the orientation of the arcs of the goniometer head relative to the primary beam. However, the solution of the general case not always can be avoided. Especially this is true, if the crystal originally was mounted on the goniometer head for other purposes, but now should be used for taking precession photographs.

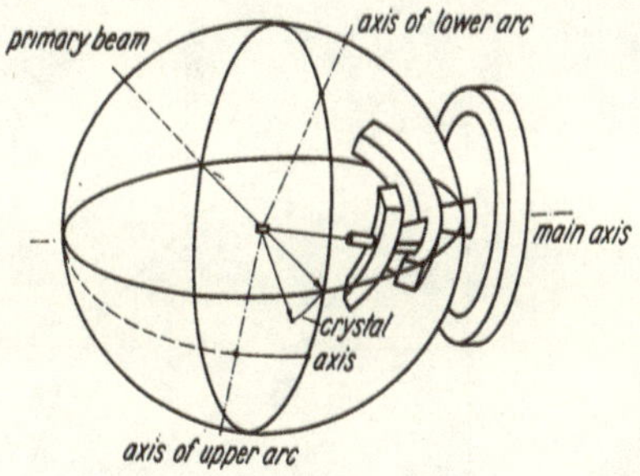

Fig. 1. Unit sphere, surrounding the crystal

First of all the misalinement of the crystal is to be determined in the known way by means of precession photographs at small angle $\bar{\mu}$. The misalinement will be characterized by the tilt δ of the crystal axis against the primary beam at $\bar{\mu} = 0°$ and the angle ψ between the equatorial plane to the main axis of the camera and the plane formed by the primary beam and the crystal axis (Fig. 1). Main axis is called the axis which bears the goniometer head. In order to calculate the necessary corrections on the goniometer head and the main axis, the crystal is thought of to be surrounded by a unit sphere. On this sphere the points are marked in which it is hit by the

axis of the lower arc of the goniometer head
axis of the upper arc of the goniometer head
main axis of the camera
crystal axis which has to be set.

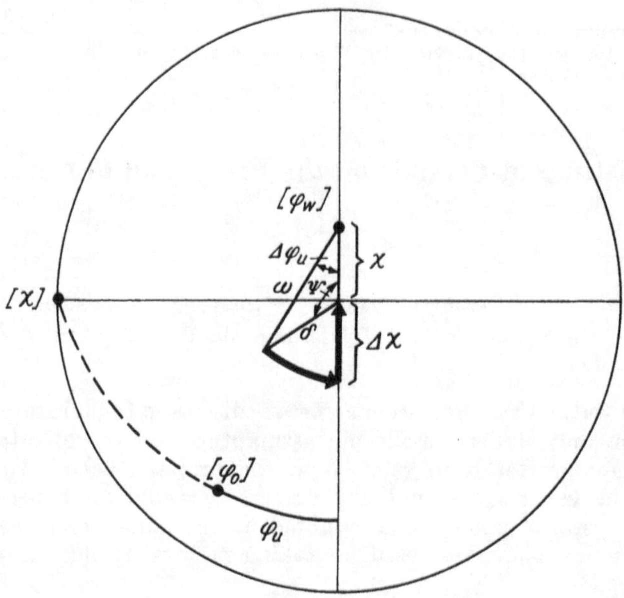

a)

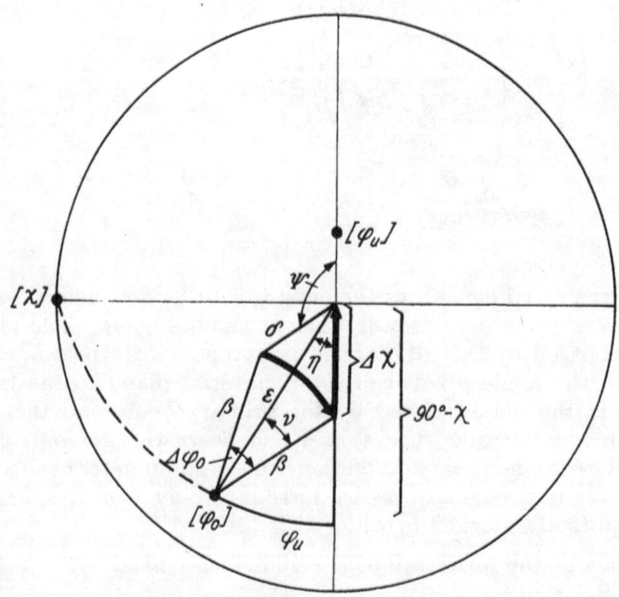

b)

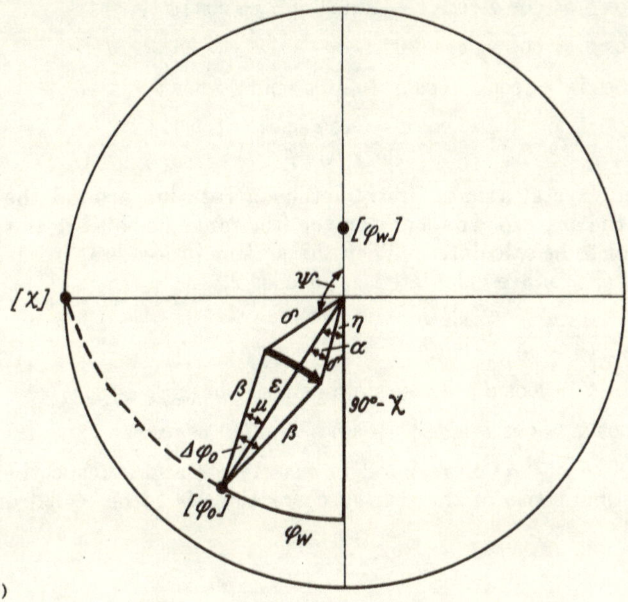

c)

Fig. 2. View on the front half of the unit sphere shown in Figure 1, looking against the primary beam. The beam hits the figures at their centers.
Thick lines: the way in which the crystal axis is oriented $[\varphi_u]$, $[\varphi_0]$, $[\varkappa]$: the points at which the axes of the lower and the upper arcs of the goniometer head and the main axis of the camera meet the surface of the sphere

a) Case 1
b) Case 2
c) Case 3

(in a) and c) read $[\varphi_u]$ instead of $[\varphi_w]$ and in c) φ_u instead of φ_w)

There are four possibilities of setting the crystal of which three shall be dealt with. In these three cases the crystal is set by rotation around the main axis, too. Which of these methods may be applied in an actual case depends on the setting ranges on the arcs of the goniometer head.

Case 1: The crystal axis is set by a rotation of angle $\Delta\varphi_u$ on the lower arc into the equatorial plane of the main axis, followed by a rotation of angle $\Delta\varkappa$ around the main axis. Figure 2a shows how the corrections are to be calculated.

$$\cos\omega = \cos\delta \cdot \cos\varkappa + \sin\delta \cdot \sin\varkappa \cdot \cos\psi$$

$$\sin\Delta\varphi_u = \sin\psi \frac{\sin\delta}{\sin\omega}$$

$$|\Delta\varkappa| = |\varkappa - \omega| .$$

Case 2: The crystal axis is set by a rotation of angle $\Delta\varphi_0$ on the upper arc into the equatorial plane, followed by a rotation of angle $\Delta\varkappa$ around the main axis (Fig. 2b).

$$\cos\varepsilon = \cos(90° - \varkappa)\cos\varphi_u$$

$$\sin\eta = \frac{\sin\varphi_u}{\sin\varepsilon}$$

$$\cos\beta = \cos\delta\cdot\cos\varepsilon - \sin\delta\cdot\sin\varepsilon\cdot\cos(\psi+\eta)$$

$$\cos\beta = \cos\varphi_\mu\cos(90^\circ - \varkappa - \Delta\varkappa)$$

$$\cos\Delta\varkappa = \cos\varepsilon\cdot\cos\beta + \sin\varepsilon\cdot\sin\beta\cdot\cos\nu$$

$$\cos(\Delta\varphi_0 - \nu) = \frac{\cos\delta - \cos\varepsilon\cdot\cos\beta}{\sin\varepsilon\cdot\sin\beta}\,.$$

Case 3: The crystal axis is firstly set by a rotation around the upper arc. The value of the angle $\Delta\varphi_0$ depends on the free range on this arc and is therefore known and not to be calculated. Then the setting proceeds as in case1 (Fig. 2c).

The angles ε, β, η are calculated as in case 2.

$$\frac{\sin\mu}{\sin(\psi+\eta)} = \frac{\sin\delta}{\sin\beta}$$

$$\cos\delta' = \cos\beta\cos\varepsilon + \sin\beta\cdot\sin\varepsilon\cdot\cos(\Delta\varphi_0 - \mu)$$

$$\cos\beta = \cos\varepsilon\cdot\cos\delta' + \sin\varepsilon\cdot\sin\delta'\cdot\cos\alpha\,.$$

δ' and $(180^\circ - \eta + \alpha)$ correspond in case 1 to δ and ψ respectively.

The directional sense of the setting corrections is to be read from a sketch like Figure 2.

References

Buerger, M. J.: The Precession Method, New York/London/Sydney 1964

(Received May 20, 1971)

Author's address:

Dr. habil. K.-H. Jost
Institut für Anorganische Chemie der Deutschen Akademie der Wissenschaften zu Berlin
1199 Berlin-Adlershof
Rudower Chaussee 5

Kristall und Technik | 7 | 1–3 | 1972 | 317–339

A. ANDREEFF, M. SCHENK

Bergakademie Freiberg, Sektion Physik
DAW, ZfK Rossendorf

Kernphysikalische Methoden in der Kristallographie und Festkörperphysik

Es wird eine Übersicht über einige „synthetische " und „analytische" kernphysikalische Methoden und ihre Anwendung zur Lösung festkörperphysikalischer und kristallographischer Aufgabenstellungen gegeben. Es wird auf teilweise neuartige Applikationsmöglichkeiten hingewiesen.

A review is given about some "synthetic" and "analytic" nuclearphysical methods and their use for studies in solid state physics and crystallography. Some new possibilities for applications are pointed out.

0. Einleitung

Eine Klassifikation kernphysikalischer Methoden in der Kristallographie und Festkörperphysik ist in vielfältiger Weise möglich und wird natürlich von subjektiven Gesichtspunkten aus bestimmt. Hier sollen diese Methoden in zwei Gruppen geteilt werden.

Die erste Gruppe bezweckt Stoffveränderungen, z. B. Struktur- und Phasenumwandlungen oder Veränderungen der chemischen Zusammensetzung. Diese Methoden sollen die „synthetischen" Verfahren genannt werden, weil durch sie neue Stoffeigenschaften zu synthetisieren sind.

Die zweite Gruppe umfaßt die Methoden, mit denen Aussagen über Stoffeigenschaften zu gewinnen sind, z. B. über die Struktur oder über Kristallfelder. Diese Methoden sollen die „analytischen" Verfahren genannt werden.

Es kommt den Autoren insbesondere darauf an, viele unterschiedliche Beispiele für Applikationsmöglichkeiten anzuführen, um Anregungen für eine möglichst große Zahl von Arbeitsgebieten zu geben.

Es wird nicht beabsichtigt, eine umfassende Auswertung der Literatur durchzuführen.

1. Synthetische Methoden

Als Grundlage für das Verständnis, die Wechselwirkung von Strahlung mit fester Materie zu nutzen, werden in einer groben Übersicht (Fig. 1) die prinzipiellen Möglichkeiten gezeigt, die beim Ablauf einer Bestrahlung in einem Festkörper auftreten können.

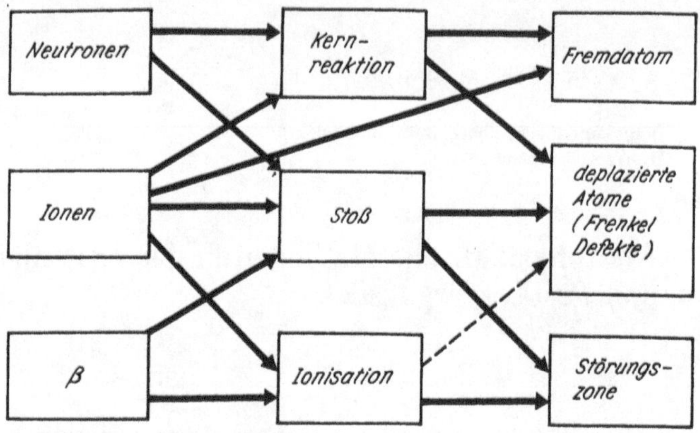

Fig. 1. Prozesse der Strahlenwirkung im festen Körper

Links sind die Strahlenarten, in der Mitte mögliche Zwischenzustände und rechts die wesentlichsten verbleibenden Strahlenschädigungen aufgetragen. Außerdem kann es zu Oberflächenrückstreuungen oder zum Durchtritt der Strahlung durch die Probe kommen. Alle Wechselwirkungen sind Variable der Strahlungsenergie, der Wirkungs- (Streu-, Reaktions-)querschnitte der Atomsorten für die applizierte Strahlenart und weiterer nicht so wesentlicher Parameter. Auf das Zusammenwirken dieser Faktoren soll hier nicht näher eingegangen werden; notwendige Hinweise werden bei der Behandlung konkreter Ergebnisse gegeben. Es wird hier nur auf den prinzipiellen Unterschied zwischen den „chemischen" Strahleneffekten (Fremdatome) und den „physikalischen" verwiesen. Die bei der Bestrahlung eingelagerten Fremdatome können nicht durch Temperaturbehandlung (Ausheilen) beseitigt werden, wie es bei den physikalischen Defekten der Fall ist. Außerdem sind viele durch Kernumwandlungen entstandene Fremdatome radioaktiv, was bei der anschließenden Behandlung der Probe zu beachten ist.

Es können hier nicht alle Varianten der Nutzung von Strahleneffekten, also alle synthetischen kernphysikalischen Methoden, behandelt werden. Die Komplexe „Implantation", „Dotierung durch Kernreaktion" und „Gezielte Gitterstörungen" sind in der Literatur noch nicht ausführlich und zusammenfassend besprochen worden, so daß nur auf sie näher eingegangen wird.

1.1. Die Implantation und ihre Erfolge in der Mikroelektronik sollen im wesentlichen als bekannt vorausgesetzt werden (s. z. B. Mayer, Eriksson, Davies). Die dabei erzielte Injektion geringer Gehalte bestimmter Fremdatome zur Erzeugung von Störleitungsphänomenen (Donatoren, Akzeptoren), die letztendlich zur Herstellung von Leitungstypänderungen und p-n-Übergängen dient, kann aber nicht nur mit Hilfe von Massentrennern oder Beschleunigern, die mit spezifischen Ionenquellen ausgerüstet sind, bewerkstelligt werden, sondern in einigen Fällen auch mittels relativ einfacher Laborgeräte. Z. B. wurde (Shimizu) mit 20···50 keV Argonionen Aluminium bzw. Antimon in Silizium eingeschossen. Die Metalle waren vorher als dünne Schicht (100···1000 Å) auf das Si-Substrat aufgebracht worden. Man nimmt an, daß die Ar-Ionen beim Auftreffen auf die Metallschicht ihre kinetische Energie wie beim Billardballstoß an die getroffenen Metallatome verlieren, welche schließlich in das Si eindringen.

Die Implantation unterscheidet sich als Vorgang nicht von der Herstellung stöchiometrischer Verbindungen durch Ioneneinschuß. Das Endprodukt zeigt jedoch prinzipiell andere Eigenschaften. Dieses Gebiet ist verhältnismäßig jung, so daß eine etwas ausführlichere Behandlung angemessen erscheint (s. auch Balarin).

Beispielsweise erreichte Foss durch eine 2 kV-HF-Gasentladung in Quecksilberdampfatmosphäre, daß Cadmiumtelluridproben (CdTe) zu einer Zusammensetzung von beispielsweise $Hg_{0.8}Cd_{0.2}Te$ verändert wurden. Auf diese Weise war es prinzipiell möglich, die spektrale Verteilung der Photoleitfähigkeit so zu beeinflussen, daß die effektive Bandbreite von 1,5 eV (für CdTe als Ausgangssubstanz) bis zu ca. 0,12 eV verringert werden konnte. Dabei wurde nur eine Schicht von etwa 100···500 Å unter der Oberfläche effektiv.

Bei ähnlichen Untersuchungen zum Beschuß von Galliumarsenid (GaAs) mit Al fand Hunsperger in Abhängigkeit von der Konzentration x in $Ga_xAl_{1-x}As$ beträchtliche Veränderungen im optischen Emissionsspektrum.

Balarin und Mitarbeiter fanden bei Untersuchungen zur Bildung von Sauerstoff-, Stickstoff- und Kohlenstoffverbindungen auf Halbleitern und Metallen (Passivierung von Oberflächen, Bildung von Widerständen und Kondensatoren im Gesamt-Vakuum-Prozeß), daß durch die Injektion entsprechender Ionen relativ niedriger Energie (1···5 keV) eine Schichtbildung auf Silizium, Kupfer, Tantal und Aluminium auftrat. Diese Oberflächenschichten waren einige 100···1000 Å dick und zeigten trotz hoher Gitterdistortion und Trapdichte gute Durchschlagsfestigkeiten bis zu einigen 10^6 V/cm. Die Schichten blieben bis 200 °C stabil. Die Dielektrizitätskonstanten stimmten mit denen der kompakten Materialien überein.

Eine in der Gleichstrom-Gasentladung durchgeführte Oberflächenoxydation wird seit einigen Jahren in der Metallographie verwendet. Hilbert (1963) gelang es als erstem, korrosionsfeste Stähle, Spaltmaterialien und andere schwierig auf andere Weise für die normalen metallographischen Verfahren zugänglichen Substanzen auf kontrollierbare Weise durch Sauerstoffbeschuß mit Oxidfilmen zu versehen. Von Korn zu Korn treten dabei starke Farbunterschiede auf, die erst eine metallographische Untersuchung ermöglichen. Die verschiedenen Farben entsprechen unterschiedlichen Oxiddicken und sind eindeutig jeweils einem Korn zuzuordnen. Die Mannigfaltigkeit der Farben und damit der Oxiddicken wird durch die unterschiedliche Orientierung der Körner und der auf diesen epitaktisch aufgewachsenen Oxidfilmen zur Oberfläche verursacht.

Schließlich ist zur Bildung stöchiometrischer Verbindungen durch Einschuß von Ionen auch noch die Entstehung von Ausscheidungen zu rechnen. Das kann z. B. bei der Härtung von Aluminium- oder Stahloberflächen von großer praktischer Bedeutung sein (Thackery, Nelson; Phillips, Seybolt; Seybolt; Hay, Martin). Es ist bekannt, daß Stahl durch Nitridierung gehärtet werden kann. Mit Stickstoffionenbeschuß konnte dieser Effekt jedoch bedeutend verstärkt werden, besonders, wenn vor dem Beschuß eine Kaltverformung durchgeführt wurde. Z. B. wurde die Härte von Eisen mit 1 Gew.-% Al von 74 auf 366 kp · mm^{-2}, im kaltverformten Zustand von 175 auf 503 kp · mm^{-2} gesteigert. Das ist auf die Ausbildung sehr feindisperser AlN-Ausscheidungen von 10···40 Å Durchmesser zurückzuführen, die zwar durch die spezifischen Bedingungen bei Ionenbeschuß, nicht aber bei normaler Diffusion erreicht werden kann.

Außerdem erfolgte die Nitridierung durch Ionenbeschuß beträchtlich schneller als die thermische Nitridierung bei der gleichen Temperatur. Ein Beispiel dafür sind die in Figur 2 dargestellten Meßkurven, die bei jeweils 1300 °C aufgenommen wurden. Die in Kurve 2 gezeigten Nitridierungstiefen nach Ionenbeschuß traten in beträchtlich kürzeren Zeiten auf als nach normaler thermischer Behandlung (Kurve 1). Das ist auf die bei der Bestrahlung gleichzeitig erzeugten Gitter-Leerstellen zurückzuführen, die zu einer strahlenbeschleunigten Diffusion führen, worauf später noch eingegangen wird.

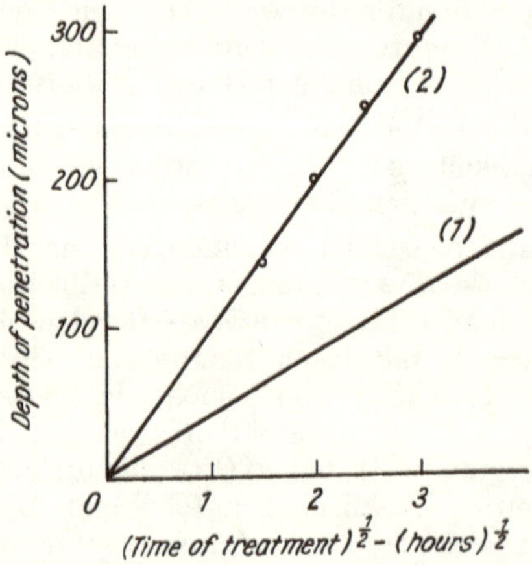

Fig. 2. Eindringtiefe von Stickstoff in Stahl bei thermischer Nitridierung (1) bzw. Hochtemperatur-Stickstoff-Ionenbeschuß (2) (nach Hay, Martin)

1.2. Nun sollen Möglichkeiten betrachtet werden, die sich durch Kernreaktionen ergeben. Unter bestimmten Bedingungen können Teilchen den Coulombwall eines Atomkerns durchdringen und mit dem Kern reagieren, wobei Folgeprodukte nach dem Schema

$$A(a, b)\, B$$

auftreten (A = Ausgangsatomkern, B = Finalkern, a = eindringendes (Beschuß-)Teilchen, b = ausgesandtes Teilchen).

Der Finalkern der Reaktion braucht nicht stabil zu sein, sondern kann unter Aussendung weiterer Teilchen oder γ-Quanten zerfallen. Die evtl. bei diesen Reaktionen auftretenden Atomdeplazierungen sollen hier nicht betrachtet werden.

Die Wahrscheinlichkeit des Zustandekommens solcher Kernreaktionen hängt u. a. wesentlich vom sog. Einfangquerschnitt der Kerne der Gitteratome und von der Energie und Ladung der Beschußteilchen ab. Da diese Parameter und Bedingungen sich um Größenordnungen bzw. prinzipiell von Experiment zu Experiment unterscheiden können, kann nicht näher auf allgemeine Gesichtspunkte eingegangen werden. Ein Hinweis erscheint jedoch erwähnenswert: die Einfangquerschnitte sind i. a. nicht für alle Beschußteilchenenergien konstant, es treten in vielen Fällen energieabhängige Resonanzen auf, d. h., daß für bestimmte Teilchenenergien die Wahrscheinlichkeit von Kernreaktionen beträchtlich ansteigt. Diesen Zusammenhang könnte man prinzipiell ausnutzen, um entsprechend Figur 3 in einer bestimmten Tiefe unterhalb der Oberfläche eine stark dotierte Schicht zu erzeugen (Hohmuth). Die Energie der Beschußteilchen nimmt mit der Eindringtiefe ab. Je nach Anfangsenergie kann die Resonanzreaktion von Ionen mit Matrixkernen in einer erwünschten Tiefe, die in gewissen Grenzen variabel ist, stattfinden.

Das gilt jedoch nicht für Experimente mit thermischen Neutronen, denn diese zeigen eine prinzipiell andere Energieverlustabhängigkeit von der Materialdicke, als es in Figur 3 schematisch angedeutet ist.

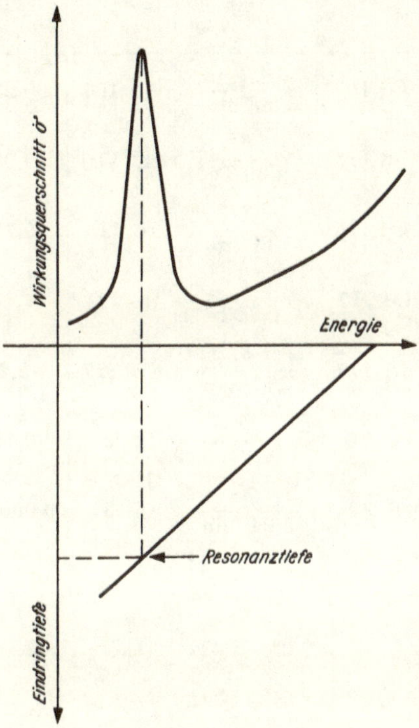

Fig. 3. Schematische Darstellung einer Ionen-Resonanzreaktion zur Erzeugung dotierter Schichten. Unteres Teilbild: Ionenenergie in Abhängigkeit von der Eindringtiefe

Mit Neutronen wurde schon eine Reihe von Kernreaktionen durchgeführt, um gezielte Dotierungseffekte in Halbleitern zu erzeugen (z. B. CLELAND). Solche Untersuchungen an Halbleitermaterialien sind deshalb besonders vielversprechend, weil es in der Folge der Reaktion häufig zu Finalkernen kommt, die im Periodensystem der Elemente direkt links oder rechts neben den Wirtsatomen liegen und deshalb als Donatoren oder Akzeptoren wirken. An einem Beispiel, der Bestrahlung von Kadmiumsulfid (CdS) mit thermischen Neutronen, soll das kurz erläutert werden. In Tabelle 1 ist die Anzahl der Fremdatome pro cm^3 nach einer Bestrahlung mit 10^{15} thermischen Neutronen/cm^2 aufgetragen (HAUSER, KÖHLER). Den größten Anteil an den insgesamt erzeugten Fremdatomen bildet Indium 115, wesentlich ist auch noch Sn 117. Hier zeigt sich ein eventueller Nachteil bei Versuchen an binären oder gar ternären Halbleitern: es entsteht eine Vielzahl von Reaktionsprodukten, die durchaus unerwünscht sein können. Das muß am konkreten Beispiel von Fall zu Fall geklärt werden. Beim hier diskutierten CdS trat im Dosisbereich $7 \cdot 10^{15} \cdots 1 \cdot 10^{17}\ n_{th}/cm$ eine beträchtliche Zunahme der Photoleitung auf (Figur 4), was interessante

Tabelle 1

Anzahl der Fremdatome pro cm^3, die bei der Bestrahlung von CdS mit 10^{15} thermischen Neutronen/cm^2 erzeugt werden (HAUSER, KÖHLER)

Ausgangsisotop	% in nat. Isotopengemisch	σ_{akt} (barn) (n, γ)-Reaktion	Folgeisotop	Zerfall / Halbwertszeit	Endprodukt	Anzahl $\cdot 10^{-12}$	Anteil an den insges. erzeugten FA (%)
Cd 106	1,215	1,0 ± 0,5	Cd 107	k, β^+ / 6,7 h	Ag 107	0,23	2,5
Cd 114	28,86	1,1 ± 0,3	Cd 115	β / 2,39 Tg	In 115	6,06	64,2
		0,14 ± 0,03	Cd 115	β / 43 Tg	In 115	0,77	8,2
Cd 116	7,58	1,5 ± 0,3	Cd 117	β / 2,72 h	In 117		
			In 117	β / 117 min	Sn 117	2,2	23,0
S 34	4,18	0,26 ± 0,05	S 35	β / 87 Tg	Cl 35	0,21	2,1
S 36	0,014	0,14 ± 0,04	S 37	β / 5,04 min	Cl 37	0,00037	$3 \cdot 10^{-3}$

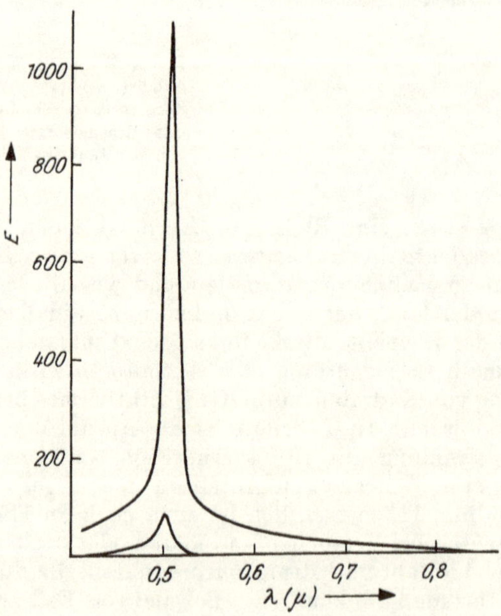

Fig. 4. CdS: spektrale Empfindlichkeit vor (untere Kurve) und nach (obere Kurve) Bestrahlung mit $7 \cdot 10^{16}$ n_{th}/cm^2 (nach HAUSER, KÖHLER)

technische Aspekte aufwirft. Die Figur 4 zeigt die spektrale Empfindlichkeit eines CdS-Kristalls vor (untere Kurve) und nach (obere Kurve) der Bestrahlung mit $7 \cdot 10^{16}\, n_{th}/cm^2$. Es erscheint also durchaus möglich, durch Reaktorbestrahlung CdS-Kristalle hoher Empfindlichkeit herstellen zu können.

1.3. Wie aus Figur 1 zu entnehmen ist, verursacht jede hochenergetische Strahlung im Festkörper eine mehr oder weniger ausgeprägte Schädigung des Gitteraufbaus durch Erzeugung von Frenkeldefekten. In den meisten Fällen ist diese Schädigung eine unerwünschte Nebenerscheinung, die es wieder zu beseitigen gilt. Das geschieht durch eine für das jeweilige Material und das Ausmaß des Schadens spezifische Temperaturerhöhung des Materials.

In einigen Fällen kann die Strahlenschädigung jedoch ausgenutzt werden, man sollte dann von „Strahlennutzung" sprechen. Einige solche Fälle sollen nun behandelt werden.

1.3.1. Ein triviales erstes Beispiel sind die Festkörperdetektoren für energiereiche Strahlung, insbesondere für hochenergetische Ionen. Hier dient die entstandene — reversible — Schädigung zum direkten Nachweis der Strahlung. In der kernphysikalischen Meßtechnik sind Halbleiterdetektoren für die Energieanalyse der untersuchten Strahlung nicht anderweitig ersetzbar.

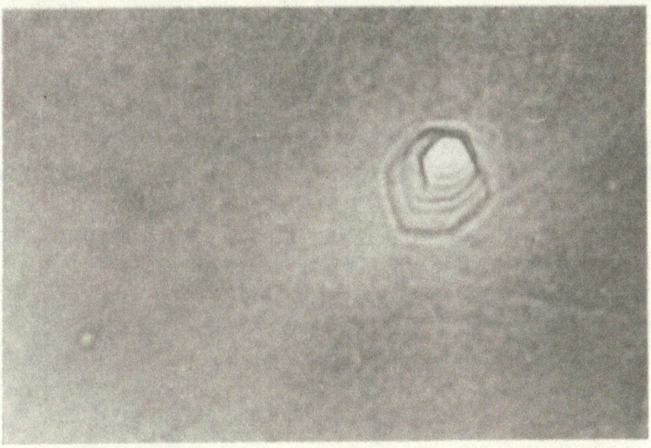

Fig. 5. Geätzter Glimmerdetektor: Spur eines Zerfallsprodukts

Eine andere Art von Detektoren, die auch in der Kernphysik verwendet wird, hat zu einem interessanten „Abfallprodukt" geführt, das auf ganz anderem Gebiet verwendet wird. Bei der Untersuchung von Kanalleitungseigenschaften einkristalliner Festkörper werden u. a. Zellulosenitratfilme (Quéré), Glimmer (Bonfiglioli, Ferro, Mojoni; Price, Walker) sowie andere Materialien benutzt, die die durch eine Einkristallfolie hindurchgeschossenen Ionen auffangen bzw. so von ihnen geschädigt werden, daß man die Teilchenspuren in ihnen elektronenmikroskopisch beobachten kann. Auf diese Analysenmethode wird später noch eingegangen; hier interessiert zunächst nur die i. a. senkrecht zur Filmoberfläche liegende Teilchenspur. Energiereiche Ionen verursachen im kristallinen Körper entlang ihres Weges eine starke Ionisation, die aufgrund von Rückstoßeffekten und intensiver thermischer Belastung zu beträchtlichen Gitterschäden führen kann. Der chemische Ätzangriff bestimmter Lösungen findet bevorzugt an solchen Gitterstörungen statt. Je nach Ätzdauer können so ganze Kanäle aus der Folie herausgeätzt werden, wie in Figur 5 für einen mit Zerfallsprodukten beschossenen Glimmerdetektor zu erkennen ist. Die hexagonale Symmetrie der Glimmerspaltflächen tritt deutlich hervor.

Verfährt man auf gleiche Weise mit α-Teilchen, die wesentlich geringere Gitterschädigungen hinterlassen als Spaltprodukte, so ist man in der Lage, Kanäle von einigen 10 Å Durchmesser durch Detektorfolien zu ätzen, und erhält auf diese Art das angedeutete Abfallprodukt dieser Experimente, nämlich Mikrosiebe, wie man sie auf keine andere Weise herstellen kann.

1.3.2. Ein weiteres Beispiel für die Nutzung einer Strahlenschädigung ist das Schweißen, Bohren und Fräsen mit Ionen- oder Elektronenstrahlen. Durch einen genügend intensiven und scharf gebündelten Strahl wird dem Material mehr Energie zugeführt, als durch Wärmeleitung wieder abgeführt werden kann: örtlich begrenzt kommt es zum Schmelzen oder sogar zum Verdampfen. Diese Möglichkeiten werden bereits industriell in der Metallurgie und Mikroelektronik angewandt, im letzteren Fall werden durch Elektronenstrahlfräsen z.B. Abgleichungen und Toleranzeichungen passiver Bauelemente vorgenommen.

1.3.3. Damit eng verwandt sind Anwendungen von Oberflächenstrahlenwirkungen in der Metallographie: bekannt dürfte das Säubern von Oberflächen durch niederenergetischen Edelgasionenbeschuß bzw. durch Glimmentladung sein. Kathodisches Oxydieren wurde bereits erwähnt; darüber hinaus bietet das kathodische Ätzen weitere Möglichkeiten, einem chemischen Ätzangriff gegenüber außerordentliche resistente Materialien zu untersuchen. Hierbei werden im Verlauf elastischer Stöße (Energieübertragung von Atom zu Atom wie beim Billardballstoß, sog. Fokusonen) in dichtgepackten Gitterreihen bevorzugt Gitterstörungen wie Korngrenzen oder Versetzungen angeätzt (HILBERT 1964).

Eine weitere Anwendungsmöglichkeit für das Ionenätzen in der Metallographie und Elektronenmikroskopie ist das Dünnen von Proben für die Durchstrahlungs-Elektronenmikroskopie (s. unter anonym). Auf diese Weise können Metalle, Gläser, Keramik und bestimmte organische Substanzen, beidseitig mit niederenergetischen Ionen beschossen, bis auf die notwendige Dicke abgeätzt werden (300···1000 Å). Das Verfahren hat gegenüber den üblichen, auf Elektropolieren und chemischer Ätzung basierenden Methoden Vorteile, die besonders in der Vermeidung von Spannungen, Verunreinigungen und Artefakten liegen.

1.3.4. Eine weitere Variante der gezielten Nutzung von Strahlenwirkungen sind Amorphisierungen. Wird einmal von allen anderen neuartigen Konsequenzen, die der amorphe Zustand mit sich zu bringen scheint, abgesehen, so fällt doch gleich die Möglichkeit ins Auge, ionen-, elektronen- oder laserstrahlerzeugte Amorphisierungen kristalliner Medien für Festwertspeicher anzuwenden. Der sog. Ovograph (nach Ovshinsky) nutzt solche Zustandsänderungen aus, um die unterschiedlichen photoelektrischen Eigenschaften der amorphen bzw. kristallinen Phase in einer dem Xerox-Verfahren ähnelnden Methode für die Vervielfältigung von Zeichnungen oder Schriften zu verwenden.

In der Mikroelektronik finden ebenfalls strahlenbedingte Amorphisierungen Interesse. Beispielsweise erzeugt intensiver Ionenbeschuß auf Si, in Abhängigkeit von der Energie der Ionen, auf der Oberfläche oder in einem Gebiet bis zu 2 μm unterhalb der Oberfläche eine amorphe Schicht, deren Leitfähigkeit um etwa 10^{-6} niedriger liegt als die der kristallinen Matrix. Eine ausheilende

Rekristallisation dieser Schicht setzt erst oberhalb 400 °C ein (Schwuttke, Brack, Gorey, Kahan, Lowe). Die amorphe Schicht könnte also als Isolator in Festkörper-Schaltkreisen eingesetzt werden.

Indirekt wird die Amorphisierung benutzt, um z. B. nach der Implantation von Bor in Si die B-Ionen auf Substitutions-Gitterplätze gelangen zu lassen (Blamires, Matthews, Nelson). Während z. B. Phosphor nach Implantation in Si und Temperung bei 630 °C (zur Ausheilung der Strahlenschädigung) zu 100% substitutionelle Gitterplätze einnimmt, sind nach der Bor-Implantation und gleicher thermischer Behandlung nur 7% der Bor-Atome substituiert eingelagert, während der Rest von 93% elektrisch nicht aktiv ist. Um einen hohen Prozentsatz der B-Implantanden auf normale Gitterplätze einzubauen, müssen Temperungen oberhalb 1000 °C durchgeführt werden, wodurch andere Elemente der Festkörperschaltkreise (z. B. die Goldkontakte) zerstört werden. Wird dagegen vor oder nach der B-Implantation durch Beschuß mit Inertgasen oder Si eine Amorphisierung der implantierten Schicht herbeigeführt, so reicht bereits die o. g. Rekristallisationstemperatur von 630 °C aus, um nahezu die Hälfte aller Bor-Atome substituiert einzulagern und auf diese Weise die erwünschte elektrische Aktivität der Dotanden zu erzielen.

1.3.5. Weiter soll auf die Möglichkeit der strahlenbedingten Phasenumwandlungen (Schenk 1970a) hingewiesen werden. In polymorphen Substanzen kann die Einlagerung von Eigenpunktdefekten zu solchen Veränderungen der thermodynamischen Parameter führen, daß Umwandlungspunktverschiebungen bis unterhalb Raumtemperatur auftreten, wie es in Figur 6 für das Beispiel Bariumtitanat gezeigt wird (Balarin, Schenk).

Das Bild zeigt die Dielektrizitätskonstante in Abhängigkeit von der Temperatur für verschiedene Neutronen-Strahlendosen. Kurve A entspricht der unbestrahlten Substanz. $BaTiO_3$ ist in seiner tetragonalen Raumtemperaturphase ferroelektrisch und besitzt eine sehr hohe DK. Auch wegen seiner piezoelektrischen Eigenschaften ist es ein technisch vielverwendetes Material.

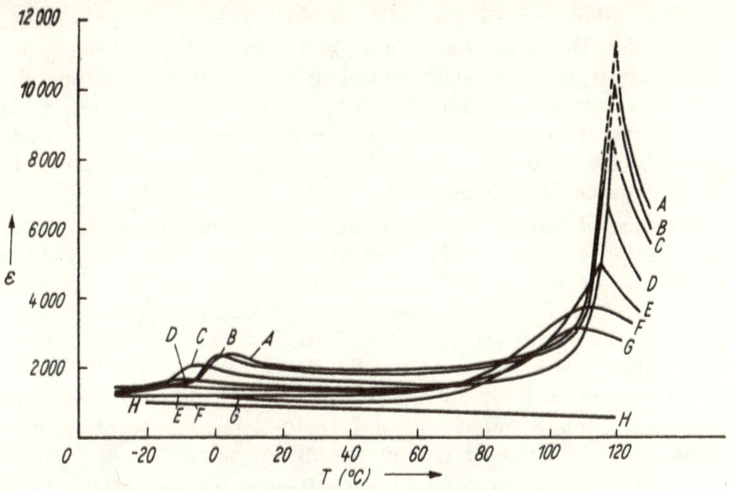

Fig. 6. $BaTiO_3$: Dielektrizitätskonstante ε in Abhängigkeit von Temperatur und Neutronenbestrahlungsdosen (s. Text) (nach Balarin, Schenk)

Störend wirkt die starke Temperaturabhängigkeit der DK in der tetragonalen Phase. Aus diesem Grunde wird die kubische Modifikation, die oberhalb 120 °C stabil ist, vielfach durch Legierungszusätze bei Raumtemperatur stabilisiert.

Bei zunehmender Strahlendosis ($B \rightarrow G$) wird die Umwandlungstemperatur tetragonal $\leftrightarrow$ kubisch, gleichbedeutend mit dem rechten Maximum der DK, zu tieferen Werten verschoben, außerdem wird die DK-Kurve geglättet. Die kubische Hochtemperaturphase ist nach einer bestimmten Neutronendosis (Kurve H) auch bei Raumtemperatur stabil, es hat also eine strahleninduzierte Phasenumwandlung tetragonal $\rightarrow$ kubisch stattgefunden. Eine Rückumwandlung in die unter normalen Bedingungen stabile tetragonale Modifikation findet erst bei Temperaturen oberhalb 1000 °C mit wahrnehmbarer Geschwindigkeit statt.

Wie Kurve H zeigt, ist die Temperaturabhängigkeit der DK der strahleninduzierten kubischen Phase nur gering und vor allem linear. Ob hier die Bestrahlung eine konkurrenzfähige Technologie anstelle der mit teilweise unerwünschten Nebenerscheinungen behafteten Dotierung bzw. Legierung mit Fremdsubstanzen darstellen könnte, ist eine Frage der Ökonomie und hängt deshalb wesentlich von der Leistungserhöhung der Kernreaktoren ab. Das Gleiche gilt natürlich für viele andere hier besprochene Verfahren.

1.3.6. Schließlich soll noch auf die strahlenbeeinflußte Diffusionsgeschwindigkeit eingegangen werden (Schenk 1970b), von der hier die strahlenbeschleunigte Diffusion interessieren soll. In den meisten Fällen verläuft bekanntermaßen die Diffusion von Legierungselementen in einer kristallinen Matrix über einen Leerstellenmechanismus. Zu den im thermischen Gleichgewicht befindlichen Leerstellen kommen durch Bestrahlung zusätzliche, so daß bei wesentlich niedrigeren Temperaturen als im Normalfall Diffusion stattfinden kann. Für neutronenbestrahltes α-Messing soll Figur 7 dies verdeutlichen ($1/\tau \sim D$). Die Gerade stellt den thermisch bedingten Diffusionskoeffizienten in Abhängigkeit von der reziproken Temperatur dar. Die Kreise sind die experimentell bestimmten Werte, die sehr gut auf einer der theoretisch abgeleiteten Kurven für strahlenbeschleunigte Diffusion liegen. Man erkennt drei Bereiche: ein Plateau, bei dem die Diffusionsgeschwindigkeit temperaturunabhängig ist (weil die Konzentration der sich im thermischen Gleichgewicht befindenden Leerstellen viel geringer ist als die der strahlenerzeugten Defekte), ein Übergangsgebiet, wo die Konzentration der Überschußleerstellen etwa gleich der der thermischen Leerstellen ist, und bei hohen Temperaturen herrscht der Diffusionsprozeß über thermische Leerstellen vor.

Die strahlenbedingte Verschiebung einer Diffusionstemperatur zu niedrigeren Werten ist für die Mikroelektronik technologisch hochinteressant. In einer Gasentladung wurde Phosphor- bzw. Bor-dotiertes Silizium mit Protonen beschossen. Der Diffusionskoeffizient war bei der Meßtemperatur von 820 °C etwa 10^5 mal so groß wie ohne Beschuß (Strack). Ein ähnlicher Effekt wurde beim Beschuß mit höherenergetischen Protonen (0,3⋯1,0 MeV) erzielt (Pfister, Baruch). Figur 8 zeigt schematisch eine so behandelte Si-Probe: unterhalb der beschossenen Streifen war die Diffusionsfront beträchtlich tiefer in das Material eingedrungen als im nur thermisch behandeltem Gebiet.

Weiterhin konnte durch Elektronen-Bestrahlung die Epitaxietemperatur von auf Steinsalz aufgedampften Goldfilmen beträchtlich erniedrigt werden (auf

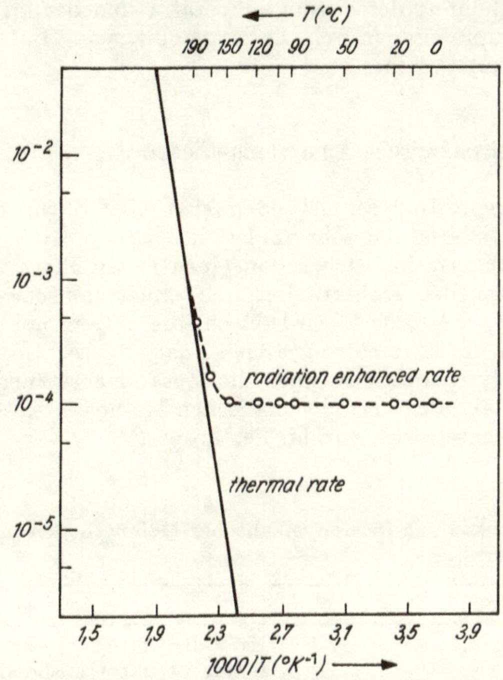

Fig. 7. α-Messing: Ordnungsrate in Abhängigkeit von der reziproken Temperatur. Durchgezogene Gerade: thermisch bedingte Geschwindigkeit; gestrichelte Kurve: theoretische Geschwindigkeit für strahlenbeschleunigte lineare Ausheilung von Defekten; Kreise: experimentelle Werte nach Neutronenbeschuß (nach Dienes, Damask)

150 °C) (Stirland). Hier führt also die strahlenbeschleunigte Diffusion zur Ausbildung der für die Epitaxie notwendigen Gitterordnung.

Bei der Herstellung von Chrom-Aufdampfschichten konnte durch gleichzeitiges Bestrahlen mit Cr-Ionen der lichtoptische Reflexionskoeffizient erhöht werden (Krimmel); es trat nämlich ein strahlenbedingtes orientiertes Wachstum auf.

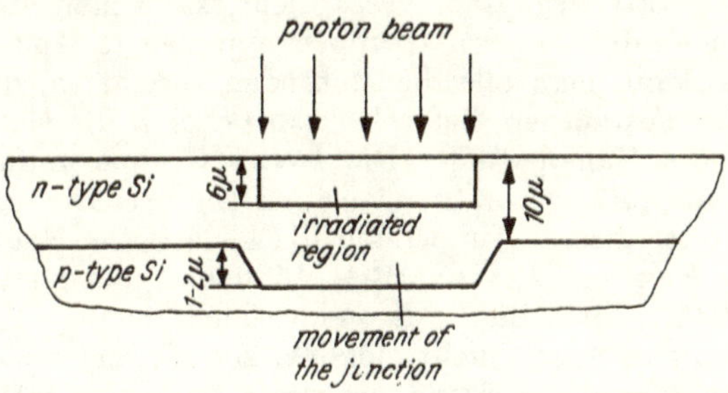

Fig. 8. Si: protonenstrahlenbeschleunigte Diffusion (nach Pfister, Baruch)

1.3.7. Weitere Beispiele der Nutzung von Strahleneffekten könnten z. B. die Erhöhung der Effektivität bestrahlter Rubin- (Flowers) und Neodym-dotierter Glas-Laser (Tittel, Kamel) (dabei erhöht sich die Zahl der Pump-niveaus) und die Erhöhung der Sprungtemperatur und der kritischen Stromdichte bestrahlter Supraleiter (durch die Erzeugung von Haftzentren für die Flußfäden) (WOHLLEBEN) sein.

2. Kernphysikalische Analysenmethoden

Die kernphysikalischen Analysenmethoden sind ebenso wie die kernphysikalisch-synthetischen Verfahren sehr zahlreich.

Die Aktivierungsanalyse hat sich schon einen festen Platz unter den chemischen Analysenmethoden erobert. Der hohe meßtechnische Aufwand, den die Aktivierungsanalyse erfordert, wird durch ihre extrem hohe Empfindlichkeit und die Möglichkeit der zerstörungsfreien Analyse gerechtfertigt. Während die Aktivierungsanalyse Aussagen über die Zusammensetzung des Körpers liefert, gestatten die anderen kernphysikalischen Methoden Aussagen über die Struktur und die Wechselwirkungen im Festkörper.

Tabelle 2

Zusammenstellung kernphysikalischer Meßmethoden in der Festkörperphysik

Methode	Aussage
1. Neutronenbeugung	Kristallstruktur
2. Rutherfordstreuung geladener Teilchen	Kristallstruktur, insbesondere Oberflächen
3. NMR 4. Mößbauerspektroskopie 5. PAC	Hyperfeinfelder
6. inelastische Neutronenstreuung	dynamische Gittereffekte, Phononen-, Magnonen-Spektren
7. IMPAC	dynamische Hyperfeinfelder
8. Positronenvernichtung	Fermiflächen; Elektronendichte

Wenn man von der Aktivierungsanalyse absieht, kann man die kernphysikalischen Analysenmethoden in zwei Gruppen einteilen (s. auch Tabelle 2). In der einen Gruppe kann man alle die Methoden vereinigen, die Aussagen über die Struktur des Festkörpers liefern. Dazu gehören die elastische Neutronenstreuung und die Kanalleitung. Die zweite Gruppe umfaßt alle Methoden, die den Zusammenhang von Struktur und Eigenschaften untersuchen. Repräsentative Vertreter dieser Gruppe sind die inelastische Neutronenstreuung, die magnetische Kernresonanz- (NMR-), Mößbauer- und gestörte Winkelkorrelationstechnik (PAC) sowie die Positronenvernichtung.

Es ist im Rahmen dieser Arbeit nicht möglich, alle Methoden zu behandeln. Deshalb wird nur eine Auswahl der bisher am wenigsten verbreiteten Methoden getroffen.

2.1. In Tabelle 3 sind die wichtigsten Eigenschaften eines thermischen Neutrons zusammengestellt. Durch diese Eigenschaften unterscheidet sich das Neutron ganz wesentlich von den für die schon klassisch gewordenen Struktur-

Tabelle 3

Wichtigste Eigenschaften thermischer Neutronen

Wellenlänge des Neutrons λ	$\approx$	Gitterkonstanten
Energieauflösung ΔE	$\approx$	Anregungsenergie elementarer Gitteranregungen
magnetisches Moment μ_N	$\neq$	0
elektrische Ladung q	$=$	0
Wirkungsquerschnitt σ_s ist keine Funktion der Ordnungszahl Z		

analysenverfahren benutzten Röntgenstrahlen bzw. Elektronen. Die Neutronen wechselwirken nicht mit den Hüllenelektronen wie die Röntgenquanten und Elektronen. Dadurch ist ihre Eindringtiefe sehr groß. Die Streuung der Neutronen geschieht vorwiegend am Kern bzw. über das magnetische Moment des Neutrons μ_N.

Während die elastische Neutronenstreuung die Kristallstruktur bzw. die magnetische Struktur eines magnetischen Materials widerspiegelt, liefert die inelastische Neutronenstreuung Aussagen über die Gitterdynamik, d. h. über das Quasiteilchenspektrum eines Festkörpers. Das ist möglich, weil die Geschwindigkeit thermischer Neutronen vergleichbar mit der Schallgeschwindigkeit ist. Die unelastische Photon-Phononenstreuung führt nur zu einer sehr geringen Frequenzverschiebung, die praktisch nicht meßbar ist, weil die Geschwindigkeit des Photons wesentlich größer ist als die Fortpflanzungsgeschwindigkeit eines Quasiteilchens. Das kann man leicht an der Kinematik des Prozesses erkennen. Ein Quasiteilchen mit dem Wellenvektor $\vec{K}$ tritt mit anderen Teilchen in Wechselwirkung, als hätte es den Impuls $\hbar\,\vec{K}$. Durch die Kinematik des Stoßprozesses ist der Wellenvektor $\vec{k}$ des stoßenden Teilchens eindeutig mit dem Wellenvektor $\vec{K}$ des Quasiteilchens verknüpft:

$$\hbar\,\vec{k} = \bar{n}\,\vec{k}' \pm \hbar\,\vec{K}\,.$$

Das Gleiche gilt für die Energie:

$$\hbar\,\omega = \hbar\,\omega' \pm \hbar\,\Omega\,.$$

Streuexperimente mit Neutronen sind deshalb vorteilhafter als Photon-Phononenstreuexperimente, trotz des erheblich höheren experimentellen Aufwandes. Sie sind allerdings nicht möglich, wenn die Neutronen von den Atomkernen zu stark absorbiert werden.

Eine experimentelle Anordnung ist schematisch in Figur 9 dargestellt (Kittel). Der aus einem Hochflußreaktor kommende Neutronenstrahl wird nach einer Kollimation durch einen Einkristall monochromatisiert. Der Monochromatorkristall ist drehbar angeordnet, dadurch kann die Energie des monochromatisierten Strahles variiert werden. Dieser Strahl fällt auf die drehbar angebrachte Probe. Der unter einem bestimmten Winkel gestreute Strahl, der nun nicht mehr monochromatisch ist, wird durch den Analysatorkristall hinsichtlich der Energie analysiert. Der Streuwinkel ist ein Maß für die Impulsänderung. Dieses Dreiachsenspektrometer hat den Nachteil, daß die Energie punktweise abgetastet werden muß. Durch den Einsatz von gepulsten Neutronenquellen, beispielsweise des Impulsreaktors im VIK Dubna, kann mit größerer Effektivität die Flugzeitmethode angewendet werden (Buras, Leciejewicz). Ein Beispiel eines solchen Quasiteilchenspektrums ist die in Figur 10 dargestellte inelastische Streuung von Neutronen in Natriumeinkristallen (Woods, Brockhouse, March, Bowers). Dabei werden akustische Gitter-

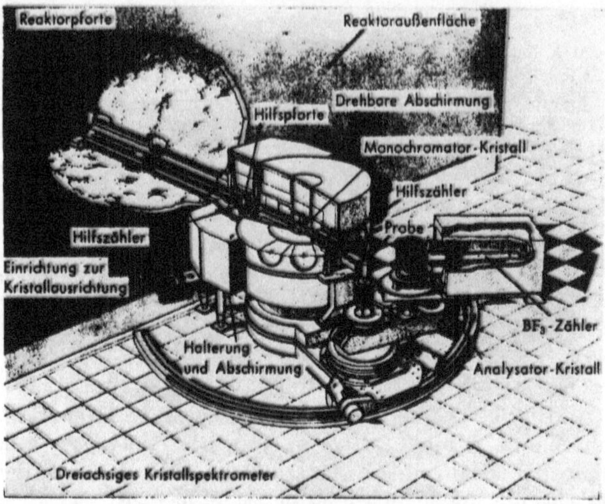

Fig. 9. BROCKHOUSE-Dreiachsen-Neutronenspektrometer (nach KITTEL)

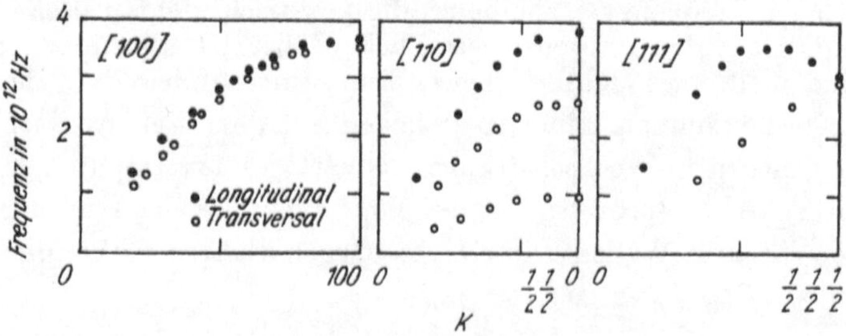

Fig. 10. Phononenspektrum eines Natrium-Einkristalls (nach WOODS, BROCKHOUSE, MARCH, BOWERS)

schwingungen erzeugt. Man erkennt deutlich den longitudinalen und transversalen Zweig der akustischen Phononen. Es zeigt sich außerdem, daß sich die Phononendispersionskurven in verschiedenen Gitterrichtungen stark unterscheiden.

Durch die Anwendung der inelastischen Neutronenstreuung läßt sich unsere Kenntnis über die Kristalldynamik wesentlich erweitern. Die inelastische Neutronenstreuung ist besonders geeignet zur Untersuchung

1. der Dynamik fester Körper,
2. der Dynamik magnetischer Systeme,
3. der Dynamik von Flüssigkeiten und
4. der Dynamik von Molekülen.

In Figur 11 sind die Quasiteilchen zusammengefaßt, die mit der inelastischen Neutronenstreuung untersucht werden können.

Aus der Kristalldynamik sind die Phononen und Magnonen als die Energiequanten der elastischen Welle bzw. der Magnetisierungswelle bekannt. In Molekülkristallen sind weitere Freiheitsgrade der Bewegung möglich, die zu anderen Quasiteilchen führen. Die Moleküle sind in der Regel Träger von Rotationsenergie; dabei kann die Rotationsenergie von Molekül zu Molekül

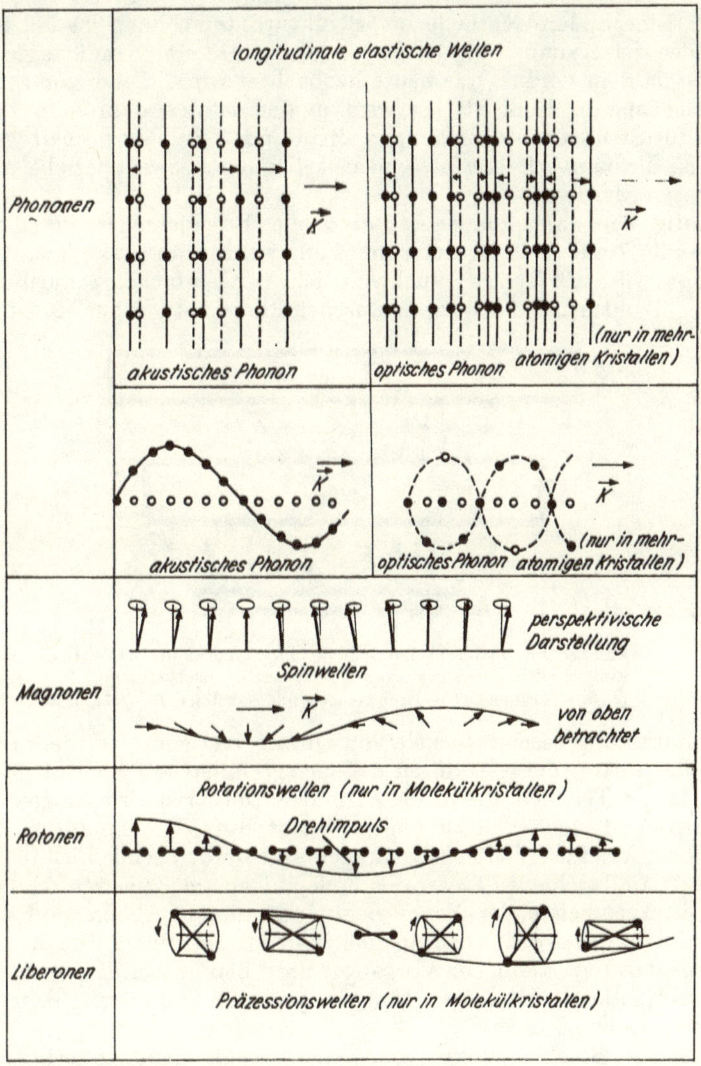

Fig. 11. Schematische Darstellung von Quasiteilchen, die mit der inelastischen Neutronenstreuung untersucht werden können

überspringen, es entsteht eine Rotationswelle. Die Energie der sogenannten Rotonen ist kleiner als die der Phononen. Bei wachsenden Temperaturen und schwächer werdenden Kristallbindungen kann es zu einer Kopplung zwischen Molekülrotation und resultierendem Kernspin kommen. Dadurch beginnt das Molekül um seine Symmetrieachse zu präzessieren. Diese Präzession kann sich im Molekülkristall als Welle fortpflanzen. Diese Quasiteilchen werden als Liberonen bezeichnet. Je komplizierter die Bausteine der Kristalle werden, um so mehr Freiheitsgrade der Bewegung stehen zur Verfügung. Die Kristalldynamik wird immer komplizierter. Gegenwärtig bietet die inelastische Neutronenstreuung die einzige Möglichkeit, solche komplizierten Vorgänge zu untersuchen.

2.2. Eine andere Methode der Strukturuntersuchung ist die Kanalleitung. Die Idee der Kanalleitung ist fast 6 Jahrzehnte alt. STARK sagte bereits 1912 vieles von dem voraus, was heute beobachtet wird. Diese von STARK erwähnt Kanalleitung in Einkristallen wird in den letzten Jahren in zunehmendem Maße für Strukturuntersuchungen angewandt. Es wird deshalb vorausgesetzt, daß das Prinzip bekannt ist, und es soll nur auf zwei spezielle Anwendungen verwiesen werden.

Häufig wird angenommen, daß große Beschleuniger für diese Methode notwendig sind. Das ist auch der Fall, wenn man maximale Informationen erhalten will. Es ist aber auch möglich, mit einfachen radioaktiven Quellen solche Strukturuntersuchungen durchzuführen. In Figur 12 ist schematisch

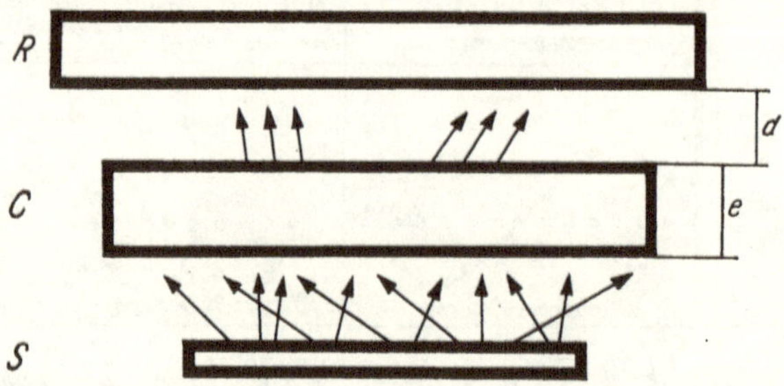

Fig. 12. Schematische Darstellung einer Anordnung zur Strukturuntersuchung von Einkristallen mit einer radioaktiven Quelle (nach QUÉRÉ). S – radioaktive Quelle (z. B. α-Strahler), C – zu untersuchende Substanz, R – Kunststoff-Folie als Detektor

eine Anordnung gezeigt, wie sie von QUÉRÉ verwendet wurde. Die von einer α-Quelle isotrop ausgestrahlten α-Teilchen fallen auf eine zu untersuchende Folie. Die α-Teilchen, die in niedrigindizierten Kristallrichtungen in die Folie eindringen, dringen nahezu ungeschwächt durch sie hindurch. Die anderen α-Teilchen erleiden Vielfachstreuungen und werden schließlich in der Folie absorbiert. Dadurch wird hinter der Folie auf einem Kunststoff-Film ein Abbild der Gittergeometrie der Folie fixiert. In diesem Abbild sind auch größere Gitterstörungen wie z. B. Stapelfehler o. ä. erkennbar. Die Kunststoff-Folie wird geätzt (ein ähnlicher Vorgang wurde bereits bei den synthetischen Methoden erwähnt); die Auswertung der geätzten Spuren geschieht letztlich mit einem einfachen Mikroskop.

Auch der Blockingeffekt kann ohne Beschleuniger zum Lokalisieren von Fremdatomen ausgenutzt werden. UGGERHØJ et al. implantierte das radioaktive Isotop ^{175}Yb in einen Siliziumkristall. Dieses ^{175}Yb emittiert ein β^--Teilchen einer Maximalenergie von 470 keV. Wenn das ^{175}Yb-Atom auf einem Gitterplatz sitzt, sollte ein Maximum in der Winkelverteilung zu beobachten sein, da die negativen Elektronen durch die positiven Kerne in Richtung der Gitterebenen fokussiert werden. Figur 13 zeigt die Impulsrate eines mit ^{175}Yb bestrahlten Siliziumkristalls. In der $\langle 111 \rangle$-Richtung ist ein Maximum zu beobachten, d. h., das Yb-Atom sitzt in den Atomreihen in dieser Richtung. Andererseits zeigt sich in $\langle 110 \rangle$-Richtung ein Minimum, was darauf hinweist, daß das Yb-Atom außerhalb der Atomreihen in dieser Richtung sitzt. Auf diese Weise kann die Lage der Yb-Atome im Siliziumkristall lokalisiert werden. Diese Methode ist sehr allgemein anwendbar, da fast alle Elemente β-strahlende Isotope besitzen.

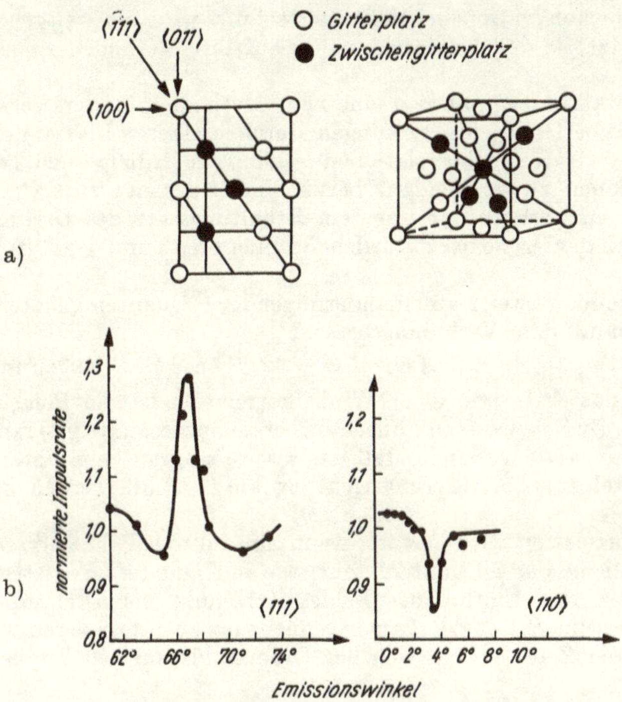

Fig. 13. Impulsratenverteilung von ^{175}Yb in einem Si-Einkristall (nach UGGERHØJ, ANDERSEN)

2.3. Zuletzt soll auf Hyperfeinwechselwirkungen eingegangen werden. Die Wechselwirkung von Atomkernen mit Materie kann auf die Wechselwirkung des magnetischen bzw. Quadrupolmomentes des Atomkerns mit dem magnetischen und elektrischen Kristallfeld zurückgeführt werden. Durch diese Wechselwirkung wird die Entartung nach der magnetischen Quantenzahl aufgehoben. Der Zustand des Kerns spaltet nach den verschiedenen Orientierungsmöglichkeiten des Kernspins relativ zu der durch das Kristallfeld ausgezeichneten Achse auf. Die Besetzung dieser Unterzustände ist unterschiedlich. Durch Energiezufuhr kann eine Umbesetzung der Unterzustände erreicht werden. Der Energieerhaltungssatz verlangt, daß dabei die zugeführte Energie E gleich dem Abstand ε der Unterzustände ist:

$$E = \varepsilon(m, m') .$$

Da die Aufspaltung von der Stärke des Kristallfeldes abhängt, kann durch Messung der zugeführten Energie E dieses Kristallfeld bestimmt werden. Die NMR und die Mößbauerspektrometrie benutzen diesen Weg zur Messung der Hyperfeinwechselwirkung. Voraussetzung für die Anwendung beider Methoden ist, daß die natürliche Linienbreite $\Gamma \approx \hbar/\tau$ kleiner oder höchstens gleich der Aufspaltung $\Delta E = \hbar\,\omega_L$ ist:

$$\frac{\Gamma}{\Delta E} = \frac{1}{\tau \cdot \omega_L} \leqq 1 . \qquad (\omega_L = \text{Lamorfrequenz})$$

Die NMR und die konventionelle Mößbauertechnik wird als weitgehend bekannt vorausgesetzt, so daß hier nicht näher auf ihre Methoden eingegangen wird.

In den letzten Jahren beginnt sich eine neue Methode, von der Kernphysik ausgehend, für Hyperfeinwechselwirkungen durchzusetzen. Dieses Verfahren der Messung von γ—γ-Winkelkorrelationen (es sind im Prinzip auch Teilchen-γ-Winkelkorrelationen zu verwenden) beruht im Gegensatz zur NMR- und Mößbauertechnik im wesentlichen auf dem Erhaltungssatz des Drehimpulses und ist deshalb von der Größe der natürlichen Linienbreite und der Aufspaltung unabhängig.

Die Winkelverteilung zweier aufeinanderfolgender γ-Quanten gehorcht nach der Quantenmechanik dem Verteilungsgesetz:

$$W(\Theta) = 1 + A_2 P_2 (\cos \Theta) + \cdots \qquad (P \cdot \cos \Theta = \text{Kugelfunktion})$$

Die Ursache für das Auftreten einer Winkelkorrelation ist die Polarisierung des Atomkerns im Zwischenzustand durch das erste ausgesandte γ-Quant. Der Drehimpulserhaltungssatz verlangt, daß das zweite γ-Quant nur unter einem bestimmten Winkel zur Polarisierungsrichtung ausgestrahlt werden kann (s. PROC.).

Diese Winkelkorrelation wird gestört, wenn ein äußeres Feld senkrecht auf der Ausbreitungsebene der γ-Quanten steht, wie in Figur 14 schematisch dargestellt ist. Unter dem Einfluß des Feldes H beginnt der Drehimpuls des Zwischenzustandes um H mit der Lamorfrequenz ω_L zu präzessieren. Erfolgt der Zerfall nach der Zeit t, so hat sich der Drehimpuls um den Winkel $\omega_L \cdot t$

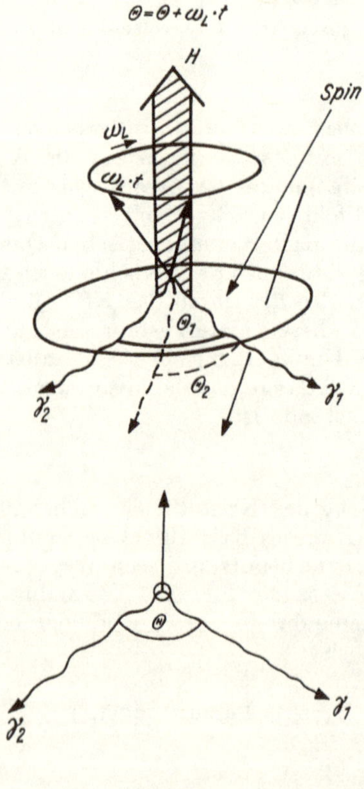

Fig. 14. Schematische Darstellung der LAMOR-Präzession des Drehimpulses im Zwischenzustand

gedreht. Das zweite γ-Quant wird deshalb unter dem Winkel $\Theta' = \Theta + \omega_L \cdot t$ ausgesandt. Die Winkelverteilung folgt deshalb dem Gesetz:

$$W(\Theta', \omega_L, t) = 1 + A_2 P_2 (\cos (\Theta + \omega_L \cdot t)) ,$$

oder, wenn man über die Zeit integriert:

$$W(\Theta', \omega_L, \tau) = 1 + A_2 P_2 (\cos (\Theta + \omega_L \cdot \tau))$$

(τ = Lebensdauer des Zwischenzustandes).
Damit ist das Prinzip der Messung vorgeschrieben: entweder man mißt unter einem festen Winkel Θ die zeitliche Verteilung der γ–γ-Koinzidenzen und eliminiert die Zeitabhängigkeit des Zerfallsgesetzes oder man mißt die gesamte Winkelverteilung. In jedem Fall erhält man das innere Magnetfeld oder den elektrischen Kristallfeldgradienten bzw. eine Überlagerung beider Felder.

Die Anwendung dieser Methode für festkörperphysikalische Fragestellungen ist noch sehr wenig verbreitet; das hat seine Ursache darin, daß der Kernphysiker in der Regel leider selten Interesse für Festkörperphysik zeigt und andererseits der Festkörperphysiker oftmals vor der komplizierten Meßtechnik und Analyse der Kernphysik zurückschreckt. Hier bietet sich also ein guter Ausgangspunkt für eine interdisziplinäre Zusammenarbeit.

Die Methode der Messung gestörter Winkelkorrelationen wird sich voraussichtlich in immer stärkerem Maße durchsetzen, da sie eine wesentliche Erweiterung der Aussagemöglichkeiten gegenüber NMR und Mößbauerspektrometrie darstellt. Einige Fragen sind nur mit dieser Technik zu klären.

Z. B. wird die NMR und Mößbauerspektrometrie durch die thermisch bedingten Atomschwingungen gestört. Die Winkelkorrelationsmessung wird dagegen nicht durch die Probentemperatur eingeschränkt. Sie kann mit gleicher Effektivität bei Heliumtemperaturen wie auch in Schmelzen, Flüssigkeiten und Gasen angewandt werden. Dieses Verfahren eignet sich deshalb besonders gut zur Untersuchung des Temperaturverhaltens der inneren Felder. So wurde beispielsweise das Hyperfeinfeld am Thuliumkern in Eisenfolien als Funktion der Temperatur gemessen (DEUTCH u. a.). Die Ergebnisse sind in Figur 15 dargestellt. Das Abweichen der experimentellen Werte von der theoretischen

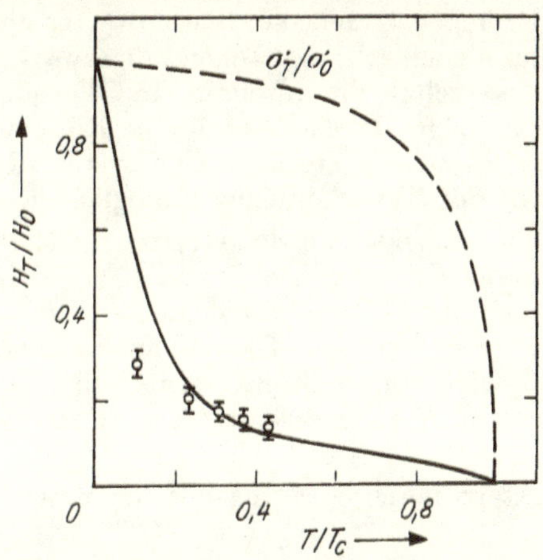

Fig. 15. Hyperfeinfeld am ^{169}Thulium in einem Eisen-Wirtsgitter als Funktion der Temperatur (nach DEUTCH, NIELSEN, BERNAS)

Kurve für ein Austauschfeld mit der gleichen Temperaturabhängigkeit wie die Suszeptibilität von reinem Eisen kann durch das Auftreten lokaler magnetischer Momente am Thuliumkern erklärt werden (ausgezogene Kurve). Dieses Experiment zeigt außerdem, daß sich die Korrelationsmessungen auch für Aussagen über die Elektronenstruktur in der Umgebung von Fremdatomen eignen.

Ein weiteres Beispiel dafür sind die Untersuchungen des Hyperfeinfeldes am Cd 111 in einem Ni-Wirtsgitter. Es wurden zwei Magnetfelder am Ort des Cd 111-Kerns gefunden (Andreeff, Unterricker, Stary).

$$B = -(65 \pm 3)\ \text{kOe}$$

und

$$B = +(170 \pm 10)\ \text{kOe}\,,$$

d. h., zwei Lamorfrequenzen. Bisher war nur das Hyperfeinfeld von -65 kOe bekannt. Es wird durch die Polarisation der Leitungselektronen erzeugt. Das zweite Feld mit positivem Vorzeichen wird dagegen wahrscheinlich durch die Wechselwirkung mit den Liganden hervorgerufen, wenn der Cd 111-Kern auf einem Zwischengitterplatz sitzt. Das positive Vorzeichen des Feldes deutet darauf hin, daß das Überlappungsintegral größer ist als der Konfigurationsmischungskoeffizient für die Elektronen aufgrund der Austauschwechselwirkung. Diese Annahme ist plausibel, da der kovalente Bindungsanteil in Kristallen mit vorwiegender Metallbindung wie Nickel sehr klein ist. Bisher wurde dieses Feld nicht gemessen, da die Sondenkerne lediglich eindiffundiert wurden und deshalb im wesentlichen substituiert oder an Gitterstörungen, wie Korngrenzen, Versetzungen usw., vorlagen. Im hier besprochenen Experiment wurden die Sondenkerne bei Raumtemperatur mit etwa 1 MeV implantiert, was ohne anschließende Ausheilung mit großer Wahrscheinlichkeit auch zum Vorhandensein von interstitiellen Fremdatomen führt. Dieses Experiment deutet auch auf Möglichkeiten für Messungen von Bindungsverhältnissen in Molekülstrukturen hin.

Es gibt theoretische Hinweise, daß sich die Eigenschaften eines Körpers bei hohen Drücken stark ändern, wenn die Elektronenhüllen benachbarter Atome sich überlagern. Solche hohen Drücke sind gegenwärtig noch nicht im Labor beherrschbar. Beim Einschießen schwerer Ionen mit etwa 1 MeV werden aber lokal extrem hohe Drücke erzeugt, die durchaus ein ,,Verquetschen" der Gitterumgebung verursachen können. Deshalb kann eine Kopplung von Implantation und gestörter Winkelkorrelationstechnik eine systematische Untersuchung der Druckabhängigkeit der Hyperfeinfelder ermöglichen.

Die NMR wie auch die Mößbauerspektrometrie gestatten die Messung lokaler zeitlicher Änderungen der inneren Felder von maximal $10^6 \cdots 10^7$ Hz. Höhere Frequenzen können nur indirekt über Interferenzphänomene und damit verbundenen Schwebungen gemessen werden. Die Winkelkorrelationstechnik erweitert diesen Meßbereich gleich um 5 Größenordnungen. Es sind z. B. Elektronenrelaxationszeiten von 10^{-12} s gemessen worden (Deutch, Nielsen, Bernas).

Eine andere Anwendung besteht in der Messung der Asymmetrieparameter der elektrischen Kristallfelder.

Der experimentelle Aufwand der Korrelationsmethode ist vergleichbar mit dem der NMR-Technik. In Figur 16 wird ein Korrelationsmeßplatz gezeigt.

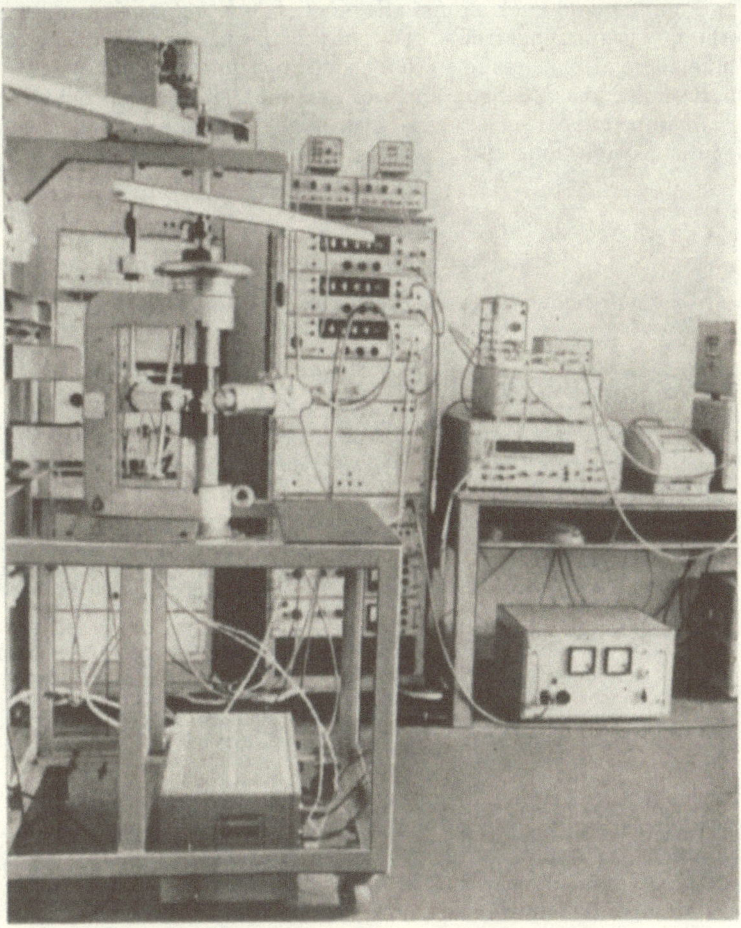

Fig. 16. Meßplatz für $\gamma - \gamma$-Winkelkorrelationen

Für die Implantationstechnik bietet die Korrelationsmethode eine Möglichkeit, den Implantationsprozeß sozusagen mikroskopisch zu verfolgen. Es können Aussagen über den Ort der implantierten Kerne gemacht werden, d. h., ob sie substitutionell oder interstitiell eingebaut werden (ANDREEFF, UNTERRICKER, STARY). Da die Hyperfeinfelder am Sondenkern stark von der Struktur der Umgebung abhängen, ist es auch möglich, Ausheil- und Diffusionsvorgänge in implantierten Systemen zu untersuchen. Die bisher verwendeten Methoden zur Untersuchung von Ausheilvorgängen sind relativ unempfindlich gegen geringe Änderungen der Lage implantierter Ionen. Ein einzelner Platzwechselvorgang kann zu einer völlig andersartigen Umgebung führen, was mit den bisherigen Methoden häufig nicht erfaßbar ist. Man könnte hier also zu einem weitergehenden Detailverständnis von Transportvorgängen oder auch Strukturumwandlungen kommen.

Eine entscheidende Bedeutung hat die Korrelationsmethode in der sogenannten „heißen" Chemie. So kann man kurzlebige hochangeregte Zustände nach der Implantation untersuchen.

In Figur 17 sind noch einmal alle möglichen Kombinationen der Korrelationsmethode zusammengestellt. Die physikalischen Größen, die man aus der Winkelkorrelationsmessung erhält, sind: die inneren magnetischen Felder B_{int}, das Produkt aus Quadrupolmoment Q und elektrischem Feldgradienten ∇E, der Asymmetrieparameter des elektrischen Feldes η, die Relaxationszeit τt oder die Korrelationszeit τ_s.

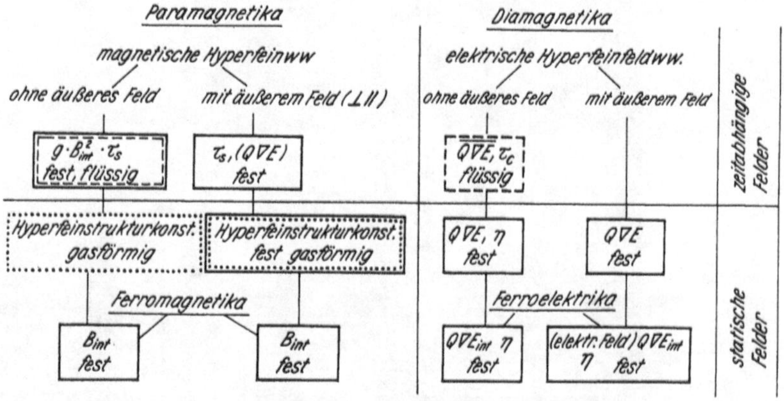

Fig. 17. Übersicht über die mit der gestörten Winkelkorrelation meßbaren inneren Feldgrößen. – Es bedeuten: B_{int} – inneres Magnetfeld; Q – Quadrupolmoment;- g – g-Faktor; ∇E – elektrischer Feldgradient; η – Asymmetrieparameter; τ_c – Korrelationszeit; τ_s – Relaxationszeit; $\perp$ – Feld senkrecht auf Meßebene; $\|$ – Feld parallel zur Meßebene; – Die Aggregatzustände der Quellen sind graphisch wie folgt dargestellt: ——— fest; – – – flüssig; ······ gasförmig

Dieser Abriß kernphysikalischer Methoden in der Kristallographie und Festkörperphysik kann aufgrund der Fülle der Verfahren nur lückenhaft sein. Die Auswahl ist natürlich durch subjektive Einflüsse bestimmt und darf deshalb nicht als Wertung aufgefaßt werden. Es kam den Autoren darauf an, vor allem weniger bekannte Methoden vorzustellen und besonders auf ihre Anwendungsmöglichkeiten hinzuweisen.

Literatur

ANDREEFF, A., UNTERRICKER, R., STARY, F.: (in Vorbereitung) anonym: Metals and Materials H. 7, 231 (1969)

BALARIN, M.: Vortrag 4. DVK-Tagung Halle 1969

BALARIN, M., OTTO, G., STORBECK, I., SCHENK, M., WAGNER, H.: Thin Solid Films **4**, 255 (1969)

BALARIN, M., SCHENK, M.: phys. stat. solidi **17**, 91 (1966)

BLAMIRES, N. G., MATTHEWS, M. D., NELSON, R. S.: Physics Letters **28 A**, 178 (1968)

BONFIGLIOLI, G., FERRO, A., MOJONI, A.: J. appl. Physics **32**, 2499 (1961)

BURAS, B., LECIEJEWICZ, J.: Nukleonika **8**, 75 (1963)

CLELAND, J. W.: in Rediconti della Scuola Internazionale di Fisica „Enrico Fermi", XVIII. Corso 1962, S. 384

DEUTCH, B. I., NIELSEN, K. B., BERNAS, H.: Physics Letters **27 B**, 209 (1968)

DIENES, G. J., DAMASK, A. C.: J. appl. Physics **29**, 1713 (1958)

FLOWERS, W.: Proc. IEEE **51**, 858 (1963)
FOSS, N. A.: J. appl. Physics **39**, 6029 (1968)
HAUSER, O., KÖHLER, V.: Tagung Festkörperphysik (14.–20. 9. 59), Balatonfüred 1961, S. 21
HAY, M. G., MARTIN, J. W.: J. Less-Common Metals **18**, 27 (1969)
HILBERT, F.: Jenaer Rdsch. **5**, 218 (1963)
HILBERT, F.: Dissertation Friedr.-Schiller-Universität Jena 1964
HOHMUTH, K.: Priv. Mitteilung
HUNSPERGER, R. G., MARSH, O. J.: Bull. Amer. Physic. Soc. II **14**, 373 (1969)
KITTEL, C.: Einführung in die Festkörperphysik. München 1968, S. 182
KRIMMEL, F.: Vakuum' Technik **19**, 187 (1970)
MAYER, J. W., ERIKSSON, L., DAVIES, J. A.: Ion Implantation in Semiconductors: Si and Ge. New York 1970
PFISTER, J. C., BARUCH, P.: J. Physic. Soc. Japan **18**, Suppl. III, 251 (1963)
PHILLIPS, V. A., SEYBOLT, A. U.: TMS-AIME **242**, 2415 (1968)
PRICE, P. B., WALKER, R. M.: J. appl. Physics **33**, 3400 (1962)
PROC. . . .: Pertubated Angular Correlations, Proc. of the Uppsala Meeting on Extranuclear Pertubations in Angular Correlations. Amsterdam 1964
QUÉRÉ, Y.: J. Physique **29**, 215 (1968)
SCHENK, M.: Habilitationsschrift Humboldt-Universität Berlin 1970a
SCHENK, M.: Radiation-Induced Diffusion in „Diffusion in metallischen Werkstoffen". Leipzig 1970b, S. 161
SCHWUTTKE, G. H., BRACK, K., GOREY, E. F., KAHAN, A., LOWE, L. F.: Int. Conf. "Ion Implantation in Semicond". Thous and Oaks 1970, C-6
SEYBOLT, A. U.: TMS-AIME **245**, 769 (1969)
SHIMIZU, K.: Denki Shikensho Iho **33**, 1383 (1969)
STARK, J.: Physik. Z. **13**, 973 (1912)
STIRLAND, D. J.: Appl. Physics Letters **8**, 326 (1966)
STRACK, H.: J. appl. Physics **34**, 2405 (1963)
THACKERY, P. A., NELSON, R. S.: Phil. Mag. **19**, 169 (1969)
TITTEL, F., KAMEL, N.: in Interaction of Radiation with Solids, New York 1967, S. 261
UGGERHØJ, E., ANDERSEN, J. U.: Canadian J. Physics **46**, 543 (1968)
WOHLLEBEN, K.: Z. angew. Physik **27**, 92 (1969)
WOODS, A. D. W., BROCKHOUSE, B. N., MARCH, R. H., BOWERS, R.: Proc. Phys. Soc. London **79**, 440 (1962)

(Eingegangen am 20. 5. 71)

Anschrift der Verfasser:

Prof. Dr. A. ANDREEFF
Bergakademie Freiberg
Sektion Physik
92 Freiberg

Dr. habil. M. SCHENK
DAW, ZfK Rossendorf
8051 Dresden
PF 19

| Kristall und Technik | 7 | 1–3 | 1972 | 341–347 |

J. Heinze

VEB Narva
Berliner Glühlampenwerk

Untersuchung des Arbeitsprozesses bei elektronenmikroskopischen Arbeiten durch eine Multimoment-Häufigkeitsanalyse

Der prozentuale Anteil aller Arbeitsfolgen bei elektronenmikroskopischen Untersuchungen wird durch eine Multimoment-Häufigkeitsanalyse ermittelt. Die Ausgangswerte wurden durch Selbstnotierung der Mitarbeiter erhalten. Die Endergebnisse werden hisichtlich ihrer Aussagekraft sowie der möglichen Verbesserung der Arbeitsproduktivität im Labor diskutiert.

The percentage of all operations sequences with electromicroscopical investigations is determined by a multimoment frequency analysis. The starting values were obtained by self-record of the laboratory staff. Final results are discussed with regard to their propositional power and the possible improvement of labour productivity in the laboratory.

Die Ausrüstung von Forschungs- und Entwicklungsabteilungen mit immer teureren Geräten zwingt zur möglichst rationellen bzw. kostengünstigen Arbeitsweise. Um dafür objektive Ausgangswerte zu erhalten, schien im Fall der elektronenmikroskopischen Arbeiten eine detaillierte Arbeitsstudie erforderlich.

Auf Grund des günstigen Aufwand-Nutzen-Verhältnisses sowie der anderen Vorzüge des Multimoment-Verfahrens ist dafür die Multimoment-Häufigkeitsanalyse (MMH) besonders geeignet.

Bekanntlich ist ja die MMH ein Stichprobenverfahren, das verbindliche Aussagen über die Häufigkeit von unregelmäßig auftretenden Vorgängen oder Größen beliebiger Art für jede gewünschte Genauigkeit bei einer vorgegebenen statistischen Sicherheit von 95% gibt (Haller-Wedel).

Die Aufgabe in unserem Fall bestand darin, die pronzentualen Anteile aller Arbeitsfolgen beim elektronenmikroskopischen Arbeiten zu ermitteln. Davon ausgehend sollte nach apparativen oder organisatorischen Maßnahmen gesucht werden, die eine möglichst effektive Ausnutzung des Gerätes sichern. Darüber hinaus gibt diese Studie Hinweise auf die Verwendung der MMH im Laborbetrieb, zeigt also eine Möglichkeit der Objektivierung geistig-experimenteller Tätigkeit in der Produktionsvorbereitung.

An Literatur lagen keine speziellen Veröffentlichungen vor. Lediglich die Monographie von Haller-Wedel, eine Lehrbrief-Reihe (. . .) sowie allgemeine statistische Grundlagen (Schindowski, Schürz) wurden verwertet.

Das in Frage stehende Elektronenmikroskopie-Labor war zur Zeit der Untersuchung aus betrieblichen Gründen mit einer halben Hochschulkraft, einer Fachschulingenieurin und einer Laborantin besetzt. Außerdem konnte für fotografische Arbeiten (Herstellen von Abzügen und Vergrößerungen) auf ein Foto-Labor zurückgegriffen werden.

An Geräten steht die Elektronenoptische Anlage EF4/EF5/EF6 des VEB C. Zeiß, Jena, sowie zwei kommerzielle Bedampfungsanlagen neben anderem üblichem Zubehör zur Verfügung. Die zu bearbeitenden Probleme umfassen Korngrößen- und Kornformanalysen an Metall-, Oxid- und Leuchtstoffpulvern; Glasoberflächenuntersuchungen; Gefüge- und Bruchgefügeuntersuchungen an Drähten und Stäben aus hochschmelzenden Metallen und Legierungen sowie Diffusionsvorgänge.

Das Labor existiert seit einem Jahr.

Auf Grund dieser Ausgangssituation wurden die Mitarbeiter bei der Aufnahme mit einbezogen und die Werte durch Selbstnotierung erhalten.

Anhand der Multimoment-Hauptformel (HALLER-WEDEL) wurde vor Beginn der Aufnahme die Anzahl von Notierungen N bei einem vorgegebenen relativen Fehler ε von 10, 15 und 20% ermittelt. Unter der vorab geschätzten Annahme, daß der kleinste Ergebnisanteil $p' = 5\%$ beträgt, wären 7300, 4100 bzw. 1820 Notierungen erforderlich gewesen. Bei Beschränkung auf etwa einen Kalendermonat für die Erhebung sind demnach pro Arbeitstag und Arbeitskraft 120, 70 bzw. 30 Notierungen erforderlich oder durchschnittlich alle 4, 7 bzw. 16 Minuten. Durch geeignetes Zusammenlegen von mehreren Arbeitsgängen bei der Auswertung lassen sich die tatsächlichen Ergebnisanteile p erhöhen. Damit sinkt — bei gleichbleibender statistischer Sicherheit und Genauigkeit — der Notierungsaufwand. Die wirkliche Aufnahme umfaßt pro Arbeitskraft und Arbeitstag 35 Notierungen, eine durchaus zumutbare zusätzliche Belastung für die Beteiligten. Abweichend von der sonst üblichen Praxis wurden auch die gesetzlichen Pausenzeiten (Frühstück und Mittagessen) mit erfaßt. Dies war notwendig, weil im Labor arbeitsbedingte Pausenverschiebungen die Regel sind. Um aber der Forderung nach gleichbleibender statistischer Belegung aller möglichen Arbeitszeiten zu genügen bzw. um Überbewertungen von einzelnen Tageszeiten zu verhindern, wurde diese Art der Notierung gewählt. Deshalb wird innerhalb dieser Analyse mit einer „Arbeitszeit" von 565 Minuten pro Arbeitstag gerechnet.

Alle anfallenden Arbeiten wurden vor Beginn der Studie in Übereinstimmung mit den Mitarbeitern in folgende Notierungskategorien eingeteilt:

1. Präparation: Dazu gehört die gesamte Probenvorbereitung einschließlich der Bedampfung, jedoch ohne Schliffherstellung bei Arbeiten in Emission. Ferner zählen dazu die Geräteeinstellung bei Wechsel von Emission auf Transmission bzw. umgekehrt, das Platteneinschleusen sowie das Ein- und Ausschalten bzw. das Warmlaufen des Gerätes.

2. Wartung, Pflege: Hierzu rechnen notwendige Säuberungen des Mikroskops und der Bedampfungsanlage, sowie das Auswechseln von schadhaften Teilen und die nachfolgende Neujustage.

3. Mikroskopieren: Diese Kategorie umfaßt das Durchmustern der Proben, Fotografieren und die Probenwechsel.

4. Fotografische Negativarbeiten: Darunter fällt das Entwickeln und Fixieren der Platten einschließlich der Vorbereitung und die Säuberung der entsprechenden Gerätschaften.

5. Fotografische Positivarbeiten: Damit ist das Herstellen von Abzügen und Vergrößerungen sowie das Beschneiden, Aussortieren und Aufkleben der Fotografien gemeint. Ein Großteil dieser Arbeiten wird im Foto-Labor vorgenommen und wird bei dieser Studie nicht berücksichtigt.

6. Auswertung: Zu dieser Kategorie zählt das Anfertigen des Abschlußberichts einschließlich der erforderlichen Berechnungen und Auswertungen.

7. Andere Arbeiten: Hervorgerufen durch den Umstand, daß die Mitarbeiter noch teilweise an anderen Aufgaben arbeiten, ist eine derartige Kategorie erforderlich. Außerdem werden hier alle arbeitsbedingten Pausen sowie die Abwesenheit durch Krankheit, Arbeitszeitverkürzung usw. gewertet. Schließlich sind auch Frühstücks- und Mittagspausen enthalten; auf die Notwendigkeit dieser Werte ist weiter oben verwiesen.

Die Ermittlung der Ausgangswerte erfolgte durch Ankreuzen der entsprechenden Spalte zu festgelegten Zeiten in üblichen Aufnahmebögen. Die dafür erforderlichen Zufallszeiten wurden einer Tabelle mit Zufallszahlen (Schindowski, Schürz) entnommen. Jede Arbeitskraft erhielt pro Arbeitstag einen einzigen Bogen, der am nächsten Morgen eingesammelt und ausgewertet wurde. Die Anzahl der Notierungen pro Tag beträgt zwischen 20 und 74 mit Maxima um $N = 35$ und $N = 70$, die Summe aller Notierungen betrug 1447. Die Ursachen für das Schwanken lag darin, daß die Laborantin während der

Tabelle 1

lfd. Nr.		Tätigkeit	Aufwand (%) $p \pm f$		Aufwandsbild (%)
1	Vorbereitung	Präparation, Vorbereitung	24,3 ± 2,2	p = 32,0; f = ± 2,4; ε = ± 7,5	
2		Wartung, Pflege, Säuberung	7,7		
3	Operative Tätigkeit	Mikroskopieren	18,3 ± 2,0	p = 35,4; f = ± 2,4; ε = ± 6,8	
4		Photographische Negativarbeiten	5,6		
5		Photographische Positivarbeiten	3,4		
6		Auswertung, Bericht	8,1 ± 1,4		
7	Verlust	Andere Arbeiten, Abwesenheit, Frühstücks- und Mittagspausen		p = 32,6; f = ± 2,4; ε = ± 7,4	
			1447 Notierungen = 100,0 %		

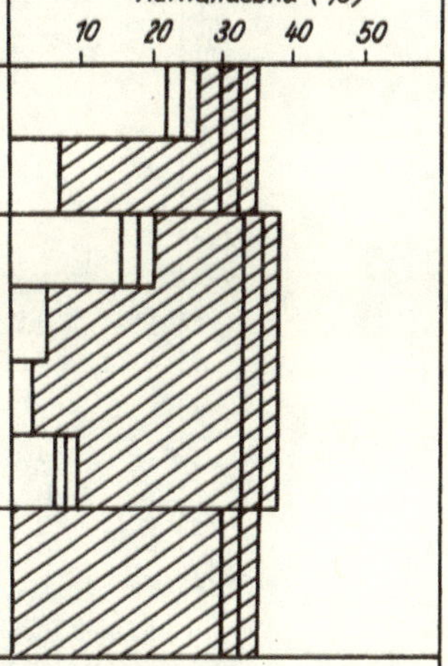

Aufnahmezeit bis auf drei Tage krank und die beiden anderen Mitarbeiter nicht ständig im Elektronenmikroskopie-Labor tätig waren. Tage, an denen eine Arbeitskraft überhaupt nicht elektronenmikroskopisch arbeitete, wurden für die betreffende Kraft nicht mitgezählt. Diese Einschränkung scheint insofern gerechtfertigt, als die Fragestellung auf die Gesamtkapazität des Labors zielte und nicht auf den jeweiligen Anteil der Mitarbeiter.

Um während der laufenden Untersuchung einen Überblick über die Genauigkeit der Notierungen zu erhalten und evtl. auftretende Fehler auszumerzen,

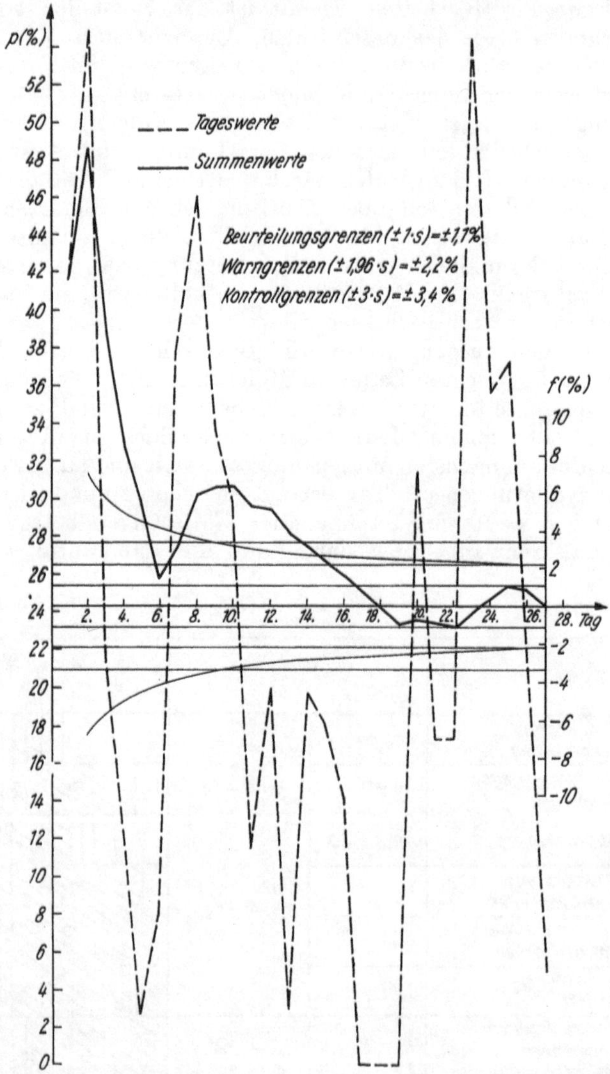

Fig. 1. Tages- und Summenkontrollkarte für die Arbeitsfolge 1 „Präparation" (Erläuterung s. Text)

wurde eines Tages- undSummenkontrollkarte (HALLER-WEDEL) für die Arbeitsfolge 1 „Präparation" angelegt und laufend vervollständigt. Nach Beendigung der Aufzeichnung wurden die gleichen Kontrollen für die Arbeitsfolge 3 „Mikroskopieren" und die Kategorie „Operative Tätigkeit" aus Vergleichsgründen vorgenommen (hier nicht veröffentlicht).

Die Ergebnisse der MMH sind der Tabelle 1 und dem Diagramm (Figur 1) zu entnehmen.

Die Tabelle 1 zeigt die prozentualen Anteile des Zeitaufwandes, bezogen auf die einzelnen Kategorien. Neben dem Ergebnisanteil p werden der absolute

Fehler f sowie der relative Fehler ε in Prozent angeführt. Die notwendigen Berechnungen wurden mit Hilfe des Sonderrechenstabes (Haller-Wedel) vorgenommen. Teilweise nicht aufgeführte f-Werte in der 4. Spalte bedeuten, daß bei vorgegebenem N und p die Meßwerte im statistisch ungesicherten Bereich (Haller-Wedel) liegen und deshalb eine Angabe des Fehlers unmöglich ist.

Um die bei der MMH in der Praxis übliche Randbedingung $-2{,}5\% < f < +2{,}5\%$ und $-10\% < \varepsilon < +10\%$ einzuhalten, wurden gleichwertige Arbeitsfolgen (Vorbereitung, operative Tätigkeit, Verluste) zusammengefaßt ausgewiesen. Die Zahlenwerte für f und ε zeigen, daß diese zusammengefaßten Werte innerhalb des Fehlerbereiches liegen, also formal korrekt sind. Um die Anschaulichkeit zu erhöhen, enthält die Tabelle 1 außerdem noch ein Aufwandsbild der Arbeitsfolgen und Kategorien mit eingezeichneten Fehlern.

Die Tage- und Summenkontrollkarte der Figur 1 zeigt grafisch den Anteil der Arbeitsfolge 1 „Präparation" über der Zeit in Einzel- und Summenschreibweise. Dabei sind die Summenkurve auf die rechte Ordinate — geteilt in Beträge des absoluten Fehlers f — und die Tageskurve auf die linke Ordinate — geteilt in Beträge des Ergebnisanteils p — zu beziehen. Der durch die trompetenförmigen Grenzlinien umschlossene Bereich gehört zur Summenkurve, die drei durch jeweils zwei Parallelen begrenzten Bereiche zur Tageskurve.

Für korrekte Aufnahmen wird gefordert (Haller-Wedel), daß die Summenkurve niemals die Grenzen der trompetenförmigen Kurve überschreitet und wenigstens einmal die Mittellinie (d. h. die grafische Repräsentation des Ergebnisanteils $p = 24{,}3\%$) schneidet. Während die letzte Forderung erfüllt ist, kann die erste Bedingung nicht eingehalten werden.

Die Ursache dafür kann leicht der Tageskurve entnommen werden. Hier wird gefordert, daß alle p_T-Werte ($\triangleq$ jeweiliger Ergebnisanteil p pro Aufnahmetag) innerhalb der Kontrollgrenze $\pm 3 \cdot s$ ($\triangleq$ dreifache Standardabweichung) liegen, einige p_T zu beiden Seiten nahe der Mittellinie p zu finden sind und nur einige wenige p_T-Werte nahe außerhalb der Warngrenzen $\pm 1{,}96 \cdot s$ auftreten. Man sieht jedoch, daß der Ergebnisanteil zwischen 0 und 56% liegt. Mit anderen Worten: An bestimmten Tagen werden keine Präparationsarbeiten ausgeführt, während an anderen Tagen mehr als die Hälfte der Arbeitszeit für Präparationen benötigt wird.

Im Sinne dieser Forderungen muß deshalb diese Verteilung als anomal bezeichnet werden. Da ähnliche Kurvenverläufe auch für die anderen Arbeitsfolgen bzw. Kategorien gefunden wurden, muß dieses Verhalten als typisch für diese Aufnahme gelten.

Die Ursache für die Anomalität liegt darin, daß die Arbeitsfolgen nicht parallel, sondern nacheinander ablaufen. (Maxima der einen Tätigkeit stehen dementsprechend Minima der anderen Arbeitsfolge gegenüber.) Damit sind aber die Arbeitsgänge nicht mehr unabhängig voneinander, sondern bedingen einander im Sinne des normalen Arbeitsrhythmus bei elektronenmikroskopischen Arbeiten. Die erforderliche statistische Zufälligkeit wird hierbei durch einen Tages- oder Wochenrhythmus überlagert, wie er in der Tageskurve auch gut zum Ausdruck kommt. Dieser Trend wurde bei der Aufnahme noch durch den Umstand verstärkt, daß die Laborantin wegen Krankheit längere Zeit fehlte und ihre Arbeit von den beiden anderen Arbeitskräften mit erledigt werden mußte. Ferner beteiligte sich die Hochschulkraft wegen anderer Aufgaben ebenfalls nur zeitweilig an elektronenmikroskopischen Arbeiten. Infolgedessen kam es nur selten zur sonst üblichen Parallelarbeit mit zeitlicher Verschiebung der Arbeitsgänge.

Es muß ferner brücksichtigt werden, daß die Intervalle gleichartiger Tätigkeit bei dieser Labortätigkeit meist einige Stunden, manchmal auch einige Tage umfassen. Die durchschnittlichen Notierungsintervalle von 16 Minuten sind dafür offensichtlich zu klein. Hier wäre eine Vergrößerung der Zeiten zwischen jeweils zwei Notierungen und Verlängerung der Gesamtaufnahme auf etwa zwei Monate ratsam.

Unter Berücksichtigung dieser kritischen Anmerkung muß gefolgert werden, daß zwar die Aufnahme formal korrekt ist, die Tätigkeit bei elektronenmikroskopischen Arbeiten jedoch gewisse Änderungen und Anpassungen erfordert. Die von de Jong (Haller-Wedel) vorgeschlagene Berücksichtigung der gegenseitigen Abhängigkeit durch Vermehrung der Anzahl von Notierungen scheint hier nicht zweckmäßig. Es leuchtet nämlich kaum ein, inwiefern ein Verlust an Zufälligkeit durch Vergrößerung der Notierungszahl ausgeglichen werden kann.

Wir beabsichtigen dagegen, bei der vorgesehenen Wiederholung der MMH im elektronenmikroskopischen Labor folgende Änderung einzuführen:

1. Erweiterung der Aufnahmezeit auf etwa zwei Monate bei gleichzeitiger Vergrößerung der Notierungsintervalle auf etwa 30 Minuten
2. Einbeziehung aller dort Beschäftigten (einschl. der Fotolaborantin) bzw. Aussetzen aller Notierungen, sobald eine Arbeitskraft fehlt
3. Verwendung von jeweils drei p_T zur Aufstellung der Tages- und Summenkontrollkarte.

Natürlich ist die Aussage der MMH auf Grund der eben skizzierten Anomalität eingeschränkt. Verzichtet man deshalb auf die angegebenen Fehlergrenzen und betrachtet man die Analyse als erste Näherung, dann lassen sich folgende Schlüsse ziehen:

- für vorbereitende Arbeiten, operative Tätigkeit und Verlustzeiten wurden jeweils rund ein Drittel der zur Verfügung stehenden Zeit verbraucht;
- Das Elektronenmikroskop war — einschließlich der Zeit für Wartung und Pflege sowie eines geschätzten Anteils aus der Arbeitsfolge 1 — im Einschichtbetrieb zu etwa 30% in Betireb;
- Elektronenmikroskopische Arbeiten, die nur von einer Hoch- oder Fachschulkraft ausgeführt werden können, machten etwa 30% der zur Verfügung stehenden Zeit aus;
- Die Kapazität des Labors ließe sich durch Einsatz einer weiteren Arbeitsgruppe (1 Mikroskopiker, 1 techn. Kraft) steigern, ohne daß neue Geräte erforderlich sind. Die zusätzliche Bereitstellung von Kapazitäten für fotografische Positivarbeiten muß gesondert berücksichtigt werden. Eine andere Alternative wäre die ausschließliche Mitarbeit der drei Kräfte im Elektronenmikroskopie-Labor bei voller Arbeitszeit.

Eine maximale Besetzung im Zweischichtbetrieb erfordert:

4 Mikroskopiker,
3 technische Kräfte für Präparations- und sonstige Arbeiten,
1 Fotolaborantin.

Dem Leiter der Forschungsstelle für hochschmelzende Metalle im VEB NARVA — Berliner Glühlampenwerk, Herrn Dr. Kaiser, wird für die Zustimmung und Förderung gedankt, ebenso wie den Mitarbeitern des Elektronenmikroskopie-Labors für ihre Mitwirkung.

Literatur

HALLER-WEDEL, E.: Multimoment-Aufnahmen in Theorie und Praxis. München 1962

....: Lehrbriefe „Arbeitsstudium, Arbeitsgestaltung, Arbeitsnormung", Nr. 14 „Mathematische Entscheidungsvorbereitung (II)", Nr. 18 „Analysenverfahren im Arbeitsstudium (I)". Berlin, 1. Aufl.

SCHINDOWSKI, E., SCHÜRZ, O.: Statistische Qualitätskontrolle. Berlin 1965, 3. fufl.

(Eingegangen am 20. 5. 71)

Anschrift des Verfassers:

Dr. Joachim HEINZE
VEB Narva
Berliner Glühlampenwerk
1017 Berlin
Ehrenbergstr. 11—14

Kristall und Technik	7	1–3	1972	349–355

Jörg Adam

Kriminalistisches Institut der DVP, Berlin

Mikroanalyse mit kombinierten spektroskopischen Methoden in der Kriminaltechnik

Am Beispiel der Untersuchung eines Lacksplitters (ca. 100 µg), der in Zusammenhang mit einem schweren Verkehrsunfall zu untersuchen war, werden Kombinationsmöglichkeiten von IR- mit UV-VIS Spektralfotometrie und Laser-Mikroanalyse für die Mikroanalytik dargestellt.

Als Beweis für die Anwendbarkeit der beschriebenen Verfahren für andere Probleme wird das von dem Cu-Phthalocyaninkomplex nach der KBr-Preßtechnik erhaltenen UV-VIS-Spektrum ausführlich diskutiert.

Folgende Arbeitsweise für eine Mikroanalyse mit kombinierten spektroskopischen Methoden wird beschrieben:

1. Registerung des IR-Spektrums mit Hilfe einer 4 mm-KBr-Mikrotablette (Substanzbedarf ca. 100 µg)
2. Verwendung der KBr-Tablette für die Laser-Mikrospektralanalyse mit gütegesteuertem Resonator
3. Verpressen der Reste der KBr- zu einer 12 mm-KBr-Tablette und Registrierung des UV-VIS-Spektrums.

Investigating a lac splinter (100 µg) from a traffic accident the effective possibilities for combining IR-, UV-VIS-spectral fotometry and laser-microanalysis are shown. For the purpose to illustrate the applicability of these methods to other problems the obtained UV-VIS-spectra are discussed. The combined spectroscopical methods for the microanalysis are described.

1. Einleitung

Zur Aufklärung krimineller Delikte ist häufig die Analyse von Substanzmengen im Halbmikro- bis Mikrobereich erforderlich. Dabei liegen die meist technischen Produkte als „Spuren“ im kriminalistischen Sinne vor (Pharmazeutika, Glassplitter, Metallreste, Textilfasern, Staub, Anstrichstoffe u. ä.).

Das Ziel der Untersuchungen ist es, eine eindeutige Aussage darüber zu treffen, ob eine gesicherte „Spur“ (Sp.) die gleiche stoffliche Zusammensetzung aufweist wie eine entsprechende Vergleichsprobe (Vgl.), oder weitergehend, ob die Spur ehemals einen Teil des Vergleichs darstellte.

Neben vielen anderen Verfahren zur Lösung dieser Fragen werden auch ständig verbesserte spektralanalytische Methoden eingesetzt.

In der vorliegenden Arbeit soll ein von uns entwickelter Analysengang am Beispiel einer Lacksplitteruntersuchung dargestellt werden. Folgende Geräte des VEB Carl Zeiss Jena standen dazu zur Verfügung:

— selbstregistrierendes Zweistrahl-Ultrarotspektralfotometer UR 10
— Laser-Mikrospektralanalysator LMA 1 mit gütegesteuertem Resonator(Q-switch) in Verbindung mit einem Quarzspektrografen Q 24
— selbstregistrierendes Zweistrahl-Spektralfotometer für den sichtbaren und ultravioletten Bereich Specord UV-VIS

2. Problemstellung

Im Zusammenhang mit der Aufklärung eines Verkehrsunfalls bestand die Aufgabe darin zu ermitteln, ob ein kleiner Lackpartikel (ca. 0.5×1.0 mm — etwa 100 μg, Sp.) aus einer Schürfwunde die gleiche Zusammensetzung aufwies wie die Lackierung eines in Frage kommenden PkW. Dazu wurden in unmittelbarer Umgebung der Deformationen am rechten Kotflügel des PkW Lacksplitter als Vergleichsprobe (Vgl.) gesichert.

3. IR-Spektralfotometrie

Die einleitende auflichtmikroskopische Untersuchung zeigte Übereinstimmung in Farbton und Struktur der einzelnen Schichten der Lacksplitter von Sp. und Vgl.

Für die nachfolgende IR-spektralfotometrische Untersuchung wurde der Lackpartikel der Sp. sowie ein gleich großes Teilchen des Vgl. mit KBr vermahlen und zu 4 mm-Tabletten verpreßt. Den mit Hilfe des UR 10 erhaltenen Spektren war zu entnehmen, daß es sich in beiden Fällen um eine Lackfarbe auf Alkydaminharz-Basis handelte, die TiO_2 als Pigment enthielt. Hinweise auf ein Blaupigment waren in diesen Spektren trotz des dunkelblauen Farbtons der Deckschichten nicht vorhanden. Sehr intensive für $CaCO_3$ charakteristische Absorptionsbanden waren den unteren Schichten der Splitter (Spachtelmasse) zuzuordnen.

Da die IR-Spektralfotometrie eine sehr weit verbreitete Analysenmethode ist, soll in der vorliegenden Arbeit nicht näher auf dieses Verfahren eingegangen werden.

4. Laser-Mikrospektralanalyse

Eine endgültige und sichere Aussage über bestehende Artgleichheit zwischen einer Spur und entsprechendem Vergleichsmaterial ist bei Anwendung nur einer Analysenmethode nicht immer möglich.

Ein zweites Verfahren stellt die Emissionsspektralanalyse dar. Neben der klassischen Anregung im Wechselstromabreißbogen hat in den letzten Jahren vor allem zur Analyse geringster Probenmengen die Anregung der Substanzen durch Laserimpulse Anwendung gefunden. Durch die Entwicklung eines gütegesteuerten Rubinresonators (Q-switch) als Zubehör zum LMA 1 wurden neue Untersuchungsmöglichkeiten geschaffen. Die mit den bisher verwendeten Nd^{3+}-Glasresonatoren nicht anregbaren transparenten Materialien (Glas, farblose Minerale, Plaste u. ä.) sind somit ebenfalls der Laser-Emissionsspektralanalyse zugänglich. Die 4 mm-KBr-Mikrotabletten werden durch die Registrierung der IR-Spektren in ihrer

Zusammensetzung nicht verändert. Da zum Erzielen qualitativ guter IR-Spektren eine völlige Homogenisierung durch intensives Vermahlen der Substanz mit dem KBr Voraussetzung ist, bilden diese Tabletten eine ideale Grundlage für die Emissionsspektralanalyse.

Um den Substanzverlust bei der Analyse möglichst gering zu halten, erfolgt die Anregung der KBr-Tabletten mit Hilfe des gütegesteuerten Rubinlasers im LMA 1 (in Verbindung mit dem Quarzspektrografen Q 24).

In der folgenden Tabelle 1 sind die Ergebnisse dieser Analyse von Sp. und Vgl. (Lacksplitter vom PKW „Moskwitsch") gegenübergestellt. Die letzte Spalte enthält Werte, die direkt vom in größeren Mengen vorliegenden Vergleichsmaterial (kompakte Lacksplitter in Paraffin eingebettet) durch Laseranregung erhalten wurden.

Die Zahlen 1 bis 5 geben mit steigender Größe ein Maß für die Intensität der Spektrallinien an; Sp bedeutet, daß die charakteristischen Linien dieses Elementes auf der Fotoplatte visuell noch deutlich erkennbar sind.

Tabelle 1

	Ba	Ca	Mg	Ti	Cu	Al	Si	Fe	Mn	K
Sp. (KBr-Matrix)	1	4	2	1	Sp	Sp	1	1	Sp	5
Vgl. (KBr-Matrix)	1	4	1	1	Sp	Sp	Sp	1	Sp	5
Lacksplitter	1	5	2	2	Sp	1	Sp	2	Sp	—

Es zeigt sich, daß in den KBr-Tabletten der Sp. und der Vgl. die gleichen Elemente in übereinstimmender Konzentration (halbquantitatives Verfahren) enthalten sind. Die geringen Abweichungen in den Mg- und Si-Gehalten lassen sich aus den größeren Verunreinigungen der aus der Schürfwunde gesicherten Sp. gegenüber der Vgl. erklären.

Das indirekte Verfahren der Untersuchung von KBr-Tabletten weist einige wesentliche Vorteile auf, obwohl die Intensitäten der Spektrallinien meist etwas geringer sind als bei der Analyse von kompakten Materialien (vgl. oben angeführte Tabelle)

- Bei geringen Substanzmengen in der Größenordnung von 100 μg reicht bei vorhergehender Laser-Mikroanalyse der verbleibende Rest der Probe nicht mehr aus, um noch deutliche und gut auswertbare IR-Spektren zu erhalten.
- Zur Laser-Mikroanalyse derartig geringer Mengen ist eine Präparation auf Mikroskopobjektträgern mit Hilfe von Einbettungsmitteln (Polystyrol, Paraffin u. ä.) nötig. Bei der nachfolgenden IR-Spektralfotometrie können Reste dieser Substanzen die Spektren erheblich verfälschen.

Da alle wesentlichen Elemente mit hinreichender Sicherheit auch mit dem indirekten Verfahren der Emissionsspektralanalyse unter Verwendung von KBr-Tabletten bei Laser-Anregung nachweisbar sind, werden die Nachteile etwas geringerer Intensitäten zu Gunsten der oben angeführten Vorteile in Kauf genommen.

5. UV-VIS-Spektralfotometrie

Die Ergebnisse der Laser-Mikrospektralanalyse lassen durch den Nachweis von Cu Vermutungen auf das in Sp. und Vgl. vorhandene Blaupigment zu. Es könnte sich dabei um das Cu-Ion des Phtalocyaninkomplexes handeln, denn diese Verbindung weist eine intensive blaue Färbung auf und findet teilweise als Pigment für hochwertige Lackfarben Verwendung. Ein derartiger Nachweis von Cu ist jedoch als Beweis für diese Annahme bei weitem nicht ausreichend.

Die Laser-Mikroanalyse erfordert so geringe Substanzmengen, daß von den 4 mm-KBr-Tabletten noch genügend Material vorliegt, um die UV-VIS-Spektralfotometrie als sehr empfindliche Methode heranzuziehen.

Für die Analyse im UV-VIS-Bereich liegen die Substanzen im allgemeinen in Lösung vor. Die Herstellung einer Lösung durch Extraktion der KBr-Tabletten mit Äthanol u. ä. hat jedoch gewisse Grenzen. So nimmt z. B. mit steigender Qualität der organischen Pigmente deren Löslichkeit stark ab. Anorganische Farbstoffe sind mit dieser Methode in den meisten Fällen gar nicht erfaßbar.

Aus diesen Gründen findet hier ein anderes Verfahren Anwendung. Die Reste der 4 mm-Tabletten werden mit zusätzlichem KBr vermahlen, zu klaren 12 mm-Tabletten verpreßt und diese unmittelbar in das selbstregistrierende Zweistrahlfotometer Specord UV-VIS eingesetzt. Im Vergleichsstrahlengang erfolgt die Kompensation mit einer reinen KBr-Tablette.

Das UV-VIS-Spektrum des verwendeten KBr (Fa. Merck, KBr für spektroskopische Zwecke) zeigt zunehmende Durchlässigkeit von hohen zu tiefen Wellenzahlen. Eine geringintensive Absorptionsbande befindet sich bei ca. 33000 cm^{-1}, die Durchlässigkeit beginnt ab ca. 47000 cm^{-1}.

In den Figuren 1 und 2 sind die Spektren des Azofarbstoffes Permanentrot in Lösung und in KBr-Matrix zum Vergleich gegenübergestellt. Auffallend ist die ausgezeichnete, reproduzierbare Auflösung des UV-Bereiches bei Anwendung der KBr-Preßtechnik (Fig. 2). Die Absorptionsmaxima erfahren gegenüber den Lösungen eine mehr oder weniger starke Verschiebung nach größeren Wellenzahlen und nähern sich dadurch den Spektren von Festkörpern an. Damit in Übereinstimmung ist auf Grund von Dispersionseffekten eine Verbreiterung der Banden festzustellen.

In Figur 3 ist als Beispiel für einen unlöslichen anorganischen Farbstoff das nach der KBr-Preßtechnik erhaltene Spektrum von Bleimennige dargestellt. Die intensiven UV-Absorptionen treten mit gleichen Lagen auch bei anderen Pb-Verbindungen auf (z. B. Bleichromat).

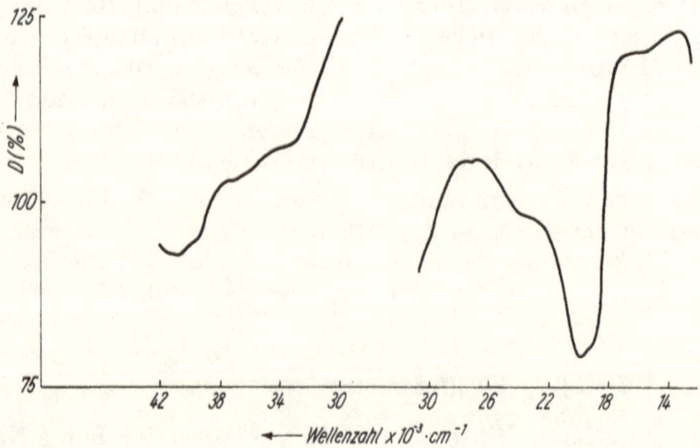

Fig. 1. Absorptionsspektrum von Permanentrot F2R (Chemapol ČSSR) (in Äthanol). Registrierzeit 11 min

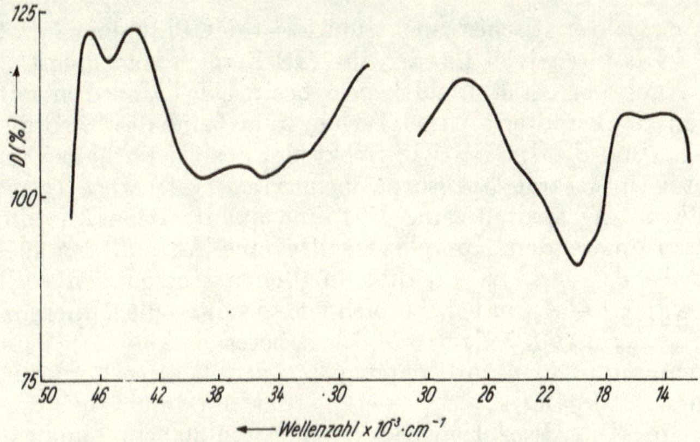

Fig. 2. Absorptionsspektrum von Permanentrot F2R (Chemapol ČSSR) (in KBr-Tablette 12 mm). Registrierzeit 11 min

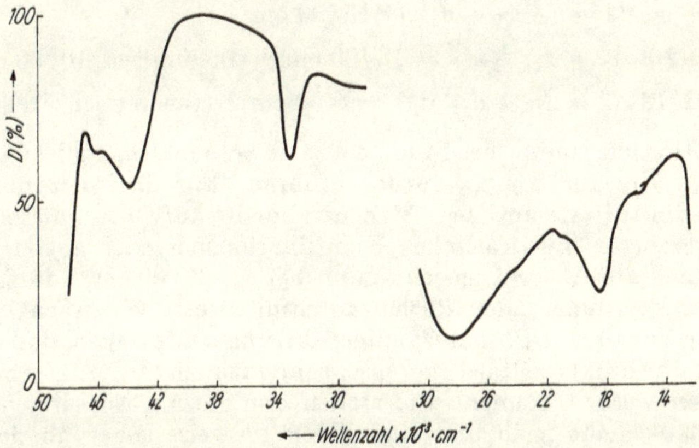

Fig. 3. Absorptionsspektrum von Bleimennige (Ohrdruff) (in KBr-Tablette 12 mm). Registrierzeit 11 min

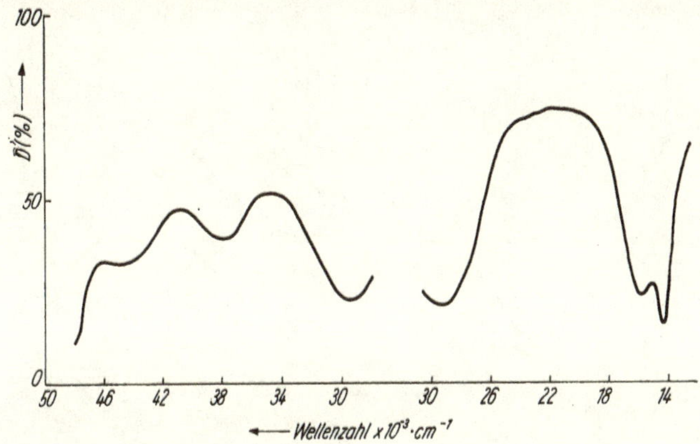

Fig. 4. Absorptionsspektrum von Phthalocyanin-blau („Horna", Maastricht) (in KBr-Tablette 12 mm). Registrierzeit 11 min

Ein sehr charakteristisches Spektrum weist der Phthalocyanin-Cu-Komplex auf (Fig. 4). Die intensiven Banden im VIS-Bereich sind dem Cu^{2+}-Ion zuzuschreiben. Nach der Kristallfeldtheorie besitzt das Cu^{2+}-Ion mit seinem d^9-Niveau einen ^2D-Grundterm. Das Termsystem beinhaltet also die Terme d_{xy}, d_{xz}, d_{yz}, $d_{x^2-y^2}$ und d_{z^2}. Im UV-VIS-Spektrum des Cu-Phthalocyaninkomplexes (Fig. 4) treten insgesamt 5 Absorptionsmaxima auf. Das Termsystem muß also sehr stark aufgespalten sein. Für eine quadratische Anordnung der das Cu^{2+}-Ion koordinierenden Komplexe sollte eine Aufspaltung in die Niveaus $d_{xz}, d_{yz} - d_{z^2} - d_{xy} - d_{x^2-y^2}$ mit drei möglichen Übergängen vorliegen. Wird die Differenz $d_{x^2-y^2} - d_{xy}$ mit Δ_0 bezeichnet, so sollten die Differenzen $d_{xy} - d_{z^2}$ und $d_{z^2} - d_{xz}, d_{yz}$ etwa 2/3 Δ_0 bzw. 1/12 Δ_0 betragen.

Die Berechnung für den vorliegenden Fall erfüllt diese Forderung recht gut, wenn für den Übergang $d_{xz}, d_{yz} \rightarrow d_{x^2-y^2}$ das Maximum bei ca. 45000 cm^{-1} betrachtet wird. Die Absorptionsbanden bei ca. 29200 cm^{-1} und ca. 15800 cm^{-1} sind dann den Übergängen $d_{xz}, d_{yz} \rightarrow d_{xy}$ bzw. $d_{xz}, d_{yz} \rightarrow d_{z^2}$ zuzuschreiben. Aus dem Spektrum sind folgende Werte zu entnehmen:

$$\Delta_0 = d_{x^2-y^2} - d_{xy} = 15800 \text{ cm}^{-1} ,$$

$$2/3\ \Delta_0 = d_{xy} - d_{z^2} = 13400 \text{ cm}^{-1} \text{ (theoretisch } 10600 \text{ cm}^{-1}) ,$$

$$1/12\ \Delta_0 = d_{z^2} - d_{xz}, d_{yz} = 1400 \text{ cm}^{-1} \text{ (theoretisch } 1300 \text{ cm}^{-1}) .$$

Das Absorptionsmaximum bei 14400 cm^{-1} ist auf eine zusätzliche geringe Aufspaltung des Niveaus d_{xz}, d_{yz} zurückzuführen. Für die Anordnung der vier Phthalsäuredinitrilreste um das Cu^{2+}-Zentralion ist auf Grund dieser Ergebnisse ein stark verzerrtes quadratisches Koordinationspolyeder anzunehmen. Das bisher unbeachtete Absorptionsmaximum bei ca. 37800 cm^{-1} dürfte durch die Struktur der koordinierenden Phthalsäuredinitrilreste verursacht werden.

Die Diskussion des Cu-Phthalocyaninspektrums sollte zeigen, daß die mit Hilfe der KBr-Preßtechnik erzielten Ergebnisse sehr gut auch für Strukturaufklärungen herangezogen werden können. Sie stehen den Lösungsspektren an Aussagekraft durchaus nicht nach und sind ihnen teilweise sogar überlegen. Nicht durchführbar sind mit diesem Verfahren quantitative Analysen. Gründe dafür

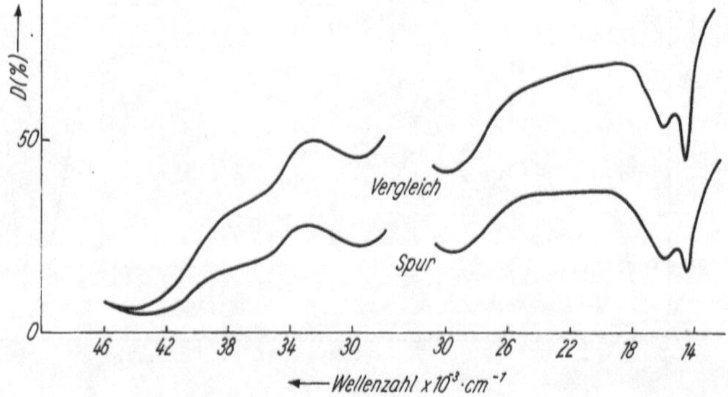

Fig. 5. Absorptionsspektren von „Spur" (Sp.) und „Vergleich" (Vgl.) (in KBr-Tablette 12 mm). Registrierzeit 11 min

sind Reflexionsverluste an den relativ unebenen Oberflächen der KBr-Tabletten, Streulichtverluste, kaum reproduzierbare Korngrößen der Substanzen sowie Dispersionseffekte. Befriedigende (qualitative) Spektren sind mit dieser Methode auch von Metalloxiden u. ä. zu erhalten. Z. B. konnten die Spektren von Cr_2O_3, Fe_2O_3 und der Granatvarietät Uwarowit registriert werden, wobei die Auflösung jedoch nicht so groß wie in den oben geschilderten Fällen war.

Die KBr-Preßtechnik für die UV-VIS-Spektralfotometrie fand, wie schon eingangs erwähnt, für die weitere Analyse der von dem Verkehrsunfall stammenden Proben (Sp. und Vgl.) Anwendung. Die erhaltenen Spektren sind in Figur 5 dargestellt. Sie zeigen eindeutig, daß sowohl in der Sp. wie auch im Vgl. als Pigment der Farbstoff Cu-Phthalocyanin (Heliogenblau) vorliegt. Das Fehlen von Absorptionsmaxima oberhalb 30000 cm^{-1} ist darauf zurückzuführen, daß kein reines Pigment (wie vergleichweise in Figur 4) vorliegt. Besonders im UV-Bereich haben zusätzliche Bestandteile wie Bindemittel (in diesem Fall Alkydaminharz), Füllstoffe usw. einen großen Einfluß auf den Verlauf der Spektren. Außerdem ist zu berücksichtigen, daß Pigmente nur in einer Konzentration von etwa 1% in Lackfarben enthalten sind.

Im Zusammenhang mit dem Verkehrsunfall stellte die angefertigte Expertise auf Grund übereinstimmender Ergebnisse von Mikroskopie, IR- und UV-VIS-Spektralfotometrie sowie Laser-Mikrospektralanalyse Artgleichheit zwischen Sp. und Vgl. fest.

Literatur

HATFIELD, W. E. et al.: Inorg. Chem. **3**, 841 (1964)
KARYAKIN, A. V. et al.: J. angew. Spektroskopie **2**, 364 (1965)
NEUNINGER, H.: Jenaer Rundschau **4**, 235 (1970)
SCHICHT, H.: Forum Krim. 8 (1968)
WICHITILL, J. et al.: Forum Krim. Sonderh. **1** (1969)

(Eingegangen am 23. Juni 1971)

Anschrift des Verfassers:

Dipl.-Min. Jörg ADAM
Kriminalistisches Institut der DVP
Berlin

Kristall und Technik	7	1–3	1972	357–358

Buchbesprechung

WILL KLEBER

Einführung in die Kristallographie, 11. Aufl.
408 Seiten, 361 Abbildungen, 49 Tabellen.
VEB Verlag Technik. Berlin 1971

Seit dem Erscheinen der 1. Auflage vor nunmehr 15 Jahren stellt die „Einführung in die Kristallographie" von W. Kleber nicht nur ein Standardwerk dar, sondern ist einer ganzen Generation von Kristallographen zum unentbehrlichen Begleiter und Ratgeber geworden. In rascher Folge erlebte das Buch seine Auflagen; zu den vorliegenden 11 deutschsprachigen gesellte sich kürzlich noch eine englische (vergl. unten).

In einem immer noch sehr handlichen – und nicht zuletzt auch preiswerten – Bande wird eine Einführung in das gesamte Gebiet der Kristallographie gegeben. Es ist das große Verdienst von W. Kleber, das stofflich umfangreiche und für den Anfänger zuweilen störrische Gebiet in didaktisch vorbildlicher und dabei inhaltlich vollgültiger Weise erschlossen zu haben. So gewinnt das Werk seinen besonderen Wert nicht nur für die Studierenden und als methodische Anleitung für die Lehrenden, sondern in besonderem Maße auch für Fachleute der angrenzenden Disziplinen (Mathematik, Physik, Chemie, Mineralogie, Geologie, Metallurgie und Materialwissenschaften), die mit kristallographischen Fragen in Berührung kommen, woraus sich seine schnelle Verbreitung erklärt.

Das Buch, das mit zahlreichen instruktiven Abbildungen versehen ist, gliedert sich in die Kapitel Kristallstrukturlehre und Kristallmorphologie, Physikalisch-chemische Kristallographie, Kristallchemie sowie Kristallphysik. Sehr ausführlich und instruktiv werden die klassischen Grundlagen der Kristallographie, die Kristallgitter, die Symmetrieoperationen, die Kristallklassen und die Raumgruppen dargestellt. Das Kapitel über Kristallchemie bringt auf 80 Seiten das Erforderliche zu den kristallchemischen Bindungszuständen und eine vollständige Systematik der Kristallstrukturen und kann bereits für sich als kleines Standardwerk angesprochen werden. Bei der Kristallphysik erfolgen Hinweise auf alle wesentlichen Effekte; sehr hohen Ansprüchen genügt der Abschnitt über Kristalloptik (ca. 40 Seiten), schließlich wird auch die röntgenographische Untersuchung von Kristallen behandelt.

Der stürmischen Entwicklung, die verschiedene Gebiete der Kristallographie in letzter Zeit genommen haben, wurde ständig Rechnung getragen, indem von Auflage zu Auflage die neuen, allgemeingültigen, gesicherten Ergebnisse eingearbeitet wurden. Gerade hierbei bewährte sich das Konzept, die Erscheinungen am Objekt Kristall durch eine morphologisch-strukturelle Betrachtungsweise zu erfassen, und das Buch ist immer lebendig und modern geblieben. In diesem Zusammenhang hätte man es sich gewünscht, daß als nächstes vielleicht einzelne Abschnitte über die Kristallbildung (Wachstum, Keimbildung, Züchtung) und die Realstrukturen vertieft und einige neuere kristallphysikalische Ergebnisse (Versetzungsmodell der plastischen Deformation, Effekte und Anwendungen von pyro- und ferroelektrischen Kristallen) stärker berücksichtigt worden wären. Um so mehr ist der Verlust zu bedauern, den die Fachwelt auch in dieser Hinsicht durch das Hinscheiden von W. Kleber betroffen hat. Bei der Vorbereitung dieser 11. Auflage konnte W. KLEber noch selbst Hand anlegen. Für die Zukunft bleibt zu wünschen, daß der Verlag Wege findet, die „Einführung in die Kristallographie“ im Kleberschen Sinn fortzuführen.

Wie erwähnt, erschien vor einiger Zeit auch eine englischsprachige Auflage: "An Introduction to Crystallography" von W. Kleber, VEB Verlag Technik, Berlin 1970; 368 Seiten, 361 Abbildungen, 49 Tabellen; Preis 36,— M.

Die Übersetzung ins Englische erfolgte durch W. A. Wooster und A. M. Wooster nach der 10. (deutschsprachigen) Auflage. Mit dieser Edition unternimmt der Verlag den lobenswerten Versuch, den „Kleber“ auch dem englischsprachigen Raum zu erschließen.

J. Bohm

Kurze Originalmitteilungen

Kristall und Technik	7	1-3	1972	K 1 - K 3

Kurze Originalmitteilungen

W. BRÜCKNER, W. HASE, E. FRITZSCH, C. PIETZSCH

Zentralinstitut für Festkörperphysik und Werkstofforschung
der Deutschen Akademie der Wissenschaften zu Berlin,
Institutsteil für metallische Spezialwerkstoffe, Dresden
und Bergakademie Freiberg, Sektion Physik

Neutronographische Untersuchungen an Chromeisenstein nach mechanischer Aktivierung

Im Zusammenhang mit Untersuchungen über die Aufschließbarkeit von Chromeisenstein nach mechanischer Aktivierung haben SCHRADER und HOFFMANN röntgenographische Untersuchungen durchgeführt, um Aussagen über den Unordnungszustand des Kristallgitters nach langzeitiger Mahlung zu erhalten. Die untersuchten Proben stammten von der Lagerstätte Aktjubinsk (UdSSR). Sie hatten die Zusammensetzung $(Fe, Mg)(Cr, Fe, Al)_2O_4$ und waren bis zu 192 h in einer Laborschwingmühle aktiviert worden.

Die angeführten röntgenographischen Untersuchungen ergaben, daß der durch langzeitige Mahlung erreichte Unordnungszustand des Kristallgitters als mehr oder weniger gestörte dichteste Anionenpackung, in deren Lücken sich die Kationen statistisch verteilt befinden, beschrieben werden kann. Zur Prüfung dieser Aussage wurden am Rossendorfer Forschungsreaktor an den identischen Proben Neutronenbeugungsmessungen ausgeführt, wobei das neutronographisch günstige Verhältnis der Streuamplituden von Sauerstoff zu Metall ausgenutzt werden sollte.

Die neutronographischen Ergebnisse sind folgende:

1. Geringe Zunahme der Reflexbreite mit der Aktivierungsdauer wie bei den röntgenographischen Untersuchungen.

2. Verringerung der Reflexintensität mit steigender Mahldauer. Wie bei SCHRADER und HOFFMANN ist die Intensitätsabnahme nicht für alle Reflexe gleich. Es ergab sich, daß diese Abnahme bei den in einem kubisch-flächenzentrierten Gitter erlaubten Indextripeln $\left(\frac{h}{2}\,\frac{k}{2}\,\frac{l}{2}\right)$ größer ist als bei den übrigen, ansonsten aber unabhängig vom Streuwinkel. In Tabelle 1 ist für verschiedene Mahldauer T_M die auf die Ausgangsprobe bezogene Intensität von zwei Reflexen angegeben. Zum Vergleich sind die analogen röntgenographischen Ergebnisse von HOFFMANN mit eingetragen. Die 96 h aktivierte Probe ist praktisch vollständig amorph.

3. Zusätzlicher Untergrund mit diffusen Maxima bei $\sin\Theta/\lambda = 0{,}23$; $0{,}35$ und evtl. bei $0{,}42\ \text{Å}^{-1}$. Mit steigender Mahldauer nehmen sowohl Mittelwert als auch Modulation dieses Untergrundes zu.

4. Geringe Vergrößerung der Gitterkonstante a mit zunehmender Aktivierungsdauer. Bei 48 h Mahlzeit beträgt $\Delta a/a$ etwa $1.0 \cdot 10^{-3}$.

Tabelle 1

T_M/h	I_{rel} Eigene Messungen (333) und (511)	(440)	HOFFMANN (333) und (511)	(440)
0	1,00	1,00	1,00	1,00
12	0,54 ± 0,09	0,68 ± 0,09	0,57	0,68
24	0,36 ± 0,07	0,47 ± 0,07	0,38	0,53
48	0,18 ± 0,04	0,26 ± 0,04	0,18	0,33

Diese neutronographischen Ergebnisse lassen im Vergleich zu den zitierten röntgenographischen Aussagen folgende präzisierte Vorstellung zu:

1. Mit zunehmender Aktivierungsdauer wird ein wachsender Teil der Probe amorphisiert. Das oben erwähnte erste diffuse Maximum entspricht einem Peak der Paardichteverteilung bei $r \approx 2,7$ Å, also dem Durchmesser von O^{2-}-Ionen. Es ist daher denkbar, daß es sich bei diesem amorphen Anteil um eine Art "eingefrorene Flüssigkeit" von O^{2-}-Ionen mit Metallionen in ihren Lücken handelt. Von einem "Gitter" sollte man bei diesem Anteil nicht sprechen. Für die Mengenanteile der amorphen Substanz erhält man etwa dieselben Werte wie röntgenographisch ermittelt.

2. In dem verbleibenden kristallinen Anteil werden in zunehmendem Maße solche Tetraeder- und Oktaederlücken, die im Spinellgitter unbesetzt sind, von Metallionen statistisch gefüllt. Nach 24 h Aktivierungsdauer sind etwa 10 % der Metallionen auf solche Plätze übergegangen. Dadurch nähert sich der kristalline Anteil einem kubisch-flächenzentrierten Gitter mit der Hälfte der Spinell-Gitterkonstante unter gleichzeitiger Verringerung seiner Menge sowie unter Verringerung der Primärteilchengröße und Zunahme der Gitterverzerrungen.

   Die von SCHRADER und HOFFMANN aus der mit der Mahldauer zunehmenden Intensitätsverringerung gefolgerten wärmeschwingungsähnlichen Gitterstörungen muß man nicht annehmen, wenn man die gesamte Intensitätsverringerung auf Amorphisierung zurückführt.

Außer den bisher besprochenen Messungen wurde die 96 h aktivierte Probe nach 48-stündiger Glühung bei 350 °C nochmals neutronographisch gemessen, um eine Aussage zu dem von HOFFMANN festgestellten Peak der DTA-Kurve bei 300 °C zu erhalten. Die Neutronenbeugungsuntersuchung ergab, daß nach der Glühung der kristalline Anteil von nahezu Null auf etwa 20 % gestiegen war, wobei sich in diesem Anteil etwa 50 % der Metallionen auf Plätzen befinden, die im Spinellgitter unbesetzt sind. Auch im amorphen Anteil hat durch die Glühung eine strukturelle Änderung stattgefunden, wie die gewachsene Modulation des diffusen Untergrundes zeigt. Nach Glühung bei 1000 °C erscheint qualitativ wieder das Beugungsbild der ungemahlenen Probe. Es ist also eine mehr oder weniger vollständige Kristallisation erfolgt.

Wir danken den Herren Prof. Dr. Schrader und Dr. Hoffmann für die Überlassung des Probenmaterials.

LITERATUR

HOFFMANN, K.: Über die mechanische Aktivierung von Chromeisenstein. Dissertation, Freiberg (Sachsen) 1969

SCHRADER, R., HOFFMANN, K.: Chemie-Ingenieur-Technik 42, 38 (1970)

(Eingegangen am 8. November 1971)

Anschrift der Verfasser:

Dr. W. BRÜCKNER
Zentralinstitut für Festkörperphysik und Werkstofforschung der DAW zu Berlin
Institutsteil für metallische Spezialwerkstoffe Dresden
8027 Dresden, Helmholtzstraße 20

Dr. W. HASE
Ingenieurhochschule Dresden
8019 Dresden
Hans-Grundig-Straße 24

Dr. E. FRITZSCH
Dr. C. PIETZSCH
Bergakademie Freiberg, Sektion Physik
92 Freiberg/Sa.
Silbermannstraße 1

Kristall und Technik	7	1-3	1972	K 5 - K 7

P. FISCHER, W. SCHMIDT, H.-G. BRÜHL, G. KÜHN

Sektion Chemie, Fachbereich Kristallographie
Sektion Physik
Karl-Marx-Universität Leipzig
Arbeitsgemeinschaft AIII-BV-Halbleiter

Lattice constants of $Al_{0.8}Ga_{0.2}As$ between -110 °C and +90 °C

The high value of the band gap (E_g = 2.25 eV at 0 °K) makes AlAs an interesting material for high-temperature semiconductor devices and for the production of light in the visible spectral range. However, the corrosion of AlAs due to atmospheric air hinders its application. The deposition of AlAs on GaAs substrates by gaseous-phase epitaxy enables to obtain more stable layers (ETTENBERG, GILBERT). It is more favorable to use Al-rich $Al_xGa_{1-x}As$ crystals applying the method of KRESSEL and WOODALL (KRESSEL, NICOLL, HAWRYLO, LOCKWOOD; WOODALL, RUPPRECHT, REUTER) since the band gap decreases only by about 2 meV per mole percent incorporated GaAs. For the realisation of p-n junctions by means of epitaxial deposition of $Al_xGa_{1-x}As$ on GaAs or vice versa the thermal behaviour of $Al_xGa_{1-x}As$ and a comparison with that of GaAs is of interest.

In this paper lattice constants and the linear expansion coefficient of $Al_xGa_{1-x}As$ crystals (x = 0.80 ± 0.03) between -100 °C and + 90 °C are presented.

A mass-spectroscopical analysis of $Al_xGa_{1-x}As$ single crystals reveals a total impurity of 0.06 atom percent, in particular Si, Ca, and K. The Al fraction distributed homogeneously has been detected by polarographical determination and confirmed by electron micro-probe investigations.

The lattice constants were determined following the procedure indicated by Buerger using a Weissenberg-goniometer with a thermostatic supplement and a film cylinder with a diameter of 114 mm (BRÜHL, SCHMIDT). The X-ray interferences corresponding to different temperatures were exposed on the same film in a displaced manner. In order to have a comparison with the lattice parameters of GaAs, a bar-like GaAs crystal was the interface lying perpendicular to the rod axis, tied to an equal-shaped $Al_xGa_{1-x}As$ crystal. Both crystals were oriented in a [211] direction. The lattice constants of each temperature measured are derived from five reflexes each, they were extrapolated to 90° according to NELSON and RILEY to increase the reliability of the results.

At 10 °C the lattice constant of $Al_{0.80}Ga_{0.20}As$ yields a = (5.6599 ± 0.0002) Å. Figure 1 shows the lattice constants of the system $Al_xGa_{1-x}As$ measured at room temperature by ETTENBERG, PAFF (△); BLACK, KU (x), and PIERRON, PARKER, McNEELY (□), GIESECKE, PFISTER (○) and by the authors (●). Figure 2 lists the lattice constants of GaAs (+ : by STRAUMANIS, KIM, ○ : by PIERRON, PARKER, McNEELY), of AlAs (△: by ETTENBERG, PFAFF), and of $Al_{0.80}Ga_{0.20}As$ (●) in dependence on the temperature. The full circle near the GaAs-line is the authors' value for the GaAs lattice parameter at +10.5 °C. On account of tendencies in the Periodic Table the lattice constant of AlAs should be smaller than that of GaAs. However, also phosphides and antimonides of aluminium and gallium demonstrate that the aluminium compound lattice constants are almost equal or larger (GIESECKE, PFISTER; VAN

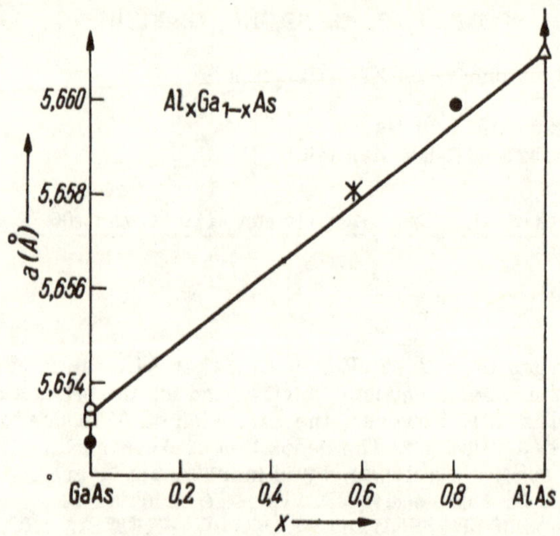

Fig. 1. Lattice parameters of $Al_xGa_{1-x}As$ at room temperature

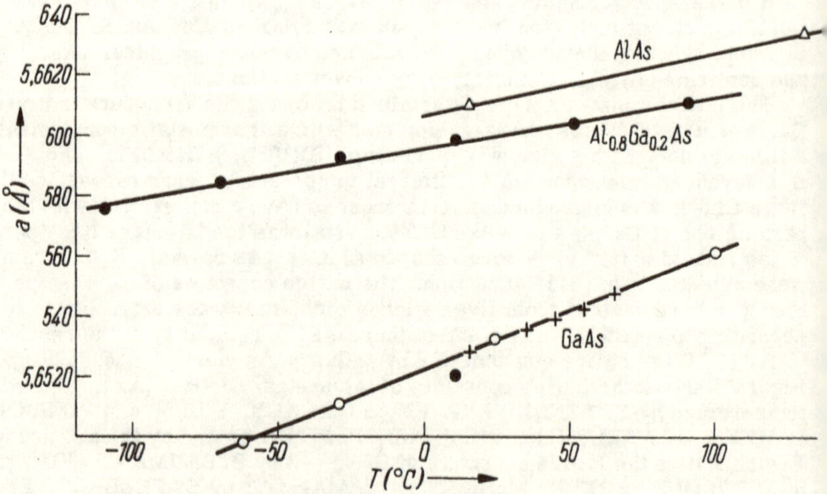

Fig. 2. Lattice parameters and their temperature dependence

VECHTEN, PHILLIPS). On the other hand, indium compounds have-with respect to gallium compounds-larger lattice constants.

For $Al_{0.80}Ga_{0.20}As$ the lattice parameters indicated between -110° and +90 °C yield a linear expansion coefficient $\alpha = (3.1 \pm 0.5) \cdot 10^{-6}$ degree^{-1}.

Taking into account this fact the lattice constant of $Al_{0.80}Ga_{0.20}As$ and GaAs should be consistent at temperatures of about 500 °C. At 20 °C occurs a misfit of both lattices of 0.12 %.

REFERENCES

BLACK, J.F., S.M. KU: J. Electrochem. Soc. 113, 349 (1966)

BRÜHL, H.-G., W. SCHMIDT: Kristall und Technik 6, K17 (1971)

ETTENBERG, M., S.L. GILBERT: National Meeting Electrochem. Soc., Los Angeles, May 1970

ETTENBERG, M., R.J. PAFF: J. appl. Physics 41, 3926 (1970)

GIESECKE, G., H. PFISTER: Acta Crystallogr. 11, 369 (1958)

KRESSEL, H., F.H. NICOLL, F.Z. HAWRYLO, H.F. LOCKWOOD: J. appl. Physics 41, 4692 (1970)

NELSON, J.B., D.P. RILEY: Proc. Physic. Soc. 57, 160 (1945)

PIERRON, E.D., D.L. PARKER, J.B. McNEELY: J. appl. Physics 38, 4669 (1967)

STRAUMANIS, M.E., C.D. KIM: Acta Crystallogr. 19, 256 (1965)

VAN VECHTEN, J.A., J.C. PHILLIPS: Physic. Rev. B2, 2160 (1970)

WOODALL, J.M., H. RUPPRECHT, W. REUTER: J. Electrochem. Soc. 116, 899 (1969)

(Received November 16, 1971)

Authors' address:

Dr. P. FISCHER
Dr. G. KÜHN
Sektion Chemie, Fachbereich Kristallographie der
Karl-Marx-Universität
703 Leipzig
Scharnhorststr. 20

Dozent Dr. W. SCHMIDT
Dipl. Phys. H.G. BRÜHL
Karl-Marx-Universität, Sektion Physik
701 Leipzig
Linnéstr. 5

Kristall und Technik	7	1-3	1972	K 9 - K 10

G. DANEV, R. STOYCHEVA

Institute of Physical Chemistry
Central Laboratory of Photographic Processes
Bulgarian Academy of Sciences, Sofia

The Melting Point of Silver Bromide

In spite of the wide application of silver bromide in photography and the numerous investigations performed with large crystals different references, including those published in recent years, state values for its melting point from 410° to 434 °C (BLANC; CARRI et al.; GRUNDLAGEN; SPRAVOTCHNIK POKASATELEI; SPRAVOTCHNIK CHIMIKA). Preparing large monocrystals for their extensive studies on the photographic properties of silver bromide MALINOWSKI and PLATIKANOVA (for a review see MALINOWSKI et al.) noticed the descrepancy between the actual melting point and the values cited in a number of literature sources. Following observations and having an extremely pure, directly synthesized silver bromide (MALINOWSKI et al.) at disposal, an exact determination of its melting point was easily on hand. Extensive investigations have shown that silver bromide obtained according to MALINOWSKI's method is considerably purer as compared to that one prepared by the most refined sedimentary method developed by Mitchell (GOLDBERG et al.). Threefold electrolysed p. a. silver and bromine (p. a.) twice distilled in vacuum were used as starting materials. The synthesis was performed following the procedure described by MALINOWSKI et al. To oxydise eventually present impurities silver was heated in air in two glass apparatus at 500 °C for 3 hours, after which the hot system was pumped to vacuum better than 10^{-5} Torr for another 3 hours. After sealing off from the vacuum pump the ampoule containing the bromine was broken and the reaction between the heated silver and the bromine vapours went on very rapidly. After filtering through several capillaries in the same apparatus, the bromine was frozen with liquid nitrogen. The apparatus was again attached to a vacuum pump and evacuated for another three hours, the silver bromide being heated at 500 °C. During this procedure a considerable part of the melt was distilled off, carrying away volatile impurities. Finally the silver bromide ingot was zone refined in a quartz tube in bromine atmosphere, passing 70 zones at a speed of 3 cm per hour. The spectral analysis of the final product carried out as described by KRASNOBAEVA et al. together with the sensitivity limits are given below:

	Pb	Fe	Cu	Ni	Mg	Mn	Ca	Cd	Bi
Limits of sensitivity	0, 08	0, 05	0, 05	0, 05	0, 01	0, 03	0, 09	0, 08	0, 05
Analysis	-	0, 05	-	-	-	-	-	-	-

Preliminary experiments to determine the melting point were carried out by following the cooling rate of a platinum/rhodium thermocouple calibrated according to the international scale at the melting points of Sn, Cd, Pb, Zn, Sb. All reference metals had purity of at least 5 N. These preliminary results showed that the melting point of silver bromide is very close to that of zinc.

Precise measurements were now carried out in a Linseis apparatus for differential thermal analysis having 0.05 class of precision. Tin with purity of 5 N was used as a reference substance. The samples were sealed in vacuum in thin wall glass ampules, each containing 2-3 gr of substance, surrounding the thermocouple. The heating and cooling curves indicated that the melting point of silver bromide in vacuum is $419,5^{o} \pm 0,2\ ^{o}C$.

It seems worthwhile to note that the very pure silver bromide shows a marked tendency for supercooling, sometimes going down to 2-3 ^{o}C.

The authors acknowledge with pleasure the suggestion of Prof. J. Malinowski to make these measurements, as well as his comments.

REFERENCES

BLANC, M.: Comptes Rendus (France) 242, Nr. 3, 273 (1958)
CARRI, J. et al.: Bull. Soc. Chim. (France) 7, 2322 (1969)
GOLDBERG, G. et al.: ASD Technical Report 61-732, 52 (1962)
GRUNDLAGEN der photographischen Prozesse mit Silberhalogeniden, Vol. I Frankfurt/Main 1968, p. 137
KRASNOBAEVA, N., NEDJALKOVA N., KOSTADINOVA, V.: Materiali ot III Nazionalna Konferenzia po Spektroskopia, Varna 1969
MALINOWSKI, J., PLATIKANOVA, V., PETKANCIN, J.: Izv. Inst. Phys. Chemia BAN 3, 119 (1963)
SPRAVOTCHNIK pokasatelei katschestva chimitscheskich reaktivov Chimia, Vol. II, 1968, p. 793
SPRAVOTCHNIK chimika Vol. II, Moscow 1963

(Received January 17, 1972)

Authors' address:

G. DANEV, R. STOYCHEVA
Bulgarian Academy of Sciences
Institute of Physical Chemistry
Central Laboratory of Photographic Processes
Sofia 13, Bulgaria

Kristall und Technik	7	1-3	1972	K 11 - K 12

J. GODARD, GRUJIC DRASKO

Laboratoire de Physique-Cristalline[1]
Université PARIS-SUD - Centre d'ORSAY
ORSAY-FRANCE

Preparation of Singe Crystals of $(NH_4)_2SO_4$

In a recent paper DESVIGNES et al. have shown that growing of single crystals of $(NH_4)_2SO_4$ is a difficult operation being easier when glycerine is added to the solution.

We can report that we succeeded in obtaining good single crystals weighing 100 gr or more without particular difficulties.

We operated by cooling down the aqueous solution between 50 and 30 °C at a rate of 1/10° each two days. The apparatus used is shown Figure 1.

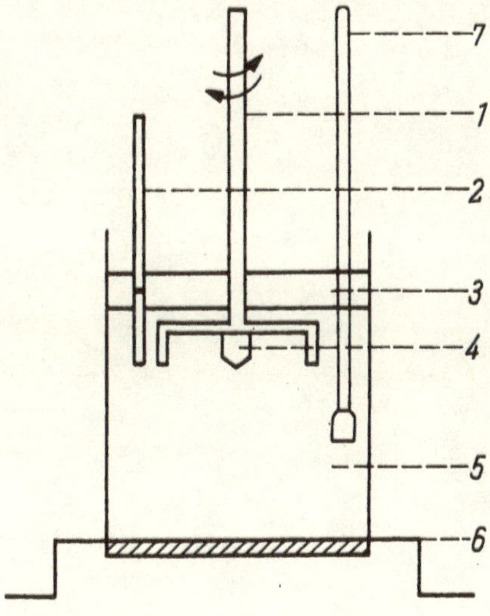

Fig. 1. Apparatus for the growing of $(NH_4)_2SO_4$

(1) Stiner
(2) Glasstube for preliminary evaporation
(3) Paraffin oil
(4) Seed
(5) Solution (3 liters)
(6) Heater
(7) Contact thermometer

---

1 Equipe de Recherche Associée au C. N. R. S. n° 13 -

In order to eliminate spurious crystals on the initial seed, it is necessary to begin the operation with a very slow evaporation, or to put on the bottom of the vessel a little crystal of $(NH_4)_2SO_4$. This crystal is dissolved, and there is a crystallization on the seed.

It is well known that the habits of $(NH_4)_2SO_4$, the twinning and the maximum supersaturation coefficient are very sensitive to cationic and anionic impurities. We obtained the best results with $MnSO_4$ as impurity (concentration 1/100 mol/lit).

REFERENCE

DESVIGNES, J. M., M. REMOISSENET, V. MARECEK: Kristall und Technik 6, 203 (1971)

(Received February 8, 1972)

Authors' adress:

J. GODARD
Laboratoire de Physique Cristalline
Université PARIS SUD - Centre d'Orsay
91 ORSAY FRANCE

GRUJIC DRASKO
Institut de Physique Nucléaire
BELGRADE - YOUGOSLAVIE

Kristall und Technik	7	1-3	1972	K 13 - K 15

Buchbesprechung

H. EHRENREICH, F. SEITZ & D. TURNBULL (Herausgeber)

Solid State Physics, Vol. 25, 420 Seiten. $ 24.00

Academic Press Inc. Publ., New York und London 1970

Der 25. Band dieser Reihe enthält drei Beiträge zu unterschiedlichen Themen: der elektronischen Struktur von Kristalloberflächen, dem Kristallwachstum und der Elastizitätstheorie von Kristallen, die alle drei das spezielle Interesse des Kristallographen beanspruchen sollten. Ferner sind ein Autoren- und ein Sachwortregister sowie ein Verzeichnis der bisher erschienenen und der demnächst erscheinenden Bände von Solid State Physics enthalten.

1. S. G. DAVISON & J. D. LEVINE: Surface States, S. 1 - 150, mit 43 Figuren, 5 Tabellen und 302 Literaturzitaten im Text.

Im Gegensatz zur Theorie des unbegrenzten Kristalls ist die quantenmechanische Theorie der Kristalloberflächen noch in einem relativ frühen Stadium der Entwicklung. Obwohl bereits wesentliche Beiträge zum Verständnis der Elektronenzustände an Kristalloberflächen geleistet wurden, behandelten zusammenfassende Arbeiten bisher immer nur bestimmte Aspekte oder Methoden. Hier wird nun erstmals eine Übersicht über die gesamte Thematik gegeben, wobei die verschiedenen Ansätze und Lösungen gegenübergestellt und in Beziehung gesetzt werden. Die verfügbaren Resultate und Daten, die Aussagen über die Eigenschaften von Oberflächenzuständen erbringen können, werden für eine Reihe von Kristallen kritisch zusammengestellt und mit dem Experiment verglichen. Grundsätzlich wird davon ausgegangen, die Einelektronen-Schrödinger-Gleichung unter den im jeweiligen Modell gewählten Randbedingungen einer "sauberen" Kristalloberfläche zu lösen. Elektronenzustände, die von einer Adschicht von Fremdatomen hervorgerufen werden, sind nur kurz gestreift, und Fragen wie Katalyse und Chemisorption bleiben außerhalb der Themenstellung.

Nach einer kurzen historischen Übersicht werden erst die (eindimensionalen) Kristallorbital-Methoden, dann die (eindimensionalen) Kristallpotential-Methoden und schließlich dreidimensionale Ansätze behandelt. Der mathematische Aufwand wird so gering, wie für eine strenge Durchführung möglich, gehalten, und das Hauptgewicht auf ein Verständnis der Grundkonzeption und eine Klarstellung der in die Modelle eingehenden Voraussetzungen und Näherungen sowie die Diskussion der Ergebnisse gelegt. Sodann werden zwei halbklassische Methoden behandelt. Für den Prototyp kovalenter Bindungen ist das Modell der unterbrochenen Bindungen ("dangling bonds") besonders für Halbleiterkristalle von Bedeutung und gestattet die Berücksichtigung von Rekonstruktionseffekten an der Oberfläche. Für den Prototyp ionogener Bindungen wird die Methode mit Madelung-Potentialen besprochen.

Ein ausführlicher Teil (80 Seiten) gilt dem Vergleich der theoretischen Modellvorstellungen mit den experimentellen Möglichkeiten. Die vorligenden experimentellen Daten werden daraufhin kritisch gesichtet, inwieweit sie Eigenschaften einer "sauberen" Oberfläche widerspiegeln oder auf sekundäre Effekte zurückgehen. Der wesentliche Durchbruch bei der experimentellen Untersuchung von Oberflächenzuständen erfolgte erst in den letzten zehn Jahren durch die Anwendung von Ultrahochvakuum. Weitaus die meisten Daten liegen für Halbleiterkristalle vor.

Das theoretische Verständnis der experimentellen Daten ist für die III-V- und die II-VI-Halbleiter schon sehr zufriedenstellend, jedoch problematisch für Si und Ge, wofür die Rekonstruktion der Oberflächenstruktur der wesentliche Grund ist. Die Autoren behandeln die experimentellen Grundphänomene, die Verteilungen der Oberflächenzustände, die experimentellen Untersuchungsmethoden (Dotierung, Einwirken von Feldern, Leitfähigkeit parallel zur Oberfläche, Photospannung, Photoemission), die Probenpräparation und die Beiträge der LEED-Technik (Beugung niedrigenergetischer Elektronen) zur Kenntnis der Oberflächenstrukturen. Schließlich wird eine Reihe von Daten und Ergebnissen für Si, Ge, sowie III-V- und II-VI-Verbindungen zusammengestellt.

2. R. L. PARKER: Crystal Growth Mechanisms: Energetics, Kinetics, and Transport, S. 151-299, mit 45 Figuren, 1 Tabelle und 319 Literaturzitaten im Text.

Der Artikel stellt eine umfassende und recht vollständige Aufnahme des im physikalischen Sinne wissenschaftlichen Bestandes zum Themenkreis des Kristallwachstums dar. Dieser Themenkreis ist ausgeprägt interdisziplinär, die Beiträge kommen nicht nur aus unterschiedlichen Wissenschaftszweigen, sondern werden auch in den verschiedensten Einrichtungen erarbeitet und in einer Vielzahl von Medien oder Veranstaltungen veröffentlicht und diskutiert. Der Autor stellt sich die Aufgabe, die zahlreichen und z. T. extrem diversen Phänomene, die hierher gehören, von einem physikalischen Standpunkt zu ordnen und zu überblicken. Das Schwergewicht liegt dabei auf quantitativ verfolgbaren Ansätzen und Modellen von wohldefinierten Systemen, sowohl in theoretischer als auch in quantitativer Hinsicht, und beinhaltet Anwendungen der statistischen Mechanik, der Thermodynamik, der kinetischen Theorie und vor allem der Transporttheorie auf Probleme der Kristallisation. Außerhalb der Aufgabenstellung bleiben qualitative und empirische Beschreibungen von Wachstums- und Züchtungsvorgängen, die sehr zahlreich vorliegen, sowie auch die eigentliche Züchtungstechnik und die -apparaturen; gleichfalls nicht behandelt werden Übergänge in festen Phasen und die Kristallisation von Polymeren.

Bereits im ersten Kapitel, der Thermodynamik von Phasenübergängen, wird deutlich, daß sich heute noch keine geschlossene Theorie der Kristallisation ohne empirische Parameter entwickeln läßt. Schon bei der Aufstellung von Phasendiagrammen bedarf es - von einigen Ansätzen für Edelgase abgesehen - solcher Parameter. Ferner wird die Thermodynamik von Grenzflächen und die irreversible Thermodynamik behandelt. In den weiteren Kapiteln wird für die Anwendung der statistischen Mechanik von der Verteilungsfunktion eines kanonischen Ensembles ausgegangen. Für die Betrachtung von Phasenübergängen wird das ursprünglich für ferromagnetische Übergänge entwickelte ISING-Modell und seine Abwandlungen als einziges vollkommen exakt gelöstes Modell, das einen Phasenübergang zeigt, hervorgehoben. Ferner werden statistische Ansätze für "rauhe" Grenzflächen behandelt. Ausführlich wird auf die mathematische Behandlung diffusionsartiger Transportvorgänge, die sogenannte Stefan-Probleme eingegangen. Die Vorgänge sind nicht stationär und führen auf schwierige Grenzwertprobleme, besonders, wenn die Randbedingungen, den Verhältnissen angepaßt, etwas komplizierter werden. Für die Keimbildung werden die klassischen Theorien und auch moderne Ansätze, z. B. für kleine Keime bei hohen Übersättigungen und im Rahmen der Cluster-Theorie aufgeführt. Das umfangreichste Kapitel gilt mit 43 Seiten den Grenzflächenstrukturen und der Grenzflächenkinetik. Nach einer thermodynamischen Betrachtung werden vor allem die atomistischen Modelle, angefangen vom Kossel-Modell bis zu den PBC-Vektoren, ausführlich diskutiert und hierbei die Kristallisation aus der Dampfphase, aus der Lösung und aus der Schmelze gesondert und soweit irgend möglich quantitativ behandelt.

Berücksichtigt werden auch die sogenannten kinematischen Ansätze, die von rein geometrischen Voraussetzungen ausgehen. Die morphologische Stabilität von wachsenden Kristallen wird wieder als mathematisches Problem behandelt, indem die Stabilität in Diffusionsfeldern gegenüber kleinen Formabweichungen betrachtet wird. Der Einfluß von Beimengungen und Verunreinigungen wird ausschließlich in seiner Wirkung auf die atomistischen Wachstumsmodelle vorgestellt und für Fragen der Trachtbeeinflussung, der Verteilung von Beimengungen usw. auf vorliegende Literatur verwiesen. Als letztes Kapitel dieser außerordentlich reichhaltigen Übersicht werden hydrodynamische Ansätze zur Beschreibung des Materialtransports durch Strömung und Konvektion verfolgt.

3. D.C. WALLACE: Thermoeleastic Theory of Stressed Crystals and Higher-Order Elastic Constants, S. 301-404, mit 9 Tabellen und 124 Literaturzitaten im Text.

Angeregt durch die in den letzten Jahren erfolgte Entwicklung bei der Untersuchung von Ultraschallwellen in Festkörpern unter mechanischer Spannung, der Bildung von akustischen Harmonischen und von Summen- und Differenzfrequenzen bei der Wechselwirkung von Ultraschallwellen, wird eine allgemeine, geschlossene Theorie der Thermoelastizität von unter mechanischen Spannungen stehenden Festkörpern zusammenfassend dargestellt und erweitert. Die spezielle Anwendung dieser Theorie erfolgt auf die nichtlinearen thermoelastischen Eigenschaften anisotroper Festkörper.

Da gleichermaßen adiabatische wie isotherme Vorgänge zu erfassen sind, hat die Theorie von den thermodynamischen Potentialen auszugehen. Während die klassische Theorie als Grundansatz infinitesimale Verformungen einer spannungsfreien Konfiguration verwendet, wird die Theorie hier für den Fall einer beliebigen Anfangsspannung entwickelt. Ferner erweist es sich für die Behandlung der nichtlinearen Effekte als angebracht, von endlichen Verformungen anstelle der infinitesimalen Verformungen auszugehen. Das umfangreichste Kapitel gilt mit 30 Seiten der allgemeinen Entwicklung der Theorie, in welcher drei verschiedene Sätze von thermoelastischen Koeffizienten zweiter Ordnung eine fundamentale Rolle spielen: die elastischen Konstanten, die Spannungs-Dehnungs-Koeffizienten und die Koeffizienten für die Wellenfortpflanzung. In einem speziellen Kapitel wird der Spezialfall kleiner Anfangsspannungen behandelt. Die Symmetrie der thermoelastischen Erscheinungen richtet sich nach den LAUE-Gruppen, für welche die Anordnungen der betreffenden Konstanten aufgeführt werden. Schließlich werden Beispiele der Anwendung der Theorie für die Bestimmung der elastischen Konstanten dritter Ordnung und der elastischen Konstanten vierter Ordnung entwickelt.

Im Hinblick auf die Interpretation von experimentellen Messungen beschränkt sich der Artikel auf einige Beispiele. Der Autor hebt ausdrücklich hervor, daß die Erfassung von veröffentlichten Ergebnissen in Tabellen oder Zusammenstellungen deswegen unterblieben ist, weil fast stets Fehler gefunden wurden. Deshalb wurde alles, was im Artikel Verwendung fand, vom Autor noch einmal neu abgeleitet oder anderweitig überprüft.

J. BOHM

Kristall und Technik	7	1-3	1972	K 17 - K 18

Buchbesprechung

R. A. LAUDISE

The Growth of Single Crystals

352 Seiten, zahlreiche Figuren, Tabellen und Literaturhinweise im Text, Autoren- und Sachverzeichnis. £ 7.25
Englewood Cliffs, New Jersey 1970, Prentice-Hall International

Das Buch hat der Autor speziell für den Kristallzüchter geschrieben. Der Buchtitel wäre treffend mit "Die Züchtung von Einkristallen" zu übertragen. Offenbar geht der Autor davon aus, daß ein Kristallzüchter keine Ausbildung oder Vorkenntnisse in der Kristallographie besitzt, und so sind die ersten drei Kapitel und damit ein Drittel des Buches einer Einführung in kristallographische Grundlagen, in die Thermodynamik und in die Kinetik des Kristallwachstums gewidmet. Das erste Kapitel behandelt vor allem Röntgenmethoden, Realstrukturen in Kristallen sowie Methoden zu ihrer Untersuchung; es kann allerdings auch für einen vorwiegend mit züchterischen Problemen Befaßten eine fundiertere Einführung in diese Gebiete nicht annähernd ersetzen. Da die Realstrukturen nun schon behandelt sind, vermißt man in einem solchen Buch, wie die Realstrukturen denn eigentlich mit der Züchtung im Zusammenhang stehen. Es ist allerdings nur das erste Kapitel, was zur Kritik herausfordert. Im zweiten Kapitel wird die Thermodynamik des Kristallwachstums anhand der verschiedenen Gleichgewichte dargestellt und eine Klassifikation der Wachstumsprozesse gegeben. Berücksichtigung finden auch die für die Züchtung so wichtigen nichtkonservativen Prozesse. Im anschließenden Kapitel werden die Kinetik des Kristallwachstums, die Diffusion, die Vorgänge an den Kristallflächen, die Keimbildung, aber auch die Erscheinungen der konstituionellen Unterkühlung, der Fazettenbildung und des Zellularwachstums dargestellt. Beide Kapitel geben dem Kristallzüchter eine sehr brauchbare und durchaus angemessene Einführung in die grundlegenden Erscheinungen des Kristallwachstums.

Der Hauptteil des Buches ist dann einer übersichtlichen Behandlung der Züchtungsmethoden gewidmet, wofür rund 200 Seiten zur Verfügung bleiben. Bemerkenswert ist ein ausführlicheres Kapitel über das Kristallwachstum durch Übergang zwischen festen (kristallisierten) Phasen. Neben den eigentlichen Züchtungsmethoden stellt der Autor auch die Modifikationsänderungen hierher, u. a. die Bildung ferroelektrischer Domänen beim Curie-Punkt und die Präparation von Eindomänenkristallen. Die weiteren Züchtungsmethoden werden in der üblichen Gliederung geboten: Züchtung aus der Schmelze, Züchtung aus der Dampfphase einschließlich Transportreaktionen und Züchtung aus Lösungen einschließlich Schmelzlösungen. Die Abhandlung erfolgt durchweg unter Bezug auf die mit den betreffenden Methoden zu züchtenden Substanzen, die apparativen Lösungen werden durch zahlreiche Prinzipskizzen erläutert. Wie der Autor selbst hervorhebt, entstammt ein großer Teil des dargestellten Materials den Ergebnissen und Erfahrungen aus den langjährigen Arbeiten zur Kristallzüchtung bei den Bell Telephone Lab.; die Sachkenntnis aus erster Hand ist in vielen Partien des Textes wohltuend zu spüren. Daneben wird zu jedem Kapitel eine größere Auswahl an Literatur präsentiert; eine

Vollständigkeit hinsichtlich der als Kristalle gezüchteten Substanzen sowie der vielen Varianten der Züchtungsmethoden wird dabei nicht angestrebt. Alles in allem ist ein sehr nützliches und dabei auch noch sehr handliches Buch geschaffen worden, was nur zu empfehlen ist und in keinem Züchtungslabor fehlen sollte.

J. BOHM

Kristall und Technik	7	1-3	1972	K 19

Buchbesprechung

G. H. BROWN, J. W. DOANE, V. D. NEFF

A Review of the Structure and Physical Properties of Liquid Crystals

94 Seiten 52 Abbildungen. £ 5,00
Butterworths, London 1971

Das vorliegende Werk ist eine Publikation im Rahmen der CRC (Chemical Rubber Company) Monoscience Series, welche als Hauptziel die Herausgabe von zusammenfassenden Schriften über den gegenwärtigen Stand einzelner, klar umrissener Gebiete verfolgt, wobei insbesondere die gegenwärtige Aktualität für deren Auswahl maßgebend ist. In der Tat kann das Gebiet der flüssigen Kristalle diese beanspruchen.

Den Autoren kann man bescheinigen, zusammenfassende (94 Seiten!) und aktuelle Darstellung gut vereint zu haben. Auch das Gewicht einzelner Abschnitte spiegelt die gegenwärtigen Hauptforschungsrichtungen wieder. Aber auch Eigenschaften der flüssigen Kristalle, die z. Zt. nur vereinzelt bearbeitet werden, finden Erwähnung.

Der einleitenden Behandlung der molekularen Geometrie der Substanzen mit kristallin-flüssigen (kr. fl.) Zuständen folgen Ausführungen über den kr. fl. Polymorphismus und die Struktur der Modifikationen. Daran schließen sich die theoretischen Vorstellungen an über flüssige Kristalle sowie die wichtige Behandlung des Einflusses äußerer Kräfte und Felder. Die statistischen Theorien der Phasenübergänge sind mit experimentellen thermodynamischen Befunden in einem Abschnitt vereinigt. Es folgen dann Eigenschaften im einzelnen, wie z. B. IR- und Raman-Spektren, Lichtstreuung, magnetische Resonanz, Doppelbrechung usw. Schließlich ist das Schlußkapitel über lyotrope kr. fl. Systeme bemerkenswert. Dadurch wird auf die Gesamtheit der kr. fl. Zustände und ihre Multivalenz aufmerksam gemacht.

Das Literaturverzeichnis ist ganz auf die Arbeiten der letzten Zeit ausgerichtet; selbst einige im Druck befindliche Publikationen wurden erfaßt.

H. SACKMANN

Buchbesprechung

H. PUPKE

Elementare Prozesse in Gasen und in Grenzschichten
Mathematisch-Naturwissenschaftliche Bibliothek, Band 37

253 Seiten mit 56 Abbildungen. 14,2 x 20,0 cm
1970. In Halbleinen M 34,50
BSB B. G. Teubner Verlagsgesellschaft, Leipzig 1970

Das im vorliegenden Buch behandelte Gebiet besitzt nicht nur als eine der wesentlichsten Grundlagen der Gasentladungsphysik große Bedeutung, sondern reicht auch in viele andere Gebiete der Physik hinein. Als Beispiel sei hier die Bedeutung der elementaren Prozesse in Grenzschichten in der modernen Festkörperphysik genannt.

Den eigentlichen Hauptteilen des Buches (Elementare Prozesse im Gas, Elementarprozesse an den Grenzflächen) sind einige für deren Verständnis notwendige Grundlagen vorangestellt, die die Eigenschaften der Gase (Statistik, Gasdruck, Stoßzahl und freie Weglänge), die Elektrodynamik der Gase, die Strahlungsgesetze und die Wechselwirkung der Teilchen (Stoßgesetze) betreffen. Dabei wird besonderer Wert auf eine klare Herausarbeitung des Begriffes und der Bedeutung des Wirkungsquerschnittes gelegt. Als Elementarprozesse im Gas werden neben Streuung, Anregung und Ionisation auch Stöße zweiter Art, Umladungs- und Dissoziationserscheinungen, Fragen der Rekombination, der Beweglichkeit und der Diffusion, sowie die Bedeutung negativer Ionen behandelt. Innerhalb der Elementarprozesse an den Grenzflächen werden zunächst die Glühemission der Metalle, die lichtelektrische Emission, die Sekundäremission und die Feldelektronenemission besprochen. Ergänzt werden die Ausführungen durch die Behandlung des Mechanismus der Entstehung positiver und negativer Ionen sowie elektromagnetischer Strahlen in Grenzschichten.

Der vom Autor geschickt gewählte Aufbau des Buches, die daraus resultierende übersichtliche Gliederung und die gegebenen klaren Formulierungen erlauben dem Leser eine schnelle Orientierung über das Gebiet und ein rasches Verständnis des gebotenen Stoffes, wozu nicht zuletzt die klaren mathematischen Ableitungen der teils klassischen, teils quantenmechanischen Rechnungen beitragen. Das nützliche Buch kann einem breiten Kreis von Physikern und Wissenschaftlern aus Nachbardisziplinen - schon vom Studenten an - bestens empfohlen werden.

J. HEYDENREICH

Kristall und Technik	7	1-3	1972	K 23

Buchbesprechung

A. P. CRACKNELL

The Fermi Surfaces of Metals. £ 3. 50

Verlag Taylor & Francis Ltd. 1971 London

Die bedeutenden Fortschritte, die auf dem Gebiet der Elektronentheorie der Metalle in den letzten 15 Jahren erzielt worden sind, beruhen zum großen Teil auf einer intensiven wechselseitigen Befruchtung experimenteller und theoretischer Forschung. Als besonders bedeutungsvoll erwies sich dabei die Bestimmung der Fermiflächen. Sie können einerseits aus der Bandstruktur der Metallelektronen und aus der Zustandsdichte berechnet und andererseits durch eine Vielzahl von Verfahren experimentell geprüft werden.

Seit der Internationalen Konferenz über Fermiflächen im Jahre 1960, deren Vorträge von Harrison und Webb herausgegeben wurden, sind die Ergebnisse so stark angewachsen - für die große Mehrzahl der metallischen Elemente liegen mehr oder weniger eingehende Untersuchungen vor -, daß eine zusammenfassende Darstellung sehr begrüßt werden muß.

Das vorliegende Buch stellt einen kaum veränderten Abdruck zweier Arbeiten dar, die der Verfasser 1969 und 1971 in den Advances of Physics veröffentlicht hat. Es besteht dementsprechend aus 2 Teilen, deren erster die sogenannten s- und p-Block-Metalle enthält (das sind die "einfachen Metalle", für welche die Vorstellung freier Elektronen eine brauchbare Näherung darstellt, also die Alkalimetalle, die Erdalkalien, sowie die Metalle der Bor-, Kohlenstoff und Stickstoff-Gruppe). Im zweiten Teil werden die übrigen Metalle (einschl. Lanthaniden und Aktiniden) besprochen.

Der überwiegende Teil des Buches referiert in kritischer Weise die vorliegenden Untersuchungen (das Literaturverzeichnis umfaßt nahezu 2. 000 Titel!), wobei im Abschnitt "Alkalimetalle" auch die Methoden kurz charakterisiert werden. Dem Ganzen vorangestellt ist eine sehr knappe Übersicht über einschlägige Grundbegriffe wie Brillouin-Zonen, Energiebänder und Fermiflächen. Der Abschnitt jeder Metallgruppe beginnt mit einer Erörterung der in ihr auftretenden Kristallstrukturen.

Das Buch gestattet jedem an seiner Thematik Interessierten, sich bequem und zuverlässig auf diesem Gebiet zu orientieren, zumal die Darstellung sehr klar ist und auf eine gute zeichnerische Wiedergabe auch komplizierter Fermiflächen großer Wert gelegt wurde. Es kann daher sehr empfohlen werden, und es wäre zu wünschen, daß dieser Übersicht über die Untersuchungsergebnisse an metallischen Elementen bald eine entsprechende über Legierungen und intermetallische Verbindungen folgt.

G. E. R. SCHULZE

Kristall und Technik	7	1-3	1972	K 25 - K 26

Buchbesprechung

Herausgegeben von N. KLEIN, D. S. TANNHAUSER, M. POLLAK

Conduction in Low-Mobility Materials

464 Seiten. £ 10.00
Taylor & Francis Ltd, London 1971

Dieser in sehr guter Aufmachung erschienene Sammelband enthält 52 Vorträge der zweiten internationalen Konferenz über Leitung in Materialien geringer Beweglichkeit, die vom 5. -8. April 1971 in Eilat, Israel, stattfand und vom Institut für den festen Zustand der Technischen Hochschule Israels organisiert wurde. An dieser Tagung nahmen 112 Wissenschaftler aus 12 Ländern teil. Das sehr komplexe Gebiet wurde in 7 Sektionen behandelt.

9 Hauptvorträge (invited papers), die nicht gleichmäßig auf alle Sektionen verteilt waren, bildeten den Kern der Tagung. Da es nicht möglich ist, alle Vorträge auch nur zu erwähnen, seien hier nur die Hauptgebiete kurz umrissen.

Die Übergangsmetalloxide werden in 13 Beiträgen behandelt. Hopping-Prozesse in Valenz-Halbleitern (Austin, Gamble), die Natur freier und gebundener Ladungsträger (van Daal) und die Morphologie von Polaronen in Vanadiumoxiden (Goodenough) sind die Hauptthemen, um die sich Berichte über IR-Untersuchungen, dielektrische und Leitungsrelaxation, Löcherbeweglichkeit u. a. gruppieren. Die Theorie der Transporterscheinungen bei niedriger Beweglichkeit (5 Vorträge) befaßt sich mit Polaronentransport und Problemen schmaler Bänder (Friedman) sowie speziell mit Gleichstrom- und optischen Eigenschaften von Übergangsmetalloxiden (Reik).

Die Probleme der Amorphen Halbleiter werden in zwei umfassenden Übersichtsvorträgen sowohl hinsichtlich der elektronischen Leitfähigkeit (Davis) als auch allgemeiner elektrischer Eigenschaften (Stuke) behandelt, während weitere 5 Vorträge sich speziell mit amorphen Silizium-, Germanium- und Kohlenstoffschichten befassen. Über Selen-Verbindungen findet man 4 Beiträge zu speziellen Untersuchungen (Hall-Effekt in Te-Se-Legierungen, trigonalen Selen-Filmen, amorphem Selen usw.)

Ein besonderes Kapitel ist Schaltprozessen vorbehalten. Zunächst gibt Fritzsche einen Überblick über Schaltmechanismen und deren Anwendungsmöglichkeiten in amorphen Halbleitern; 6 weitere Berichte untersuchen thermische und elektronische Schalteffekte, organische Filme, verschiedene Gläser, Chalkogenide usw. 11 Vorträge sind Organischen Materialien gewidmet. In dem einleitenden Vortrag geht Silver auf die Theorie und die Erscheinungsformen der Leitfähigkeit dieser Stoffgruppe ein, ehe in den folgenden einzelne Substanzen bzw. Substanzgruppen diskutiert werden.

Der letzte Problemkreis umfaßt schließlich Ionenkristalle und Flüssigkeiten. In 5 Berichten werden dünne NaCl-Filme, Injektion heißer Elektronen in flüssigen Stickstoff und Gase hoher Dichte, Beweglichkeiten in flüssigen Dielektrika und Excitonen besprochen.

Bereits diese stark verkürzte Inhaltsangabe zeigt, daß die Teilgebiete eng miteinander verbunden sind, und diese sich sehr oft überschneiden. Die Vorträge geben einen sehr guten Über- und Einblick in ein Gebiet, das seit einigen

Jahren immer mehr an Bedeutung gewinnt. Dieser Band kann daher jedem, der sich mit dieser Problematik befaßt, sei es auch nur am Rande, empfohlen werden, da man auch viele Hinweise über die eigentliche Thematik der Tagung hinaus erhält.

A. LÖSCHE

Kristall und Technik	7	1-3	1972	K 27

Buchbesprechung

E.A. KANER, V.G. SKOBOV

Plasma Effects in Metals: Helicon and Alfvèn Waves

142 Seiten, 38 Abbildungen. £ 2.50
Taylor & Francis Ltd., London 1971

Das in den letzten Jahren stark angewachsene theoretische wie auch experimentelle Interesse an der Ausbreitung hochfrequenter elektromagnetischer Wellen in Festkörpern, insbesondere in Metallen, läßt die vorliegende Monographie über Plasmaeffekte in Metallen als besonders begrüßenswert erscheinen. Ausgehend von einer allgemeinen Formulierung des Problems werden Spektrum, Dämpfung und Polarisation elektromagnetischer Wellen in Metallen berechnet. Die Theorie der Heliconwellen wird zunächst für Metalle mit sphärischer Fermifläche hergeleitet und mit experimentellen Daten für die Alkalimetalle verglichen. Es folgen theoretische Überlegungen unter Berücksichtigung der Anisotropie von Fermiflächen. Die Behandlung der Alfvènwellen erfolgt analog. Es folgen theoretische Betrachtungen zu einigen Besonderheiten, in denen sich das Elektron-Loch-Plasma eines Metalls von einem gasförmigen Plasma unterscheidet. Hierzu gehören insbesondere Quanteneffekte sowie die Kopplung elektromagnetischer Wellen mit anderen für ein Metall typischen Schwingungen (hier behandelt die Kopplung mit akustischen Phononen). Wellen mit diskretem Spektrum sowie elektromagnetische Wellen mit Frequenzen nahe der Zyklotronresonanzfrequenz werden gleichfalls behandelt. Das Schwergewicht der Monographie liegt auf theoretischem Gebiet, experimentelle Ergebnisse werden nur in beschränktem Maße an einigen wichtigen Stellen zitiert. Zu begrüßen ist die klare Herausarbeitung des Unterschiedes zwischen elektromagnetischen Wellen in Metallen und in Gasplasmen. Nicht eingegangen wird auf Supraleiter sowie auf Metalle mit magnetischen Ordnungszuständen.

Die Monographie ist vom mathematischen Formalismus her so aufgebaut, daß sie auch dem experimentell tätigen Physiker sehr empfohlen werden kann. Sie gibt einen guten Überblick über den gegenwärtigen Stand der Theorie auf diesem Gebiet.

H. NEUMANN

Kristall und Technik	7	1-3	1972	K 29 - K 30

Buchbesprechung

LANDOLT - BÖRNSTEIN

Zahlenwerte und Funktionen aus Naturwissenschaften und Technik
Neue Serie, Gruppe III: Kristall- und Festkörperphysik, Band 6:
P. ECKERLIN, Aachen und H. KANDLER, Burghausen
Strukturdaten der Elemente und intermetallischen Phasen

XXVIII, 1019 Seiten. Leinen DM 620,-
Berlin - Heidelberg - New York. Springer-Verlag 1971

Als Band 6 der Gruppe 3 (Kristall- und Festkörperphysik) der neuen Serie des Landolt-Börnstein liegt eine gut 1000 Seiten umfassende Zusammenstellung von Strukturdaten der Elemente und intermetallischen Phasen vor, soweit sie bis Ende 1967 veröffentlicht waren. Der Begriff "intermetallische Phase" ist dabei so abgegrenzt, daß darunter Phasen in Systemen, die B, C, H, N, O oder Halogene nicht enthalten, verstanden werden. Die Zusammenstellung bringt über 8.100 solche Phasen. Boride, Karbide und Hydride werden außerdem gesondert behandelt.

Die Besprechung der Phasen erfolgt in alphabetischer Reihenfolge auf Grund der chemischen Formeln, gegebenenfalls sind die Mineralnamen in Fußnoten angegeben.
Die Strukturdaten umfassen: Raumgruppe, Gitterkonstanten, Anzahl der Formeleinheiten in der Elementarzelle, Strukturtyp, Dichte, Umfang der Strukturbestimmung. Dazu kommen als zusätzliche Angaben: Existenzbereiche der Phasen, Abhängigkeit von der Zusammensetzung, Vorgeschichte des Materials. Zur angewandten Untersuchungsmethode wird mitgeteilt, ob es sich um Einkristall- oder Pulver-Untersuchungen handelt und ob sie mit Röntgen-Elektronen- oder Neutronenstrahlen durchgeführt wurden. Dem Ganzen vorangestellt sind Kristallsymmetrie-Tabellen, die Listen der Symmetrieelemente, der Symmetrieklassen und Lauegruppen, der Raumgruppensymbole sowie der Strukturtypen in verschiedenen Bezeichnungsweisen umfassen.

Das Literaturverzeichnis von Originalarbeiten erstrebt nur für die neueren Arbeiten (ab 1960) möglichste Vollständigkeit. Schon unter diesem Aspekt mußten allein für den Teil über die intermetallischen Phasen etwa 4.500 Zitate zusammengetragen werden.

Jeder Betrachter dieses Werkes wird - wie bei anderen Landolt-Börnsteinbänden auch - bewundernd vor der riesigen Arbeit stehen, die hier zu seinem Nutzen durchgeführt worden ist. Zumal, wenn die Bearbeitung mit der erwünschten Sorgfalt erfolgt ist, was im vorliegenden Fall nach Ausweis von stichprobenmäßigen Kontrollen in vollem Maße zutrifft.

Es ist sicher verständlich, daß der Benutzer sich wünschte, noch weitere Informationen so leicht zugänglich zu finden, z.B. die Atomparameter, deren Kenntnis zur Ermittlung der so wichtigen Koordinationsverhältnisse notwendig ist. Andererseits steht dem der entsprechend anwachsende Umfang und Preis entgegen. Vielleicht könnte man doch einen Mittelweg gehen, indem solche Angaben gemacht werden, wenn der Aufwand erträglich ist, und anderfalls darauf notgedrungen verzichtet wird.

K 30

Es ist zu hoffen, daß das verdienstvolle Werk durch weite Verbreitung einem möglichst großen Kreis von Interessenten nutzbar wird, wenn auch sein naturgemäß hoher Preis die Finanzkraft der Einzelnen überschreiten dürfte.

G. E. R. SCHULZE

Kristall und Technik	7	1-3	1972	K 31

Buchbesprechung

R. E. HUMMEL

Optische Eigenschaften von Metallen und Legierungen
Mit einer Einführung in die Elektronentheorie der Metalle
(Reine und angewandte Metallkunde in Einzeldarstellungen
Herausgegeben von W. KÖSTER, Band 22)

154 Abb. VIII, 222 Seiten. Gebunden DM 68,--; US $ 19,70
Springer-Verlag Berlin-Heidelberg-New-York 1971
ISBN 3-540-05234-8

Bekanntlich hat die Untersuchung der optischen Konstanten von Metallen und Legierungen im letzten Jahrzehnt einen bemerkenswerten Aufschwung genommen im Zusammenhang mit der starken Weiterentwicklung der Metallelektronentheorie. Dies ist ganz verständlich, da die optischen Messungen den direktesten Zugang zu den Kenngrößen der Metallelektronentheorie zu liefern vermögen, sofern die Schwierigkeiten überwunden werden können, die sich aus der geringen Eindringtiefe der Lichtwellen in Metalle ergeben. In welchem Maße dies bisher gelungen ist, wird im letzten Kapitel des vorliegenden Buches gezeigt.

Als Einführung für "den angehenden Metallkundler und Festkörperphysiker in verständlicher Form in die optischen Eigenschaften der Metalle und in die für das Verständnis der Metalloptik notwendigen Grundlagen der Elektronentheorie" ist der Hauptteil des Buches jedoch den theoretischen Grundlagen (5 Kapitel) und den experimentellen Methoden (1 Kapitel) gewidmet. Erst im 8. Kapitel werden Meßergebnisse an Hand elektronentheoretischer Modelle diskutiert. Der Anteil, der auf Legierungen entfällt, ist sehr klein, intermetallische Verbindungen werden nur gelegentlich erwähnt, z. B. hinsichtlich ihrer Farbe.

Obwohl er es als "heikles Unterfangen" bezeichnet, widmet der Verfasser dem Problem der Farbe von Metallen einen eigenen Abschnitt und zitiert in diesem Zusammenhang bereits in der Einleitung E. Swedenborg und Goethes Farbenlehre. Dagegen wird die optische Anisotropie hexagonaler Metalle, die immerhin ein Drittel aller Metalle bilden, nicht behandelt. Der gelegentliche Hinweis in Fußnoten, daß hier tensorielle Eigenschaften zu betrachten seien, wird einem Leser wenig nützen, dem man andererseits die Grundlagen der Bändertheorie der Metallelektronen (z. B. Durchrechnung mit Kronig-Penney-Potential) zu vermitteln für notwendig hält.

Es muß aber betont werden, daß das Kapitel über die "Quantenmechanische Behandlung der optischen Konstanten" - wie die gesamte Darstellung - außerordentlich klar und leicht verständlich geschrieben ist, durch instruktive Abbildungen wirksam unterstützt. In dieser Hinsicht ist es zu bedauern, daß viele Interessenten an einer wirklich leicht verständlichen elementaren Darstellung der Metall-Elektronentheorie sie in diesem Buch nicht finden werden, sofern sie sich nicht speziell mit Metalloptik beschäftigen.

Nach dem Gesagten ist das Buch für Metallkundler uneingeschränkt zu empfehlen. Der Festkörperphysiker, der in die Metalloptik eingeführt werden will, wird es ebenfalls mit Nutzen zu Rate ziehen, sich aber zugleich wünschen, daß anstelle von vielem ihm wohlbekanntem die Darstellung des eigentlichen Gegenstandes vertieft und in mancher Hinsicht auch erweitert worden wäre.

G. E. R. SCHULZE

Kristall und Technik
Crystal Research and Technology

Zeitschrift für experimentelle und technische Kristallographie

Herausgegeben von
H. NEELS, Leipzig

Erscheinungsweise monatlich mit einem Umfang von 128 Seiten im Format 16,7 × 24 cm, mit Abbildungen, Bezugspreis 20,– M.

BEITRÄGE DES VIERTEN HEFTES 1972

Originalbeiträge

Fortsetzungsbestellungen durch eine Buchhandlung erbeten

AKADEMIE-VERLAG · BERLIN

www.ingramcontent.com/pod-product-compliance
Lightning Source LLC
LaVergne TN
LVHW092154300926
849161LV00002B/270

Mometrix
TEST PREPARATION

ORELA®

Middle Grades Social Science Secrets Study Guide